International Series in Operations Research & Management Science

Volume 349

The book series **International Series in Operations Research and Management Science** encompasses the various areas of operations research and management science. Both theoretical and applied books are included. It describes current advances anywhere in the world that are at the cutting edge of the field. The series is aimed especially at researchers, advanced graduate students, and sophisticated practitioners.

The series features three types of books:

• Advanced expository books that extend and unify our understanding of particular areas.

• Research monographs that make substantial contributions to knowledge.

• Handbooks that define the new state of the art in particular areas. Each handbook will be edited by a leading authority in the area who will organize a team of experts on various aspects of the topic to write individual chapters. A handbook may emphasize expository surveys or completely new advances (either research or applications) or a combination of both.

The series emphasizes the following four areas:

Mathematical Programming: Including linear programming, integer programming, nonlinear programming, interior point methods, game theory, network optimization models, combinatorics, equilibrium programming, complementarity theory, multiobjective optimization, dynamic programming, stochastic programming, complexity theory, etc.

Applied Probability: Including queuing theory, simulation, renewal theory, Brownian motion and diffusion processes, decision analysis, Markov decision processes, reliability theory, forecasting, other stochastic processes motivated by applications, etc.

Production and Operations Management: Including inventory theory, production scheduling, capacity planning, facility location, supply chain management, distribution systems, materials requirements planning, just-in-time systems, flexible manufacturing systems, design of production lines, logistical planning, strategic issues, etc.

Applications of Operations Research and Management Science: Including telecommunications, health care, capital budgeting and finance, economics, marketing, public policy, military operations research, humanitarian relief and disaster mitigation, service operations, transportation systems, etc.

This book series is indexed in Scopus.

Dmitrii Lozovanu • Stefan Wolfgang Pickl

Markov Decision Processes and Stochastic Positional Games

Optimal Control on Complex Networks

Springer

Dmitrii Lozovanu
Institute of Mathematics and CS
Moldowa Academy of Science
Chisinau, Moldova

Stefan Wolfgang Pickl
Universität der Bundeswehr München
Neubiberg, München, Germany

ISSN 0884-8289 ISSN 2214-7934 (electronic)
International Series in Operations Research & Management Science
ISBN 978-3-031-40179-4 ISBN 978-3-031-40180-0 (eBook)
https://doi.org/10.1007/978-3-031-40180-0

This Springer imprint is published by the registered company Springer Nature Switzerland AG
The registered company address is: Gewerbestrasse 11, 6330 Cham, Switzerland

The work is dedicated to Murray Gell-Mann, who died in the first phase of the creation of this book. He died on May 24, 2019.

"Sometimes the probabilities are very close to certainties, but they are never really certainties. . . "

Murray Gell-Mann

Preface

Markov decision processes and stochastic games represent an important class of Operations Research models with large applications for the study of practical problems in such areas as ecology, economics, engineering, telecommunication, and biology. Because of this, there has been an increasing interest in the investigations of new classes of problems for Markov decision process models to develop efficient algorithms for solving them.

This book presents the most recent results concerned with finding the solutions to finite state space Markov decision problems and determining special Nash equilibria for stochastic games with average and expected total discounted reward payoffs. A new class of stochastic games studied in the book represents stochastic positional games that extend and generalize deterministic positional games.

The authors present their results concerning the application of quasi-monotonic programming for studying and solving Markov decision problems with average and discounted optimization criteria. Based on this, it is shown how to prove the existence of stationary Nash equilibria for stochastic positional games with average and discounted payoffs. The new conditions for the determination of pure and mixed stationary equilibria for the considered class of games are derived, and algorithms for determining the optimal stationary strategies of the players in the case of two-player zero-sum games are described.

The algorithms for Markov decision problems are extended for the control problem on networks and the corresponding algorithms to determine their solutions are presented. The most important results of the book are related to stochastic positional games for which the existence of stationary Nash equilibria is proved. Moreover, for this class of games, conditions for existing Nash equilibria in pure stationary strategies are formulated. In addition, the algorithms for deterministic positional games on networks are specified based on efficient polynomial algorithms for determining the optimal strategies of the players.

New details concerned with applications of the backward dynamic programming technique to determine the basic probabilistic characteristic in finite state Markov chains and to solve Markov decision problems are presented. At the end, the results are embedded in concrete applications.

Chapter 1 introduces the necessary definitions and notions for discrete Markov processes and presents the most recent results concerned with determining the basic probabilistic characteristic for finite state space Markov chains. New algorithms for determining the limiting and differential matrices based on the z-transform are proposed and an asymptotic behavior analysis of state-time probabilities is made. It is shown how to estimate the expected average rewards and the expected total discounted rewards in a Markov process with rewards using the z-transform. Furthermore, new dynamic programming algorithms are developed to determine the main probabilistic characteristics for stationary and non-stationary Markov processes.

In Chap. 2, models of Markov decision processes with finite and infinite time horizons are considered. For the finite horizon model, the decision problem with an expected total reward criterion is considered and the backward dynamic programming algorithm is presented. For infinite horizon decision models, two basic problems are studied: the Markov decision problem with an expected total discounted reward criterion and the Markov decision problem with an expected average reward criterion. The value and policy iteration algorithms for determining the optimal stationary policies for these problems are characterized.

The linear programming approach is discussed and analyzed in detail, and linear programming models are used to show how Markov decision problems can be formulated in terms of stationary strategies (stationary policies) as quasi-monotonic programming problems with linear constraints. New algorithms for determining the solution to stochastic control problems on networks are proposed.

In Chap. 3, the results concerned with the existence and determination of Nash equilibria for stochastic games with finite state and action spaces are presented. Mainly, stochastic games with average and discounted payoffs criteria are studied. Some classical results related to the existence and determination of Nash equilibria in such games are discussed and conditions for the existence of stationary Nash equilibria are formulated.

An important class of stochastic games, for which Nash equilibria in stationary strategies are proved, represents stochastic positional games and positional games on networks. Based on constructive proofs of these results, conditions for determining the optimal stationary strategies of the players are formulated.

Moreover, for the discounted stochastic positional games and average stochastic positional games with a unichain property, the existence of Nash equilibria in pure stationary strategies is proved. In the following, it is shown how the results for stochastic positional games can be used for studying and solving some classes of deterministic positional games on networks.

Some concrete applications of positional games for studying a class of multi-objective discrete control problems and hierarchical control problems on networks are presented. Future perspectives are described.

Multigraphs of Markov decision processes represent the special focus and spirit of this book on complex networks, which can serve as a comprehensive textbook for students but also as a starting point for further research. In particular, concerning the characterization and analysis of algorithmic decision-making and interdiction games and in the immersive field of complexity science, one could benefit from the models provided.

The authors would like to thank Prof. Dr. Maximilian Moll for carefully reading the manuscript. Furthermore, we would like to thank Dr. Andrea Ferstl, Dr. Leonhard Kunczik, Rudy Milani, Ulrike Stein, and Tino Krug for their help during the typesetting process, and also especially Verena Krüger for her outstanding help in the editing process.

Thank you very much to the colleagues at Santa Fe Institute for the inspiring discussions.

Chisinau, Moldova | Dmitrii Lozovanu
München, Germany | Stefan Wolfgang Pickl
24 May 2023

Contents

Notation

A	Set of actions in a Markov decision process
$A(x)$	Set of actions in the state x
$A^i(x)$	Set of actions in the state x of player i
a	An action for the decision maker in a Markov decision process or an action vector of the players in a stochastic game
a^i	An action of player i in a stochastic game
$\mathbb{C}$	Complex space
$C = (c_{x,y})$	Matrix of transition costs in a control problem on a network
$c_{x,y}$	Cost of the system's transition from state x to state y
$D_x(t)$	Variance of total reward during t transitions when process starts in state x
$\mathbb{E}\{\cdot\}$	Expected value
$\mathbb{E}_{\theta,\pi}\{\cdot\}$	Expectation with respect to probability measure $\mathbb{P}_{\theta,\pi}\{\cdot\}$
E	Set of edges of a graph
$e = (x, y)$	A directed edge of a graph
$E(x)$	Set of directed edges originating in vertex x
$F(z)$	z-transform for a non-decreasing discrete-time function $f(t)$
$G = (X, E)$	Graph with the set of vertices X and the set of edges E
$G_p = (X, E_p)$	Graph induced by the probability function p of a Markov chain
I	Identity matrix
$K(z)$	Characteristic polynomial of a transition probability matrix P
$\mathbb{L}$	Discrete-time system
L	Length of input data of a decision problem
$L_c^-(\psi)$	Sublevel set of function ψ for a given constant c, i.e., $L_c^-(\psi) = \{\mathbf{s} \in \mathbf{S} \mid \psi(\mathbf{s}) \leq c\}$
$L_c^+(\psi)$	Superlevel set of function ψ for a given constant c, i.e., $L_c^+(\psi) = \{\mathbf{s} \in \mathbf{S} \mid \psi(\mathbf{s}) \geq c\}$
$\mathbb{P}(\mathscr{A} \mid \mathscr{B})$	Probability of $\mathscr{A}$ for given $\mathscr{B}$
$\mathbb{P}_{\theta,\pi}\{\cdot\}$	Probability measure with respect to policy π and distribution θ on X

$P = (p_{x,y})$	Transition probability matrix of a Markov chain
P^t	t-th power of matrix P
$P^s = (p^s_{x,y})$	Transition probability matrix induced by a stationary policy (strategy) s in a Markov decision process
$P(t)$	Dynamic stochastic matrix
$P(\pi^t)$	Transition probability matrix induced by policy π^t at the moment of time t
$\overline{P}(t)$	P^t
$P_x(y, t)$	Probability to reach state y using t transitions from state x
$p_{x,y}$	Transition probability from state x to state y in a Markov chain
$p^a_{x,y}$	Transition probability from state x to state y for a given action $a \in A(x)$ in a Markov decision process (or for an action vector a of the players in a stochastic game)
$Q = (q_{x,y})$	Limiting matrix for a finite state space Markov chain
$q_{x,y}$	Limiting probability to reach y when process stats in x
$\mathbb{R}^m$	m-dimensional real space
$R = (r_{x,y})$	Matrix of rewards in a Markov process with rewards
$r_{x,y}$	Reward in a Markov process when the process makes a transition from a state x to state y with probability $p_{x,y}$
r_x	$r_x = \sum_{y \in X} p_{x,y} r_{x,y}$ – reward in the state x for a Markov process with rewards
$r_{x,a}$	Reward in the state x for an action $a \in A(x)$ in a Markov decision process (or for an action vector in a stochastic game)
$\mathbf{r}$	Reward vector for a Markov process
$\mathbb{S}$	Set of pure stationary strategies
$\mathbb{S}^i$	Set of pure stationary strategy of player i in a stochastic game
S	Set of mixed stationary strategies
S^i	Set of mixed stationary strategies of player i in a stochastic game
s	A strategy from $\mathbb{S}$ or from S
s^i	A strategy of player i from $\mathbb{S}^i$ or from S^i
$\mathbb{T}$	$\{0, 1, 2, \ldots, t, \ldots\}$
$T(t)$	$P^t - Q$
t	Time moment for the state in a discrete dynamical system
$U_t(x(t))$	Set of feasible controls in the state $x(t)$ at the time moment t
$u(t)$	Control vector from $U_t(x(t))$
X	Set of states for a discrete-time system or for a Markov chain
X_C	Set of controllable states for a control problem on a network
X_N	Set of uncontrollable states for a control problem on a network
$X(x)$	Set of neighboring vertices for vertex x in a graph
$x(t)$	State of the dynamical system at the moment of time t
Z	Set of dynamical states for the discrete control problem, i.e., $Z = \{x(t) = (x, t) \mid x \in X,\ t = 0, 1, 2, \ldots\}$
Z^C	Set of controllable dynamical states for the control problem
Z^N	Set of dynamical states for the discrete control problem
γ	Discount rate in discrete Markov decision models

γ^i	Discount rate of player i in a discounted stochastic game
ϵ	A small quantity
μ_x	Immediate cost in the state x for a Markov process with transition costs; it is expressed as $\sum_{y\in X} p_{x,y}c_{x,y}$
$\mathcal{P}$	Set of all policies
$\mathcal{P}(M)$	Set of all Markov policies
$\mathcal{P}(S)$	Set of all stationary policies
π	Policy in a Markov decision process (or a stationary distribution in an ergodic Markov chain)
$\pi(t)$	Policy at the moment of time t in a Markov decision process
π^t	$\pi^t = \pi(t)$
$\sigma_x(t)$	Expected total reward (cost) during t transitions in a Markov process when process starts in x
σ_x^γ	Expected total discounted reward (cost) with discount rate γ in a Markov process when process starts in x
$\sigma_x^\gamma(t)$	Expected total discounted reward (cost) during t transitions in a Markov process with discounted rate γ and starting state x
τ	Time counter in discrete systems
τ_e	Transition time through a directed edge $e = (x, y)$ in the graph of states' transitions for the control problem on a network
$\tau_{x(t_j)}\big(u(t_j)\big)$	Transition time from state $x(t_j)$ to state $x(t_{j+1})$ induced by the control vector $u(t) \in U_t(x(t_j))$
ω_x	Expected reward (or cost) per transition in the Markov process with rewards with starting state x
$\omega_x(t)$	Expected reward (cost) during t transitions in the Markov process with starting state x
ξ_t	Random variable representing the state at the time t
η_t	Random variable representing the action at the time t
$*$	Superscript denoting an optimal strategy, value, or decision rule in the problems
argmax	An element or a subset of elements at which the maximum of a function is obtained
argmin	An element or a subset of elements at which the minimum of a function is obtained
0	Scalar or vector zero
$\times$	Cartesian product
$\square$	Completion of the proof
$\Longrightarrow$	Implication token

Chapter 1
Discrete Markov Processes and Numerical Algorithms for Markov Chains

Abstract This chapter states the necessary classical results on discrete-time Markov processes and presents some approaches for determining the basic probabilistic characteristics of finite state space Markov chains. The main focus is on the elaboration of efficient numerical algorithms for computing the state-time probabilities, the limiting and differential matrices, as well as the average and expected total discounted rewards for discrete-time Markov processes. New algorithms for determining the limiting and differential matrices for such processes based on the z-transform are developed, and innovative procedures for calculating the state-time probabilities based on dynamic programming are substantiated.

Keywords Markov chain · Markov processes with rewards · State-time probabilities · Limiting matrix · Differential matrix · Average expected rewards · Discounted expected rewards

1.1 Definitions and Some Preliminary Results

The basic definitions and the main results related to discrete Markov processes can be found in [39, 61, 72–74, 78, 82, 83, 152, 199]. Here, we specify only some of the most important notions and results that we use in the book. In these sections, we focus on the dynamic programming technique for calculating the state-time probabilities and on algorithms for determining the limiting matrix in a finite state space Markov chain. Furthermore, we use the properties of the z-transform from [72] to asymptotically estimate the behavior of state-time probabilities, and based on this, an approach for determining the limiting matrix in a Markov chain is proposed. In the following, we show that such an approach can be used to determine the differential matrices.

D. Lozovanu, S. W. Pickl, *Markov Decision Processes and Stochastic Positional Games*, International Series in Operations Research & Management Science 349,
https://doi.org/10.1007/978-3-031-40180-0_1

1.1.1 Stochastic Processes and Markov Chains

A *stochastic process* is a collection of random variables $\{\xi_t : t \in \mathbb{T}\}$, where $\mathbb{T}$ is an index set that usually represents time. If $\mathbb{T} = \{0, 1, 2, \dots\}$, then $\{\xi_t : t \in \mathbb{T}\}$ is called a *discrete-time process*, and if $\mathbb{T} = [0, \infty)$, then $\{\xi_t : t \in \mathbb{T}\}$ is called a *continuous-time process*.

All possible values that can take ξ_t at any time $t \in \mathbb{T}$ form the *state space* of the stochastic process that we denote by X. If X is a discrete set, we have a *discrete state space process*; otherwise, we have a *continuous state space process*. Stochastic processes can be divided into the following four categories, depending on the continuous or discrete nature of the time parameter and the related random variables:

- Discrete-time stochastic processes with discrete state space
- Discrete-time stochastic processes with continuous state space
- Continuous-time stochastic processes with discrete state space
- Continuous-time stochastic processes with continuous state space

Here, we only consider discrete-time stochastic processes with discrete state spaces. An important class of such processes is represented by *Markov chains*.

A discrete-time stochastic process $\{\xi_t : t \in \mathbb{T}\}$ with a discrete state space X is called *Markov chain* (or *discrete Markov process*) if for every t $(t = 0, 1, 2, \dots)$ the following property holds:

$$\begin{aligned}\mathbb{P}(\xi_t = x_t|\, \xi_0 = x_0, \xi_1 = x_1, \xi_2 = x_2, \dots, \xi_{t-1} = x_{t-1}) \\ = \mathbb{P}(\xi_t = x_t|\, \xi_{t-1} = x_{t-1}),\end{aligned} \tag{1.1}$$

where $\mathbb{P}(\mathscr{A}|\mathscr{B})$ is the probability of $\mathscr{A}$ for given $\mathscr{B}$.

The property (1.1) is called *Markov property*. This property means that the probability of the state at time t in the process only depends on the state at time $t-1$ and does not depend on the earlier states, i.e., the future in the process depends on the past only through the present. Therefore, it is often said that Markov chains represent *memoryless processes*.

The probabilities $p_{x,y} = \mathbb{P}(\xi_t = y|\xi_{t-1} = x)$ are called *one-step transition probabilities* of the Markov chain to pass from a state $x \in X$ to states $y \in X$ in one step. If the one-step transition probabilities do not depend on time, the Markov chain is called *stationary Markov chain* (or *time-homogeneous Markov chain*). In the stationary case of a Markov chain with a finite set of states X, the one-step transition probabilities can be given by a transition probability matrix $P = (p_{x,y})$, where $p_{x,y}$ expresses the probability Markov chain to pass from state $x \in X$ to state $y \in Y$ in one step. This matrix possesses the property that $p_{x,y} \geq 0, \forall x, y \in X$ and $\sum_{y \in X} p_{x,y} = 1, \forall x \in X$, i.e., P is a *stochastic matrix*.

A Markov chain can be represented by a directed graph $G_p = (X, E_p)$ with the set of vertices X, representing the set of states and the set of the directed edges E, where $e = (x, y) \in E$ if and only if $p_{x,y} > 0$. Such a graph we call *transition graph of the Markov chain.*

1.1.2 State-Time Probabilities in a Markov Chain

Consider a *discrete-time Markov process* that models the evolution of the stochastic dynamical system $\mathbb{L}$ with the finite set of states $X = \{x_1, x_2, \ldots, x_n\}$. Let us assume that at the time moment $t = 0$, the system is in the state $x_{i_0} \in X$, and for an arbitrary state $x \in X$, the probabilities $p_{x,y}$ of the system's transitions from x to another state $y \in X$ are given, i.e.,

$$\sum_{y \in X} p_{x,y} = 1, \quad \forall x \in X; \quad p_{x,y} \geq 0, \qquad \forall x, y \in X.$$

Here, the probabilities $p_{x,y}$ do not depend on time, i.e., we have a *stationary Markov chain* determined by the *stochastic matrix* $P = (p_{x,y})$ and a given starting state x_{i_0}. So, we consider a stationary finite state space *Markov chain* where the transition time from one state to another is constant and it is equal to 1 [83, 152, 199]. For the dynamical system $\mathbb{L}$, $\mathbb{P}(\xi_t = x_{i_0} | \xi_0 = x)$ denotes the probability to reach the state x at the time moment t if it starts transitions at the time moment $t = 0$ in the state x_{i_0}. We consider the probability $\mathbb{P}(\xi_t = x_{i_0} | \xi_0 = x)$ at the discrete-time moment $t = 0, 1, 2, \ldots$. Following [71], we define and calculate $\mathbb{P}(\xi_t = x | \xi_0 = x_{i_0})$ by using the following recursive formula:

$$\mathbb{P}(\xi_{\tau+1} = x | \xi_0 = x_{i_0}) = \sum_{y \in X} \mathbb{P}(\xi_\tau = y | \xi_0 = x_{i_0}) p_{y,x}, \quad \tau = 0, 1, 2, \ldots, t-1, \tag{1.2}$$

where $\mathbb{P}(\xi_0 = x_{i_0} | \xi_0 = x_{i_0}) = 1$ and $\mathbb{P}(\xi_0 = x_{i_0} | \xi_0 = x) = 0$ for $x \in X \setminus \{x_{i_0}\}$. We call these probabilities *state-time probabilities of the system* $\mathbb{L}$. Formula (1.2) can be represented in the matrix form

$$\pi(\tau + 1) = \pi(\tau) P, \qquad \tau = 0, 1, 2, \ldots, t-1, \tag{1.3}$$

where $\pi(\tau) = (\pi_{x_1}(\tau), \pi_{x_2}(\tau), \ldots, \pi_{x_n}(\tau))$ is a vector, where an arbitrary component $\pi_{x_i}(\tau)$ expresses the probability of the system $\mathbb{L}$ to reach state x_i from state x_{i_0} using τ transitions, i.e., $\pi_{x_i}(\tau) = \mathbb{P}(\xi_\tau = x_i | \xi_0 = x_{i_0})$. At the starting moment of time $\tau = 0$, the vector $\pi(\tau)$ is given and its components are defined as follows: $\pi_{x_{i_0}}(0) = 1$ and $\pi_{x_i}(0) = 0$ for arbitrary $x_i \neq x_{i_0}$.

If for the dynamical system $\mathbb{L}$ with given starting vector $\pi(0)$ we apply formula (1.3) for $\tau = 0, 1, 2, \ldots, t-1$, then we obtain

$$\pi(t) = \pi(0)P^t, \tag{1.4}$$

where $P^t = P \times P \times \cdots \times P$. So, an arbitrary element $p^{(t)}_{x_i,x_j}$ of this matrix represents the probability of the system $\mathbb{L}$ to reach state x_j from x_i by using t transitions.

It is easy to see that for an arbitrary given starting vector $\pi(0)$, the following formula holds:

$$\sum_{i=1}^{n} \pi_i(\tau) = 1, \qquad \tau = 0, 1, 2, \ldots. \tag{1.5}$$

The correctness of this relation can be easily proved by using the induction principle with respect to τ. Indeed, for $\tau = 0$, equality (1.5) holds according to the definition. If we assume that (1.5) takes place for every $\tau \le t$, then for $\tau = t+1$, we obtain

$$\sum_{i=1}^{n} \pi_i(t+1) = \sum_{i=1}^{n} \sum_{x_j \in X} p_{x_j,x_i} \pi_j(t) = \sum_{x_j \in X} \pi_j(t) \sum_{i=1}^{n} p_{x_j,x_i} = \sum_{x_j \in X} \pi_j(t) = 1.$$

This means that P^t is a stochastic matrix.

In a more general form, formula (1.5) can be expressed as follows:

$$\mathbb{P}(\xi_t = y | \xi_0 = x) = \sum_{z \in X} \mathbb{P}(\xi_\tau = z | \xi_0 = x) \mathbb{P}(\xi_t = y | \xi_\tau = z), \ \forall x, y \in X, \tag{1.6}$$

where t and τ represent arbitrary non-negative integer values such that $t > \tau$. In the case $t = \tau + 1$, we have $\mathbb{P}(\xi_t = x | \xi_\tau = z) = p_{z,y}, \forall z, y \in X$, and formula (1.6) becomes (1.2). Relation (1.6) corresponds to the *Chapman-Kolmogorov equation* for Markov chains [158]. If we specify this equation in the matrix form, we obtain

$$p^{(t)}_{x,y} = \sum_{z \in X} p^{(\tau)}_{x,z} p^{(t-\tau)}_{z,y}, \ \forall x, y \in X; \ \text{ i.e., } \ P^t = P^\tau P^{t-\tau}.$$

1.1.3 Limiting Probabilities and Stationary Distributions

Let's take a finite state space Markov chain determined by a stochastic matrix P and assume that for P, there exists the limit

$$\lim_{t\to\infty} P^t = Q.$$

Then for a given starting state of system $\mathbb{L}$ in the case of a large number of transitions, the vector of state-time probability $\pi(t)$ can be approximated with the corresponding row vector of the matrix Q. Indeed, if in (1.4) we take the limit, then we have

$$\pi = \lim_{t\to\infty} \pi(t) = \pi(0) \lim_{t\to\infty} P^t = \pi(0)Q.$$

An arbitrary component π_{x_j} of $\pi = (\pi_{x_1}, \pi_{x_2}, \ldots, \pi_{x_n})$ can be treated as the probability that the system $\mathbb{L}$ will occupy the state x_j after a large number of transitions if it starts transitions in x_{i_0}.

The vector π is called the *vector of limiting state probabilities* for the dynamical system $\mathbb{L}$ with a given starting state x_{i_0}. Based on the property mentioned above, we may conclude that $\sum_{x\in X} \pi_x = 1$ for an arbitrary given starting vector $\pi(0)$, i.e., vector π represents a *probability distribution on* X. This means that the elements $q_{x,y}$ of the matrix Q satisfy the condition

$$\sum_{y\in X} q_{x,y} = 1, \quad \forall x \in X,$$

where $q_{x,y} \geq 0$, $\forall x, y \in X$, i.e., $Q = (q_{x,y})$ is a stochastic matrix. An arbitrary element $q_{x,y}$ of this matrix expresses the limiting probability of the system $\mathbb{L}$ to occupy the state $y \in X$ if the system starts transitions in x. The matrix Q is called the *matrix of limiting state probabilities*[71]; in short, we call this matrix the *limiting matrix*.

If the limiting matrix Q possesses the property that all its rows are the same, then the corresponding Markov process is called *Markov unichain* [6, 71]. The Markov chain for which the limiting matrix Q contains at least two different rows is called *Markov multichain* [83, 152, 199]. A more detailed classification of Markov chains with respect to the corresponding structure of matrices P and Q is given in the next two sections.

For the Markov unichain, we have $q_{x,y} = q_{z,y} = \pi_y, \quad \forall x, z, y \in X$. This means that the limiting state probabilities $\pi_y, \ y \in X$ do not depend on the state in which system $\mathbb{L}$ starts transitions. In this case, the vector π of limiting state probabilities determines a distribution on X that can be found by solving the system of linear equations

$$\pi(I - P) = 0; \qquad \sum_{y \in X = 1}^{n} \pi_y = 1, \tag{1.7}$$

where I is the identity matrix. The first equation in (1.7) corresponds to the condition

$$\pi = \pi P \tag{1.8}$$

that can be obtained from (1.3) if $\tau \to \infty$. The second equation in (1.7) reflects the property that after a large number of transitions, the dynamical system will occupy one of the states $x_j \in X$.

In general, the system of Eqs. (1.7) can be considered for an arbitrary Markov chain, including the case when $\lim_{t\to\infty} P^t$ does not exist. However, in the following, we can see that such a system may not have a unique solution:

A distribution π on X that satisfies the condition (1.8) is called *stationary distribution* for a Markov chain with transition probability matrix P. In terms of matrix theory, this means that π is a stationary distribution for a Markov chain with transition probability matrix P if and only if the corresponding row vector π is a left eigenvector for P corresponding to the eigenvalue 1. A stationary distribution π for a Markov chain always exists, but it is not necessarily unique.

In the following, we can see that the rank of the matrix $(I - P)$ for a Markov unichain is equal to $n-1$, and the system (1.7) has a unique solution [137, 199]. In the case of a Markov multichain, the rank of the matrix $(I - P)$ is less than $n - 1$, and the solution to system (1.7) is not unique.

For the Markov chain with the following transition probability matrix

$$P = \begin{pmatrix} \frac{1}{2} & \frac{1}{2} \\ \frac{2}{5} & \frac{3}{5} \end{pmatrix},$$

it is easy to check that there exists $\lim_{t\to\infty} P^t = Q$ and the rank of $(I - P)$ is equal to $n - 1$. Therefore, we can determine the unique stationary distribution $\pi = (\pi_1, \pi_2)$ for this Markov chain by solving the corresponding system (1.7)

$$\begin{cases} \frac{1}{2}\pi_1 - \frac{2}{5}\pi_2 = 0, \\ -\frac{1}{2}\pi_1 + \frac{2}{5}\pi_2 = 0, \\ \pi_1 + \pi_2 = 1, \end{cases}$$

and we obtain $\pi_1 = 4/9, \pi_2 = 5/9$, i.e., $\pi = (4/9,\ 5/9)$.

In this case, we have the following limiting matrix:

$$Q = \begin{pmatrix} \frac{4}{9} & \frac{5}{9} \\ \frac{4}{9} & \frac{5}{9} \end{pmatrix}.$$

For this example, a component π_{x_j} of the vector π that satisfies (1.7) can be treated as the probability of the system $\mathbb{L}$ to occupy the state x_j after a large number of transitions.

The Markov chain with the following transition probability matrix

$$P = \begin{pmatrix} 0 & 1 \\ 1 & 0 \end{pmatrix}$$

represents an example for which $\lim_{t\to\infty} P^t$ does not exist and the system (1.7) has a unique stationary distribution $\pi = (\pi_1, \pi_2)$.

It is easy to check that

$$P^{2t} = \begin{pmatrix} 1 & 0 \\ 0 & 1 \end{pmatrix}, \quad P^{2t+1} = \begin{pmatrix} 0 & 1 \\ 1 & 0 \end{pmatrix}, \quad \forall t \geq 0,$$

i.e., the limit $\lim_{t\to\infty} P^t$ does not exist. However, the rank of $(I - P)$ is equal to $n - 1$, and the corresponding system of linear equations (1.7)

$$\pi_1 - \pi_2 = 0, \quad -\pi_1 + \pi_2 = 0, \quad \pi_1 + \pi_2 = 1,$$

has a unique solution $\pi_1 = 1/2, \pi_2 = 1/2$, that determines the stationary distribution $\pi = (1/2, 1/2)$, i.e., in this case, we can define the limiting matrix as

$$Q = \begin{pmatrix} \frac{1}{2} & \frac{1}{2} \\ \frac{1}{2} & \frac{1}{2} \end{pmatrix}.$$

For this example, a component π_{x_j} of the vector π can be treated as the probability of the system $\mathbb{L}$ to occupy the state x_j at random moments of times during a large number of transitions.

The first example of a Markov chain presented above refers to *aperiodic Markov chains*, and the second example of a Markov chain refers to *periodic Markov chains*. A more detailed characterization of periodic and aperiodic Markov chains and how to define the limiting matrix for an arbitrary Markov chain are discussed in the next sections.

1.1.4 Definition of the Limiting Matrix

The limiting matrix Q for a Markov chain with an arbitrary transition probability matrix P can be defined by using the following limit:

$$\lim_{t\to\infty} \frac{1}{t} \sum_{\tau=0}^{t-1} P^{\tau}.$$

This limit is called *Cesàro limit* [31, 48, 152, 171, 199]. In [30] it was proved that this limit exists for an arbitrary stochastic matrix P.

So, for an arbitrary Markov chain, the limiting matrix Q is defined by

$$Q = \lim_{t\to\infty} \frac{1}{t} \sum_{\tau=0}^{t-1} P^{\tau}, \tag{1.9}$$

where $P^0 = I$ is the identity matrix. In a component notation, this means that for arbitrary $x, y \in X$,

$$q_{x,y} = \lim_{t\to\infty} \frac{1}{t} \sum_{\tau=0}^{t-1} p_{x,y}^{(\tau)},$$

where $p_{x,y}^{(\tau)}$ is a component of P^{τ}. Additionally, in [30] it was shown that if $\lim_{t\to\infty} P^t$ exists, then

$$\lim_{t\to\infty} P^t = \lim_{t\to\infty} \frac{1}{t} \sum_{\tau=0}^{t-1} P^{\tau} = Q.$$

It is easy to show that for an arbitrary stochastic matrix P, the limiting matrix Q satisfies the following equalities:

$$PQ = QP = QQ = Q. \tag{1.10}$$

Using definition (1.9) of limiting matrix Q, we can interpret an element $q_{x,y}$ of this matrix as the long-run fraction of time that the system occupies in the state y when it starts transitions in x. In the case when $\lim_{t\to\infty} P^t$ exists, where $\lim_{t\to\infty} P^t = Q$, an element $q_{x,y}$ of Q can be treated as the steady-state probability that the system is in state y when it starts in x.

1.1.5 Classification of States for a Markov Chain

We present a classification of states for a Markov chain that we use for the asymptotic analysis of state-time probabilities and for the determination of the limiting matrix Q. For this, we need some definitions:

A state y is said to be *accessible* from a state x, written as $x \to y$, if there exists $t > 0$ such that $p^{(t)}_{x,y} > 0$. So, y is accessible from x if y can be reached from x with a positive probability. Note that each $x \in X$ is accessible from itself, since $p^{(0)}_{x,y} = 1$ if $x = y$ and $p^{(0)}_{x,y} = 0$ if $x \neq y$.

A state x is said to *communicate* with state y, written as $x \leftrightarrow y$, if $x \to y$ and $y \to x$. *Communication relation* is an *equivalence* relation, and therefore, the set of states of a Markov chain can be partitioned into communicating classes such that only members of the same class communicate with each other: that is, two states x and y belong to the same class if $x \leftrightarrow y$. If we use the graph interpretation of a Markov chain, then each communicating class corresponds to a set of vertices of strongly connected component $G_i = (Y_i, E_i)$ of transition graph $G = (X, E)$.

A subset Y of X is called *closed set* if no state outside Y is accessible from any state in Y. If a closed set X does not contain a proper closed subset, then it is called *irreducible set*. An irreducible set in X induces a distinct Markov chain that we call *irreducible Markov chain*. The irreducible set consisting of a single state is called *absorbing state*. It is easy to observe that a state $x \in X$ is absorbing if and only if $p_{x,x} = 1$ and $p_{x,y} = 0$ for $y \neq X$.

We say that state x has *period* k if any return to state x must occur in multiples of k times steps. So, the period of state $x \in X$ can be defined as

$$k = gcd\{t > 0 : \mathbb{P}(\xi_t = x | \xi_0 = x)\},$$

where gcd is the greatest common divisor. If $k = 1$, the state x is *aperiodic*. If $k > 1$, the state x is said to be *periodic with period* k. A *Markov chain is aperiodic* if all its states are aperiodic.

A state $x \in X$ is called *transient* if the probability of never returning to this state is larger than zero. Formally, the transient property of state x is defined as follows: Let the random variable T_x represent the hitting time, which is the first (earliest) return time to x, i.e.,

$$T_x = \inf\{t \geq 1 : \xi_t = x | \xi_0 = x\}.$$

Then state x is transient if

$$\mathbb{P}(T_x < \infty) = \sum_{t=1}^{\infty} \mathbb{P}(T_x = t) < 1.$$

This means that state x is transient if $\mathbb{P}(T_x = \infty) > 0$.

A state $x \in X$ is *recurrent* if it is not transient. So, a recurrent state has a finite hitting time with probability 1. It can be shown that a state x is recurrent if and only if the expected number of visits is infinite, i.e., $\sum_{t=0}^{\infty} p_{x,x}^{(t)} = \infty$. Even if the hitting time of a state x is finite with probability 1, it does not need to have a finite expectation. The *mean recurrence time* at state x is the expected return time $\mathbb{E}(T_x) = \sum_{t=1}^{\infty} \mathbb{P}(T_x = t)$. A recurrent state x is called *positive recurrent state* if $\mathbb{E}(T_x)$ is finite; otherwise, state x is called *null recurrent state*.

A state $x \in X$ is said to be *ergodic* if it is aperiodic and positive recurrent. This means that state x is ergodic if it is recurrent, has period $k = 1$, and has mean recurrence time. If all states in an irreducible Markov chain are ergodic, then the chain is called *ergodic Markov chain*. It can be easily shown that a finite state irreducible Markov chain is ergodic if it has an aperiodic state. In general, a Markov chain is ergodic if there is a number N such that any state can be reached from many other states in any number of steps greater than or equal to N.

Thus, we can make the following conclusions:

The set of recurrent states of finite state space Markov chains can be partitioned into disjoint closed irreducible sets X_i, $i = 1, 2, \ldots, m$ such that we can write X as $X = X_1 \cup X_2 \cup \cdots \cup X_m \cup X_{m+1}^0$, where X_{m+1}^0 represents the set of transient states that do not belong to any closed set. After that, by relabeling the states, we can express the transition probability matrix P and Q as follows:

$$P = \begin{pmatrix} P_1 & 0 & \ldots 0 & 0 \\ 0 & P_2 & \ldots 0 & 0 \\ . & . & \ldots. & . \\ . & . & \ldots. & . \\ . & . & \ldots. & . \\ 0 & 0 & \ldots P_{k'} & 0 \\ W_1 & W_2 & \ldots W_{k'} & W_{k'+1} \end{pmatrix}, \qquad Q = \begin{pmatrix} Q_1 & 0 & \ldots 0 & 0 \\ 0 & Q_2 & \ldots 0 & 0 \\ . & . & \ldots. & . \\ . & . & \ldots. & . \\ . & . & \ldots. & . \\ 0 & 0 & \ldots Q_{k'} & 0 \\ R_1 & R_2 & \ldots R_{k'} & 0 \end{pmatrix},$$

where P_i and Q_i correspond to transitions between states in X_i; the matrices W_i and R_i correspond to transitions from states in X_{m+1}^0 to the state in X_i, and $W_{k'+1}$ is the matrix that corresponds to transitions between states in X^0. The matrix P represented in such a form is called the *transition probability matrix in canonical form.*

The Markov chain consisting of a single closed irreducible set we call *irreducible Markov chain*. If the Markov chain contains a single irreducible set and a set of transient states (possibly empty), then we call it *unichain*.

Otherwise, we say that the Markov chain is a *multichain*. For a Markov unichain, all rows of matrix Q are the same and correspond to a stationary distribution π that can be determined by solving system (1.7); in the case of a Markov multichain, the matrix Q contains at least two different rows.

1.1.6 An Algorithm for Determining the Limiting Matrix

If the matrix of transition probabilities P is given in canonical form, then in order to determine the limiting matrix Q, the following decomposition algorithm from [58, 152] can be applied: Each component P_i corresponds to an *irreducible Markov chain*; therefore, all rows of the matrix Q_i are identical. Each matrix Q_i is determined according to the formula

$$Q_i = \overline{e}_i \pi^i,$$

where $\overline{e}_i$ is the column vector, where each component is equal to one, and π^i is a row vector with the component that represents the solution to the system

$$\pi^i = \pi^i P_i, \quad \sum_j \pi^i_j = 1.$$

The matrices R_i can be calculated by using the formula

$$R_i = (I - W_{k'+1})^{-1} W_i Q_i, \quad i = 1, 2, \ldots, k'. \tag{1.11}$$

The correctness of this formula can be proved if we represent the transition matrix P as

$$P = \begin{pmatrix} \overline{P} & 0 \\ \overline{W} & W_{k'+1} \end{pmatrix}$$

and the limiting matrix Q as

$$Q = \begin{pmatrix} \overline{Q} & 0 \\ \overline{R} & 0 \end{pmatrix},$$

and then apply the property (1.10). Indeed, since $PQ = Q$, we obtain

$$\overline{W}\,\overline{Q} + W_{k'+1}\overline{R} = \overline{R}.$$

Thus,

$$W_i Q_i + W_{k'+1} R_i = R_i, \quad i = 1, 2, \ldots, k'.$$

Taking into account that, in the case of a Markov multichain, the inverse matrix $(I - W_{k'+1})^{-1}$ for $(I - W_{k'+1})$ exists (see [14, 31]), then we obtain (1.11) from this formula. So, to determine the limiting matrix Q, it is sufficient to represent the matrix P in canonical form and then to determine $Q_1, Q_2, \ldots, Q_{k'}$ and

$R_1, R_2, \ldots, R_{k'}$. An algorithm for relabeling the states in Markov chains, which transforms the matrix P of probability transitions into its canonical form, was described in [58, 90, 152]. This algorithm uses $O(n^2)$ elementary operations. In general, to determine the irreducible sets X_i, we can use the *graph of transition probabilities* $G = (X, E)$ for Markov chains. The *adjacency matrix* of this graph is obtained from the matrix P by changing the non-zero probabilities with units. Then the strongly connected components that do not contain any directed edges to other strongly connected components correspond to the irreducible sets X_i. Therefore, if we apply an algorithm for finding the strongly connected components of G_p, we obtain an algorithm for finding the irreducible sets X_i.

Example Consider the problem of determining the limiting matrix Q of the Markov chain with the transition probability matrix

$$P = \begin{pmatrix} 1 & 0 & 0 & 0 \\ 0 & \frac{1}{2} & \frac{1}{2} & 0 \\ 0 & \frac{2}{5} & \frac{3}{5} & 0 \\ \frac{1}{4} & \frac{1}{4} & \frac{1}{4} & \frac{1}{4} \end{pmatrix}.$$

In Fig. 1.1, the transition graph $G_p = (X, E_p)$ of the Markov chain with transition probability matrix P is given.

The strongly connected components of this graph are determined by the subsets

$$X_1 = \{1\}, \quad X_2 = \{2, 3\}, \quad X_3 = \{4\}.$$

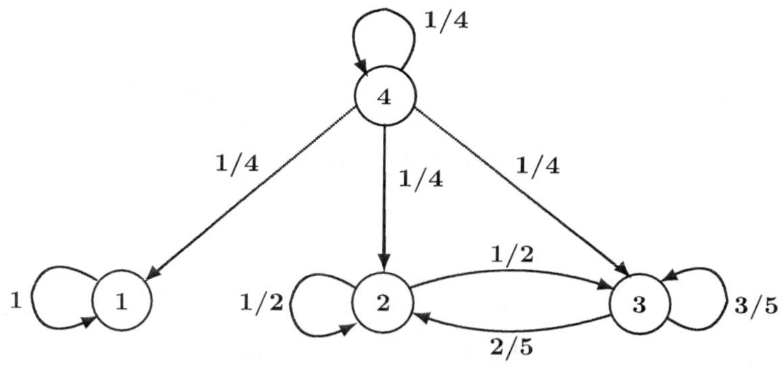

Fig. 1.1 Graph $G_p = (X, E_p)$ of transition probabilities

The irreducible sets correspond to strongly connected components that do not contain outgoing directed edges to other strongly connected components. Thus, we have two irreducible sets $X_1 = \{1\}$ and $X_2 = \{2, 3\}$. Therefore, the *transition probability matrix* mentioned above has a canonical form, where

$$P_1 = (1),\ P_2 = \begin{pmatrix} \frac{1}{2} & \frac{1}{2} \\ \frac{2}{5} & \frac{3}{5} \end{pmatrix},\ W_1 = \begin{pmatrix} \frac{1}{4} \end{pmatrix},\ W_2 = \begin{pmatrix} \frac{1}{4} & \frac{1}{4} \end{pmatrix},\ W_3 = \begin{pmatrix} \frac{1}{4} \end{pmatrix}.$$

Therefore, we solve the system of linear equations $\pi^i = \pi^i P_i$, $\sum_j \pi^i_j = 1$ for $i = 1, 2$ and determine $\pi^1 = (1)$, $\pi^2 = (4/9\ \ 5/9)$. This means that

$$Q_1 = (1), \quad Q_2 = \begin{pmatrix} \frac{4}{9} & \frac{5}{9} \\ \frac{4}{9} & \frac{5}{9} \end{pmatrix}.$$

Now we calculate $R_i = (I - W_3)^{-1} W_i Q_i$, $i = 1, 2$ and obtain

$$R_1 = \begin{pmatrix} \frac{3}{4} \end{pmatrix}^{-1} (1) \begin{pmatrix} \frac{1}{4} \end{pmatrix} = \begin{pmatrix} \frac{1}{3} \end{pmatrix};$$

$$R_2 = \begin{pmatrix} \frac{3}{4} \end{pmatrix}^{-1} \begin{pmatrix} \frac{1}{4} & \frac{1}{4} \end{pmatrix} \begin{pmatrix} \frac{4}{9} & \frac{5}{9} \\ \frac{4}{9} & \frac{5}{9} \end{pmatrix} = \begin{pmatrix} \frac{8}{27} & \frac{10}{27} \end{pmatrix}.$$

In such a way, we obtain the limiting matrix

$$Q = \begin{pmatrix} 1 & 0 & 0 & 0 \\ 0 & \frac{4}{9} & \frac{5}{9} & 0 \\ 0 & \frac{4}{9} & \frac{5}{9} & 0 \\ \frac{1}{3} & \frac{8}{27} & \frac{10}{27} & 0 \end{pmatrix}.$$

It is easy to check that in the worst case, this algorithm uses $O(|X|^3)$ elementary operations. We can see that along with this algorithm, other variations for determining the limiting matrix in a Markov chain can be used. In the following, we analyze an approach for determining the limiting matrix that allows us to study the asymptotic behavior of the state-time probability in Markov chains.

This approach is based on the properties of the generating functions [48, 71]. In [71], the generating functions are called z-transforms.

1.1.7 An Approximation Algorithm for Limiting Probabilities Based on the Ergodicity Condition

The main idea of the algorithm described in this subsection is based on the ergodicity condition for Markov chains and on the structure properties of graph G_p of probability transitions. We consider the problem of finding the limiting probabilities $q_{i_0,j}$, $j = 1, 2, \ldots, n$ in the Markov chain with a fixed starting state $x(0) = x_{i_0}$ and a given stochastic matrix P.

We show that this problem can be reduced to an auxiliary problem of finding the limiting probabilities in a new Markov chain for which the corresponding graph of probability transitions $G^0 = (X, E^0)$ is strongly connected. Finally, for a new Markov chain, we obtain the problem of determining the limiting probability

$$q^0_{i_0,j} = \pi^0_j, \quad j = 1, 2, \ldots, n,$$

where $\pi^0 = (\pi^0_1, \pi'_2, \ldots, \pi^0_n)$ can be found by solving the system of linear equations

$$\pi^0 = \pi^0 P^0, \quad \sum_{j=1}^{n} \pi^0_j = 1. \tag{1.12}$$

Here, P^0 is the stochastic matrix of the auxiliary irreducible Markov chain with the graph of probability transitions $G^0 = (X, E^0)$. Afterward, we show that each component $q_{i_0,j}$ can be obtained from the corresponding π^0_j, $j \in \{1, 2, \ldots, n\}$ using a special approximation procedure from [85]. We define the auxiliary graph $G^0 = (X, E^0)$ and the corresponding auxiliary Markov chain in the following way:

Let $G^1 = (X^1, E^1)$, $G^2 = (X^2, E^2)$, $\ldots$, $G^k = (X^k, E^k)$ be the strongly connected components of graph $G_p = (X, E_p)$, where $\bigcup_{i=1}^{k} X^i = X$ and $k > 1$. Denote by $G^{i_r} = (X^{i_r}, E^{i'_n})$, $r = 1, 2, \ldots, k'$ the deadlock components of graph Gp and assume that $x_{i_0} \in X \setminus \bigcup_{r=1}^{k'} X^{i_r}$ and G_p satisfy the condition that each vertex $x \in X$ is attainable from x_{i_0}. If G_p contains vertices x_j, which could not be reached from x_{i_0}, then we set $q_{i_0,j} = 0$ and delete these vertices from G_p. Our problem can easily be solved, also in the case that x_{i_0} belongs to a deadlock component G^{i_r}. In this case, the limiting probabilities can also be determined easily:

We put $q_{i_0,j} = 0$ for $x_j \in X \setminus X^{i_r}$ and determine the non-zero components $q_{i_0,j}$ by solving the system of linear equations $\pi^{i_r} = \pi^{i_p} P^{(i_r)}$, $\sum_{y \in X^{i_r}} \pi^{i_r}_y = 1$ (see Procedure 1 of the algorithm from the previous subsection).

The strongly connected graph $G^0 = (X, E^0)$ of the auxiliary Markov unichain is obtained from the graph $G_p = (X, E_p)$, using the following construction:

- Graph G^0 contains the same set of vertices X as graph G_p.
- The set of directed edges E^0 of graph G^0 consists of the set of edges E and the new directed edges $e = (x, x_{i_0})$, oriented from $x \in \bigcup_{r=1}^{k'} X^{i_r}$ to x_{i_0}, i.e., $E^0 = E \cup E'$, where $E' = \left\{(x, x_{i_0}) \,\middle|\, x \in \bigcup_{r=1}^{k'} X^{i_r}\right\}$.
- We define the probabilities $p^0_{x,y}$ for $(x, y) \in E^0 (x \in X_2)$ in G^0 using the following rules:

  (a) $p^0_{x,y} = p_{x,y}$ for $(x, y) \in E^0$ if $x \in X \setminus \bigcup_{r=1}^{k'} X^{i_r}$.
  (b) $p^0_{x,x_{i_0}} = \epsilon$ for $(x, x_{i_0}) \in E'$, $x \in \bigcup_{r=1}^{k'} X^{i_r}$, where ϵ is a small positive value in comparison with p_e for $e \in E$.
  (c) $p^0_{x,y} = p_{x,y}(1 - \epsilon)$ for $(x, y) \in E$ if $x, y \in X \setminus \bigcup_{r=1}^{k'} X^{i_r}$.

Graph G^0 is strongly connected, and therefore, for the corresponding Markov chain with the new probability transition matrix P^0, there exists the limiting matrix Q^0 whose rows are all identical. The vector $\pi^0 = (\pi^0_1, \pi^0_2, \ldots, \pi^0_n)$ with the components $\pi^0_j = q^0_{i,j}$, $i, j = 1, 2, \ldots, n$ can be found by solving the system of linear equations (1.12). Moreover, on the basis of the algorithm from the previous section, we may conclude that for a small ϵ, the solution to this system (1.12) represents the approximation values for the components of the vector of the limiting probabilities in the initial Markov chain with a given starting state i_0 and probability matrix P. The probabilities $p^0_{x,y}$ for $(x, y) \in E^0$ are defined according to the items (a)–(c) that allow us to preserve in the auxiliary Markov chain the small values of the limiting state probabilities π^0_j for the states $x_j \in X \setminus \bigcup_{r=1}^{k'} X^{i_r}$. In addition, we can see that condition c) in the auxiliary problem allows us to preserve an appropriate proportion of the probabilities of the limiting states in each irreducible set X^{i_r} and between different ergodic classes as the proportion of the limiting probabilities in the initial problem.

Using the results from [85], we can see that the exact values of the limiting probabilities $\pi_j = q_{i_0,j}$, $j = 1, 2, \ldots, n$ can be found from the corresponding π^0_j using a special approximation procedure from [84, 85] if ϵ is a suitable small value.

Indeed, let us assume that the probabilities $p_{i,j}$ are given in the form of irreducible decimal fractions $p_{i,j} = a_{i,j}/b_{i,j}$, $i, j = 1, 2, \ldots, n$, where the numerators as well as the denominators are integer numbers. Then the values π_j can be found from π^0_j using the roundoff procedure from [85] if ϵ satisfies the following condition:

$$\epsilon \leq 2^{-L},$$

where

$$L = \sum_{(x,y)\in E} \log(a_{x,y}+1) + \sum_{(x,y)\in E} \log(b_{x,y}+1) + 2\log(n) + 1.$$

Here, L is the length of the binary-coded representation of the elements of the matrix P, where each probability $p_{x,y}$ is given by the integer couple $a_{x,y}, b_{x,y}$. If for given ϵ the values π_j^0, $j = 1, 2, \ldots, n$ are known, then each π_j^0 can be represented as a convergent continued fraction, and we may find a unique rational fraction A_j/B_j that satisfies the condition $\left|\pi_j^0 - A_j/B_j\right| \le 2^{-2L-2}$. After that, we fix $\pi_j = A_j/B_j$. In such a way, we find the exact limiting probabilities π_j, $j = 1, 2, \ldots, n$.

In general, we can see that the probabilities π_j^0 can be expressed as functions that depend on ϵ, i.e.,

$$\pi_j^0 = \pi_j^0(\epsilon), \quad j = 1, 2, \ldots, n.$$

This means that the limiting probabilities π_j for our initial problem can be found as follows:

$$\pi_j = \lim_{\epsilon\to 0} \pi_j^0(\epsilon), \quad j = 1, 2, \ldots, n.$$

Below there is an example that illustrates the algorithm described above.

Example Again, we use the data for a Markov chain from the previous example and consider the problem of determining the limiting state probabilities in the case if the starting state of the system is $x_{i_1} = 2$ and the stochastic matrix of probabilities is

$$P = \begin{pmatrix} 1 & 0 & 0 & 0 \\ 0.25 & 0.25 & 0.25 & 0.25 \\ 0 & 0 & 0.5 & 0.5 \\ 0 & 0 & 0.5 & 0.5 \end{pmatrix}.$$

The corresponding graph $G_p = (X, E_p)$ of this Markov chain is represented in Fig. 1.6.

We apply the algorithm described above to determine the vector of limiting state probabilities $\pi = (\pi_1, \pi_2, \pi_3)$, where $\pi_1 = q_{2,1}$, $\pi_2 = q_{2,2}$, $\pi_3 = q_{2,3}$, $\pi_4 = q_{2,4}$. The corresponding strongly connected auxiliary graph $G^0 = (X, E^0)$ is represented in Fig. 1.2, and the corresponding matrix of probabilities P' for the auxiliary Markov

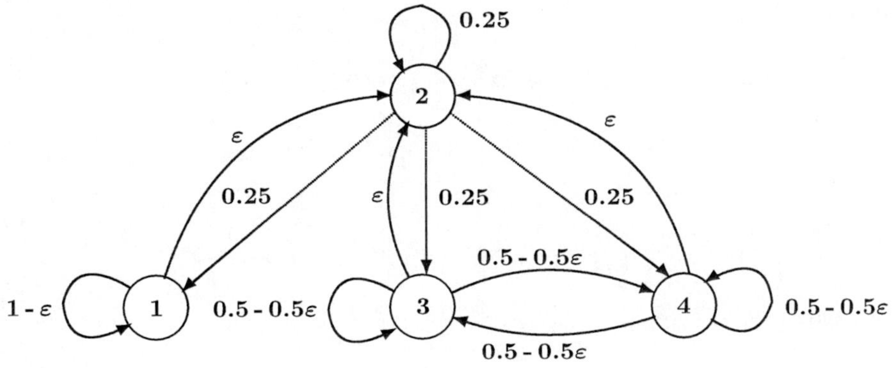

Fig. 1.2 Auxiliary graph $G^0 = (X, E^0)$

chain is given below:

$$P^o = \begin{pmatrix} 1-\epsilon & \epsilon & 0 & 0 \\ 0.25 & 0.25 & 0.25 & 0.25 \\ 0 & \epsilon & 0.5-0.5\epsilon & 0.5-0.5\epsilon \\ 0 & \epsilon & 0.5-0.5\epsilon & 0.5-0.5\epsilon \end{pmatrix}.$$

For the matrix P, we have $p_{1,1} = 1$, $p_{1,2} = p_{1,3} = p_{1,4} = 0$; $p_{2,1} = p_{2,2} = p_{2,3} = p_{2,4} = 1/4$; $p_{3,1} = p_{3,2} = 0$, $p_{3,3} = p_{3,4} = 1/2$; $p_{4,1} = p_{4,2} = 0$, $p_{4,3} = p_{4,4} = 1/2$.

Therefore, we can take $L \geq 35$. This implies that we can set $\epsilon = 0.0001$.

We consider the system of linear equations

$$\pi^0 = \pi^0 P^0; \qquad \sum_{j=1}^{4} \pi_j^0 = 1$$

and determine the solution to this system. So, if for $\epsilon = 0.0001$, we solve the system of linear equations

$$\begin{cases} \pi_1^0 = (1-\epsilon)\pi_1^0 + 0.25\pi_2^0; \\ \pi_2^0 = \epsilon\pi_1^0 + 0.25\pi_2^0 + \epsilon\pi_3^0 + \epsilon\pi_4^0; \\ \pi_3^0 = 0.25\pi_2^0 + (0.5-0.5\epsilon)\pi_3^0 + (0.5-0.5\epsilon)\pi_4^0; \\ \pi_4^0 = 0.25\pi_2^0 + (0.5-0.5\epsilon)\pi_3^0 + (0.5-0.5\epsilon)\pi_4^0; \\ 1 = \pi_1^0 + \pi_2^0 + \pi_3^0 + \pi_4^0; \end{cases}$$

then, we obtain the solution

$$\pi_1^0 = 0.3333; \quad \pi_2^0 = 0.0001; \quad \pi_3^0 = 0.3333; \quad \pi_4^0 = 0.3333.$$

If for each π_j^0, $j = 1, 2, 3, 4$ we find the approximate rational fraction, then we determine

$$\pi_1 = \frac{1}{3}; \quad \pi_2 = 0; \quad \pi_3 = \frac{1}{3}; \quad \pi_4 = \frac{1}{3}$$

that satisfy the conditions

$$\left|\pi_1^0 - \frac{1}{3}\right| \le 0.0001; \quad |\pi_2^0 - 0| \le 0.0001;$$
$$\left|\pi_3^0 - \frac{1}{3}\right| \le 0.0001; \quad \left|\pi_4^0 - \frac{1}{3}\right| \le 0.0003.$$

Therefore, finally we obtain

$$q_{2,1} = \frac{1}{3}; \quad q_{2,2} = 0; \quad q_{2,3} = \frac{1}{3}; \quad q_{2,4} = \frac{1}{3}.$$

The system above can be solved in general form with respect to ϵ, and the following representation of the solution in parametrical form can be obtained:

$$\pi_1^0(\epsilon) = \frac{1}{3+4\epsilon}; \qquad \pi_2^0(\epsilon) = \frac{4\epsilon}{3+4\epsilon};$$
$$\pi_3^0(\epsilon) = \frac{2+3\epsilon}{(2+2\epsilon)(3+4\epsilon)}; \quad \pi_4^0(\epsilon) = \frac{2+\epsilon}{(2+2\epsilon)(3+4\epsilon)}.$$

If after that we find the corresponding limits when $\epsilon \to 0$, then we obtain

$$\pi_1 = \lim_{\epsilon \to 0} \pi_1^0(\epsilon) = \frac{1}{3}; \quad \pi_2 = \lim_{\epsilon \to 0} \pi_2^0(\epsilon) = 0;$$
$$\pi_3 = \lim_{\epsilon \to 0} \pi_3^0(\epsilon) = \frac{1}{3}; \quad \pi_4 = \lim_{\epsilon \to 0} \pi_4^0(\epsilon) = \frac{1}{3},$$

i.e., we obtain the limiting probabilities for our problem.

1.2 Asymptotic Behavior of State-Time Probabilities

The results of this section are concerned with the applications of the generating functions for studying the asymptotic behavior of state-time probabilities in a Markov chain. In the following, we show that such an approach can be used for the estimation of the expected total rewards in a discrete Markov process with rewards (see Sects. 1.9 and 1.10). We use the generating functions in terms of the z-transform from [71]. So, we consider the z-transform for a discrete-time function $f(t),\ t = 0, 1, 2, \ldots$ with the values $f(0), f(1), (2), \ldots$ that do not increase in magnitude faster than a geometric sequence. For an arbitrary function, the z-transform is defined as

$$F(z) = \sum_{t=0}^{\infty} f(t)z^t.$$

If the function $f(t)$ holds the property mentioned above, then its z-transform $F(z)$ is unique. This means that each time function $f(t)$ has only one z-transform $F(z)$, and the inverse of its transformation produces only the original time function. The probability transients in a Markov chain are geometric sequences; therefore, the z-transforms can be used for studying the behavior of state-time probabilities. Below we present the z-transforms of some typical time function $f(t)$ that we use in the following:

If $f(t)$ is the step function

$$f(t) = \begin{cases} 1, & \text{if } \ t = 0, 1, 2, \ldots \\ 0, & \text{if } \ t < 0, (x, y), \end{cases}$$

then $F(z) = \sum_{t=0}^{\infty} f(t)z^t = 1 + z + z^2 + z^3 + \ldots$, and we have $F(z) = \dfrac{1}{1-z}$.

For a geometric sequence $f(t) = a^t, \geq 0$, we obtain $F(z) = \dfrac{1}{1-az}$ because

$$F(z) = \sum_{t=0}^{\infty} f(t)z^t = \sum_{t=0}^{\infty} a^t z^t = \sum_{t=0}^{\infty} (az)^t = \frac{1}{1-az}.$$

If $f(t) = ta^t$, we can show that $F(z) = \dfrac{az}{(1-az)^2}$.

Indeed,

$$F(z) = \sum_{t=0}^{\infty} ta^t z^t = z\frac{d}{dz}\left(\sum_{t=0}^{\infty} a^t z^t\right) = z\frac{d}{dz}\left(\frac{1}{1-az}\right) = \frac{az}{(1-az)^2}.$$

An important property of the z-transform that we use is the following: If a time function $f(t)$ with the z-transform $F(z)$ is shifted to the right with one unit so as to become $f(t+1)$, then the z-transform of the shifted function is

$$\sum_{t=0}^{\infty} f(t+1)z^t = \sum_{\tau=1}^{\infty} f(\tau)z^{\tau-1} = z^{-1}\big(F(z) - f(0)\big).$$

Thus, based on the definition of the z-transform, we can easily obtain the following elementary properties:

(1) $F(z) = F_1(z) + F_2(z)$ if $f(t) = f_1(t) + f_2(t)$,

where $F_1(z) = \sum_{t=0}^{\infty} f_1(t)z^t$ and $F_2(z) = \sum_{t=0}^{\infty} f_2(t)z^t$;

(2) $F(z) = kF'(z)$ if $f(t) = kf'(t)$,

where $F'(z) = \sum_{t=0}^{\infty} f'(t)z^t$;

(3) $F(z) = z^{-1}[F'(z) - f'(0)]$ if $f(t) = f'(t+1)$,

where $F'(z) = \sum_{t=0}^{\infty} f'(t)z^t$;

(4) $F(z) = zF'(z)$ if $f(t) = f'(t-1)$,

where $F'(z) = \sum_{t=0}^{\infty} f'(t)z^t$;

(5) $F(z) = \dfrac{1}{1-z}$ if $f(t) = \begin{cases} 1, & t = 0, 1, 2, \ldots, \\ 0, & t < 0; \end{cases}$

(6) $F(z) = \dfrac{1}{1-\alpha z}$ if $f(t) = \alpha^t$;

(7) $F(z) = \dfrac{\alpha z}{(1-\alpha z)^2}$ if $f(t) = t\alpha^t$;

(8) $F(z) = \dfrac{z}{(1-z)^2}$ if $f(t) = t$;

(9) $F(z) = F'(\alpha z)$ if $f(t) = \alpha^t f'(t)$,

where $F'(z) = \sum_{t=0}^{\infty} f'(t)z^t$.

In a more detailed form, the properties of the z-transform for different classes of discrete-time functions were described in [48, 71, 83, 171]. Note that the z-transform can be applied to the vectors and matrices by applying the z-transform to each component of the array. If we denote the z-transform of the vector $\pi(t)$ by $F(z)$, and if we apply the z-transform to the relationship

$$\pi(t+1) = \pi(t)P,$$

then we obtain

$$F(z+1) = F(z)P.$$

Based on property (3), this formula can be written as follows:

$$z^{-1}(F(z) - F(0)) = F(z)P,$$

i.e.,

$$F(z)(I - zP) = F(0),$$

where I is the identity matrix. Taking into account that $F(0) = \pi(0)$, we finally obtain

$$F(z) = \pi(0)(I - zP)^{-1}.$$

So, the z-transform of the vector of state-time probabilities is equal to the initial vector of probabilities $\pi(0)$ multiplied by the matrix $(I - zP)^{-1}$. This means that the solution to the transient problem is contained in the matrix $(I - zP)^{-1}$. We can obtain the probabilities of our transient problem in an analytical form if we weigh the rows of the matrix $(I - zP)^{-1}$ by the initial state probabilities, sum them up, and then take the inverse transformation. In such a way, we can obtain the matrix $P(t)$ that represents the inverse transformation of the matrix $(I - zP)^{-1}$.

An example that illustrates how to obtain the matrix $P(t)$ for the Markov process with the following stochastic matrix of probability transitions is given below:

$$P = \begin{pmatrix} \frac{1}{2} & \frac{1}{2} \\ \frac{2}{5} & \frac{3}{5} \end{pmatrix}.$$

In order to determine the matrix $P(t)$, we form the matrix

$$(I - zP) = \begin{pmatrix} 1 - \frac{1}{2}z & -\frac{1}{2}z \\ -\frac{2}{5}z & 1 - \frac{3}{5}z \end{pmatrix}$$

and find the inverse matrix

$$(I - zP)^{-1} = \frac{1}{\Delta(z)} \begin{pmatrix} 1 - \frac{3}{5}z & \frac{1}{2}z \\ \frac{2}{5}z & 1 - \frac{1}{2}z \end{pmatrix},$$

where $\Delta(z) = det(I - zP) = (1 - z)(1 - z/10)$.

So,

$$(I - zP)^{-1} = \begin{pmatrix} \dfrac{1 - \frac{3}{5}z}{(1 - z)(1 - \frac{1}{10}z)} & \dfrac{\frac{1}{2}z}{(1 - z)(1 - \frac{1}{10}z)} \\ \dfrac{\frac{2}{5}z}{(1 - z)(1 - \frac{1}{10}z)} & \dfrac{1 - \frac{1}{2}z}{(1 - z)(1 - \frac{1}{10}z)} \end{pmatrix}.$$

Each element of this matrix is a function of z with a factorable denominator $(1 - z)(1 - \frac{1}{10}z)$. By partial fraction expansion, each element of this matrix can be expressed as the sum of two terms, where one is with denominator $1 - z$ and another is with denominator $1 - \frac{1}{10}z$.

After such transformations, we obtain

$$(I - zP)^{-1} = \begin{pmatrix} \dfrac{\frac{4}{9}}{1 - z} + \dfrac{\frac{5}{9}}{1 - \frac{1}{10}z} & \dfrac{\frac{5}{9}}{1 - z} + \dfrac{\frac{-5}{9}}{1 - \frac{1}{10}z} \\ \dfrac{\frac{4}{9}}{1 - z} + \dfrac{\frac{-4}{9}}{1 - \frac{1}{10}z} & \dfrac{\frac{5}{9}}{1 - z} + \dfrac{\frac{4}{9}}{1 - \frac{1}{10}z} \end{pmatrix}.$$

Finally, we obtain the inverse matrix and it is represented as follows:

$$(I - zP)^{-1} = \frac{1}{1 - z} \begin{pmatrix} \frac{4}{9} & \frac{5}{9} \\ \frac{4}{9} & \frac{5}{9} \end{pmatrix} + \frac{1}{1 - \frac{1}{10}z} \begin{pmatrix} \frac{5}{9} & -\frac{5}{9} \\ -\frac{4}{9} & \frac{4}{9} \end{pmatrix}.$$

If we take the inverse transform of $(I - zP)^{-1}$, then we obtain

$$P(t) = \begin{pmatrix} \frac{4}{9} & \frac{5}{9} \\ \frac{4}{9} & \frac{5}{9} \end{pmatrix} + \frac{1}{10^t} \begin{pmatrix} \frac{5}{9} & -\frac{5}{9} \\ -\frac{4}{9} & \frac{4}{9} \end{pmatrix}.$$

From this formula in the case $t \to \infty$, we obtain

$$P(t) \to Q = \begin{pmatrix} \frac{4}{9} & \frac{5}{9} \\ \frac{4}{9} & \frac{5}{9} \end{pmatrix}.$$

Note that the example given above corresponds to an aperiodic Markov chain, and for the matrix P, there exists the limit $\lim_{t\to\infty} P^t$. An example for which this limit does not exist is given by the following matrix:

$$P = \begin{pmatrix} 0 & 1 \\ 1 & 0 \end{pmatrix}.$$

For this example, we have

$$(I - zP) = \begin{pmatrix} 1 & -z \\ -z & 1 \end{pmatrix},$$

where $\Delta(z) = det(I - zP) = 1 - z^2 = (1 - z)(1 - z)$.

So,

$$(I - zP)^{-1} = \begin{pmatrix} \frac{1}{(1-z)(1+z)} & \frac{z}{(1-z)(1+z)} \\ \frac{z}{(1-z)(1+z)} & \frac{1}{(1-z)(1+z)} \end{pmatrix}.$$

Using similar transformations as in the previous example, we find

$$(I - zP)^{-1} = \frac{1}{1-z} \begin{pmatrix} \frac{1}{2} & \frac{1}{2} \\ \frac{1}{2} & \frac{1}{2} \end{pmatrix} + \frac{1}{1+z} \begin{pmatrix} \frac{1}{2} & -\frac{1}{2} \\ -\frac{1}{2} & \frac{1}{2} \end{pmatrix}.$$

Based on property 6) of the z-transform, we obtain the matrix

$$P(t) = \begin{pmatrix} \frac{1}{2} & \frac{1}{2} \\ \frac{1}{2} & \frac{1}{2} \end{pmatrix} + (-1)^t \begin{pmatrix} \frac{1}{2} & -\frac{1}{2} \\ -\frac{1}{2} & \frac{1}{2} \end{pmatrix}.$$

We can see that in this case, for the matrix P, the limit $\lim_{t\to\infty} P(t)$ does not exist. However, the limiting matrix Q can be found by using the z-transform.

Thus, the vector $\pi(t)$ can be calculated by using the following formula:

$$\pi(t) = \pi(0)P(t).$$

By comparison of this formula with formula $\pi(t) = \pi(0)P^t$, we conclude that $P(t) = P^t$. This formula can be used for the asymptotic analysis of the transient problem. An important property of the matrix $P(t)$, which can be obtained on the basis of the z-transform, is the following: Among the component matrices in the representation of $\overline{P}(t)$, there exists at least one matrix that is stochastic, and it corresponds in the term $(I - zP)^{-1}$ to the matrix with the coefficient $1/(1 - z)$. This follows from the fact that the determinant of $(I - zP)$ is equal to zero for $z = 1$, i.e., the stochastic matrix P always has at least one *eigenvalue* (characteristic value) equal to 1. In the ergodic case, the Markov process $P(t)$ contains exactly such a stochastic matrix. Moreover, in the ergodic case, the rows of the stochastic components are identical and correspond to the vector of limiting state probabilities of the process.

This component does not depend on time and we denote it by Q. The remaining terms of $P(t)$ express the transient behavior of the process and represent matrices multiplied by coefficients of the form $\lambda_k^t, t\lambda_k^t, t^2\lambda_k^t, \ldots$, where $|\lambda_k| \leq 1$ and for the ergodic processes $|\lambda_k| < 1$. In the following, we can see that λ_k represents the eigenvalues (proper values) of the characteristic polynomial of the matrix P. The transient component of $P(t)$, which depends on t, is denoted by $T(t)$; this component vanishes as t becomes very large. The matrices of the component $T(t)$ possess the property that the sum of the elements in each row is equal to zero. This property for transient components holds because it expresses the perturbations from the limiting state probabilities. The matrices with such properties are called *differential matrices*. So, the matrix $P(t)$ for the Markov process can be represented as follows:

$$P(t) = Q + T(t),$$

where Q is a stochastic matrix where the rows represent the vector of limiting state probabilities, and $T(t)$ is the sum of the differential matrices with the geometric coefficients which for the ergodic case tends to zero if $t \to \infty$.

The z-transform shows that the matrix Q exists for an arbitrary transition probability matrix P, and $T(t)$ can be expressed as the sum of differential matrices multiplied by the coefficients $t^i \lambda_k^t$, where λ_k are the roots of the characteristic polynomial of the matrix P.

The problem of determining the limiting matrix Q and the differential matrices for a Markov chain with a given transition probability matrix P based on the z-transform was studied in [125]. In the general case, this problem is complicated from a computational point of view, because the algorithms for determining these matrices need the calculation of the roots of the characteristic polynomial of the matrix P. In [125], efficient algorithms for determining the limiting matrix P^* were proposed, as well as the differential matrices in the case when the roots of the characteristic polynomial of the matrix P are known.

1.3 Determining the Limiting Matrix Based on the z-Transform

In this section, we develop a new approach for determining the limiting matrix in discrete Markov processes based on the z-transform and classical numerical methods [92, 105]. Based on this approach, we substantiate an algorithm to determine the limiting matrix Q. We can see that the running time of the algorithm is $O(n^4)$, where n is the number of the states of the Markov chain. However, in the following, we show that the proposed approach can be extended to calculate the differential matrices.

1.3.1 Main Results

Let $\mathbb{C}$ be a complex space and denote by $\mathbb{M}(\mathbb{C})$ the set of complex matrices with n rows and n columns. We consider the function $K : \mathbb{C} \mapsto \mathbb{M}(\mathbb{C})$, where

$$K(z) = I - zP, \quad z \in \mathbb{C}.$$

We denote the elements of the matrix $K(z)$ by $a_{i,j}(z)$, $i, j = 1, 2, \ldots, n$, i.e.,

$$a_{i,j}(z) = \delta_{i,j} - zp_{i,j} \in \mathbb{C}[z],$$

where

$$\delta_{i,j} = \begin{cases} 1 & \text{if} \quad i = j, \\ 0 & \text{if} \quad i \neq j, \end{cases} \qquad i, j = 1, 2, \ldots, n.$$

It is evident that the determinant $\Delta(z)$ of the matrix $K(z)$ is a polynomial of degree less than or equal to n, $(\deg(\Delta(z)) \leq n, \Delta(z) \in \mathbb{C}[z])$.

Therefore, if we denote $\mathcal{D} = \{z \in \mathbb{C} \mid \Delta(z) \neq 0\}$, then we obtain $|\mathbb{C}\backslash\mathcal{D}| \leq \deg(\Delta(z)) \leq n$, and for an arbitrary $z \in \mathcal{D}$, there exists the inverse matrix of $K(z)$.

So, we can define the function $F : \mathcal{D} \to \mathbb{M}(\mathbb{C})$, where

$$F(z) = (K(z))^{-1}.$$

Then the elements $F_{i,j}(z)$, $i, j = 1, 2, \ldots, n$ of $F(z)$ can be found as follows:

$$F_{i,j}(z) = \frac{M_{j,i}(z)}{\Delta(z)}, \quad i, j = 1, 2, \ldots, n,$$

where

$$M_{i,j}(z) = (-1)^{i+j} K_{i,j}(z),$$

and $A_{i,j}(z)$ is a determinant of the matrix obtained from $K(z)$ by deleting the row i and the column j, $i, j = 1, 2, \ldots, n$. Therefore,

$$M_{j,i}(z) \in \mathbb{C}[z], \ \deg(M_{j,i}(z)) \leq n - 1, \quad i, j = 1, 2, \ldots, n.$$

Note that $\Delta(1) = 0$, because for the matrix $K(1)$, the following property holds:

$$\sum_{j=1}^{n}(\delta_{i,j} - p_{i,j}) = \sum_{j=1}^{n}\delta_{i,j} - \sum_{j=1}^{n}p_{i,j} = \delta_{i,i} - 1 = 0, \quad i = 1, 2, \ldots, n.$$

This means that $1 \in \mathbb{C}\backslash\mathcal{D}$ and $\Delta(z)$ can be factorized by $z - 1$.

Taking into account that $F_{i,j}(z)$ is a rational fraction with the denominator $\Delta(z)$, we can uniquely represent $F_{i,j}(z)$ in the following form:

$$F_{i,j}(z) = B_{i,j}(z) + \sum_{y \in \mathbb{C}\backslash\mathcal{D}} \sum_{k=1}^{m(y)} \frac{\alpha_{i,j,k}(y)}{(z - y)^k}, \quad i, j = 1, 2, \ldots, n, \tag{1.13}$$

where $m(z)$ is the order of the root of the polynomial $\Delta(z)$, $z \in \mathbb{C}\backslash\mathcal{D}$, $\alpha_{i,j,k}(y) \in \mathbb{C}$, $\forall y \in \mathbb{C}\backslash\mathcal{D}$, $k = 1, 2, \ldots, m(y)$, $i, j = 1, 2, \ldots, n$.

In the representation of $F_{i,j}(z)$ given above, the degree of the polynomial $B_{i,j}(z) \in \mathbb{C}[z]$ satisfies the condition

$$\deg(B_{i,j}(z)) = \deg(M_{j,i}(z)) - \deg(\Delta(z)),$$

where $\deg(M_{j,i}(z)) \geq \deg(\Delta(z))$; otherwise, $B_{i,j}(z) = 0$.

To represent (1.13) in a more convenient form, we use the expressions of series expansion for the functions $\nu_k(z) = 1/(1-z)^k$, $k = 1, 2, \ldots$. First of all, we observe that for these functions, there exists a series expansion. In particular for $k = 1$, the function $\nu_1(z)$ can be represented by $\nu_1(z) = \sum_{t=0}^{\infty} z^t$. In the general case (for an arbitrary $k > 1$), the following recursive relation holds: $\nu_{k+1}(z) = \mathrm{d}\nu_k(z)/(k\mathrm{d}z)$, $k = 1, 2, \ldots$. Using these properties and the induction principle, we obtain the series expansion of the function $\nu_k(z)$, $\forall k \geq 1$: $\nu_k(z) = \sum_{t=0}^{\infty} H_{k-1}(t)z^t$, where $H_{k-1}(t)$ is a polynomial of degree less than or equal to $(k-1)$.

Based on the properties mentioned above, we can make the following transformation in (1.13):

$$
\begin{aligned}
F_{i,j}(z) &= B_{i,j}(z) + \sum_{y\in\mathbb{C}\setminus\mathcal{D}} \sum_{k=1}^{m(y)} \frac{\left(-\frac{1}{y}\right)^k \alpha_{i,j,k}(y)}{\left(1-\frac{1}{y}z\right)^k} \\
&= B_{i,j}(z) + \sum_{y\in\mathbb{C}\setminus\mathcal{D}} \sum_{k=1}^{m(y)} \left(-\frac{1}{y}\right)^k \alpha_{i,j,k}(y)\nu_k\left(\frac{z}{y}\right) \\
&= B_{i,j}(z) + \sum_{y\in\mathbb{C}\setminus\mathcal{D}} \sum_{k=1}^{m(y)} \left(-\frac{1}{y}\right)^k \alpha_{i,j,k}(y) \sum_{t=0}^{\infty} H_{k-1}(t)\left(\frac{z}{y}\right)^t \\
&= B_{i,j}(z) + \sum_{t=0}^{\infty} \left(\frac{z}{y}\right)^t \sum_{y\in\mathbb{C}\setminus\mathcal{D}} \sum_{k=0}^{m(y)-1} \left(-\frac{1}{y}\right)^{k+1} \alpha_{i,j,k+1}(y)H_k(t).
\end{aligned}
$$

We can observe that in the relation above, the expression

$$
\sum_{k=0}^{m(y)-1} \left(-\frac{1}{y}\right)^{k+1} \alpha_{i,j,k+1}(y)H_k(t)
$$

represents a polynomial of degree less than or equal to $m(y)-1$, and we can write it in the form $\sum_{k=0}^{m(y)-1} \beta_{i,j,k}(y)t^k$, where $\beta_{i,j,k}(y)$ represent the corresponding coefficients of this polynomial. Therefore, if in this expression we substitute $\sum_{k=0}^{m(y)-1}(-1/y)^{k+1}\alpha_{i,j,k+1}(y)H_k(t)$ with $\sum_{k=0}^{m(y)-1} \beta_{i,j,k}(y)t^k$, then we obtain

$$
\begin{aligned}
F_{i,j}(z) &= B_{i,j}(z) + \sum_{t=0}^{\infty} z^t \sum_{y\in\mathbb{C}\setminus\mathcal{D}} \sum_{k=0}^{m(y)-1} \frac{t^k}{y^t}\beta_{i,j,k}(y) \\
&= W_{i,j}(z) + \sum_{t=1+\deg(B_{i,j}(z))}^{\infty} z^t \sum_{y\in\mathbb{C}\setminus\mathcal{D}} \sum_{k=0}^{m(y)-1} \frac{t^k}{y^t}\beta_{i,j,k}(y),
\end{aligned}
$$

$$
i, j = 1, 2, \ldots, n, \quad (1.14)
$$

where $\beta_{i,j,k}(y) \in \mathbb{C}$, $\forall y \in \mathbb{C}\backslash\mathcal{D}$, $k = 0, 1, 2, \ldots, m(y) - 1$, $i, j = 1, 2, \ldots, n$, and $W_{i,j}(z) \in \mathbb{C}[z]$ is a polynomial, where the degree satisfies the condition $\deg(W_{i,j}(z)) = \deg(B_{i,j}(z))$, $i, j = 1, 2, \ldots, n$.

In addition, we observe that the norm of the matrix P satisfies the condition $\|P\| = \max_{i=1,2,\ldots,n} \sum_{j=1}^{n} p_{i,j} = 1$, and therefore, $\|zP\| = |z|\|P\| = |z|$.

Let $|z| < 1$. Thus, for $F(z)$, we have

$$F(z) = (I - zP)^{-1} = \sum_{t=0}^{\infty} P^t z^t.$$

This means that

$$F_{i,j}(z) = \sum_{t=0}^{\infty} p_{i,j}(t) z^t, \quad i, j = 1, 2, \ldots, n. \tag{1.15}$$

From the definition of the z-transform and from (1.14) and (1.15), we obtain

$$p_{i,j}(t) = \sum_{y \in \mathbb{C}\backslash\mathcal{D}} \sum_{k=0}^{m(y)-1} \frac{t^k}{y^t} \beta_{i,j,k}(y), \quad \forall t > \deg(B_{i,j}(z)), \quad i, j = 1, 2, \ldots, n.$$

Since $0 \le \overline{p}_{i,j}(t) \le 1$, $i, j = 1, 2, \ldots, n$, $\forall t \ge 0$, we have

$$|y| \ge 1, \ \forall y \in \mathbb{C}\backslash\mathcal{D}, \quad \beta_{i,j,k}(1) = 0, \ \forall k \ge 1.$$

This implies $\alpha_{i,j,k}(1) = 0$, $\forall k \ge 2$.

Now let us assume that $\Delta(z) = (z - 1)^{m(1)} H(z)$, where $H(1) \ne 0$.

Then relation (1.13) is represented as follows:

$$\begin{aligned} F_{i,j}(z) &= \frac{\alpha_{i,j,1}(1)}{z-1} + B_{i,j}(z) + \sum_{y \in (\mathbb{C}\backslash\mathcal{D})\backslash\{1\}} \sum_{k=1}^{m(y)} \frac{\alpha_{i,j,k}(y)}{(z-y)^k} \\ &= \frac{\alpha_{i,j,1}(1)}{z-1} + \frac{Y_{i,j}(z)}{H(z)}, \quad i, j = 1, 2, \ldots, n, \end{aligned}$$

where $Y_{i,j}(z) \in \mathbb{C}[z]$ and

$$\begin{aligned} \deg(Y_{i,j}(z)) &= \deg(B_{i,j}(z)) + \deg(T(z)) = \deg(B_{i,j}(z)) + \deg(\Delta(z)) - m(1) \\ &= \deg(M_{j,i}(z)) - m(1) \le n - 1 - m(1) \le n - 2, \quad i, j = 1, 2, \ldots, n. \end{aligned}$$

If we denote

$$Y(z) = (Y_{i,j}(z))_{i,j=1,2,\ldots,n}, \quad \alpha_1(1) = (\alpha_{i,j,1}(1))_{i,j=1,2,\ldots,n},$$

then the matrix $F(z)$ can be represented as follows:

$$F(z) = \frac{1}{z-1}\alpha_1(1) + \frac{1}{H(z)}Y(z). \tag{1.16}$$

From this formula and from the definition of the limiting state matrix Q, we have

$$Q = -\alpha_1(1), \tag{1.17}$$

i.e., in the representation of the inverse matrix of $(I - zP)$, the limiting matrix Q will correspond to the term with the coefficient $1/(1-z)$.

From (1.16) and (1.17), we obtain the formula

$$Q = \lim_{z\to 1}(1-z)(I-zP)^{-1}.$$

In the following, we show how to determine the polynomial $\Delta(z)$ and the function $F(z)$ in the matrix form.

1.3.2 Constructing the Characteristic Polynomial

Let us consider the *characteristic polynomial*

$$K(\lambda) = |P - \lambda I| = \sum_{k=0}^{n} \nu_k \lambda^k.$$

It is easy to observe that $\nu_n = |-I| = (-1)^n \neq 0$. This means that $\deg(K(\lambda)) = n$, and the characteristic polynomial can be written in the following form:

$$K(\lambda) = (-1)^n(\lambda^n - \alpha_1\lambda^{n-1} - \alpha_2\lambda^{n-2} - \ldots - \alpha_n).$$

If we assume that $\alpha_0 = -1$, then the coefficients ν_k can be represented as follows:

$$\nu_k = (-1)^{n+1}\alpha_{n-k}, \quad k = 0, 1, 2, \ldots, n.$$

In [68] it was shown that the coefficients α_k can be calculated using $O(n^3)$ elementary operations based on Leverrier's method. This method can be applied to determine the coefficients α_k in the following way:

(1) Determine the matrices

$$P^k = \left(p_{i,j}^{(k)}\right), \quad i, j = 1, 2, \ldots, n, \quad k = 1, 2, \ldots, n,$$

where $P^k = P \times P \times \cdots \times P$.

(2) Find the traces of the matrices P^k:

$$\overline{s}_k = \operatorname{tr}(P^k) = \sum_{j=1}^{n} p_{j,j}^{(k)}, \quad k = 1, 2, \ldots, n.$$

(3) Calculate the coefficients

$$\alpha_k = \frac{1}{k}\left(\overline{s}_k - \sum_{j=1}^{k-1} \alpha_j \overline{s}_{k-j}\right), \quad k = 1, 2, \ldots, n.$$

Thus, if the coefficients α_k are known, then we can determine the coefficients of the polynomial $\Delta(z) = \sum_{k=0}^{n} \beta_k z^k$. Indeed, if $z \in \mathbb{C}\backslash\{0\}$, then

$$\begin{aligned}
\Delta(z) = |I - zP| &= (-z)^n \left|P - \frac{1}{z} I\right| = (-1)^n z^n K\left(\frac{1}{z}\right) \\
&= (-1)^n z^n \sum_{k=0}^{n} \nu_k \frac{1}{z^k} = (-1)^n \sum_{k=0}^{n} \nu_k z^{n-k} = \sum_{k=0}^{n} (-1)^n \nu_{n-k} z^k \\
&= \sum_{k=0}^{n} (-1)^n (-1)^{n+1} \alpha_k z^k = \sum_{k=0}^{n} (-\alpha_k) z^k.
\end{aligned}$$

For $z = 0$, we have

$$\Delta(0) = |I| = 1 = -\alpha_0.$$

Therefore, finally we obtain

$$\Delta(z) = \sum_{k=0}^{n} (-\alpha_k) z^k, \quad \forall z \in \mathbb{C}.$$

This means that

$$\beta_k = -\alpha_k, \quad k = 0, 1, 2, \ldots, n.$$

So, the coefficients β_k, $k = 0, 1, 2, \ldots, n$ can be calculated using a similar recursive formula

$$\beta_k = -\alpha_k = -\frac{1}{k}\left(\overline{s}_k - \sum_{j=1}^{k-1} \alpha_j s_{k-j}\right) = -\frac{1}{k}\left(\overline{s}_k + \sum_{j=1}^{k-1} \beta_j \overline{s}_{k-j}\right), \quad k=1, 2, \ldots, n,$$

$$\beta_0 = -\alpha_0 = 1.$$

Based on the result described above, we can propose the following algorithm to determine the coefficients β_k:

Algorithm 1.1 Determining the Coefficients of the Characteristic Polynomial

1. Calculate the matrices $P^k = \left(p_{i,j}^{(k)}\right), \quad i, j = 1, 2, \ldots, n, \ k = 1, 2, \ldots, n.$
2. Determine the traces of the matrices P^k:

$$\overline{s}_k = \operatorname{tr}(P^k) = \sum_{j=1}^{n} p_{j,j}^{(k)}, \quad k = 1, 2, \ldots, n.$$

3. Find the coefficients

$$\beta_0 = 1, \ \beta_k = -\frac{1}{k}\left(\overline{s}_k + \sum_{j=1}^{k-1} \beta_j \overline{s}_{k-j}\right), \quad k = 1, 2, \ldots, n.$$

1.3.3 Determining the z-Transform Function

Now let us show how to determine the function $F(z)$. Consider

$$H'(z) = (z-1)H(z)$$

and denote $N = \deg(H'(z)) = n - (m(1) - 1)$. We have already shown that the function $F(z)$ can be represented in the following matrix form:

$$F(z) = \frac{1}{\delta(z)} \sum_{k=0}^{N-1} z^k R^{(k)},$$

where

$$(z-1)^{m(1)-1}\sum_{k=0}^{N-1} z^k R_{i,j}^{(k)} = M_{j,i}, \quad i, j = 1, 2, \ldots, n.$$

We use the identity relation $I = (I - zP)(I - zP)^{-1}$ and include some elementary transformations:

$$\delta(z)I = (I - zP)\sum_{k=0}^{N-1} z^k R^{(k)} = \sum_{k=0}^{N-1} z^k R^{(k)} - \sum_{k=0}^{N-1} z^{k+1}(PR^{(k)})$$

$$= R^{(0)} + \sum_{k=1}^{N-1} z^k (R^{(k)} - PR^{(k-1)}) - z^N (PR^{(N-1)}).$$

Let $H'(z) = \sum_{k=0}^{N} \beta_k^* z^k$ and let us substitute this expression in the formula above. Then we obtain the following formula to determine the matrices $R^{(k)}$, $k = 0, 1, 2, \ldots, N-1$:

$$R^{(0)} = \beta_0^* I; \quad R^{(k)} = \beta_k^* I + PR^{(k-1)}, \quad k = 1, 2, \ldots, N-1. \tag{1.18}$$

So, we have

$$F(z) = \left(\frac{V_{i,j}(z)}{\delta(z)}\right)_{i,j=1,2,\ldots,n},$$

where

$$V_{i,j}(z) = \sum_{k=0}^{N-1} R_{i,j}^{(k)} z^k, \quad i, j = 1, 2, \ldots, n.$$

Based on these formulae, we can propose the following algorithm to determine the matrix Q.

1.3.4 The Algorithm for Calculating the Limiting Matrix

Consider

$$H(z) = \sum_{k=0}^{N-1} \gamma_k z^k; \quad Y(z) = \sum_{k=0}^{N-2} y^{(k)} z^k; \quad y^* = \alpha_1(1).$$

Then according to relation (1.16), we obtain

$$\frac{V_{i,j}(z)}{\delta(z)} = F_{i,j}(z) = \frac{y^*_{i,j}}{z-1} + \frac{\sum_{k=0}^{N-2} y^{(k)}_{i,j} z^k}{H(z)}, \quad i,j = 1,2,\dots,n.$$

This implies

$$\begin{aligned}
V_{i,j}(z) &= \sum_{k=0}^{N-1} R^{(k)}_{i,j} z^k = y^*_{i,j} H(z) + (z-1)\sum_{k=0}^{N-2} y^{(k)}_{i,j} z^k \\
&= y^*_{i,j}\sum_{k=0}^{N-1}\gamma_k z^k + \sum_{k=0}^{N-2} y^{(k)}_{i,j} z^{k+1} - \sum_{k=0}^{N-2} y^{(k)}_{i,j} z^k \\
&= \sum_{k=0}^{N-1}\gamma_k y^*_{i,j} z^k + \sum_{k=1}^{N-1} y^{(k-1)}_{i,j} z^k - \sum_{k=0}^{N-2} y^{(k)}_{i,j} z^k \\
&= \left(\gamma_0 y^*_{i,j} - y^{(0)}_{i,j}\right) + \sum_{k=1}^{N-2}\left(\gamma_k y^*_{i,j} + y^{(k-1)}_{i,j} - y^{(k)}_{i,j}\right) z^k \\
&\quad + \left(\gamma_{N-1} y^*_{i,j} + y^{(N-2)}_{i,j}\right) z^{N-1}, \qquad i,j = 1,2,\dots,n.
\end{aligned}$$

If we equate the corresponding coefficients of the variable z with the same exponents, then we obtain the following system of linear equations:

$$\begin{cases}
R^{(0)}_{i,j} = \gamma_0 y^*_{i,j} - y^{(0)}_{i,j}, \\
R^{(k)}_{i,j} = \gamma_k y^*_{i,j} + y^{(k-1)}_{i,j} - y^{(k)}_{i,j}, \quad k = 1,2,\dots,N-2, \quad i,j = 1,2,\dots,n, \\
R^{(N-1)}_{i,j} = \gamma_{N-1} y^*_{i,j} + y^{(N-2)}_{i,j}.
\end{cases}$$

This system is equivalent to the following system:

$$\begin{cases}
y^{(0)}_{i,j} = \gamma_0 y^*_{i,j} - R^{(0)}_{i,j}, \\
y^{(k)}_{i,j} = \gamma_k y^*_{i,j} + y^{(k-1)}_{i,j} - R^{(k)}_{i,j}, \quad k = 1,2,\dots,N-2, \quad i,j = 1,2,\dots,n, \\
y^{(N-2)}_{i,j} = -\gamma_{N-1} y^*_{i,j} + R^{(N-1)}_{i,j}.
\end{cases}$$

Here, we can observe that there exist the coefficients $u^{(k)}_{i,j}, v^{(k)}_{i,j} \in \mathbb{C}$, $k = 0,1,2,\dots,N-2$, $i,j = 1,2,\dots,n$, such that

$$y^{(k)}_{i,j} = u^{(k)}_{i,j} y^*_{i,j} + v^{(k)}_{i,j}, \quad k = 0,1,2,\dots,N-2, \quad i,j = 1,2,\dots,n.$$

From the first equation, we obtain

$$u_{i,j}^{(0)} = \gamma_0, \quad v_{i,j}^{(0)} = -R_{i,j}^{(0)}, \quad i, j = 1, 2, \ldots, n.$$

From the next $N - 2$ equations, we obtain

$$\begin{aligned} y_{i,j}^{(k)} &= \gamma_k y_{i,j}^{*} + y_{i,j}^{(k-1)} - R_{i,j}^{(k)} \\ &= \gamma_k y_{i,j}^{*} + u_{i,j}^{(k-1)} y_{i,j}^{*} + v_{i,j}^{(k-1)} - R_{i,j}^{(k)} \\ &= \left(\gamma_k + u_{i,j}^{(k-1)}\right) y_{i,j}^{*} + \left(v_{i,j}^{(k-1)} - R_{i,j}^{(k)}\right), \\ &\qquad k = 1, 2, \ldots, N-2, \quad i, j = 1, 2, \ldots, n, \end{aligned}$$

which involve the recursive equations

$$\begin{aligned} u_{i,j}^{(k)} &= u_{i,j}^{(k-1)} + \gamma_k, \quad v_{i,j}^{(k)} = v_{i,j}^{(k-1)} - R_{i,j}^{(k)}, \quad k = 1, 2, \ldots, N-2, \\ i, j &= 1, 2, \ldots, n. \end{aligned}$$

We obtain the direct formula for the calculation of the coefficients:

$$u_{i,j}^{(k)} = \sum_{r=0}^{k} \gamma_r, \quad v_{i,j}^{(k)} = -\sum_{r=0}^{k} R_{i,j}^{(r)}, \quad k=0, 1, 2, \ldots, N-2, \quad i, j=1, 2, \ldots, n.$$

If we introduce these coefficients into the last equation of the system, then we obtain

$$\begin{aligned} & u_{i,j}^{(N-2)} y_{i,j}^{*} + v_{i,j}^{(N-2)} = -\gamma_{N-1} y_{i,j}^{*} + R_{i,j}^{(N-1)}, \quad i, j = 1, 2, \ldots, n \\ \Leftrightarrow\ & y_{i,j}^{*} \sum_{r=0}^{N-1} \gamma_r = \sum_{r=0}^{N-1} R_{i,j}^{(r)}, \quad i, j = 1, 2, \ldots, n \\ \Leftrightarrow\ & y_{i,j}^{*} = \frac{\sum_{r=0}^{N-1} R_{i,j}^{(r)}}{\sum_{r=0}^{N-1} \gamma_r} = \frac{R_{i,j}}{H(1)}, \quad i, j = 1, 2, \ldots, n, \end{aligned}$$

where $R_{i,j} = \sum_{r=0}^{N-1} R_{i,j}^{(r)}$, $i, j = 1, 2, \ldots, n$. Finally, if we denote $R = (R_{i,j})_{n \times n}$, then we obtain:

$$Q = -\frac{1}{H(1)} R. \tag{1.19}$$

Based on the result above, we can describe the algorithm to determine the matrix Q.

Algorithm 1.2 Determining the Limiting Matrix Q

1. Find the coefficients of the characteristic polynomial $\Delta(z) = \sum_{k=0}^{n} \beta_k z^k$ using Algorithm 1.1.
2. Divide $m(1)$ times the polynomial $\Delta(z)$ by $z - 1$, using Horner's scheme, and find the polynomial $H(z)$ that satisfies the condition $H(1) \neq 0$. At the same time, we preserve the coefficients β_k^*, $k = 0, 1, 2, \ldots, N$ of the polynomial $(z-1)H(z)$ obtained at the previous step of Horner's scheme.
3. Determine $H(1)$ according to the rule described above.
4. Find the matrices $R^{(k)}$, $k = 0, 1, 2, \ldots, N-1$, according to (1.18).
5. Determine the matrix $R = \sum_{k=0}^{N-1} R^{(k)}$.
6. Calculate the matrix Q according to formula (1.19).

It is easy to check that the running time of Algorithm 1.2 is $O(|X|^4)$. Indeed, steps (1) and (4) of the algorithm use $O(|X|^4)$ elementary operations, and each of the remaining steps uses in the worst case $O(|X|^3)$ elementary operations.

Numerical Examples
Below we give some examples which illustrate the main details of the algorithm described above.

Example 1 Consider the discrete Markov process with the stochastic matrix of probability transitions

$$P = \begin{pmatrix} 0 & 1 \\ 1 & 0 \end{pmatrix}.$$

We have already noted that

$$P^{2t} = \begin{pmatrix} 1 & 0 \\ 0 & 1 \end{pmatrix}, \quad P^{2t+1} = \begin{pmatrix} 0 & 1 \\ 1 & 0 \end{pmatrix}, \quad \forall t \geq 0,$$

i.e., this Markov chain is 2−periodic.
So, for the considered example, $\lim_{t\to\infty} P^t$ does not exist.

Nevertheless, we can see that the matrix Q exists and it can be determined by using the algorithm described above. If we apply this algorithm, then we obtain:

(1) Find

$$P = \begin{pmatrix} 0 & 1 \\ 1 & 0 \end{pmatrix}, \quad P^2 = \begin{pmatrix} 1 & 0 \\ 0 & 1 \end{pmatrix};$$

$$\overline{s}_1 = \mathrm{tr}(P) = 0, \qquad \overline{s}_2 = \mathrm{tr}\big(P^2\big) = 2;$$

$$\beta_0 = 1, \ \ \beta_1 = -\overline{s}_1 = 0, \qquad \beta_2 = -\frac{1}{2}(\overline{s}_2 + \beta_1 \overline{s}_1) = -1.$$

(2) Divide the polynomial $\beta_2 z^2 + \beta_1 z + \beta_0$ by $z-1$ using Horner's scheme

	-1	0	1
1	-1	-1	0
1	-1	-2	

and obtain $m(1) = 1$, $N = 2$; $\beta_0^* = 1$, $\beta_1^* = 0$, $\beta_2^* = -1$; $\gamma_0 = -1$, $\gamma_1 = -1$.

(3) Determine

$$H(1) = \gamma_0 + \gamma_1 = -2.$$

(4) Calculate

$$R^{(0)} = \beta_0^* I = \begin{pmatrix} 1 & 0 \\ 0 & 1 \end{pmatrix}, \quad R^{(1)} = \beta_1^* I + P R^{(0)} = \begin{pmatrix} 0 & 1 \\ 1 & 0 \end{pmatrix}.$$

(5) Find

$$R = R^{(0)} + R^{(1)} = \begin{pmatrix} 1 & 0 \\ 0 & 1 \end{pmatrix} + \begin{pmatrix} 0 & 1 \\ 1 & 0 \end{pmatrix} = \begin{pmatrix} 1 & 1 \\ 1 & 1 \end{pmatrix}.$$

(6) Determine

$$Q = -\frac{1}{H(1)} R = \frac{1}{2} \begin{pmatrix} 1 & 1 \\ 1 & 1 \end{pmatrix} = \begin{pmatrix} 0.5 & 0.5 \\ 0.5 & 0.5 \end{pmatrix}.$$

In such a way, we obtain the limiting matrix

$$Q = \begin{pmatrix} 0.5 & 0.5 \\ 0.5 & 0.5 \end{pmatrix}.$$

Note that the considered process is not ergodic in the sense of the definition from [58, 83] because the matrix P^t contains zero elements for all $t \geq 0$.

However, the limiting matrix exists and the rows of this matrix are identical. As we have already shown, the vector of the limiting probabilities $\pi^* = (0.5,\ 0.5)$ can also be found by solving the system of linear equations (1.7).

Example 2 Consider a Markov process with the stochastic matrix

$$P = \begin{pmatrix} 0.5 & 0.5 \\ 0.4 & 0.6 \end{pmatrix}.$$

It is easy to observe that here, we have an ergodic Markov chain.

We can determine the matrix Q by using our algorithm.

In the same way, we determine

$$P = \begin{pmatrix} 0.5 & 0.5 \\ 0.4 & 0.6 \end{pmatrix}, \qquad P^2 = \begin{pmatrix} 0.45 & 0.55 \\ 0.44 & 0.56 \end{pmatrix};$$

$$\overline{s}_1 = \mathrm{tr}(P) = 0.5 + 0.6 = 1.1, \qquad \overline{s}_2 = \mathrm{tr}(P^2) = 0.45 + 0.56 = 1.01;$$

$$\beta_0 = 1, \ \beta_1 = -\overline{s}_1 = -1.1, \qquad \beta_2 = -\frac{1}{2}(\overline{s}_2 + \beta_1 \overline{s}_1) = 0.1.$$

Using Horner's scheme

	0.1	−1.1	1
1	0.1	−1	0
1	0.1	−0.9	

we obtain

$$\beta_0^* = 1, \ \beta_1^* = -1.1, \ \beta_2^* = 0.1; \ \gamma_0 = -1, \ \gamma_1 = 0.1; \ H(1) = \gamma_0 + \gamma_1 = -0.9;$$

$$R^{(0)} = \beta_0^* I = \begin{pmatrix} 1 & 0 \\ 0 & 1 \end{pmatrix}, \quad R^{(1)} = \beta_1^* I + P R^{(0)} = \begin{pmatrix} -0.6 & 0.5 \\ 0.4 & -0.5 \end{pmatrix};$$

$$R = R^{(0)} + R^{(1)} = \begin{pmatrix} 0.4 & 0.5 \\ 0.4 & 0.5 \end{pmatrix}; \quad Q = -\frac{1}{H(1)} R = \frac{1}{9} \begin{pmatrix} 4 & 5 \\ 4 & 5 \end{pmatrix}.$$

Finally, we obtain

$$Q = \begin{pmatrix} \frac{4}{9} & \frac{5}{9} \\ \frac{4}{9} & \frac{5}{9} \end{pmatrix}.$$

The rows of this matrix are the same and all elements of the matrix P^t are non-zero if $t \to \infty$.

So, this is an ergodic Markov chain with the vector of limiting probabilities $\pi^* = (4/9,\ 5/9)$. As we have shown, this vector can be found by solving system (1.7).

Example 3 We consider a Markov multichain with the stochastic matrix of probability transitions

$$P = \begin{pmatrix} 1 & 0 & 0 \\ 0 & 1 & 0 \\ \frac{1}{3} & \frac{1}{3} & \frac{1}{3} \end{pmatrix}.$$

In this case, the solution to the system of linear equations (1.7) is not unique.

If we apply the proposed algorithm, we can determine the matrix Q.

According to this algorithm, we obtain:

$$P = \begin{pmatrix} 1 & 0 & 0 \\ 0 & 1 & 0 \\ \frac{1}{3} & \frac{1}{3} & \frac{1}{3} \end{pmatrix}, \quad P^2 = \begin{pmatrix} 1 & 0 & 0 \\ 0 & 1 & 0 \\ \frac{4}{9} & \frac{4}{9} & \frac{1}{9} \end{pmatrix}, \quad P^3 = \begin{pmatrix} 1 & 0 & 0 \\ 0 & 1 & 0 \\ \frac{13}{27} & \frac{13}{27} & \frac{1}{27} \end{pmatrix};$$

$$\overline{s}_1 = \mathrm{tr}(P) = \frac{7}{3}, \quad \overline{s}_2 = \mathrm{tr}(P^2) = \frac{19}{9}, \quad \overline{s}_3 = \mathrm{tr}(P^3) = \frac{55}{27}; \quad \beta_0 = 1,$$

$$\beta_1 = -\overline{s}_1 = -\frac{7}{3}, \quad \beta_2 = -\frac{\overline{s}_2 + \beta_1 \overline{s}_1}{2} = \frac{5}{3}, \quad \beta_3 = -\frac{\overline{s}_3 + \beta_1 \overline{s}_2 + \beta_2 \overline{s}_1}{3} = -\frac{1}{3}.$$

If we apply Horner's scheme

	$-\frac{1}{3}$	$\frac{5}{3}$	$-\frac{7}{3}$	1
1	$-\frac{1}{3}$	$\frac{4}{3}$	-1	0
1	$-\frac{1}{3}$	1	0	
1	$-\frac{1}{3}$	$\frac{2}{3}$		

,

then we obtain

$$\beta_0^* = -1, \ \beta_1^* = \frac{4}{3}, \ \beta_2^* = -\frac{1}{3}; \ \gamma_0 = 1, \ \gamma_1 = -1/3; \ H(1) = \gamma_0 + \gamma_1 = \frac{2}{3};$$

$$R^{(0)} = \beta_0^* I = \begin{pmatrix} -1 & 0 & 0 \\ 0 & -1 & 0 \\ 0 & 0 & -1 \end{pmatrix}, \quad R^{(1)} = \beta_1^* I + P R^{(0)} = \begin{pmatrix} 1/3 & 0 & 0 \\ 0 & 1/3 & 0 \\ -1/3 & -1/3 & 1 \end{pmatrix};$$

$$R = R^{(0)} + R^{(1)} = \begin{pmatrix} -2/3 & 0 & 0 \\ 0 & -2/3 & 0 \\ -1/3 & -1/3 & 0 \end{pmatrix};$$

$$Q = -\frac{1}{H(1)} R = \begin{pmatrix} 1 & 0 & 0 \\ 0 & 1 & 0 \\ 1/2 & 1/2 & 0 \end{pmatrix}.$$

So, finally we have

$$Q = \begin{pmatrix} 1 & 0 & 0 \\ 0 & 1 & 0 \\ 1/2 & 1/2 & 0 \end{pmatrix}.$$

In this case, all rows of the matrix Q are different. It is easy to observe that for the considered example, $\lim_{t\to\infty} P^t = Q$ exists.

Finally, we can remark that the proposed approach also allows us to determine the differential components of the matrix $T(t)$.

1.4 An Approach for Finding the Differential Matrices

Now we show how the approach from the previous section can be used to determine the differential components of the matrix $T(t)$. We propose a simple modification of this approach that allows us to calculate all components of the transient matrix in case the roots of the characteristic polynomial are known. So, if there exist efficient algorithms for determining the roots of the characteristic polynomial, then the differential components of the matrix $T(t)$ can be determined efficiently by using algorithms similar to the algorithms from the previous section. We show that in this case, the differential matrices of the matrix $T(t)$ can be calculated by using $O(|X|^4)$ elementary operations.

1.4.1 The General Scheme of the Algorithm

To formulate the algorithm for determining the differential matrices, we use the relationship between the coefficients of the matrix $P(t) = P^t$ and the corresponding coefficients of $F(z)$ in formulae (1.14) and (1.15). An arbitrary element $p_{i,j}(t)$ of the matrix $P(t) = P^t$ represents the probability of the system to reach the state x_j from x_i using t transitions. The corresponding coefficients in formulae (1.14) and (1.15) have the same meaning, and therefore, we obtain an arbitrary element $p_{i,j}(t)$ of the matrix $P(t)$, which can be expressed by the formula

$$p_{i,j}(t) = \sum_{y \in \mathbb{C} \setminus \mathcal{D}} \sum_{k=0}^{m(y)-1} \frac{t^k}{y^t} \beta_{i,j,k}(y), \quad \forall t > \deg(B_{i,j}(z)), \ i, j = 1, 2, \ldots, n,$$

where $\mathcal{D} = \{z \in \mathbb{C} \,|\, |I - zP| \neq 0\}$, $\beta_{i,j,k}(y) \in \mathbb{C}$, $\forall y \in \mathbb{C} \setminus \mathcal{D}$, $k = 0, 1, 2, \ldots, m(y) - 1$, $m(y)$ is the order of the root y of the polynomial $\Delta(z) = |I - zP|$, and $B_{i,j}(z)$ is a polynomial of degree less than or equal to $n - 1$, $i, j = 1, 2, \ldots, n$.

If we denote $\beta_k(y) = (\beta_{i,j,k}(y))_{i,j=\overline{1,n}}$, $\forall y \in \mathbb{C} \setminus \mathcal{D}$, $k = 0, 1, 2, \ldots, m(y) - 1$, then

$$P(t) = \sum_{y \in \mathbb{C} \setminus \mathcal{D}} \sum_{k=0}^{m(y)-1} \frac{t^k}{y^t} \beta_k(y), \quad \forall t \geq n. \tag{1.20}$$

As we have shown, $\mathbb{C} \setminus \mathcal{D}$ consists of the set of inverses of the non-zero proper values of the matrix P, where the order of each element is the same as the order of the corresponding proper value. This means that (1.20) gives the components of the matrix $P(t)$ with respect to the proper values of the matrix P. Thus, if we determine these components, then we find the limit and differential matrices in a Markov chain.

We have already described the algorithm for determining the stationary components of the matrix $P(t)$ in the previous section. In the following, we show how to determine all matrices $\beta_k(y)$ from (1.20). In order to develop such an algorithm, we use some properties of the linear recurrent equations.

1.4.2 Linear Recurrent Equations and Their Main Properties

Consider an arbitrary set K on which the operations of summation and multiplication are defined. On this set we consider the following relation:

$$a_n = \sum_{k=0}^{m-1} q_k a_{n-1-k}, \quad \forall n \geq m, \tag{1.21}$$

where q_k are given elements from K. A sequence $a = \{a_n\}_{n=0}^{\infty}$ is called the *linear m-recurrence on K* if there exists a vector $q = (q_k)_{k=0}^{m-1} \in K^m$ such that (1.21) holds. Here, we call q the *generating vector*, and we call $I_m^{[a]} = (a_n)_{n=0}^{m-1}$ the *initial value of the sequence a*. The sequence a is called the *linear recurrence on K* if there exists $m \in \mathbb{N}^*$ ($\mathbb{N}^* = \{1, 2, \dots\}$) such that the sequence a is a linear m-recurrence on K. If $q_{m-1} \neq 0$, then the sequence a is called non-degenerated; otherwise, it is called degenerated.

Denote:

$Rol[K][m]$	the set of non-degenerated linear m-recurrences on K
$Rol[K]$	the set of non-degenerated recurrences on K
$G[K][m](a)$	the set of generating vectors of length m of the sequence $a \in Rol[K][m]$
$G[K](a)$	the set of generating vectors of the sequence $a \in Rol[K]$

In the following, we consider K as a subfield of the field of complex numbers $\mathbb{C}$ and $a = \{a_n\}_{n=0}^{\infty} \subseteq \mathbb{C}$.

We call the function $G^{[a]} : \mathbb{C} \mapsto \mathbb{C}, G^{[a]}(z) = \sum_{n=0}^{\infty} a_n z^n$ the *generating function of the sequence* $a = (a_n)_{n=0}^{\infty} \subseteq \mathbb{C}$, and we call the function $G_t^{[a]} : \mathbb{C} \to \mathbb{C}, G_t^{[a]}(z) = \sum_{n=0}^{t-1} a_n z^n$ the *partial generating function of order* t *of the sequence* $a = (a_n)_{n=0}^{\infty} \subseteq \mathbb{C}$.

Let $a \in Rol[K][m]$ be $q \in G[K][m](a)$. For this sequence, we consider the *characteristic polynomial* $\Delta_m^{[q]}(z) = 1 - zG_m^{[q]}(z)$ and the characteristic equation $\Delta_m^{[q]}(z) = 0$.

For an arbitrary $\alpha \in K^*$, we also call the polynomial $\Delta_{m,\alpha}^{[q]}(z) = \alpha\Delta_m^{[q]}(z)$ characteristic polynomial of the sequence a. We introduce the following notations:

$\Delta[K][m](a)$	the set of characteristic polynomials of degree m of the sequence $a \in Rol[K]$
$\Delta[K](a)$	the set of characteristic polynomials of the sequence $a \in Rol[K]$

If we operate with arbitrary recurrence (not obligatory, non-degenerated), then for the corresponding set, we will use similar notations and will specify these with the mark "*", i.e., we will denote: $Rol^*[K][m]$, $Rol^*[K]$, $G^*[K][m](a)$, $G^*[K](a)$, $\Delta^*[K][m](a)$, $\Delta^*[K](a)$.

We use the following properties:

(1) Let $a \in Rol[K][m]$, $q \in G[K][m](a)$, $\Delta_{m,\alpha}^{[q]}(z) = \prod_{k=0}^{p-1}(z - z_k)^{s_k}$, $z_i \neq z_j \; \forall i \neq j$.
Then $a_n = I_m^{[a]} \cdot ((B^{[a]})^T)^{-1} \cdot (\beta_n^{[a]})^T, \;\; \forall n \in \mathbb{N} \;\; (\mathbb{N} = \{0, 1, 2, \dots\})$,
$$\text{where } \beta_i^{[a]} = \left(\frac{\tau_{i,j}}{z_k^i}\right)_{k=\overline{0,p-1},\; j=\overline{0,s_k-1}},$$
$$\tau_{i,j} = \begin{cases} i^j \text{ if } i^2 + j^2 \neq 0, \\ 1 \;\text{ if } i = j = 0, \end{cases} \quad i \in \mathbb{N}, \quad B^{[a]} = (\beta_i^{[a]})_{i=0}^{m-1}.$$

(2) If a is a matrix sequence, $a \in Rol[M_n(K)][m]$ and $q \in G[M_n(K)][m](a)$, then $a \in Rol^*[K][mn]$ and $|I - zG_m^{[q]}(z)| \in \Delta^*[K][mn](a)$.

1.4.3 The Main Results and the Algorithm

Consider a matrix sequence $a = (P(t))_{t=0}^{\infty}$. Then it is easy to observe that the following recurrent relation $a_t = Pa_{t-1}$, $\forall t \geq 1$ holds. So, a belongs to $Rol[M_n(\mathbb{R})][1]$ with the generating vector $q = (P) \in G[M_n(\mathbb{R})][1](a)$.

Therefore, according to property (2) from Sect. 1.4.2, we have $a \in Rol^*[\mathbb{R}][n]$ and $\Delta(z) \in \Delta^*[\mathbb{R}][n](a)$.

Let $r = \deg \Delta(z)$ and consider the subsequence $\overline{a} = (P(t))_{t=n-r}^{\infty}$ of the sequence a. We have $\overline{a} \in Rol[\mathbb{R}][r]$ and $\Delta(z) \in \Delta[\mathbb{R}][r](\overline{a})$.

For the corresponding elements, this relation can be expressed as follows: $\overline{a}_{i,j} \in Rol[\mathbb{R}][r]$, $\Delta(z) \in \Delta[\mathbb{R}][r](\overline{a}_{i,j})$, $i, j = 1, 2, \dots, n$.

According to property 1) from Sect. 1.4.2, we obtain

$$\begin{aligned} p_{i,j}(t) &= a_{i,j}(t) = \overline{a}_{i,j}(t-n+r) \\ &= I_r^{[\overline{a}_{i,j}]}(B^T)^{-1}(\beta_{t-n+r})^T, \quad i, j = 1, 2, \dots, n, \ \forall t \geq n-r, \quad (1.22) \end{aligned}$$

where

$$\beta_t = \left(\frac{t^k}{y^t}\right)_{k=\overline{0,m(y)-1};\, y \in \mathbb{C}\setminus\mathcal{D}}, \quad B = (\beta_j)_{j=\overline{0,r-1}}, \quad 0^0 \equiv 1 \qquad (1.23)$$

for $t \geq 0$.

Therefore, we can determine the initial values of the subsequences $\overline{a}_{i,j}$, $i, j = 1, 2, \dots, n$:

$$I_r^{[\overline{a}_{i,j}]} = (\overline{a}_{i,j}(t))_{t=0}^{r-1} = (a_{i,j}(t))_{t=n-r}^{n-1} = (\overline{p}_{i,j}(t))_{t=n-r}^{n-1}, \quad i, j = 1, 2, \dots, n. \quad (1.24)$$

If for $y \in \mathbb{C}\setminus\mathcal{D}$ we denote

$$I_r^{[\overline{a}_{i,j}]}(B^T)^{-1} = (\gamma_{i,j,l}(y))_{l=\overline{0,m(y)-1}}, \quad i, j = 1, 2, \dots, n, \qquad (1.25)$$

then formula (1.22) can be represented in the following form:

$$\begin{aligned} p_{i,j}(t) &= \sum_{y \in \mathbb{C}\setminus\mathcal{D}} \sum_{s=0}^{m(y)-1} \frac{(t-n+r)^l}{y^{t-n+r}} \gamma_{i,j,s}(y) \\ &= \sum_{y \in \mathbb{C}\setminus\mathcal{D}} \sum_{s=0}^{m(y)-1} \sum_{k=0}^{l} C_l^k (r-n)^{l-k} y^{n-r} \gamma_{i,j,l}(y) \frac{t^k}{y^t} \end{aligned}$$

$$= \sum_{y\in\mathbb{C}\setminus\mathcal{D}} \sum_{k=0}^{m(y)-1} \frac{t^k}{y^t} \sum_{l=k}^{m(y)-1} C_l^k (r-n)^{l-k} y^{n-r} \gamma_{i,j,l}(y)$$

$$= \sum_{y\in\mathbb{C}\setminus\mathcal{D}} \sum_{k=0}^{m(y)-1} \frac{t^k}{y^t} \beta_{i,j,k}(y), \quad i, j = 1, 2, \ldots, n, \ \forall t \geq n-r, \quad (1.26)$$

where

$$\beta_{i,j,k}(y) = y^{n-r} \sum_{l=k}^{m(y)-1} C_l^k (r-n)^{l-k} \gamma_{i,j,l}(y),$$

$$\forall y \in \mathbb{C}\setminus\mathcal{D}, \ k = 0, 1, 2, \ldots, m(y)-1, \ i, j = 1, 2, \ldots, n. \quad (1.27)$$

If we rewrite the relations (1.26) in the matrix form, then we obtain the representation (1.20) of the matrices $\beta_k(y)$ $(y \in \mathbb{C}\setminus\mathcal{D}, k = 0, 1, 2, \ldots, m(y)-1)$, i.e., these matrices can be determined according to formula (1.27).

Based on the results above, we can describe the following algorithm for the decomposition of the transient matrix:

Algorithm 1.3 The Decomposition of the Transient Matrix

Input Data: *The matrix of the transition probability* P.

Output Data: *The matrices* $\beta_k(y)$ $(y \in \mathbb{C}\setminus\mathcal{D}, k = 0, 1, 2, m(y)-1)$.

1. Calculate the coefficients of the polynomial $\Delta(z)$ for the matrix P, using the algorithm from Sect. 1.3 (the algorithm based on Leverrier's method [68]).
2. Solve the equation $\Delta(z) = 0$ and find all roots of this equation in $\mathbb{C}$; then determine $\mathbb{C}\setminus\mathcal{D}$.
3. Determine the order of each root $m(y)$ of the polynomial $\Delta(z)$ (the order of each root can be found using Horner's scheme; $m(y)$ is equal to the number of successive factorizations of the polynomial $\Delta(z)$ by $(z-y)$, $\forall y \in \mathbb{C}\setminus\mathcal{D}$).
4. Calculate the matrix B, using formula (1.23).
5. Determine the matrix $(B^T)^{-1}$.
6. Calculate the values C_l^k, $l = 0, 1, 2, \ldots, \max_{y\in\mathbb{C}\setminus\mathcal{D}} m(y)-1$, $k = 0, 1, 2, \ldots, l$, according to Pascal's triangle rule: $C_l^0 = C_l^l = 1$, $C_l^k = C_{l-1}^{k-1} + C_{l-1}^k$ $(k = 1, 2, \ldots, l-1)$.
7. Find recursively $(r-n)^l$, $l = 0, 1, 2, \ldots, \max_{y\in\mathbb{C}\setminus\mathcal{D}} m(y)-1$.
8. For every $i, j = 1, 2, \ldots, n$, take the following steps:

   a. Find the initial value $I_r^{[\overline{a}_{i,j}]}$ according to formula (1.24).
   b. Calculate the values $\gamma_{i,j,l}(y)$, $y \in \mathbb{C}\setminus\mathcal{D}$, $l = 0, 1, 2, \ldots, m(y)-1$, according to (1.25).

c. For arbitrary $y \in \mathbb{C}\backslash\mathcal{D}, k = 0, 1, 2, \ldots, m(y) - 1$, determine the coefficients $\beta_{i,j,k}(y)$ of the matrix $\beta_k(y)$, using formula (1.27) and the parameters calculated in steps 6 and 7.

1.4.4 Comments on the Complexity of the Algorithm

The proposed algorithm can be used to determine the differential matrices in case the characteristic values of the matrix P are known. Therefore, the computational complexity of the algorithm depends on the computational complexity of the algorithm for determining the characteristic values of the matrix P.

If the set of characteristic values of the matrix P is known, then it is easy to observe that the running time of the algorithm for determining the differential matrices is $O(n^4)$. We obtain this estimation of the algorithm if we estimate the number of elementary operations at steps 3–8 in the worst case.

Note that the matrix $\beta_0(1)$ corresponds to the limiting probability matrix Q of the Markov process, and therefore, this matrix can be calculated using $O(n^4)$ elementary operations.

So, based on the results described above, we may conclude that the matrix $\overline{P}(t)$ can be represented as follows:

$$P(t) = \sum_{y \in \mathbb{C}\backslash\mathcal{D}} \sum_{k=0}^{m(y)-1} \beta_k(y) \frac{t^k}{y^t}, \quad \forall t \geq n - r.$$

For $t = 0, 1, 2, \ldots, n - r - 1$, this formula can be expressed in the form

$$P(t) = L(t) + \sum_{y \in \mathbb{C}\backslash\mathcal{D}} \sum_{k=0}^{m(y)-1} \beta_k(y) \frac{t^k}{y^t}, \tag{1.28}$$

where $L(t)$ is a matrix that depends only on t. If the matrices $\beta_k(y)$, $\forall y \in \mathbb{C}\backslash\mathcal{D}$, $k = 0, 1, 2, m(y) - 1$ are known, then we can determine the matrices $L(t)$ from (1.28), taking into account that $P(t) = P^t$, $\forall t \geq 0$.

In [71] it was noted that the matrices $L(t)$, $t = 0, 1, 2, \ldots, n - r - 1$, and $\beta_k(y)$ for each $y \in (\mathbb{C}\backslash\mathcal{D})\backslash\{1\}, k = 0, 1, 2, \ldots, m(y) - 1$ are differential matrices, i.e., the sum of elements of each row is equal to zero. The unique non-differential component in the representation (1.28) is the matrix $\beta_0(1)$; the remaining matrices $\beta_k(1)$, $k = 1, 2, \ldots, m(1) - 1$ are zero (see [105]).

1.5 An Algorithm to Find the Limiting and Differential Matrices

The results from the previous sections can be used for a simultaneous calculation of the limiting and the differential matrices in a Markov process. We propose a modification of the algorithms from Sects. 1.3 and 1.4.3 that allows us to determine the limiting and the differential matrices in the case if the roots of the characteristic polynomial are known.

1.5.1 The Representation of the z-Transform

Let us consider the method for determining the matrix $F(z) = (I - zP)^{-1}$ from the previous section with a simple modification:

In the calculation procedure, we will not divide $F(z)$ by $(z-1)^{m(1)-1}$.

Then it easy to observe that

$$F(z) = \frac{1}{\Delta(z)} \sum_{k=0}^{n-1} R^{(k)} z^k, \tag{1.29}$$

where the matrix-coefficients $R^{(k)}$, $k = 0, 1, 2, \ldots, n-1$ are determined recursively according to the formula

$$R^{(0)} = \beta_0 I; \quad R^{(k)} = \beta_k I + P R^{(k-1)}, \quad k = 1, 2, \ldots, n-1, \tag{1.30}$$

and the values β_k, $k = 0, 1, 2, \ldots, n$ represent the coefficients of the polynomial $\Delta(z)$ calculated according to the algorithm described above.

As we have shown in Sect. 1.3, the elements of the matrix $F(z)$ can be represented by the following formula:

$$F_{i,j}(z) = B_{i,j}(z) + \sum_{y \in \mathbb{C} \setminus \mathcal{D}} \sum_{k=1}^{m(y)} \frac{\alpha_{i,j,k}(y)}{(z-y)^k}, \quad i, j = 1, 2, \ldots, n. \tag{1.31}$$

In the general form, relation (1.31) can be written as follows:

$$F_{i,j}(z) = B_{i,j}(z) + \frac{Q_{i,j}(z)}{\Delta(z)}, \quad i, j = 1, 2, \ldots, n, \tag{1.32}$$

where $Q_{i,j}(z) \in \mathbb{C}[z]$ and $\deg(Q_{i,j}(z)) < \deg(\Delta(z)) = r$, $i, j = 1, 2, \ldots, n$.

If we write equality (1.29) for each element and after that, we make the corresponding substitutions in (1.32), then we obtain the formula

$$\sum_{k=0}^{n-1} R_{i,j}^{(k)} z^k = B_{i,j}(z)\Delta(z) + Q_{i,j}(z), \quad i, j = 1, 2, \dots, n.$$

So, $B_{i,j}(z) = \sum_{k=0}^{n-1-r} b_{i,j,k} z^k$ and $Q_{i,j}(z) = \sum_{k=0}^{r-1} q_{i,j,k} z^k$ represent the quotient and the rest, respectively, after the division of the polynomial $\sum_{k=0}^{n-1} R_{i,j}^{(k)} z^k$ by $\Delta(z)$. Therefore, the polynomials $B_{i,j}(z)$ and $Q_{i,j}(z)$ can be found by using the algorithm described below.

Algorithm 1.4 Determining the Polynomials $B_{i,j}(z)$ and $Q_{i,j}(z)$

- For $i, j = 1, 2, \dots, n$, calculate:
  - $q_{i,j,k} = R_{i,j}^{(k)}, \; k = 0, 1, 2, \dots, n-1.$
- For $k = n-1, n-2, \dots, r$, calculate:
  - $b_{i,j,k-r} = \dfrac{q_{i,j,k}}{\beta_r};$
  - $q_{i,j,k-t} = q_{i,j,k-t} - b_{i,j,k-r}\beta_{r-t}, \quad t = 0, 1, 2, \dots, r.$

1.5.2 *Expansion of the z-Transform*

Let $\mu \in \mathbb{C}\backslash\mathcal{D}$, $m(\mu) = m$ (μ^{-1} is a non-zero characteristic value of the matrix P and assume that the order of this characteristic value is m). According to formulae (1.31) and (1.32) for the separated root μ, we have

$$\frac{Q_{i,j}(z)}{\Delta(z)} = \sum_{k=1}^{m} \frac{\alpha_{i,j,k}(\mu)}{(z-\mu)^k} + \sum_{y \in (\mathbb{C}\backslash\mathcal{D})\backslash\{\mu\}} \sum_{k=1}^{m(y)} \frac{\alpha_{i,j,k}(y)}{(z-y)^k}, \quad i, j = 1, 2, \dots, n. \tag{1.33}$$

Let $\Delta(z) = (z-\mu)^m D(z)$, $D(z) = \sum_{k=0}^{r-m} d_k z^k$ and denote $\deg(D(z)) = M$. Relation (1.33) can be written as follows:

$$\frac{Q_{i,j}(z)}{\Delta(z)} = \frac{G_{i,j}(z)}{(z-\mu)^m} + \frac{E_{i,j}(z)}{D(z)}, \quad i, j = 1, 2, \dots, n,$$

where $E_{i,j}(z) = \sum_{k=0}^{M-1} e_{i,j,k} z^k$, $G_{i,j}(z) = \sum_{k=0}^{m-1} g_{i,j,k} z^k \in \mathbb{C}[z]$, $i, j = 1, 2, \dots, n$.

Making an elementary transformation, we obtain

$$Q_{i,j}(z) = G_{i,j}(z)D(z) + E_{i,j}(z)(z-\mu)^m, \quad i,j = 1,2,\ldots,n.$$

By expanding $(z-\mu)^m = \sum_{k=0}^{m} C_m^k(-\mu)^{m-k}z^k$ and then introducing the notation $\xi(k) = C_m^k(-\mu)^{-k}$, $k = 0,1,2,\ldots,m$, we have

$$(z-\mu)^m = \sum_{k=0}^{m} C_m^k(-\mu)^{m-k}z^k = (\xi(m))^{-1}\sum_{k=0}^{m}\xi(k)z^k.$$

Now for our relation, we make the following transformations:

$$\begin{aligned}
\sum_{t=0}^{r-1} q_{i,j,t}z^t &= \sum_{k=0}^{m-1} g_{i,j,k}z^k \sum_{l=0}^{M} d_l z^l + (\xi(m))^{-1}\sum_{l=0}^{M-1} e_{i,j,l}z^l \sum_{k=0}^{m}\xi(k)z^k \\
&= \sum_{k=0}^{m-1}\sum_{l=0}^{M} g_{i,j,k}d_l z^{k+l} + (\xi(m))^{-1}\sum_{k=0}^{m}\sum_{l=0}^{M-1}\xi(k)e_{i,j,l}z^{k+l} \\
&= \sum_{t=0}^{r-1} z^t \left[\sum_{\substack{k+l=t\\ 0\le k\le m-1\\ 0\le l\le M}} g_{i,j,k}d_l + (\xi(m))^{-1} \sum_{\substack{k+l=t\\ 0\le k\le m\\ 0\le l\le M-1}} \xi(k)e_{i,j,l} \right].
\end{aligned}$$

Equating the corresponding coefficients in this formula, we obtain

$$\begin{aligned}
q_{i,j,t} &= \sum_{\substack{0\le k\le m-1\\ 0\le t-k\le M}} d_{t-k}g_{i,j,k} + (\xi(m))^{-1} \sum_{\substack{0\le l\le t\\ t-m\le l\le M-1}} \xi(t-l)e_{i,j,l} \\
&= \sum_{k=0}^{m-1} d_{t-k}I_{\{0\le x\le M\}}(t-k)g_{i,j,k} + \\
&\qquad +(\xi(m))^{-1}\sum_{l=0}^{t}\xi(t-l)I_{\{t-m\le x\le M-1\}}(l)e_{i,j,l},
\end{aligned}$$

where $I_A(x)$ is the index of the set A: $I_A(x) = 1$, $\forall x \in A$ and $I_A(x) = 0$, $\forall x \notin A$.

Now we observe that for $t \le M-1$, the formula above can be written in the following form:

$$q_{i,j,t} = \sum_{k=0}^{m-1} d_{t-k}I_{\{x\le t\}}(k)g_{i,j,k} + (\xi(m))^{-1}\sum_{l=0}^{t}\xi(t-l)I_{\{x\ge t-m\}}(l)e_{i,j,l}$$

$$= \sum_{k=0}^{m-1} d_{t-k} I_{\{x \le t\}}(k) g_{i,j,k} +$$

$$+(\xi(m))^{-1}\left(e_{i,j,t} + \sum_{l=0}^{t-1} \xi(t-l) I_{\{x \ge t-m\}}(l) e_{i,j,l}\right).$$

This involves

$$e_{i,j,t} = \xi(m)\Bigg[q_{i,j,t} - \sum_{k=0}^{m-1} d_{t-k} I_{\{x \le t\}}(k) g_{i,j,k} +$$

$$-(\xi(m))^{-1} \sum_{l=0}^{t-1} \xi(t-l) I_{\{x \ge t-m\}}(l) e_{i,j,l}\Bigg].$$

So, finally we obtain the following expression:

$$e_{i,j,t} = w_{i,j,t} + \sum_{k=0}^{m-1} x_{t,k}\, g_{i,j,k}, \quad t = 0, 1, 2, \ldots, M-1, \;\; i, j = 1, 2, \ldots, n.$$

In the following, we determine the coefficients $w_{i,j,t}$ and $x_{t,k}$ from the expression above.

Then we have

$$e_{i,j,t} = w_{i,j,t} + \sum_{k=0}^{m-1} x_{t,k}\, g_{i,j,k} = \xi(m) q_{i,j,t} - \sum_{k=0}^{m-1} \xi(m) d_{t-k} I_{\{x \le t\}}(k) g_{i,j,k}$$

$$- \sum_{l=0}^{t-1} \xi(t-l) I_{\{x \ge t-m\}}(l) \left[w_{i,j,l} + \sum_{k=0}^{m-1} x_{l,k}\, g_{i,j,k} \right]$$

$$= \left[\xi(m) q_{i,j,t} - \sum_{l=0}^{t-1} \xi(t-l) I_{\{x \ge t-m\}}(l) w_{i,j,l}\right]$$

$$- \sum_{k=0}^{m-1} g_{i,j,k} \left[\xi(m) d_{t-k} I_{\{x \le t\}}(k) + \sum_{l=0}^{t-1} \xi(t-l) I_{\{x \ge t-m\}}(l) x_{s,k}\right].$$

So, we obtain

$$\begin{cases} x_{t,k} = -\xi(m)d_{t-k}I_{\{x\le t\}}(k) - \sum\limits_{l=\max\{0,\, t-m\}}^{t-1} \xi(t-l)x_{l,k}, \\ \qquad\qquad k = 0, 1, 2, \dots, m-1; \\ w_{i,j,t} = \xi(m)q_{i,j,t} - \sum\limits_{l=\max\{0,\, t-m\}}^{t-1} \xi(t-l)w_{i,j,l}, \\ \qquad\qquad t = 0, 1, 2, \dots, M-1; \quad i, j = 1, 2, \dots, n. \end{cases} \tag{1.34}$$

In the case $t \ge M$, we obtain the following transformations:

$$\begin{aligned} q_{i,j,t} &= \sum_{k=0}^{m-1} d_{t-k}I_{\{0\le x\le M\}}(t-k)g_{i,j,k} + \\ &\quad +(\xi(m))^{-1}\sum_{l=0}^{M-1} \xi(t-l)I_{\{x\ge t-m\}}(l)\left[w_{i,j,l} + \sum_{k=0}^{m-1} x_{l,k}\, g_{i,j,k}\right] \\ &= (\xi(m))^{-1}\sum_{l=0}^{M-1} \xi(t-l)I_{\{x\ge t-m\}}(l)w_{i,j,l} + \\ &\quad +\sum_{k=0}^{m-1} g_{i,j,k}\Bigg[d_{t-k}I_{\{0\le x\le M\}}(t-k) + \\ &\quad +(\xi(m))^{-1}\sum_{l=0}^{M-1} \xi(t-l)I_{\{x\ge t-m\}}(l)x_{l,k}\Bigg]. \end{aligned}$$

This involves

$$\sum_{k=0}^{m-1} r_{t,k}\, g_{i,j,k} = s_{i,j,t}, \quad t = M, M+1, \dots, r-1, \quad i, j = 1, 2, \dots, n, \tag{1.35}$$

where

$$\begin{cases} r_{t,k} = d_{t-k}I_{\{0\le x\le M\}}(t-k) + (\xi(m))^{-1} \sum\limits_{l=\max\{0,\ t-m\}}^{M-1} \xi(t-l)x_{l,k}; \\ s_{i,j,t} = q_{i,j,t} - (\xi(m))^{-1} \sum\limits_{l=\max\{0,\, t-m\}}^{M-1} \xi(t-l)w_{i,j,l}, \ k = 0, 1, 2, \dots, m-1. \end{cases} \tag{1.36}$$

The results described above allow us to determine the values $\alpha_{i,j,k}(\mu)$, $k = 1, 2, \ldots, m$; $i, j = 1, 2, \ldots, n$. Indeed, according to formula (1.33), we have

$$\frac{G_{i,j}(z)}{(z-\mu)^m} = \sum_{k=1}^{m} \frac{\alpha_{i,j,k}(\mu)}{(z-\mu)^k} = \frac{1}{(z-\mu)^m} \sum_{k=1}^{m} \alpha_{i,j,k}(\mu)(z-\mu)^{m-k}.$$

If we express this formula in a more detailed form, then we obtain

$$\begin{aligned}
\sum_{l=0}^{m-1} g_{i,j,l} z^l &= \sum_{k=1}^{m} \alpha_{i,j,k}(\mu)(z-\mu)^{m-k} = \sum_{k=0}^{m-1} \alpha_{i,j,m-k}(\mu)(z-\mu)^k \\
&= \sum_{k=0}^{m-1} \alpha_{i,j,m-k}(\mu) \sum_{l=0}^{k} C_k^l (-\mu)^{k-l} z^l \\
&= \sum_{l=0}^{m-1} z^l \sum_{k=l}^{m-1} \alpha_{i,j,m-k}(\mu) C_k^l (-\mu)^{k-l}.
\end{aligned}$$

So, we have

$$g_{i,j,l} = \sum_{k=s}^{m-1} C_k^l (-\mu)^{k-l} \alpha_{i,j,m-k}(\mu), \quad l = 0, 1, 2, \ldots, m-1, \quad i, j = 1, 2, \ldots, n.$$

If we substitute the expression $g_{i,j,l}$ in (1.35), then we obtain

$$\begin{aligned}
s_{i,j,t} &= \sum_{k=0}^{m-1} r_{t,k} \sum_{l=k}^{m-1} C_l^k (-\mu)^{l-k} \alpha_{i,j,m-l}(\mu) \\
&= \sum_{l=0}^{m-1} \alpha_{i,j,m-l}(\mu) \sum_{k=0}^{l} C_l^k (-\mu)^{l-k} r_{t,k} \\
&= \sum_{l=1}^{m} \alpha_{i,j,l}(\mu) \sum_{k=0}^{m-l} C_{m-l}^k (-\mu)^{m-l-k} r_{t,k} \\
&= \sum_{l=1}^{m} r_{t,l}^* \alpha_{i,j,l}(\mu), \qquad t = M, M+1, \ldots, r-1, \quad i, j = 1, 2, \ldots, n,
\end{aligned}$$

where

$$r_{t,l}^{*}=\sum_{k=0}^{m-l} C_{m-l}^{k}(-\mu)^{m-l-k} r_{t,k},\ t=M, M+1, \ldots, r-1,\ s=1,2, \ldots, m. \tag{1.37}$$

The solution to the system is

$$\alpha_{i,j}(\mu)=(R^{*})^{-1} S_{i,j}, \quad i, j=1,2, \ldots, n, \tag{1.38}$$

where

$$\alpha_{i,j}(\mu)=((\alpha_{i,j,l}(\mu))_{l=\overline{1,m}})^{T}, \quad S_{i,j}=((s_{i,j,t})_{t=\overline{M,r-1}})^{T}$$

and

$$R^{*}=(r_{t,l}^{*})_{t=\overline{M,r-1},\quad l=\overline{1,m}}.$$

1.5.3 The Main Conclusion and the Algorithm

The complex functions $\nu_k(z)=(1-z)^{-k}$, $\forall k \geq 1$ introduced in Sect. 1.3.1 satisfy the recurrent relation $\nu_{k+1}(z) = \mathrm{d}\nu_k(z)/(k\mathrm{d}z)$, $\forall k \geq 1$. In addition, we have shown that $\nu_k(z) = \sum_{t=0}^{\infty} H_{k-1}(t) z^t$, $\forall k \geq 1$, where the coefficient $H_{k-1}(t)$ is a polynomial of degree less than or equal to $k-1$. Moreover, the calculation formula for the elements $\beta_{i,j,k}(y)$ and the corresponding matrices

$$W_{i,j}(y,t)=\sum_{k=0}^{m(y)-1}(-y)^{-k-1} \alpha_{i,j,k+1}(y) H_k(t), \quad \forall y \in \mathbb{C} \backslash \mathcal{D},\ i, j=1,2, \ldots, n$$

have been obtained.

Let $H_k(t)=\sum_{l=0}^{k} u_s^{(k)} t^l$, $\forall k \geq 0$.

Then

$$\nu_{k+1}(z)=\frac{1}{k} \frac{\mathrm{d}}{\mathrm{d}z} \sum_{t=0}^{\infty} H_{k-1}(t) z^t=\frac{1}{k} \sum_{t=1}^{\infty} t H_{k-1}(t) z^{t-1}=\frac{1}{k} \sum_{t=0}^{\infty}(t+1) H_{k-1}(t+1) z^t$$

and we obtain

$$\begin{aligned} H_k(t) &= \frac{1}{k}(t+1) H_{k-1}(t+1) = \frac{1}{k}(t+1) \sum_{l=0}^{k-1} u_l^{(k-1)}(t+1)^l \\ &= \frac{1}{k} \sum_{l=0}^{k-1} u_l^{(k-1)}(t+1)^{l+1} = \frac{1}{k} \sum_{l=0}^{k-1} u_l^{(k-1)} \sum_{l=0}^{l+1} C_{l+1}^{l} t^l \end{aligned}$$

$$= \frac{1}{k}\sum_{l=0}^{k-1} u_l^{(k-1)}\left(1 + \sum_{l=1}^{l+1} C_{l+1}^{l} t^{l}\right)$$

$$= \frac{1}{k}\sum_{l=0}^{k-1} u_l^{(k-1)} + \frac{1}{k}\sum_{l=1}^{k} t^{l} \sum_{l=l-1}^{k-1} u_l^{(k-1)} C_{l+1}^{l}.$$

This means that

$$\begin{cases} u_0^{(0)} = 1, \\ u_0^{(k)} = \frac{1}{k}\sum\limits_{l=0}^{k-1} u_l^{(k-1)}, \ u_j^{(k)} = \frac{1}{k}\sum\limits_{l=j-1}^{k-1} C_{l+1}^{r} u_l^{(k-1)}, \ \forall k \geq 1, \ j = 1, 2, \ldots, k. \end{cases} \tag{1.39}$$

We obtain a formula for calculating the elements of the matrices for the expressions

$$\begin{cases} \beta_{i,j,k}(y) = \sum\limits_{l=k+1}^{m(y)} (-y)^{-l} \alpha_{i,j,l}(y) u_k^{(l-1)}, \\ y \in \mathbb{C}\backslash\mathcal{D}, \ k = 0, 1, 2, \ldots, m(y) - 1, \ i, j = 1, 2, \ldots, n. \end{cases} \tag{1.40}$$

Based on the results described above, we can use the following algorithm to determine the limiting and the differential matrices in a Markov chain:

Algorithm 1.5 Determining the Limiting and Differential Matrices

Input Data: *The matrix of the probability transitions* P.

Output Data: *The matrices* $\beta_k(y)$ $(y \in \mathbb{C}\backslash\mathcal{D}, \ k = 0, 1, 2, \ldots, m(y)-1)$.

1. Calculate the coefficients of the polynomial $\Delta(z)$ for the matrix P, using the algorithm from Sect. 1.3 (the algorithm based on Leverrier's method [68]).
2. Solve the equation $\Delta(z) = 0$ and find all roots of this equation in $\mathbb{C}$; then determine $\mathbb{C}\backslash\mathcal{D}$.
3. Determine the order of each root $m(y)$ of the polynomial $\Delta(z)$ (the order of each root can be found using Horner's scheme; $m(y)$ is equal to the number of successive factorizations of the polynomial $\Delta(z)$ by $(z - y), \ \forall y \in \mathbb{C}\backslash\mathcal{D}$).
4. Calculate the matrices $R^{(k)}, k = 0, 1, 2, \ldots, n-1$, according to formula (1.30).
5. Find the values $q_{i,j,k}$, $k = 0, 1, 2, \ldots, r-1$, $i, j = 1, 2, \ldots, n$, using the calculation procedure described in Sect. 1.5.1.
6. Calculate C_s^k, $s = 1, 2, \ldots, \max_{y\in\mathbb{C}\backslash\mathcal{D}} m(y)$, $k = 0, 1, 2, \ldots, s$, using Pascal's triangle rule.
7. Determine $u_j^{(k)}$, $k = 0, 1, 2, \ldots, \max_{y\in\mathbb{C}\backslash\mathcal{D}} m(y)-1$, $j = 0, 1, 2, \ldots, k$, using formula (1.39).

8. For every $\mu \in \mathbb{C}\backslash\mathcal{D}$, follow items a–g:

   a. Determine the values $\xi(k) = C_m^k(-\mu)^{-k}$, $k = 0, 1, 2, \ldots, m$ $(m = m(\mu))$.

   b. Determine the coefficients d_k, $k = 0, 1, 2, \ldots, r - m$, using Horner's scheme.

   c. Calculate the values $x_{t,k}$, $t = 0, 1, 2, \ldots, M-1$, $k = 0, 1, 2, \ldots, m-1$, according to (1.34).

   d. Calculate the values $r_{t,k}$, $t = M, M+1, \ldots, r-1$, $k = 0, 1, 2, \ldots, m-1$, using formula (1.36).

   e. Determine the elements of the matrix R^* according to the relation (1.37).

   f. Determine the matrix $(R^*)^{-1}$ using known numerical algorithms.

   g. For $i, j = 1, 2, \ldots, n$, apply items g_1–g_4:

      g_1. Calculate the values $w_{i,j,t}$, $t = 0, 1, 2, \ldots, M{-}1$, according to formula (1.34).

      g_2. Calculate the values $s_{i,j,t}$, $t = M, M{+}1, \ldots, r{-}1$, using formula (1.36).

      g_3. Determine the vector $\alpha_{i,j}(\mu)$ according to relation (1.38).

      g_4. Calculate the elements $\beta_{i,j,k}(\mu)$ of the matrix $\beta_k(\mu)$, $k = 0, 1, 2, \ldots, m(\mu){-}1$, according to formula (1.40).

It is easy to observe that the computational complexity of this algorithm is similar to the computational complexity of the algorithm from the previous section. If the characteristic values of the matrix P are known, then the algor finds the limit and the differential matrices, using $O(n^4)$ elementary operations. Note that this algorithm can be applied if the set (or a subset) of characteristic values of the matrix P is known. In this case, we use the set (or the subset) $\mathbb{C} \setminus \mathbb{D}$ of the inverse non-zero characteristic values; the algor determines the corresponding matrices, which correspond to these characteristic values. The computational complexity of the algorithm in the case of unknown characteristic values depends on the complexity of the algorithm for determining the roots of the characteristic polynomial.

Numerical Examples
The following illustrates the details of the proposed algorithms for periodic as well as for aperiodic Markov chains.

Example 1 Let a Markov chain with the following transition probability matrix be given:

$$P = \begin{pmatrix} 1 & 0 & 0 \\ 0 & 0.5 & 0.5 \\ 0.5 & 0 & 0.5 \end{pmatrix}$$

and consider the problem of determining the limiting matrix and the differential components of the matrix $\overline{P}(t)$.

First, we apply Algorithm 1.3.

Step 1.
Calculate the coefficients of the characteristic polynomial. Thus, we find

$$P^2 = \begin{pmatrix} 1 & 0 & 0 \\ 0.25 & 0.25 & 0.5 \\ 0.75 & 0 & 0.25 \end{pmatrix}, \qquad P^3 = \begin{pmatrix} 1 & 0 & 0 \\ 0.5 & 0.125 & 0.375 \\ 0.875 & 0 & 0.125 \end{pmatrix};$$

$$\overline{s}_1 = \mathrm{tr} P = 2, \quad \overline{s}_2 = \mathrm{tr} P^2 = 1.5, \quad \overline{s}_3 = \mathrm{tr} P^3 = 1.25$$

and determine

$$\beta_0 = 1, \quad \beta_1 = -\overline{s}_1 = -2, \quad \beta_2 = -(\overline{s}_2 + \overline{\beta}_1 \overline{s}_1)/2 = 1.25,$$
$$\beta_3 = -(\overline{s}_3 + \beta_1 \overline{s}_2 + \beta_2 \overline{s}_1)/3 = -0.25.$$

Steps 2–3.
Find the roots of the equation $\Delta(z) = 0$ and the set $\mathbb{C} \backslash \mathcal{D}$:

$$\Delta(z) = \sum_{k=0}^{3} \beta_k z^k = 1 - 2z + 1.25z^2 - 0.25z^3 = (1 - z)(1 - 0.5z)^2,$$
$$\mathbb{C} \backslash \mathcal{D} = \{z \in \mathbb{C} \mid \Delta(z) = 0\} = \{1, 2\}, \quad m(1) = 1, \ m(2) = 2, \ r = n = 3.$$

Step 4.
Find the matrix B:

$$\overline{\beta}_0 = (1, 1, 0), \quad \overline{\beta}_1 = (1, 0.5, 0.5),$$
$$\overline{\beta}_2 = (1, 0.25, 0.5) \implies B = \begin{pmatrix} 1 & 1 & 0 \\ 1 & 0.5 & 0.5 \\ 1 & 0.25 & 0.5 \end{pmatrix}.$$

Step 5.
Calculate $(B^T)^{-1}$:

$$(B^T)^{-1} = \begin{pmatrix} 1 & 0 & -2 \\ -4 & 4 & 6 \\ 4 & -4 & -4 \end{pmatrix}.$$

Steps 6–7.

Find the coefficients C_s^k, using Pascal's triangle rule:

$$C_0^0 = C_1^0 = C_1^1 = 1, \quad (r-n)^0 = 0^0 = 1, \quad (r-n)^1 = 0^1 = 0.$$

Steps 8*a*–8*b*.

Determine $I_r^{[\overline{a}_{i,j}]}$ and $\gamma_{i,j,l}(y)$:

$$\Gamma_{1,1} = I_3^{[\overline{a}_{1,1}]}(B^T)^{-1} = (1,1,1)\begin{pmatrix} 1 & 0 & -2 \\ -4 & 4 & 6 \\ 4 & -4 & -4 \end{pmatrix} = (1,0,0),$$

$$\Gamma_{1,2} = (0,0,0)(B^T)^{-1} = (0,0,0),$$

$$\Gamma_{1,3} = (0,0,0)(B^T)^{-1} = (0,0,0),$$

$$\Gamma_{2,1} = (0,0,0.25)(B^T)^{-1} = (1,-1,-1),$$

$$\Gamma_{2,2} = (1,0.5,0.25)(B^T)^{-1} = (0,1,0),$$

$$\Gamma_{2,3} = (0,0.5,0.5)(B^T)^{-1} = (0,0,1),$$

$$\Gamma_{3,1} = (0,0.5,0.75)(B^T)^{-1} = (1,-1,0),$$

$$\Gamma_{3,2} = (0,0,0)(B^T)^{-1} = (0,0,0),$$

$$\Gamma_{3,3} = (1,0.5,0.25)(B^T)^{-1} = (0,1,0).$$

Step 8*c*.

Find the coefficients $\beta_{i,j,k}(y)$ for the limiting and the differential matrices by using the formula:

$$\beta_{i,j,k}(y) = \sum_{l=k}^{m(y)-1} 0^{l-k}\gamma_{i,j,l}(y) = \gamma_{i,j,k}(y)$$

for $y \in \mathbb{C}\backslash\mathcal{D}$, $k = 0, 1, 2, \ldots, m(y)-1$, $i, j = 1, 2, 3$.

Based on this formula, we obtain

$$\beta_0(1) = \begin{pmatrix} 1 & 0 & 0 \\ 1 & 0 & 0 \\ 1 & 0 & 0 \end{pmatrix}, \quad \beta_0(2) = \begin{pmatrix} 0 & 0 & 0 \\ -1 & 1 & 0 \\ -1 & 0 & 1 \end{pmatrix}, \quad \beta_1(2) = \begin{pmatrix} 0 & 0 & 0 \\ -1 & 0 & 1 \\ 0 & 0 & 0 \end{pmatrix}.$$

So, the matrix $\overline{P}(t)$ can be represented as follows:

$$P(t) = \begin{pmatrix} 1\,0\,0 \\ 1\,0\,0 \\ 1\,0\,0 \end{pmatrix} + \begin{pmatrix} 0\,0\,0 \\ -1\,1\,0 \\ -1\,0\,1 \end{pmatrix} \left(\frac{1}{2}\right)^t + \begin{pmatrix} 0\,0\,0 \\ -1\,0\,1 \\ 0\,0\,0 \end{pmatrix} t \left(\frac{1}{2}\right)^t, \tag{1.41}$$

because $\lambda = 1/2$.

If we apply Algorithm 1.5 for Example 1, then we obtain:
Steps 1–3.
Repeat steps 1–3 of Algorithm 1.3 and find:

$$P^2 = \begin{pmatrix} 1 & 0 & 0 \\ 0.25 & 0.25 & 0.5 \\ 0.75 & 0 & 0.25 \end{pmatrix}, \quad P^3 = \begin{pmatrix} 1 & 0 & 0 \\ 0.5 & 0.125 & 0.375 \\ 0.875 & 0 & 0.125 \end{pmatrix};$$

$$\beta_0 = 1, \quad \beta_1 = -2, \quad \beta_2 = 1.25, \quad \beta_3 = -0.25;$$

$$\mathbb{C}\backslash\mathcal{D} = \{1, 2\}, \quad m(1) = 1, \quad m(2) = 2, \quad r = n = 3.$$

Steps 4–5.
Find $R^{(k)}$ and $q_{i,j,k}$:

$$R^{(0)} = \beta_0 I = \begin{pmatrix} 1 & 0 & 0 \\ 0 & 1 & 0 \\ 0 & 0 & 1 \end{pmatrix}, \quad R^{(1)} = \beta_1 I + PR^{(0)} = \begin{pmatrix} -1 & 0 & 0 \\ 0 & -1.5 & 0.5 \\ 0.5 & 0 & -1.5 \end{pmatrix},$$

$$R^{(2)} = \beta_2 I + PR^{(1)} = \begin{pmatrix} 0.25 & 0 & 0 \\ 0.25 & 0.5 & -0.5 \\ -0.25 & 0 & 0.5 \end{pmatrix};$$

$$q_{i,j,k} = R^{(k)}_{i,j}, \quad i, j = 1, 2, 3, \quad k = 0, 1, 2.$$

Steps 6–7.
Calculate C^k_s and $u^{(k)}_l$:

$$C^0_1 = C^1_1 = C^0_2 = C^2_2 = 1, \quad C^1_2 = C^0_1 + C^1_1 = 2; \quad u^{(0)}_0 = u^{(1)}_0 = u^{(1)}_1 = 1.$$

Step 8.
For $\mu = 1$ and $\mu = 2$, follow the items a–g: If we fix $\mu = 1$, then we have

8′) $m = m(\mu) = 1, \quad M = r - m = 2, \quad \xi(0) = 1, \quad \xi(1) = -1.$

Based on Horner's scheme, we obtain

	−0.25	1.25	−2	1
1	−0.25	1	−1	0

$\Rightarrow d_0 = -1,\quad d_1 = 1,\quad d_2 = -0.25.$

Then we calculate

$$x_{0,0} = -\xi(1)d_0 = -1,\quad x_{1,0} = -\xi(1)d_1 - \xi(1)x_{0,0} = 0;$$
$$r_{2,0} = d_2 - \xi(1)x_{1,0} = -0.25,\quad r^*_{2,1} = r_{2,0} = -0.25;$$
$$R^* = (-0.25);\quad (R^*)^{-1} = (-4);$$
$$w_{i,j,0} = -q_{i,j,0} = -R^{(0)}_{i,j},\ w_{i,j,1} = -q_{i,j,1} + w_{i,j,0} = -R^{(0)}_{i,j} - R^{(1)}_{i,j},\ i, j=1,2,3;$$
$$(s_{i,j,2})_{3\times3} = (q_{i,j,2} - w_{i,j,1})_{3\times3} = (R^{(0)}_{i,j} + R^{(1)}_{i,j} + R^{(2)}_{i,j})_{3\times3} = \begin{pmatrix} 0.25 & 0 & 0 \\ 0.25 & 0 & 0 \\ 0.25 & 0 & 0 \end{pmatrix};$$
$$(\alpha_{i,j}(1))_{3\times3} = -4(s_{i,j,2})_{3\times3} = \begin{pmatrix} -1 & 0 & 0 \\ -1 & 0 & 0 \\ -1 & 0 & 0 \end{pmatrix};$$
$$(\beta_{i,j,0}(1))_{3\times3} = (-\alpha_{i,j}(1))_{3\times3} = \begin{pmatrix} 1 & 0 & 0 \\ 1 & 0 & 0 \\ 1 & 0 & 0 \end{pmatrix} \Rightarrow \beta_0(1) = \begin{pmatrix} 1 & 0 & 0 \\ 1 & 0 & 0 \\ 1 & 0 & 0 \end{pmatrix}.$$

If we fix $\mu = 2$, then we have

8″) $m = m(\mu) = 2,\quad M = r - m = 1;\ \xi(0) = 1,\ \xi(1) = -1,\ \xi(2) = 0.25\,.$

Find the coefficients d_i, using Horner's scheme

	−0.25	1.25	−2	1
2	−0.25	0.75	−0.5	0
2	−0.25	0.25	0	

$\Rightarrow d_0 = 0.25,\ d_1 = -0.25.$

Then we calculate

$$x_{0,0} = -\xi(2)d_0 = -0.0625,\ x_{0,1} = 0;\ r_{1,0} = 0,\ r_{1,1} = 0.25,\ r_{2,0} = -0.0625,$$
$$r_{2,1} = -0.25 \Rightarrow r^*_{1,1} = 0.25,\ r^*_{1,2} = 0,\ r^*_{2,1} = -0.125,\ r^*_{2,2} = -0.0625$$

$$\Longrightarrow R^* = \begin{pmatrix} 0.25 & 0 \\ -0.125 & -0.0625 \end{pmatrix}$$

$$\Longrightarrow \quad (R^*)^{-1} = \begin{pmatrix} 4 & 0 \\ -8 & -16 \end{pmatrix};$$

$$w_{i,j,0} = 0.25 q_{i,j,0} = 0.25 R_{i,j}^{(0)}; \;\; s_{i,j,1} = q_{i,j,1} + 4 w_{i,j,0} = R_{i,j}^{(0)} + R_{i,j}^{(1)},$$

$$s_{i,j,2} = q_{i,j,2} - w_{i,j,0} = R_{i,j}^{(2)} - 0.25 R_{i,j}^{(0)} \quad \Longrightarrow \quad S_{i,j} = \begin{pmatrix} R_{i,j}^{(0)} + R_{i,j}^{(1)} \\ R_{i,j}^{(2)} - 0.25 R_{i,j}^{(0)} \end{pmatrix}$$

$$\Longrightarrow \quad \alpha_{i,j}(2) = (R^*)^{-1} S_{i,j} = \begin{pmatrix} 4R_{i,j}^{(0)} + 4R_{i,j}^{(1)} \\ -4R_{i,j}^{(0)} - 8R_{i,j}^{(1)} - 16\, R_{i,j}^{(2)} \end{pmatrix}, \; i, j = 1, 2, 3;$$

$$\beta_{i,j,0}(2) = -0.5 \alpha_{i,j,1}(2) + 0.25 \alpha_{i,j,2}(2) = -3R_{i,j}^{(0)} - 4R_{i,j}^{(1)} - 4R_{i,j}^{(2)},$$

$$\beta_{i,j,1}(2) = 0.25 \alpha_{i,j,2}(2) = -R_{i,j}^{(0)} - 2R_{i,j}^{(1)} - 4R_{i,j}^{(2)}, \; i, j = 1, 2, 3;$$

$$\Longrightarrow \quad \beta_0(2) = \begin{pmatrix} 0 & 0 & 0 \\ -1 & 1 & 0 \\ -1 & 0 & 1 \end{pmatrix}, \quad \beta_1(2) = \begin{pmatrix} 0 & 0 & 0 \\ -1 & 0 & 1 \\ 0 & 0 & 0 \end{pmatrix}.$$

So, we obtain formula (1.41).

Example 2 Let a 2-periodic Markov process determined by the matrix of probability transition

$$P = \begin{pmatrix} 0 & 1 \\ 1 & 0 \end{pmatrix}$$

be given, and consider the problem of determining the limiting and the differential components of the matrix $P(t)$.

First, we apply Algorithm 1.3.

Steps 1–3.

Find the characteristic polynomial $\Delta(z) = 0$ and the set $\mathbb{C} \backslash \mathcal{D}$ in a similar way as in Example 1:

$$P^2 = \begin{pmatrix} 1 & 0 \\ 0 & 1 \end{pmatrix}, \;\; \overline{s}_1 = \mathrm{tr} P = 0, \;\; \overline{s}_2 = \mathrm{tr} P^2 = 2 \Longrightarrow \beta_0 = 1, \;\; \beta_1 = -\overline{s}_1 = 0,$$

$$\beta_2 = -(\overline{s}_2 + \beta_1 \overline{s}_1)/2 = -1 \Longrightarrow \Delta(z) = \sum_{k=0}^{2} \beta_k z^k = 1 - z^2 = (1 - z)(1 + z),$$

i.e.,

$$\mathbb{C}\backslash\mathcal{D} = \{z \in \mathbb{C} \mid \Delta(z) = 0\} = \{1, -1\}, \ \ m(1) = m(-1) = 1, \ r = n = 2.$$

Steps 4–5.

Find the matrices B and $(B^T)^{-1}$:

$$\overline{\beta}_0 = (1, 1), \ \ \overline{\beta}_1 = (1, -1) \Rightarrow B = \begin{pmatrix} 1 & 1 \\ 1 & -1 \end{pmatrix},$$

$$(B^T)^{-1} = \begin{pmatrix} 0.5 & 0.5 \\ 0.5 & -0.5 \end{pmatrix}.$$

Steps 6–8.

Determine C_s^k, $I_r^{[\overline{a}_{i,j}]}$ and $\gamma_{i,j,l}(y)$:

$$C_0^0 = 1, \ \ (r-n)^0 = 0^0 = 1; \ \Gamma_{1,1} = \Gamma_{2,2} = (1, 0)\begin{pmatrix} 0.5 & 0.5 \\ 0.5 & -0.5 \end{pmatrix} = (0.5,\ 0.5),$$

$$\Gamma_{1,2} = \Gamma_{2,1} = (0, 1)\begin{pmatrix} 0.5 & 0.5 \\ 0.5 & -0.5 \end{pmatrix} = (0.5,\ -0.5)$$

$$\Longrightarrow \ \beta_0(1) = \begin{pmatrix} 0.5 & 0.5 \\ 0.5 & 0.5 \end{pmatrix}, \ \ \beta_0(-1) = \begin{pmatrix} 0.5 & -0.5 \\ -0.5 & 0.5 \end{pmatrix}.$$

So, we obtain the following formula for the matrix $P(t)$:

$$\overline{P}(t) = \begin{pmatrix} 0.5 & 0.5 \\ 0.5 & 0.5 \end{pmatrix} + \begin{pmatrix} 0.5 & -0.5 \\ -0.5 & 0.5 \end{pmatrix}(-1)^t. \tag{1.42}$$

Now we apply Algorithm 1.5 for Example 2.

Steps 1–3.

Repeat the steps 1–3 of Algorithm 1.3 and find

$$P^2 = \begin{pmatrix} 1 & 0 \\ 0 & 1 \end{pmatrix}, \ \ \beta_0 = 1, \ \ \beta_1 = 0, \ \ \beta_2 = -1$$

$$\Longrightarrow \ \Delta(z) = 1 - z^2 = (1-z)(1+z)$$

$$\Longrightarrow \ \mathbb{C}\backslash\mathcal{D} = \{1, -1\}, \ \ m(1) = m(-1) = 1, \ \ r = n = 2.$$

Step 4.
Calculate

$$R^{(0)} = \beta_0 I = \begin{pmatrix} 1 & 0 \\ 0 & 1 \end{pmatrix}, \quad R^{(1)} = \beta_1 I + P R^{(0)} = \begin{pmatrix} 0 & 1 \\ 1 & 0 \end{pmatrix}.$$

Steps 5–7.
Find

$$q_{i,j,k} = R_{i,j}^{(k)}, \; i, j = 1, 2, \; k = 0, 1; \quad C_1^0 = C_1^1 = 1, \; u_0^0 = 1.$$

Step 8.
For $\mu = 1$ and $\mu = -1$, follow the items a–g:
If we fix $\mu = 1$, then we have

$$8') \qquad m = m(\mu) = 1, \; M = r - m = 1, \; \xi(0) = 1, \; \xi(1) = -1;$$

then determine d_0 and d_1, using Horner's scheme

	-1	0	1
1	-1	-1	0

$\Rightarrow d_0 = -1, \; d_1 = -1.$

This means that

$$x_{0,0} = -\xi(1)d_0 = -1; \; r_{1,0} = d_1 - \xi(1)x_{0,0} = -2$$

$$\Longrightarrow r_{1,1}^* = -2$$

$$\Longrightarrow R^* = (-2), \; (R^*)^{-1} = (-0.5); \; w_{i,j,0} = \xi(1)q_{i,j,0} = -R_{i,j}^{(0)}, \; i, j = 1, 2;$$

$$s_{i,j,1} = q_{i,j,1} + \xi(1)w_{i,j,0} = R_{i,j}^{(0)} + R_{i,j}^{(1)}$$

$$\Longrightarrow \alpha_{i,j}(1) = (-0.5)(s_{i,j,1}) = -0.5R_{i,j}^{(0)} - 0.5R_{i,j}^{(1)}, \; i, j = 1, 2;$$

$$\beta_{i,j,0}(1) = -\alpha_{i,j,1}(1) = 0.5R_{i,j}^{(0)} + 0.5R_{i,j}^{(1)}, \; i, j = 1, 2;$$

$$\Longrightarrow \beta_0(1) = \begin{pmatrix} 0.5 & 0.5 \\ 0.5 & 0.5 \end{pmatrix}.$$

If we fix $\mu = -1$, then we have

$$8'') \qquad m = m(\mu) = 1, \; M = r - m = 1, \; \xi(0) = 1, \; \xi(1) = 1;$$

then determine d_0 and d_1, using Horner's scheme

	−1	0	1
−1	−1	1	0

$\Rightarrow d_0 = 1,\ d_1 = -1.$

This means that

$$x_{0,0} = -\xi(1)d_0 = -1;\ \ r_{1,0} = d_1 + \xi(1)x_{0,0} = -2$$

$$\Longrightarrow r_{11}^* = -2$$

$$\Longrightarrow R^* = (-2),\ (R^*)^{-1} = (-0.5);\ \ w_{i,j,0} = \xi(1)q_{i,j,0} = R_{i,j}^{(0)},\ i,j = 1,2;$$

$$s_{i,j,1} = q_{i,j,1} - \xi(1)w_{i,j,0} = R_{i,j}^{(1)} - R_{i,j}^{(0)}$$

$$\Longrightarrow \alpha_{i,j}(-1) = (-0.5)(s_{i,j,1}) = -0.5R_{i,j}^{(1)} + 0.5R_{i,j}^{(0)},\ i,j = 1,2;$$

$$\beta_{i,j,0}(-1) = \alpha_{i,j,0}(-1) = -0.5R_{i,j}^{(1)} + 0.5R_{i,j}^{(0)},\ i,j = 1,2;$$

$$\Longrightarrow \beta_0(-1) = \begin{pmatrix} 0.5 & -0.5 \\ -0.5 & 0.5 \end{pmatrix}.$$

So, we obtain formula (1.42).

In the examples given above, the roots of the characteristic polynomials are real values. Below we consider an example where the characteristic polynomial contains complex roots. For this example, the calculations in the algorithms are similar as in the case with real roots; however, at the final stage of the algorithms, in order to obtain the real representation of $P(t)$, it is necessary to make some additional elementary transformations that eliminate the imaginary component of $T(t)$. We illustrate these transformations in the example below.

Example 3 Let a Markov chain with the matrix of probability transitions

$$P = \begin{pmatrix} \frac{1}{2} & 0 & \frac{1}{2} \\ \frac{1}{2} & \frac{1}{4} & \frac{1}{4} \\ \frac{1}{4} & \frac{1}{2} & \frac{1}{4} \end{pmatrix}$$

be given, and consider the problem of determining the limiting matrix and the differential components of the matrix $P(t)$.

If we apply Algorithm 1.3, then we obtain:

Step 1.

Calculate the coefficients of the characteristic polynomial. Thus, we find

$$P^2 = \begin{pmatrix} \frac{3}{8} & \frac{1}{4} & \frac{3}{8} \\ \frac{7}{16} & \frac{3}{16} & \frac{3}{8} \\ \frac{7}{16} & \frac{1}{4} & \frac{5}{16} \end{pmatrix}, \quad P^3 = \begin{pmatrix} \frac{13}{32} & \frac{1}{4} & \frac{11}{32} \\ \frac{13}{32} & \frac{15}{64} & \frac{23}{64} \\ \frac{27}{64} & \frac{7}{32} & \frac{23}{64} \end{pmatrix};$$

$$\overline{s}_1 = \operatorname{tr} P = 1, \quad \overline{s}_2 = \operatorname{tr} P^2 = \frac{7}{8}, \quad \overline{s}_3 = \operatorname{tr} P^3 = 1$$

and determine

$$\beta_0 = 1, \quad \beta_1 = -\overline{s}_1 = -1, \quad \beta_2 = -(\overline{s}_2 + \overline{\beta}_1 \overline{s}_1)/2 = \frac{1}{16},$$

$$\beta_3 = -(\overline{s}_3 + \beta_1 \overline{s}_2 + \beta_2 \overline{s}_1)/3 = -\frac{1}{16}.$$

Steps 2–3.

Find the roots of the equation $\Delta(z) = 0$ and the set $\mathbb{C}\backslash\mathcal{D}$:

$$\Delta(z) = \sum_{k=0}^{3} \beta_k z^k = 1 - z + \frac{1}{16} z^2 - \frac{1}{16} z^3 = -\frac{1}{16}(z-1)(z-4i)(z+4i),$$

$\mathbb{C}\backslash\mathcal{D} = \{z \in \mathbb{C} \,|\, \Delta(z) = 0\} = \{1, -4i, 4i\}$; $m(1) = m(-4i) = m(4i) = 1$, $r = n = 3$.

Step 4.

Find the matrix B: $\overline{\beta}_0 = (1, 1, 1)$, $\overline{\beta}_1 = (1, i/4, -i/4)$, $\overline{\beta}_2 = (1, -1/16, -1/16)$, i.e.,

$$B = \begin{pmatrix} 1 & 1 & 1 \\ 1 & \frac{i}{4} & -\frac{i}{4} \\ 1 & -\frac{1}{16} & -\frac{1}{16} \end{pmatrix}.$$

Step 5.
Calculate $(B^T)^{-1}$:

$$(B^T)^{-1} = \begin{pmatrix} \frac{1}{17} & \frac{8}{17} + \frac{2}{17}i & \frac{8}{17} - \frac{2}{17}i \\ 0 & -2i & 2i \\ \frac{16}{17} & -\frac{8}{17} + \frac{32}{17}i & -\frac{8}{17} - \frac{32}{17}i \end{pmatrix}.$$

Steps 6–7.
Find the coefficients C_s^k, using Pascal's triangle rule:

$$C_0^0 = C_1^0 = C_1^1 = 1, \quad (r-n)^0 = 0^0 = 1, \quad (r-n)^1 = 0^1 = 0.$$

Steps 8a–8b.
Determine $I_r^{[\overline{a}_{i,j}]}$ and $\gamma_{i,j,l}(y)$:

$$\Gamma_{1,1} = I_3^{[\overline{a}_{1,1}]}(B^T)^{-1} = \left(1, \frac{1}{2}, \frac{3}{8}\right)(B^T)^{-1} = \left(\frac{7}{17}, \frac{5}{17} - \frac{3}{17}i, \frac{5}{17} + \frac{3}{17}i\right),$$

$$\Gamma_{1,2} = \left(0, 0, \frac{1}{4}\right)(B^T)^{-1} = \left(\frac{4}{17}, -\frac{2}{17} + \frac{8}{17}i, -\frac{2}{17} - \frac{8}{17}i\right),$$

$$\Gamma_{1,3} = \left(0, \frac{1}{2}, \frac{3}{8}\right)(B^T)^{-1} = \left(\frac{6}{17}, -\frac{3}{17} - \frac{5}{17}i, -\frac{3}{17} + \frac{5}{17}i\right),$$

$$\Gamma_{2,1} = \left(0, \frac{1}{2}, \frac{7}{16}\right)(B^T)^{-1} = \left(\frac{7}{17}, -\frac{7}{34} - \frac{3}{17}i, -\frac{7}{34} + \frac{3}{17}i\right),$$

$$\Gamma_{2,2} = \left(1, \frac{1}{4}, \frac{3}{16}\right)(B^T)^{-1} = \left(\frac{4}{17}, \frac{13}{34} - \frac{i}{34}, \frac{13}{34} + \frac{i}{34}\right),$$

$$\Gamma_{2,3} = \left(0, \frac{1}{4}, \frac{3}{8}\right)(B^T)^{-1} = \left(\frac{6}{17}, -\frac{3}{17} + \frac{7}{34}i, -\frac{3}{17} - \frac{7}{34}i\right),$$

$$\Gamma_{3,1} = \left(0, \frac{1}{4}, \frac{7}{16}\right)(B^T)^{-1} = \left(\frac{7}{17}, -\frac{7}{34} + \frac{11}{34}i, -\frac{7}{17} - \frac{11}{34}i\right),$$

$$\Gamma_{3,2} = \left(0, \frac{1}{2}, \frac{1}{4}\right)(B^T)^{-1} = \left(\frac{4}{17}, -\frac{2}{17} - \frac{9}{17}i, -\frac{2}{17} + \frac{9}{17}i\right),$$

$$\Gamma_{3,3} = \left(1, \frac{1}{4}, \frac{5}{16}\right)(B^T)^{-1} = \left(\frac{6}{17}, \frac{11}{34} + \frac{7}{17}i, \frac{11}{17} - \frac{7}{17}i\right).$$

Step 8c.

Find the coefficients $\beta_{i,j,k}(y)$ for the limiting and the differential matrices, using the formula

$$\beta_{i,j,k}(y) = \sum_{l=k}^{m(y)-1} 0^{l-k} \gamma_{i,j,l}(y) = \gamma_{i,j,k}(y)$$

for $y \in \mathbb{C}\backslash\mathcal{D}$, $k = 0, 1, 2, \ldots, m(y)-1$, $i, j = 1, 2, 3$.

Based on this formula, we obtain

$$\beta_0(1) = \begin{pmatrix} \frac{7}{17} & \frac{4}{17} & \frac{6}{17} \\ \frac{7}{17} & \frac{4}{17} & \frac{6}{17} \\ \frac{7}{17} & \frac{4}{17} & \frac{6}{17} \end{pmatrix},$$

$$\beta_0(-4i) = \begin{pmatrix} \frac{5}{17} - \frac{3}{17}i & -\frac{2}{17} + \frac{8}{17}i & -\frac{3}{17} - \frac{5}{17}i \\ -\frac{7}{34} - \frac{3}{17}i & \frac{13}{34} - \frac{1}{34}i & -\frac{3}{17} + \frac{7}{34}i \\ -\frac{7}{34} + \frac{11}{34}i & -\frac{2}{17} - \frac{9}{17}i & \frac{11}{34} + \frac{7}{34}i \end{pmatrix},$$

$$\beta_0(4i) = \begin{pmatrix} \frac{5}{17} + \frac{3}{17}i & -\frac{2}{17} - \frac{8}{17}i & -\frac{3}{17} + \frac{5}{17}i \\ -\frac{7}{34} + \frac{3}{17}i & \frac{13}{34} + \frac{1}{34}i & -\frac{3}{17} - \frac{7}{34}i \\ -\frac{7}{34} - \frac{11}{34}i & -\frac{2}{17} + \frac{9}{17}i & \frac{11}{34} - \frac{7}{34}i \end{pmatrix}.$$

So, the matrix $P(t)$ can be represented as follows:

$$\overline{P}(t) = \beta_0(1) + \lambda^t \beta_0(-4i) + \overline{\lambda}^t \beta_0(4t), \tag{1.43}$$

where $\lambda = 1/(-4i) = i/4$ and $\overline{\lambda} = 1/(4i) = -i/4$.

If we set $i = \cos(\pi/2) + i\sin(\pi/2)$, then

$$\lambda^t = \left(\frac{i}{4}\right)^t = \left(\frac{1}{4}\right)^t \left(\cos\frac{t\pi}{2} + i\sin\frac{t\pi}{2}\right).$$

Here, $\beta_0(4i)$ is the conjugate matrix of $\beta_0(-4i)$, i.e., $\beta_0(4i) = \overline{\beta}_0(-4i)$.

Therefore,

$$\lambda^t \beta_0(-4i) + \overline{\lambda}^t \beta_0(4t) = 2\Big(\mathrm{Re}(\beta_0(-4i))\mathrm{Re}(\lambda^t) - \mathrm{Im}(\beta_0(-4i))\mathrm{Im}(\lambda^t)\Big).$$

If we take these properties into account, then we obtain

$$P(t) = \begin{pmatrix} \frac{7}{17} & \frac{4}{17} & \frac{6}{17} \\ \frac{7}{17} & \frac{4}{17} & \frac{6}{17} \\ \frac{7}{17} & \frac{4}{17} & \frac{6}{17} \end{pmatrix} + 2\left(\frac{1}{4}\right)^t \cos\frac{t\pi}{2} \begin{pmatrix} \frac{5}{17} & -\frac{2}{17} & -\frac{3}{17} \\ -\frac{7}{34} & \frac{13}{34} & -\frac{3}{17} \\ -\frac{7}{34} & -\frac{2}{17} & \frac{11}{34} \end{pmatrix} +$$

$$+ 2\left(\frac{1}{4}\right)^t \sin\frac{t\pi}{2} \begin{pmatrix} \frac{3}{17} & -\frac{8}{17} & \frac{5}{17} \\ \frac{3}{17} & \frac{1}{34} & -\frac{7}{34} \\ -\frac{11}{34} & \frac{9}{17} & -\frac{7}{34} \end{pmatrix}.$$

Formula (1.43) can be obtained by using Algorithm 1.5.

The calculation procedure according to this algorithm is similar to the calculation procedures in the previous examples.

1.6 Fast Computing Schemes for Limiting and Differential Matrices

The algorithms from Sects. 1.3–1.5 can be improved if the results from [33, 35, 81, 94, 95, 145, 173] concerned with fast matrix multiplication, efficient computation of the characteristic polynomial, and resuming matrix polynomials are applied. Based on these results, we substantiate an algorithm for finding the limiting matrix with complexity $O(n^3)$ and an algorithm for calculating differential matrices with complexity $O(n^{\omega+1})$, where $O(n^{\omega})$ is the complexity of the used matrix multiplication algorithm [93]. The theoretical computational complexity estimation of the algorithm is governed by the fastest known matrix multiplication algorithm for which $\omega < 2.372864$.

1.6.1 Fast Matrix Multiplication and Matrix Inversion

The classical method that multiplies two $n \times n$ matrices (the naive algorithm) needs $O(n^3)$ elementary operations. Actually, to multiply two matrices, more efficient algorithms can be used, especially in the case of large n. The first algorithm of matrix multiplication with computation complexity $O(n^\omega)$, where $\omega < 3$, was proposed by Strassen [173] in 1969.

He developed an algorithm with $\omega \approx 2.802$. Later, this algorithm was improved by others (see references from [5, 200]), and a series of algorithms with $\omega \approx 2,796$ (Pan 1978), $\omega \approx 2,2522$ (Bini 1981), and $\omega \approx 2.479$ (Strassen 1986) were obtained. The last more efficient algorithms were elaborated by Coppersmith-Winograd [33] with $\omega < 2.376$ in 1990 and Williams with $\omega < 2.372873$ in 2014. The fastest known algorithm at the moment is the Le Gall version [94], developed in 2014 as an improvement of the Coppersmith-Winograd algorithm. He proved that $\omega = 2.3728639\ldots$. The fastest known matrix multiplication algorithms (Coppersmith-Winograd, Williams, Le Gall) are frequently used as a building block in other algorithms to prove theoretical time bounds.

However, unlike the Strassen algorithm, they are considered galactic algorithms and are not used in practice due to their advantage only for very large matrices that cannot be processed by modern hardware. Despite the disadvantages mentioned above, there exist parallel algorithms for computing the product of two $n \times n$ matrices. D'Alberto and Nicolau studied the adaptive Winograd's matrix multiplications in [35], and Ballard, Demmel, Holtz, Lipshitz, and Schwartz described a communication-optimal parallel algorithm for Strassen's matrix multiplication in [1]. Similar to the matrix multiplication operation, the matrix inversion is one of the most basic problems in mathematics and computer science. There are multiple algorithms for finding the inverse of an invertible matrix: the Gauss-Jordan elimination, LU decomposition, Newton's iterative method, Cayley-Hamilton method, eigen decomposition, Cholesky decomposition, reciprocal basis vectors method, blockwise inversion, and other methods. The simplest method to inverse an invertible matrix is the Gauss-Jordan elimination. According to this method, the identity matrix is augmented by an identity matrix to the right of a given matrix, and after that, through the application of elementary row operations, the reduced echelon form is found, the obtained left block is the identity matrix, and the right block is the inverse of the given invertible matrix. If the algorithm is unable to reduce the left block to the identity matrix, then the initial matrix is not invertible. The disadvantage of this method is its $O(n^3)$ computational complexity, which is the same as for a naive matrix multiplication algorithm. However, due to its simplicity, it can be considered to be a part of more complex algorithms, at least in the case when there are also other algorithm parts with a complexity bigger than $O(n^3)$ since the entire algorithm complexity is not so much affected by this method.

In contrast to the Gauss-Jordan elimination, the blockwise inversion method [13, 34] is a divide-and-conquer algorithm, which allows reducing the complexity of the algorithm due to its relationship with matrix multiplication complexity.

It was shown in [34] that the blockwise inversion method runs with the same time complexity as the matrix multiplication algorithm that is used internally. Since the best known matrix multiplication complexity is $O(n^{\omega})$, this means that the fastest known matrix inversion algorithm runs with the same time complexity $O(n^{\omega})$.

1.6.2 Determining the Characteristic Polynomial and Resuming the Matrix Polynomial

The characteristic polynomial for a given matrix can be found by using the algorithms from [81, 145]. In [145], there were several combined ideas to get a new Las Vegas randomized algorithm for computing the characteristic polynomial. The complexity of this randomized algorithm is $O(n^{\omega})$. The Keller Gehrig's deterministic algorithm [81] works for all inputs and in the worst case has the complexity $O(n^{\omega}lg(n))$.

The problem of resuming the matrix polynomial is the following: For a given $n \times n$ matrix P and a polynomial $T(z) = \sum_{k=0}^{m} a_k z^k$ with numeric coefficients to compute the matrix $T(P) = \sum_{k=0}^{m} a_k P^k$.

This problem can be solved by using the naive algorithm recursively, computing the matrices $P^0 = I_n,\ P^1 = P^0 P,\ P^2 = P^1 P,\ \ldots\ P^m = P^{m-1} P$, performing m matrix multiplications, multiplying each matrix with the corresponding coefficient a_k, $k \in \{1, 2, \ldots, m\}$, and summing over k. The complexity of such an algorithm is $O(mn^{\omega})$.

The $r \times r$ scheme presented in [95] shows that we can use only $O(r)$ matrix multiplications where $r = [\sqrt{m+1}]$. This means that we have to store the matrices $P^0, P^1, \ldots, P^r$ and reuse them along a polynomial subdivision into a set of sub-polynomials of degree less than r or equal to r. So, the complexity of the algorithm is $O(\max\{mn^2, rn^{\omega}\})$. On the conditions $m \approx n$ and $\omega <= 2.5$, the running time of this algorithm is $O(n^3)$.

For resuming matrix polynomials, there exist also parallel algorithms. For example, Alonso, Boratto, Peinado, Ibáñez, and Sastre in [1] evaluated matrix polynomials, using several GPGPUs.

1.6.3 A Modified Algorithm to Find the Limiting Matrix

Let $K(z)$ be the characteristic polynomial for the transition probability matrix P. Then, based on the results from Sects. 1.3.2–1.3.4, the following equalities

$$\Delta(z) = (-1)^n z^n K(z^{-1}) = (z-1)^{m(1)} V(z)$$

hold. Substituting z with z^{-1}, we obtain

$$(-1)^n z^{-n} K(z) = (z^{-1} - 1)^{m(1)} V(z^{-1}).$$

This implies the relation

$$V(z^{-1}) = \frac{K(z)}{(-z)^n (z^{-1} - 1)^{m(1)}}. \tag{1.44}$$

Additionally, from Sects. 1.3.2–1.3.4, we have

$$S^{(k)} = \sum_{j=0}^{k} \beta^*_{k-j} P^j, \; k = \overline{0, N-1}. \tag{1.45}$$

Taking into account that

$$V'(z) = (z-1)V(z), \; V(z) = \sum_{k=0}^{N-1} \gamma_k z^k \; and \; V'(z) = \sum_{k=0}^{N} \beta^*_k z^k,$$

and dividing the polynomial $V'(z)$ by $z - 1$, using Horner's scheme, we obtain

$$\gamma_k = \sum_{j=k+1}^{N} \beta^*_j, \; k = \overline{0, N-1}.$$

Since $\sum_{j=0}^{N} \beta^*_j = V'(1) = 0$, we have

$$\gamma_k = -\sum_{j=0}^{k} \beta^*_j, \; k = \overline{0, N-1}. \tag{1.46}$$

So, by using equalities (1.45) and (1.46), we obtain the following formula:

$$S = \sum_{k=0}^{N-1} S^{(k)} = -\sum_{k=0}^{N-1} \gamma_{N-1-k} P^k. \tag{1.47}$$

Now, using the equalities

$$V(z^{-1}) = \sum_{k=0}^{N-1} \gamma_k z^{-k} = z^{1-N} \sum_{k=0}^{N-1} \gamma_k z^{N-1-k} = z^{1-N} \sum_{k=0}^{N-1} \gamma_{N-1-k} z^k$$

and applying relation (1.44), we obtain

$$\sum_{k=0}^{N-1} \gamma_{N-1-k} z^k = z^{N-1} V(z^{-1}) = z^{n-m(1)} V(z^{-1}) =$$

$$= \frac{z^{n-m(1)} K(z)}{(-z)^n (z^{-1} - 1)^{m(1)}} = (-1)^{N-1} \frac{K(z)}{(z-1)^{m(1)}}.$$

Taking into account that $K(z) \vdots (z-1)^{m(1)}$, we can represent $K(z)$ as follows: $K(z) = (z-1)^{m(1)} T(z)$. This means that the previous relation becomes

$$\sum_{k=0}^{N-1} \gamma_{N-1-k} z^k = (-1)^{N-1} T(z). \tag{1.48}$$

Formula (1.48) represents an identity for two polynomials. Substituting z with P, we obtain an equality of two matrix polynomials

$$\sum_{k=0}^{N-1} \gamma_{N-1-k} P^k = (-1)^{N-1} T(P). \tag{1.49}$$

Formula (1.47) becomes

$$S = (-1)^N T(P), \tag{1.50}$$

i.e., the limiting matrix is

$$Q = -\frac{S}{V(1)} = \frac{(-1)^{N-1}}{V(1)} T(P). \tag{1.51}$$

So, the limiting matrix Q represents the product of the matrix $R = T(P)$ with a constant $c = \dfrac{(-1)^{N-1}}{V(1)}$. Because the limiting matrix Q is a stochastic matrix, we can determine the constant c through normalization operation, i.e., c^{-1} represents the sum of the elements from an arbitrary row of the matrix R.

The results presented above substantiate the following modified version of Algorithm 1.2 for determining the limiting matrix:

Algorithm 1.6 Determining the Limiting Matrix

Input: *The transition matrix* $P \in \mathcal{M}_n(\mathbb{R})$.
Output: *The limiting matrix* $Q \in \mathcal{M}_n(\mathbb{R})$.

1. Determine the characteristic polynomial $K(z) = |P - zI_n|$.
2. Divide the polynomial $K(z)$ by $(z-1)^{m(1)}$, where $m(1)$ is the multiplicity of the eigenvalue $z_0 = 1$, obtaining the quotient $T(z)$.
3. Compute the matrix $R = T(P)$.
4. Determine the limiting matrix Q by dividing the matrix R by the sum of the elements of an arbitrary row.

Note that for a better performance of Algorithm 1.6, it is necessary to use the fast methods from Sects. 1.6.1 and 1.6.2. It is easy to observe that if we apply the algorithm mentioned in these sections, then we obtain an algorithm for determining the limiting matrix Q with complexity $O(n^3)$.

Numerical Examples

We present some numerical examples that illustrate the main details of the elaborated algorithm from Sect. 1.6.3.

Example 1 Consider the Markov process with the transition probability matrix $P = \begin{pmatrix} 0 & 1 \\ 1 & 0 \end{pmatrix}$. We can see that $P^{2t} = \begin{pmatrix} 1 & 0 \\ 0 & 1 \end{pmatrix}$ and $P^{2t+1} = \begin{pmatrix} 0 & 1 \\ 1 & 0 \end{pmatrix}$, $\forall t \geq 0$, i.e., the Markov chain is 2-periodic.

So, in this case, the limit $\lim_{t\to\infty} P^t$ does not exist, but there exists the limiting matrix Q, which can be found by using Algorithm 1.6. If we apply this algorithm, then we have:

$$K(z) = |P - zI_n| = \begin{vmatrix} -z & 1 \\ 1 & -z \end{vmatrix} = (z-1)(z+1) \Rightarrow$$

$$T(z) = z + 1 \Rightarrow T(P) = I_2 + P = \begin{pmatrix} 1 & 1 \\ 1 & 1 \end{pmatrix} \Rightarrow$$

$$Q = \frac{1}{1+1} T(P) = \begin{pmatrix} 1/2 & 1/2 \\ 1/2 & 1/2 \end{pmatrix}.$$

So, we obtain the same result as in numerical example 1 from [92, 125]. Note that the Markov process for the considered process is not ergodic; however, the rows of the limiting matrix are the same, and the vector of limiting probabilities is $\pi^* = (0.5, 0.5)$, i.e., regardless of the initial state of the Markov process, it will occupy both states with the same probability 0.5 after a large number of transitions.

Example 2 Consider the Markov process with the transition probability matrix $P = \begin{pmatrix} 0.5 & 0.5 \\ 0.4 & 0.6 \end{pmatrix}$. We find the matrix Q by applying Algorithm 1.6:

$$K(z) = |P - zI_n| = \begin{vmatrix} 0.5 - z & 0.5 \\ 0.4 & 0.6 - z \end{vmatrix} = (z-1)(z-0.1) \Rightarrow$$

$$T(z) = z - 0.1 \Rightarrow T(P) = -0.1 \cdot I_2 + P = \begin{pmatrix} 0.4 & 0.5 \\ 0.4 & 0.5 \end{pmatrix} \Rightarrow$$

$$Q = \frac{1}{0.4 + 0.5} T(P) = \begin{pmatrix} 4/9 & 5/9 \\ 4/9 & 5/9 \end{pmatrix}.$$

So, we obtain the same result as in numerical example 2 from [92, 125]. The rows of the limiting matrix are the same, and the Markov process is ergodic, with the vector of limiting probabilities $\pi^* = (4/9,\ 5/9)$.

Example 3 We consider a non-ergodic Markov process with the transition probability matrix $P = \begin{pmatrix} 1 & 0 & 0 \\ 0 & 1 & 0 \\ 1/3 & 1/3 & 1/3 \end{pmatrix}$. If for this example we apply Algorithm 1.6 to determine the limiting matrix, then we obtain:

$$K(z) = |P - zI_n| = \begin{vmatrix} 1-z & 0 & 0 \\ 0 & 1-z & 0 \\ 1/3 & 1/3 & 1/3 - z \end{vmatrix} = (z-1)^2(1/3 - z) \Rightarrow$$

$$T(z) = 1/3 - z \Rightarrow T(P) = (1/3)I_3 - P = \begin{pmatrix} -2/3 & 0 & 0 \\ 0 & -2/3 & 0 \\ -1/3 & -1/3 & 0 \end{pmatrix} \Rightarrow$$

$$Q = \frac{1}{-\frac{2}{3} + 0 + 0} \cdot T(P) = \begin{pmatrix} 1 & 0 & 0 \\ 0 & 1 & 0 \\ 1/2 & 1/2 & 0 \end{pmatrix}.$$

So, we obtain the same result as in numerical example 3 from [92, 125]. In this case, all rows of the matrix Q are different. It is easy to observe that for the considered example, there exists the limit $\lim_{t\to\infty} P^t$, and this limit is Q.

1.6.4 A Modified Algorithm to Find Differential Matrices

Now we substantiate an algorithm to determine the differential matrices with computational complexity of $O(n^{\omega+1})$. We show that such an algorithm can be elaborated if the roots of the characteristic polynomial of matrix P are known.

As we have shown in Sect. 1.3.2, the decomposition

$$P(t) = \sum_{y\in\mathbb{C}\setminus\mathcal{D}} \sum_{k=0}^{m(y)-1} \frac{t^k}{y^t}\beta_k(y), \ \forall t \geq n-r \tag{1.52}$$

holds and the coefficients

$$\beta_k(y), \ y \in (\mathbb{C}\setminus\mathcal{D})\setminus\{1\}, \ k = \overline{0, m(y)-1}$$

represent the differential matrices of the Markov process. Since the set $\mathbb{C}\setminus\mathcal{D}$ consists of all inverses of eigenvalues of the transition probability matrix P, decomposition (1.52) is equivalent to the decomposition

$$P(t) = \sum_{y\in\sigma(P)} \sum_{k=0}^{m(y)-1} t^k y^t \overline{\beta}_k(y), \ \forall t \geq n-r,$$

where $\sigma(P) = \{z \in \mathbb{C} \mid K(z) = 0\}$ is the spectrum of the transition probability matrix P and $K(z)$ is the characteristic polynomial of matrix P. Recall that $r = deg(K_P(z))$, $m(y)$ is the multiplicity of the root y of the polynomial $K_P(z)$, and $\overline{\beta}_k(y) = \beta_k(y^{-1})$, $\forall y \in \sigma(P)$. In particular, from this decomposition, we have

$$P(t) = \sum_{y\in\sigma(P)} \sum_{k=0}^{m(y)-1} t^k y^t \overline{\beta}_k(y), \ t = \overline{n-r, n-1}.$$

In components, this relation can be written as follows:

$$p_{i,j}(t) = \sum_{y\in\sigma(P)} \sum_{k=0}^{m(y)-1} t^k y^t \overline{\beta}_{i,j,k}(y), \ t = \overline{n-r, n-1}, \ i, j = \overline{1, n}.$$

If we denote

$$\beta_t = (t^k y^t)_{(y,k)\in\sigma(P)\times\{0,\ldots,m(y)-1\}}, \ \forall t \geq 0, \ B = (\beta_j)_{j=\overline{0,r-1}}, \ 0^0 \equiv 1 \tag{1.53}$$

and

$$p_{i,j} = (p_{i,j}(t))_{t=n-r}^{n-1}, \ \overline{\beta}_{i,j} = (\overline{\beta}_{i,j,k}(y))_{(y,k)\in\sigma(P)\times\{0,\ldots,m(y)-1\}}, \ i, j = \overline{1, n},$$

then we obtain

$$p_{i,j} = \overline{\beta}_{i,j} B^T, \; i, j = \overline{1, n},$$

where B^T is the transpose of matrix B. Since the vectors $\beta_j, \; j = \overline{1, r}$ are linearly independent, the matrix B is invertible. As a consequence,

$$\overline{\beta}_{i,j} = p_{i,j}(B^T)^{-1}, \; i, j = \overline{1, n}.$$

Finally, if we consider the matrices $\Delta = (\overline{\beta}_{i,j})_{(i,j)\in\{1,\ldots,n\}\times\{1,\ldots,n\}}$ and $\Pi = (p_{i,j})_{(i,j)\in\{1,\ldots,n\}\times\{1,\ldots,n\}}$, then we obtain

$$\Delta = \Pi(B^T)^{-1}. \tag{1.54}$$

Since Π is an $n^2 \times r$ matrix and B is an $r \times r$ matrix, then the result Δ is an $n^2 \times r$ matrix, which contains all elements of $rn \times n$ coefficient matrices $\overline{\beta}_k(y), \; y \in \sigma(P)$. A row of Δ corresponding to a pair index (i, j) represents the elements from position (i, j) of the limiting and differential matrices. This means that the columns of Δ represent the flattened version of the limiting and differential matrices. In a similar way, we can refer this to the matrix Π whose columns represent the flattened version of the powers $P^t, \; t = \overline{n - r, n - 1}$.

Now, if we analyze the computational complexity of calculating Δ according to (1.54), then we can notice the following: We can obtain a better performance for calculating (1.54) if the set of rows of Δ and Π are split into about $\left\lceil \frac{n^2}{r} \right\rceil$ parts of dimension r. After that, Eq. (1.54) can be regarded as about $\left\lceil \frac{n^2}{r} \right\rceil$ matrix multiplications of dimension $r \times r$. This means that we have the complexity $O(r^{\omega-1}n^2)$ to determine Δ. Since in the worst case, we have $r = n$, the complexity of Δ according to formula (1.54) is $O(n^{\omega+1})$.

In addition to this complexity, we need to take into account the complexity of calculating the matrices Π, B, B^T, and $(B^T)^{-1}$. In order to obtain the matrix Π, we need to have all matrices $P^t, t = \overline{n - r, n - 1}$, i.e., in the worst case, we need $O(n^{\omega+1})$ operations. If the eigenvalues of the matrix P are known, then the matrices B and B^T can be computed by using $O(n^2)$ elementary operations and the computation of $(B^T)^{-1}$ takes $O(n^{\omega})$ arithmetic operations. Taking into account all of the above remarks, we can conclude that the entire algorithm complexity is $O(n^{\omega+1})$.

So, having the eigenvalues of the transition probability matrix P, we can compute the matrix of limiting state probabilities and the differential matrices, using $O(n^{\omega+1})$ arithmetic operations.

Thus, based on the results mentioned above, we can propose the following algorithm to determine the differential matrices:

Algorithm 1.7 Determining the Limiting and Differential Matrices

Input: *The transition probability matrix* $P \in \mathcal{M}_n(\mathbb{R})$.
Output: *Limiting and differential matrices.*

1. Calculate $P^0 = I_n,\ P^1 = P^0P,\ P^2 = P^1P,\ \ldots\ P^{n-1} = P^{n-2}P$.
2. Build the matrix Π by collecting elements from the powers of matrix P.
3. Determine the matrix B according to (1.53).
4. Compute the matrix $(B^T)^{-1}$.
5. Determine matrix $\triangle$ according to (1.54).
6. Build the limiting and differential matrices by collecting elements from Δ.

The proposed algorithms represent a modification of the algorithms from Sect. 1.5, based on fast matrix multiplication, matrix inversion, finding of characteristic polynomial, and resuming the polynomial mentioned in Sects. 1.6.1 and 1.6.2. Therefore, the computational complexity of finding the limiting matrix is $O(n^3)$, and the complexity of calculating differential matrices is $O(n^{\omega+1})$, where n is the number of the states of the Markov chain, and $O(n^\omega)$ is the complexity of the used fast matrix multiplication algorithm.

Numerical Examples

We present numerical examples that illustrate the details of Algorithm 1.7 for periodic and aperiodic Markov chains.

Example 1 Let the Markov chain with transition probability matrix $P = \begin{pmatrix} 1 & 0 & 0 \\ 0 & 0.5 & 0.5 \\ 0.5 & 0 & 0.5 \end{pmatrix}$ be given and consider the problem of determining the differential components. We apply Algorithm 1.7:

$$(1)\ \ P^0 = \begin{pmatrix} 1 & 0 & 0 \\ 0 & 1 & 0 \\ 0 & 0 & 1 \end{pmatrix},\ P = \begin{pmatrix} 1 & 0 & 0 \\ 0 & 0.5 & 0.5 \\ 0.5 & 0 & 0.5 \end{pmatrix},\ P^2 = \begin{pmatrix} 1 & 0 & 0 \\ 0.25 & 0.25 & 0.5 \\ 0.75 & 0 & 0.25 \end{pmatrix}.$$

$$(2\text{–}4)\ \ K(z) = |P - zI_n| = \begin{vmatrix} 1-z & 0 & 0 \\ 0 & 0.5-z & 0.5 \\ 0.5 & 0 & 0.5-z \end{vmatrix} = (1-z)(0.5-z)^2 \Rightarrow$$

$$\Rightarrow r = deg(K(z)) = 3,\ \sigma(P) = \{1, 0.5\},\ m(1) = 1,\ m(0.5) = 2$$
$$\Rightarrow \beta_0 = (1, 1, 0),\ \beta_1 = (1, 0.5, 0.5),\ \beta_2 = (1, 0.25, 0.5) \Rightarrow$$
$$B = \begin{pmatrix} 1 & 1 & 0 \\ 1 & 0.5 & 0.5 \\ 1 & 0.25 & 0.5 \end{pmatrix}$$

$$\Rightarrow \quad (B^T)^{-1} = \begin{pmatrix} 1 & 0 & -2 \\ -4 & 4 & 6 \\ 4 & -4 & -4 \end{pmatrix}; \qquad \Pi = \begin{pmatrix} 1 & 1 & 1 \\ 0 & 0 & 0 \\ 0 & 0 & 0 \\ -- & -- & -- \\ 0 & 0 & 0.25 \\ 1 & 0.5 & 0.25 \\ 0 & 0.5 & 0.5 \\ -- & -- & -- \\ 0 & 0.5 & 0.75 \\ 0 & 0 & 0 \\ 1 & 0.5 & 0.25 \end{pmatrix}.$$

$$(5)\ \Delta = \begin{pmatrix} 1 & 1 & 1 \\ 0 & 0 & 0 \\ 0 & 0 & 0 \\ -- & -- & -- \\ 0 & 0 & 0.25 \\ 1 & 0.5 & 0.25 \\ 0 & 0.5 & 0.5 \\ -- & -- & -- \\ 0 & 0.5 & 0.75 \\ 0 & 0 & 0 \\ 1 & 0.5 & 0.25 \end{pmatrix} \cdot \begin{pmatrix} 1 & 0 & -2 \\ -4 & 4 & 6 \\ 4 & -4 & -4 \end{pmatrix} = \begin{pmatrix} 1 & 0 & 0 \\ 0 & 0 & 0 \\ 0 & 0 & 0 \\ -- & -- & -- \\ 1 & -1 & -1 \\ 0 & 1 & 0 \\ 0 & 0 & 1 \\ -- & -- & -- \\ 1 & -1 & 0 \\ 0 & 0 & 0 \\ 0 & 1 & 0 \end{pmatrix}.$$

$$(6)\ \overline{\beta}_0(1) = \begin{pmatrix} 1 & 0 & 0 \\ 1 & 0 & 0 \\ 1 & 0 & 0 \end{pmatrix},\ \overline{\beta}_0(0.5) = \begin{pmatrix} 0 & 0 & 0 \\ -1 & 1 & 0 \\ -1 & 0 & 1 \end{pmatrix},\ \overline{\beta}_1(0.5) = \begin{pmatrix} 0 & 0 & 0 \\ -1 & 0 & 1 \\ 0 & 0 & 0 \end{pmatrix}.$$

So, the t-step transition probability matrix can be represented as follows:

$$P(t) = \begin{pmatrix} 1 & 0 & 0 \\ 1 & 0 & 0 \\ 1 & 0 & 0 \end{pmatrix} + \begin{pmatrix} 0 & 0 & 0 \\ -1 & 1 & 0 \\ -1 & 0 & 1 \end{pmatrix} \left(\frac{1}{2}\right)^t + \begin{pmatrix} 0 & 0 & 0 \\ -1 & 0 & 1 \\ 0 & 0 & 0 \end{pmatrix} t \left(\frac{1}{2}\right)^t, \ \forall t \geq 0.$$

Example 2 Let the 2-periodic Markov process determined by the matrix of probability transition $P = \begin{pmatrix} 0 & 1 \\ 1 & 0 \end{pmatrix}$ be given, and consider the problem of determining the limiting and differential components. If we apply Algorithm 1.7, then we obtain:

$$(1)\ P^0 = \begin{pmatrix} 1 & 0 \\ 0 & 1 \end{pmatrix},\ P = \begin{pmatrix} 0 & 1 \\ 1 & 0 \end{pmatrix}.$$

$$(2\text{–}4)\ K(z) = |P - zI_n| = \begin{vmatrix} -z & 1 \\ 1 & -z \end{vmatrix} = z^2 - 1 = (z-1)(z+1)$$
$$\Rightarrow r = deg(K(z)) = 2,\ \sigma(P) = \{1, -1\},\ m(1) = m(-1) = 1$$

$$\Rightarrow \beta_0 = (1, 1),\ \beta_1 = (1, -1) \Rightarrow B = \begin{pmatrix} 1 & 1 \\ 1 & -1 \end{pmatrix}$$

$$\Rightarrow \quad (B^T)^{-1} = \begin{pmatrix} 0.5 & 0.5 \\ 0.5 & -0.5 \end{pmatrix}; \qquad \Pi = \begin{pmatrix} 1 & 0 \\ 0 & 1 \\ -- & -- \\ 0 & 1 \\ 1 & 0 \end{pmatrix}.$$

(5) $$\Delta = \begin{pmatrix} 1 & 0 \\ 0 & 1 \\ -- & -- \\ 0 & 1 \\ 1 & 0 \end{pmatrix} \cdot \begin{pmatrix} 0.5 & 0.5 \\ 0.5 & -0.5 \end{pmatrix} = \begin{pmatrix} 0.5 & 0.5 \\ 0.5 & -0.5 \\ -- & -- \\ 0.5 & -0.5 \\ 0.5 & 0.5 \end{pmatrix}.$$

(6) $$\overline{\beta}_0(1) = \begin{pmatrix} 0.5 & 0.5 \\ 0.5 & 0.5 \end{pmatrix}, \qquad \overline{\beta}_0(-1) = \begin{pmatrix} 0.5 & -0.5 \\ -0.5 & 0.5 \end{pmatrix}.$$

So, the t-step transition probability matrix can be represented as follows:

$$P(t) = \begin{pmatrix} 0.5 & 0.5 \\ 0.5 & 0.5 \end{pmatrix} + \begin{pmatrix} 0.5 & -0.5 \\ -0.5 & 0.5 \end{pmatrix} (-1)^t, \ \forall t \geq 0.$$

1.7 Dynamic Programming Algorithms for Markov Chains

In this section, we develop backward dynamic programming algorithms for finite state space Markov chains. We show how such algorithms can be efficiently used for calculation and analyze some of the main probabilistic characteristics of a Markov chain. First, we consider the problem of determining the probability of the transitions of the systems from a given starting state to a given final state if the final state should be reached at a moment of time from a given interval. We show that the asymptotic behavior analysis of the proposed backward dynamic programming calculation procedures for this problem allows us to substantiate several polynomial-time algorithms for calculating the limiting matrix in Markov chains [121, 148], including also the algorithms from Sects. 1.1.6 and 1.5.

1.7.1 Determining the State-Time Probabilities with Restrictions on the Number of Transitions

We consider the problem of determining the probability of systems' transitions from state x_{i_0} to state x with the condition that state x should be reached at the time moment $t(x)$ such that $t_1 \leq t(x) \leq t_2$, where t_1 and t_2 are given.

So, we need to calculate the probability that the system $\mathbb{L}$ will reach the state x at least at one of the moments of time $t_1, t_1 + 1, \ldots, t_2$. We denote this probability by $P_{x_{i_0}}(x,\ t_1 \le t(x) \le t_2)$, i.e.,

$$P_{x_{i_0}}(x,\ t_1 \le t(x) \le t_2) = \mathbb{P}(\xi_t = x \mid \xi_0 = x_{i_0}, t_1 \le t \le t_2).$$

Some reflections on this definition allow us to write the following formula:

$$P_{x_{i_0}}(x,\ t_1 \le t(x) \le t_2) = \\ = P_{x_{i_0}}(x,\ 0 \le t(x) \le t_2) - P_{x_{i_0}}(x,\ 0 \le t(x) \le t_1 - 1). \quad (1.55)$$

Furthermore, we describe some results that can be used to calculate the probability $P_x(y,\ 0 \le t(y) \le t)$ for $x, y \in X$ and $t = 1, 2, \ldots$. For this reason, we give the graphical interpretation of the Markov processes, using the graph of transition probabilities $G_p = (X, E_p)$. Recall that in such a graph, each vertex $x \in X$ corresponds to a state of the dynamical system and a possible transition of the system from one state x to another state y with positive probability $p_{x,y}$ represented by the directed edge $e = (x, y) \in E_p$ from x to y; the corresponding probabilities $p_{x,y}$ can be associated with directed edges $(x, y) \in E_p$ in G_p. It is evident that on graph G_p, each vertex x contains at least one leaving edge (x, y) and $\sum_{y \in X} p_{x,y} = 1$. In general, this graph can be treated as a random graph [20] with some additional properties. As an example, the graph of states' transitions $G_p = (X, E_p)$ for the Markov process with the stochastic matrix

$$P = \begin{pmatrix} 0.3 & 0.3 & 0.4 & 0 \\ 0.5 & 0 & 0.5 & 0 \\ 0 & 0.4 & 0 & 0.6 \\ 0 & 0.3 & 0.5 & 0.2 \end{pmatrix}$$

is represented in Fig. 1.3.

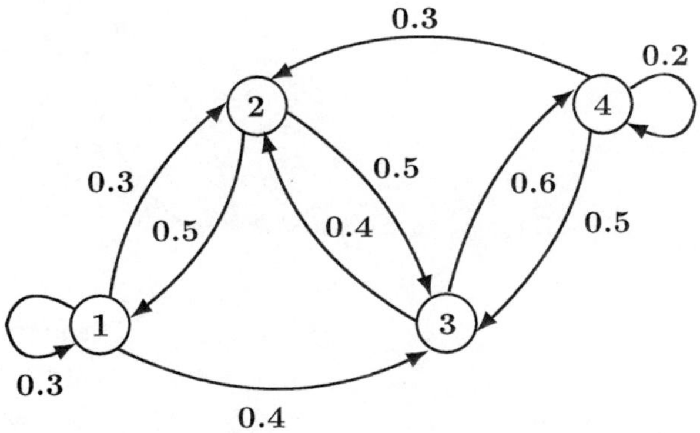

Fig. 1.3 Graph $G_p = (X, E_p)$ of the Markov chain

In the following, we also consider the stochastic process that may stop if one of the states from a given subset of states of the dynamical system is reached. This means that the graph of such a process may contain deadlock vertices.

So, we consider the stochastic process for which the graph may contain the *deadlock vertices* $y \in X$ and $\sum_{z \in X} p_{x,z} = 1$ for the vertices $x \in X$, which contain at least one leaving directed edge. For example, in Fig. 1.4, a graph $G_p = (X, E_p)$ containing deadlock vertices is represented. This graph corresponds to the stochastic process with the following matrix of states' transitions:

$$P = \begin{pmatrix} 0.3 & 0.3 & 0.4 & 0 \\ 0.5 & 0 & 0.3 & 0.2 \\ 0 & 0.6 & 0 & 0.4 \\ 0 & 0 & 0 & 0 \end{pmatrix}.$$

Such a graph does not correspond to a Markov process and the matrix of the transition probability P contains rows with zero components. Nevertheless, in this case, the probabilities $P_{x_0}(x, t)$ can be calculated based on recursive formula (1.3). Note that the matrix P can easily be transformed into a stochastic matrix, replacing the probabilities $p_{y,y} = 0$ for the deadlock states $y \in X$ with the probabilities $p_{y,y} = 1$. This transformation leads to a new graph, which corresponds to a Markov process because the obtained graph contains a new directed edge $e = (y, y)$ with $p_e = 1$ for $y \in X$. In this graph, the vertices $y \in X$ contain the loops and the corresponding states of the dynamical system in a new Markov process that represent the absorbing states. So, the stochastic process that may stop in a given set of states can be represented either by the graph with deadlock vertices or by a graph with absorbing vertices. In Fig. 1.5, the graph is represented with

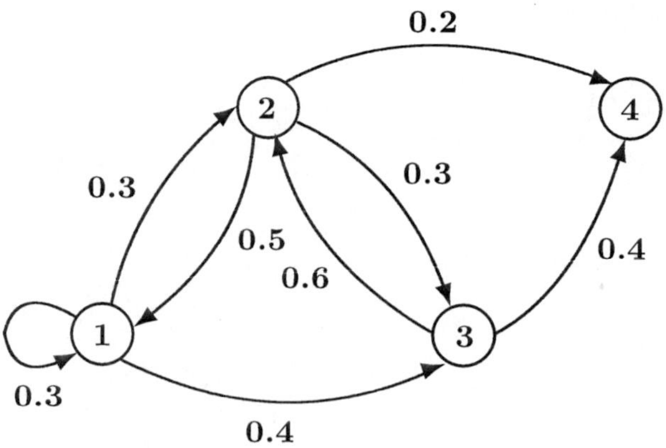

Fig. 1.4 Graph $G_p = (X, E_p)$ with deadlock vertex $y = 4$

Fig. 1.5 Graph $G_p = (X, E_p)$ with absorbing state $y = 4$

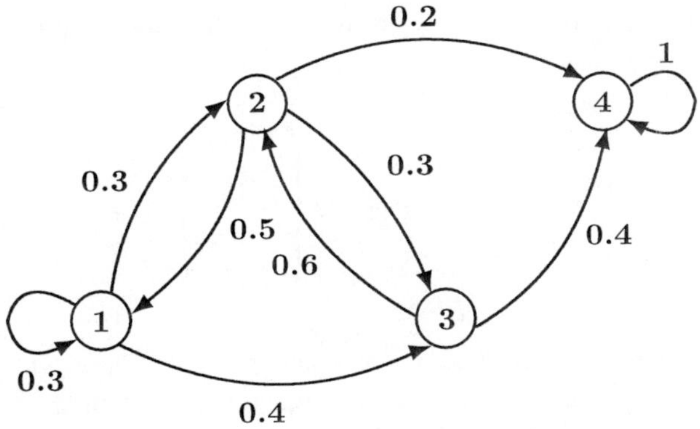

absorbing vertex $y = 4$ for the Markov process defined by the matrix P given below.

$$P = \begin{pmatrix} 0.3 & 0.3 & 0.4 & 0 \\ 0.5 & 0 & 0.3 & 0.2 \\ 0 & 0.6 & 0 & 0.4 \\ 0 & 0 & 0 & 1.0 \end{pmatrix}.$$

It is easy to see that the stochastic matrix P in this example is obtained from the previous one by replacing $p_{4,4} = 0$ with $p_{4,4} = 1$. In this case, the corresponding graph with the absorbing vertex $y = 4$ is obtained from the graph in Fig. 1.4 by adding the directed edge $e = (4, 4)$ with $p_{4,4} = 1$.

We use the graph with absorbing vertices for the calculation of the probabilities $P_x(y, 0 \le t(y) \le t)$.

Lemma 1.8 *Let a Markov process for which graph $G_p = (X, E_p)$ contains an absorbing vertex $y \in X$ be given. Then for an arbitrary state $x \in X$, the following recurrence formula holds:*

$$P_x(y, 0 \le t(y) \le \tau + 1) = \sum_{z \in X} p_{x,z} P_z(y, 0 \le t(y) \le \tau),$$
$$\tau = 0, 1, 2, \ldots, \qquad (1.56)$$

where $P_x(y, 0 \le t(y) \le 0) = 0$ if $x \ne y$ and $P_y(y, 0 \le t(y) \le 0) = 1$.

Proof It is easy to observe that for $\tau = 0$, the lemma holds. Moreover, we can see that here, the condition that y is an absorbing state is essential; otherwise, for $x = y$, recursive formula (1.56) fails to hold. For $t \ge 1$, the correctness of formula (1.56) follows from the definition of the probabilities $P_x(y, 0 \le t(y) \le t + 1)$, $P_z(y, 0 \le t(z) \le t)$, and the induction principle on τ. □

The recursive formula from this lemma can be written in matrix form as follows:

$$\pi'(\tau + 1) = P\pi'(\tau), \quad \tau = 0, 1, 2, \ldots. \qquad (1.57)$$

Here, P is the stochastic matrix of the Markov process with the absorbing state $y \in X$ and

$$\pi'(\tau) = \begin{pmatrix} \pi'_1(\tau) \\ \pi'_2(\tau) \\ \vdots \\ \pi'_n(\tau) \end{pmatrix}, \qquad \tau = 0, 1, 2, \dots$$

are the column vectors, where an arbitrary component $\pi'_i(\tau)$ expresses the probability of the dynamical system to reach state y from x_i by using not more than τ units of time, i.e., $\pi'_i(\tau) = P_{x_i}(y,\ 0 \le t(y) \le \tau)$. At the starting moment of time $\tau = 0$, the vector $\pi'(0)$ is given. All components of this vector are equal to zero except the component corresponding to the absorbing vertex which is equal to one, i.e.,

$$\pi'_i(0) = \begin{cases} 0, & \text{if } x_i \ne y, \\ 1, & \text{if } x_i = y. \end{cases}$$

If we apply this formula for $\tau = 0, 1, 2, \dots, t-1$, then we obtain

$$\pi'(t) = P^t \pi'(0), \quad t = 1, 2, \dots, \tag{1.58}$$

where P^t is the t-th power of the matrix P. So, if we denote by j_y the index of the column of the matrix P^t, which corresponds to the absorbing state y, then an arbitrary element $p^{(t)}_{i,j_y}$ of this column expresses the probability of the system $\mathbb{L}$ to reach state y from x_i by using not more than t units of time, i.e., $p^{(t)}_{i,j_y} = P_{x_i}(y,\ 0 \le t(x) \le t)$. This allows us to formulate the following lemma:

Lemma 1.9 *Let a discrete Markov process with the absorbing state $y \in X$ be given. Then:*

(a) $P_{x_i}(y, t) = P_{x_i}(y,\ 0 \le t(y) \le t), \quad for \quad x_i \in X \setminus \{y\},\ t = 1, 2, \dots.$
(b) $P_{x_i}(y,\ t_1 \le t(y) \le t_2) = p^{(t_2)}_{i,j_y} - p^{(t_1-1)}_{i,j_y},$

where $p^{(t_2)}_{i,j_y}$, $p^{(t_1-1)}_{i,j_y}$ represent the corresponding elements of the matrices P^{t_2} and P^{t_1-1}.

Proof Condition (a) in this lemma holds because

$$P_{x_i}(y, t) = p^{(t)}_{i,j_y} = P_{x_i}(y,\ 0 \le t(y) \le t).$$

Condition (b) follows from Lemma 1.8 and the following properties:

$$P_{x_i}(y,\ 0 \le t(y) \le t_2) = p^{(t_2)}_{i,j_y}, \quad P_{x_i}(y,\ 0 \le t(y) \le t_1 - 1) = p^{(t_1-1)}_{i,j_y}. \qquad \square$$

So, to calculate $P_{x_i}(y,\ t_1 \le t(y) \le t_2)$, it is sufficient to find the matrices P^{t_1-1}, P^{t_2} and then to apply the formula from Lemma 1.9.

Below we give an example that illustrates the calculation procedure of the state probabilities based on the recursive formula above for the stationary case of the Markov process.

Example 1 Let the Markov process with the stochastic matrix P, which corresponds to the graph of transition probabilities represented in Fig. 1.4, be given. It is easy to see that the state $y = 4$ is an absorbing state. We consider the problem of finding the probabilities $P_{x_i}(y, 4)$ and $P_{x_i}(y,\ 2 \le t(x) \le 4)$, where $x_i = 2$. To determine this probability, we use the probability matrices $P^1 = P$ and P^4:

$$P^1 = \begin{pmatrix} 0.3 & 0.3 & 0.4 & 0 \\ 0.5 & 0 & 0.3 & 0.2 \\ 0 & 0.6 & 0 & 0.4 \\ 0 & 0 & 0 & 1.0 \end{pmatrix}.$$

All probabilities $P_{x_i}(y, 4)$, $i = 1, 2, 3$ can be derived from the last column of the following matrix:

$$P^4 = \begin{pmatrix} 0.1701 & 0.1881 & 0.1542 & 0.4876 \\ 0.1455 & 0.1584 & 0.1335 & 0.5626 \\ 0.1260 & 0.0990 & 0.0954 & 0.6796 \\ 0 & 0 & 0 & 1.0 \end{pmatrix}.$$

The probability $P_{x_i}(y,\ 2 \le t(x) \le 4)$ for $x_i = 2$ we find on the basis of Lemma 1.9. According to condition (a) of this lemma, we have

$$P_2(4,\ 0 \le t(4) \le 4) = P_2(4, 4) = p_{2,4}^{(4)} = 0.5626;$$

$$P_2(4,\ 0 \le t(4) \le 1) = P_2(4, 1) = p_{2,4}^{(1)} = p_{2,4} = 0.2,$$

and according to condition (b), we obtain

$$P_2(4,\ 2 \le t(4) \le 4) = p_{2,4}^{(4)} - p_{2,4}^{(1)} = 0.5626 - 0.2 = 0.3626\,.$$

The procedure of calculating the probabilities $P_x(y,\ 0 \le t(y) \le t)$ in the case of the Markov process without absorbing states can easily be reduced to the procedure of calculating the probabilities in the Markov process with the absorbing state y by using the following transformation of the stochastic matrix P: We put $p_{i_y,j} = 0$ if $j \ne i_y$ and $p_{i_y,i_y} = 1$. It is easy to see that such a transformation of the matrix P does not change the probabilities $P_x(y,\ 0 \le t(y) \le t)$. After such a transformation, we obtain a new stochastic matrix to which the recursive formula from the lemma above can be applied.

In general, for the Markov processes without absorbing states, these probabilities can be calculated by using the algorithm that operates with the original matrix P without changing its elements. Below, such an algorithm is described.

Algorithm 1.10 Determining the State-Time Probabilities of the System with a Restriction on the Number of Transitions

Preliminary step (Step 0): Put $P_x(y,\ 0 \le t(y) \le 0) = 0$ for every $x \in X \setminus \{y\}$ and $P_y(y,\ 0 \le t(x) \le 0) = 1$.

General step (Step $\tau + 1,\ \tau \ge 0$): For every $x \in X$, calculate

$$P_x(y,\ 0 \le t(x) \le \tau + 1) = \sum_{z \in X} p_{x,z} P_z(y,\ 0 \le t(y) \le \tau) \tag{1.59}$$

and then put

$$P_y(y,\ 0 \le t(y) \le \tau + 1) = 1. \tag{1.60}$$

If $\tau < t - 1$, then go to the next step, i.e., $\tau = \tau + 1$; otherwise, stop.

Theorem 1.11 *Algorithm 1.10 correctly finds the probabilities $P_x(y,\ 0 \le t(x) \le \tau)$ for $x \in X$, $\tau = 1, 2, \ldots, t$. The running time of the algorithm is $O(|X|^2 t)$.*

Proof It is easy to see that at each general step of the algorithm, the probabilities $P_x(y,\ 0 \le t(x) \le \tau + 1)$ are calculated on the basis of formula (1.59) that takes condition (1.60) into account. This calculation procedure is equivalent to the calculation of the probabilities $P_x(y,\ 0 \le t(x) \le \tau + 1)$ with the condition that the state y is an absorbing state. So, the algorithm correctly finds the probabilities $P_x(y,\ 0 \le t(x) \le \tau)$ for $x \in X$, $\tau = 1, 2, \ldots, t$. In order to estimate the running time of the algorithm, it is sufficient to estimate the number of elementary operations at the general step of the algorithm. It is easy to see that at iteration τ, the algorithm uses $O(|X|^2)$ elementary operations. So, the running time of the algorithm is $O(|X|^2 t)$. □

If we use in Algorithm 1.10 the same notation $\pi'_i(\tau) = P_{x_i}(y,\ 0 \le t(y) \le \tau)$, $\pi_{i_y}(\tau) = P_y(y,\ 0 \le t(y) \le \tau)$ as in formula (1.56), then we obtain the following simple description in matrix form:

Algorithm 1.12 Calculation of the State-Time Probabilities of the System in Matrix Form (Stationary Case)

Preliminary step (Step 0): Fix the vector $\pi'(0) = (\pi'_1(0), \pi'_2(0), \ldots, \pi'_n(0))$, where $\pi'_i(0) = 0$ for $i \ne i_y$ and $\pi'_{i_y}(0) = 1$.

General step (Step$\tau + 1,\ \tau \ge 0$): For a given τ, calculate

$$\pi'(\tau + 1) = P\pi'(\tau)$$

and then put

$$\pi'_{i_y}(\tau + 1) = 1. \tag{1.61}$$

If $\tau < t - 1$, then go to the next step, i.e., $\tau = \tau + 1$; otherwise, stop.

Note that in the algorithm, condition (1.61) allows us to preserve the value $\pi'_{i_y}(t) = 1$ at every time moment t in the calculation process. This condition reflects the property that the system remains in state y at every time step t if state y is reached. We can modify this algorithm to determine the probability $P_x(y,\ 0 \le t(y) \le 0)$ in a more general case if we assume that the system will remain at every time step t in the state y with the probability $\pi'_{i_y}(t) = q(y)$, where $q(y)$ may differ from 1, i.e., $q(y) \le 1$. In the following, we can see that this modification allows us to elaborate a new polynomial-time algorithm for determining the matrix of the limiting probabilities in a stationary Markov process: If $q(y)$ is known, then we can use the following algorithm to calculate the state probabilities of the system with a given restriction on the number of its transitions.

Algorithm 1.13 Calculation of the State-Time Probabilities of the System with Known Probability of Its Remaining in the Final State (Stationary Case)

Preliminary step (Step 0): Fix the vector $\pi'(0) = (\pi'_1(0), \pi'_2(0), \dots, \pi'_n(0))$, where $\pi'_i(0) = 0$ for $i \ne i_y$ and $\pi'_{i_y}(0) = q(y)$.

General step (Step $\tau + 1,\ \tau \ge 0$): For a given τ, calculate

$$\pi'(\tau + 1) = P\pi'(\tau)$$

and then put

$$\pi'_{i_y}(\tau + 1) = q(y). \tag{1.62}$$

If $\tau < t - 1$, then go to the next step, i.e., $\tau = \tau + 1$; otherwise, stop.

Remark 1.14 The results and algorithms described above are valid for an arbitrary stochastic process and for an arbitrary graph for which $\sum_{z \in X} p_{x_i,z} = r(x_i) \le 1$.

An example, which illustrates the calculation procedure in this algorithm, is given below.

Example 2 Consider the Markov process with the stochastic matrix P, which corresponds to the graph of transition probabilities $G_p = (X, E_p)$ represented in Fig. 1.5. Fix $y = 4$ and let us calculate the probabilities $\pi_{i_x} = P_x(y,\ 0 \le t(y) \le t)$ for $x \in \{1, 2, 3\}$ and $t = 3$. If we use this algorithm in the case $q(y) = 1$, i.e., we apply Algorithm 1.12, then we obtain:

Step 0.

$$P_1(4,\ 0 \le t(4) \le 0) = 0; \qquad P_2(4,\ 0 \le t(4) \le 0) = 0;$$
$$P_3(4,\ 0 \le t(4) \le 0) = 0; \qquad P_4(4,\ 0 \le t(4) \le 0) = 1.$$

Step 1.

$$P_1(4,\ 0 \le t(4) \le 1) = 0; \qquad P_2(4,\ 0 \le t(4) \le 1) = 0;$$
$$P_3(4,\ 0 \le t(4) \le 1) = 0.6; \qquad P_4(4,\ 0 \le t(4) \le 1) = 1.$$

Step 2.

$$P_1(4,\ 0 \le t(4) \le 2) = 0.4 \cdot 0.6 = 0.24;$$
$$P_2(4,\ 0 \le t(4) \le 2) = 0.5 \cdot 0.6 = 0.30;$$
$$P_3(4,\ 0 \le t(4) \le 2) = 0.6;$$
$$P_4(4,\ 0 \le t(4) \le 2) = 1.$$

Step 3.

$$P_1(4,\ 0 \le t(4) \le 3) = 0.3 \cdot 0.24 + 0.3 \cdot 0.30 + 0.4 \cdot 0.6 = 0.402;$$
$$P_2(4,\ 0 \le t(4) \le 3) = 0.5 \cdot 0.24 + 0.5 \cdot 0.6 = 0.42;$$
$$P_3(4,\ 0 \le t(4) \le 3) = 0.4 \cdot 0.30 + 0.6 = 0.72;$$
$$P_4(4,\ 0 \le t(4) \le 3) = 1.$$

If we put $q(y) = 0.7$ in this algorithm, then we obtain:

Step 0.

$$P_1(4,\ 0 \le t(4) \le 0) = 0; \qquad P_2(4,\ 0 \le t(4) \le 0) = 0;$$
$$P_3(4,\ 0 \le t(4) \le 0) = 0; \qquad P_4(4,\ 0 \le t(4) \le 0) = 0.7\,.$$

Step 1.

$$P_1(4,\ 0 \le t(4) \le 1) = 0; \qquad P_2(4,\ 0 \le t(4) \le 1) = 0;$$
$$P_3(4,\ 0 \le t(4) \le 1) = 0.42; \qquad P_4(4,\ 0 \le t(4) \le 1) = 0.7\,.$$

Step 2.

$$P_1(4,\ 0 \le t(4) \le 2) = 0.4 \cdot 0.42 = 0.168;$$
$$P_2(4,\ 0 \le t(4) \le 2) = 0.5 \cdot 0.42 = 0.21;$$
$$P_3(4,\ 0 \le t(4) \le 2) = 0.6 \cdot 0.7 = 0.42;$$
$$P_4(4,\ 0 \le t(4) \le 2) = 0.7\,.$$

Step 3.

$$P_1(4,\ 0 \le t(4) \le 3) = 0.3 \cdot 0.168 + 0.3 \cdot 0.21 + 0.4 \cdot 0.42 = 0.2814;$$
$$P_2(4,\ 0 \le t(4) \le 3) = 0.5 \cdot 0.168 + 0.5 \cdot 0.42 = 0.294;$$
$$P_3(4,\ 0 \le t(4) \le 3) = 0.4 \cdot 0.21 + 0.6 \cdot 0.7 = 0.504;$$
$$P_4(4,\ 0 \le t(4) \le 3) = 0.7\,.$$

1.7.2 An Approach to Finding the Limiting Probabilities Based on Dynamic Programming

Now let us show that the results from Sect. 1.7.1 allow us to elaborate an algorithm to determine the matrix of the limiting probabilities for Markov chains which is similar to the algorithm from Sect. 1.5. To characterize this algorithm, we analyze the algorithms from Sect. 1.7.1 in the case of a large number of iterations, i.e., when $\tau \to \infty$. We can see that in the case $\tau \to \infty$, these give the conditions to determine the probabilities of the limiting states in the Markov chain.

Denote by $Q = (q_{i,j})$ the limiting matrix for the Markov chain induced by the stochastic matrix $P = (p_{x_i,x_j})$. We denote the column vectors of the matrix Q by

$$\overline{q}^j = \begin{pmatrix} q_{1,j} \\ q_{2,j} \\ \vdots \\ q_{n,j} \end{pmatrix}, \quad j = 0, 1, 2, \ldots, n,$$

and the row vectors of the matrix Q are denoted by $\overline{q}_i = (q_{i,1}, q_{i,2}, \ldots, q_{i,n})$, $i = 1, 2, \ldots, n$. To characterize algorithms for finding the limiting matrix Q for an arbitrary Markov chain, we need to analyze the structure of the corresponding graph of transition probabilities $G_p = (X, E_p)$ and to study the behavior of the algorithms from the previous subsection in the case $t \to \infty$. First of all, we note that for the ergodic Markov chain with positive recurrent states, graph G_p is strongly connected, and all row vectors $\overline{q}_i$, $i = 1, 2, \ldots, n$ are the same. In this case, the limiting state probabilities can be derived by solving the system of linear equations

$$\pi = \pi P, \quad \sum_{J=1}^{n} \pi_j = 1,$$

i.e., $\overline{q}_i = \pi$, $i = 1, 2, \ldots, n$. In general, such an approach can be used for an arbitrary Markov unichain.

If we have a Markov multichain, then graph G_p consists of several strongly connected components $G^1 = (X^1, E^1)$, $G^2 = (X^2, E^2)$, $\ldots$, $G^k = (X^k, E^k)$, where $\bigcup_{i=1}^{k} X^i = X$.

Additionally, among these components, there are such strongly connected components $G^{i_r} = (X^{i_r}, E^{i_r})$, $r = 1, 2, \ldots, k'$ $(k' < k)$ that do not contain a leaving directed edge $e = (x, y)$, where $x \in X^{i_r}$ and $y \in X \setminus X^{i_r}$. We call such components G^{i_r} *deadlock components* in G_p. In the following, we use these deadlock components for the characterization of the structure of the Markov multichains.

Lemma 1.15 *If $G^{i_r} = (X^{i_r}, E^{i_r})$ is a strongly connected deadlock component in G_p, then X^{i_r} is a positive recurrent class of Markov chains (irreducible chains); if $x \in X \setminus \bigcup_{r=1}^{k'} X^{i_r}$, then x is a transient state of the Markov chain.*

Lemma 1.15 reflects the well-known properties of Markov chains from [71, 83, 152, 171] expressed in the terms of the graphs of probability transitions.

Now we analyze some properties that can be derived from the algorithms from the previous subsection in the case $t \to \infty$.

Let a Markov process with a finite set of states X be given. For an arbitrary state $x_j \in X$, we denote by X_j the subset of states $x_i \in X$ for which in G_p there exists at least a directed path from x_i to x_j. In addition, we denote $N = \{1, 2, \ldots, n\}$, $I(X_j) = \{i \mid x_i \in X_j\}$.

Lemma 1.16 *Let a Markov process with a finite set of states X be given and let us assume that x_j is an absorbing state. Furthermore, let $\pi'(x_j)$ be a solution to the following system of linear equations:*

$$\pi'(x_j) = P\pi'(x_j); \quad \pi_{j,j} = 1; \quad \pi_{i,j} = 0 \quad for \quad i \in N \setminus I(X_j), \tag{1.63}$$

where

$$\pi'(x_j) = \begin{pmatrix} \pi_{1,j} \\ \pi_{2,j} \\ \vdots \\ \pi_{n,j} \end{pmatrix}.$$

Then $\pi'(x_j) = \overline{q}^j$, *i.e.,* $\pi_{i,j} = q_{i,j}$, $i = 1, 2, \ldots, n$. *If* x_j *is a unique absorbing state on the graph* G_p *of the Markov chain and if* x_j *in* G_p *is attainable from every* $x_i \in X$ *(i.e.,* $I(X_j) = N$*), then* $\pi_{i,j} = q_{i,j} = 1$, $i = 1, 2, \ldots, n$.

Proof We apply Algorithm 1.12 with respect to a given absorbing state x_j, $(y_j = x_j)$ if $t \to \infty$. Then $\pi'(t) \to \pi'(x_j)$, and therefore, we obtain $\pi'(x_j) = P\pi'(x_j)$, where $\pi_{j,j} = 1$ and $\pi_{i,j} = 0$ for $i \in N \setminus I(X_j)$. The correctness of the second part of the lemma corresponds to the case if $I(X_j) = N$, and therefore, we obtain the vector π^j with the components $\pi_{i,j} = 1$, $i = 1, 2, \ldots, n$, and it is the solution to the system $\pi'(x_j) = P\pi'(x_j)$, $\pi_{j,j} = 1$. So, the lemma holds. □

Remark 1.17 If x_j is not an absorbing state, then Lemma 1.16 may fail to hold.

Remark 1.18 Lemma 1.16 can be extended to the case if $\sum_{y \in X} p_{x_i,y} = q(x_i) \leq 1$ for some states $x_i \in X$. In this case, the solution to the system (1.63) can also be treated as the limiting probabilities of the system to reach the state x_j. In such processes, there exists always at least one component $\pi_{i,j}$ of the vector $\pi'(x_j)$ that is less than 1, even if $X_j = X$.

Let us show that the result formulated above allows us to find the vector of limiting probabilities $\overline{q}^j$ of the matrix Q if the diagonal elements $q_{j,j}$ of Q are known. We consider the subset of states $Y^+ = \{x_j \mid q_{j,j} \geq 0\}$. It is easy to observe that $Y^+ = \bigcup_{r=1}^{k'} X^{i_r}$, and we denote the corresponding set of indices of this set by $I(Y^+)$. For each $j \in I(Y^+)$, we define the set X_j in the same way as above.

Lemma 1.19 *If a non-zero diagonal element* $q_{j,j}$ *of the limiting matrix* Q *in the Markov multichain is known, i.e.,* $q_{j,j} = q(x_j)$*, then the corresponding vector* $\overline{q}^j$ *of the matrix* Q *can be found by solving the following system of linear equations:*

$$\overline{q}^j = P\overline{q}^j; \quad q_{j,j} = q(x_j); \quad q_{i,j} = 0 \quad for \quad i \in N \setminus I(X_j).$$

Proof We apply Algorithm 1.13 with respect to the fixed final state $y_j = x_j \in X$ with $q(y_j) = q_{j,j}$ if $t \to \infty$. Then for given $y_j = x$, we have $\pi(t)' \to \overline{q}^j$, and therefore, we obtain $\overline{q}^j = P\overline{q}^j$, where $q(y_j) = q_{j,j}$ and $q_{i,j} = 0$ for $i \in N \setminus I(X_j)$. So, the lemma holds. □

Based on this lemma and Algorithm 1.13, we can prove the following result:

Theorem 1.20 *The limiting state matrix* Q *for Markov chains can be derived by using the following algorithm:*

(1) *For each ergodic class* X^{i_r}*, solve the system of linear equations*

$$\pi^{i_r} = \pi^{i_r} P^{(i_r)}, \qquad \sum_{j \in I(X^{i_r})} \pi_j^{i_r} = 1,$$

where π^{i_r} is the row vector with components $\pi_j^{i_r}$ for $j \in I(X^{i_r})$ and $P^{(i_r)}$ is the submatrix of P induced by the class X^{i_r}. Then for every $j \in I(X^{i_r})$, put $q_{j,j} = \pi_j^{i_r}$; for each $j \in I\left(X \setminus \bigcup_{r=1}^{k'} X^{i_r}\right)$, set $q_{j,j} = 0$.

(2) *For every $j \in I(Y^+)$, $Y^+ = \bigcup_{r=1}^{k'} X^{i_r}$, solve the system of linear equations*

$$\overline{q}^j = P\overline{q}^j; \quad q_{j,j} = \pi_{j,j}; \quad q_{i,j} = 0, \quad \forall i \in N \setminus I(X_j)$$

and determine the vector $\overline{q}^j$. For every $j \in I(X \setminus Y^+)$, set $\overline{q}^j = \mathbf{0}$, where $\mathbf{0}$ is the column vector with zero components. The algorithm determines the matrix Q, using $O(n^3)$ elementary operations.

Proof Let us show that the algorithm correctly finds the limiting matrix Q. Step (1) of the algorithm determines the limiting probabilities $q_{j,j}$. This step is based on Lemma 1.15 and on the conditions that each ergodic class X^{i_r} and each transient state $x \in X \setminus Y^+$ should satisfy. So, step 1) correctly finds the limiting probabilities $q_{i,j}$ for $j \in N$. Step 2) of the algorithm is based on Lemma 1.19 and therefore correctly determines the vectors $\overline{q}^j$ of the matrix Q with the known diagonal elements $q_{j,j}$. So, the algorithm correctly determines the limiting matrix Q of the Markov multichain. The running time of the algorithm is $O(n^3)$. We obtain this estimation if we estimate the number of elementary operations at each step of the algorithm. At step (1) of the algorithm, we solve $n' \leq n$ system of equations, where each system contains $|X^{i_r}|$ variables and $\sum_{r=1}^{k'} |X_{i_r}| \leq n$. Therefore, as a whole, the solutions to these systems can be obtained using $O(n^3)$ elementary operations. At step (2) of the algorithm, we solve n systems of linear equations to determine the vectors $\overline{q}^j$. However, these systems have the same left part and can be solved simultaneously by using the Gaussian elimination. Therefore we obtain $O(n^3)$ elementary operations. □

An example that illustrates the details of the algorithm described in this theorem is given below. As we have noted, in the worst case, the running time of the algorithm is $O(n^3)$; however, intuitively it is clear that the upper bound of this estimation cannot be reached. Practically, this algorithm efficiently finds the limiting matrix Q. In the next section, we show how the proposed algorithm can be modified such that the solution can be found in a more suitable form.

Example Consider the problem of determining the limiting matrix of probabilities Q for the Markov multichain determined by the following probability transition matrix:

$$P = \begin{pmatrix} 1 & 0 & 0 & 0 \\ 0.25 & 0.25 & 0.25 & 0.25 \\ 0 & 0 & 0.5 & 0.5 \\ 0 & 0 & 0.5 & 0.5 \end{pmatrix}.$$

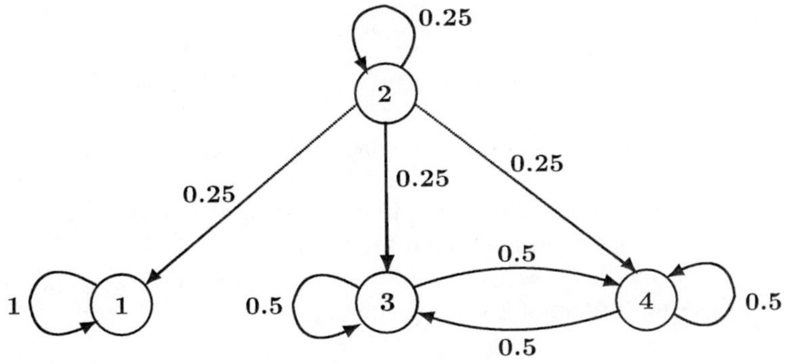

Fig. 1.6 Graph $G_p = (X, E_p)$ of a Markov multichain

The corresponding graph $G_p = (X, E_p)$ of this Markov multichain is represented in Fig. 1.6. We apply the algorithm from Theorem 1.20.

(1) Find the deadlock components X^{i_r}, $r = 1, 2, \ldots, k'$ on graph G_p, and for each of them, solve the system of linear equations

$$\pi^{i_r} = \pi^{i_r} P^{(i_r)}, \qquad \sum_{j \in I(X^{i_r})} \pi_j^{i_r} = 1.$$

In our case, we have two deadlock components $X^1 = \{1\}$, $X^2 = \{3, 4\}$, and therefore, we have to solve the following two systems of linear equations

$$\pi^1 = \pi^1 P^{(1)}, \qquad \pi_1^1 = 1;$$
$$\pi^2 = \pi^2 P^{(2)}, \qquad \pi_3^2 + \pi_4^2 = 1,$$

where

$$\pi^1 = (\pi_1^1); \qquad P^{(1)} = (1);$$
$$\pi^2 = (\pi_3^2, \pi_4^2); \quad P^{(2)} = \begin{pmatrix} 0.5 & 0.5 \\ 0.5 & 0.5 \end{pmatrix}.$$

So, we have to solve the following systems:

$$\begin{cases} \pi_1^1 = \pi_1^1, \quad \pi_1^1 = 1, \\ \pi_3^2 = 0.5\pi_3^2 + 0.5\pi_4^2, \\ \pi_4^2 = 0.5\pi_3^2 + 0.5\pi_4^2, \\ \qquad \pi_3^2 + \pi_4^2 = 1, \end{cases}$$

and we obtain $\pi_1^1 = 1$; $\pi_3^2 = 0.5$; $\pi_4^2 = 0.5$.

This means that the diagonal elements $q_{j,j}$ of the matrix Q are:

$$q_{1,1} = 1; \quad q_{3,3} = 0.5; \quad q_{4,4} = 0.5; \quad q_{2,2} = 0.$$

Here the vertex $x = 2$ corresponds to a transient state; therefore, $q_{2,2} = 0$. At the next step of the algorithm, we obtain the vectors $\overline{q}^j$, $j = 1, 2, 3, 4$.

(2) Fix $j = 1$. For this case, $q_{1,1} = 1$, $X_1 = \{1, 2\}$, $N \setminus X_1 = \{3, 4\}$. Therefore, we have to solve the system of linear equations

$$\overline{q}^1 = P\overline{q}^1; \quad q_{1,1} = 1; \quad q_{1,3} = 0; \quad q_{1,4} = 0.$$

This system can be written as follows:

$$\begin{cases} q_{1,1} = q_{1,1}; \\ q_{2,1} = 0.25q_{1,1} + 0.25q_{2,1} + 0.25q_{3,1} + 0.25q_{4,1}; \\ q_{3,1} = \qquad\qquad\qquad\qquad 0.5\ q_{3,1} + 0.5\ q_{4,1}; \\ q_{4,1} = \qquad\qquad\qquad\qquad 0.5\ q_{3,1} + 0.5\ q_{4,1}; \\ q_{1,1} = 1, \quad q_{3,1} = 0, \quad q_{4,1} = 0, \end{cases}$$

and we determine

$$\overline{q}^1 = \begin{pmatrix} 1 \\ 0.33(3) \\ 0 \\ 0 \end{pmatrix}.$$

For $j = 2$, we have $q_{2,2} = 0$; therefore,

$$\overline{q}^2 = \begin{pmatrix} 0 \\ 0 \\ 0 \\ 0 \end{pmatrix}$$

because the state $x = 2$ is a transient state.

For $j = 3$, we have $q_{3,3} = 0.5$, $X_3 = \{2, 3, 4\}$, $N \setminus X_3 = \{1\}$. In this case, it is necessary to solve the system of linear equations

$$\overline{q}^3 = P\overline{q}^3; \quad q_{3,3} = 0.5; \quad q_{1,3} = 0.$$

Thus, if we solve the system of linear equations

$$\begin{cases} q_{1,3} = q_{1,3}; \\ q_{2,3} = 0.25q_{1,3} + 0.25q_{2,3} + 0.25q_{3,3} + 0.25q_{3,4}; \\ q_{3,3} = \qquad\qquad\qquad\qquad 0.5\ q_{3,3} + 0.5\ q_{3,4}; \\ q_{4,3} = \qquad\qquad\qquad\qquad 0.5\ q_{3,3} + 0.5\ q_{3,4}; \\ q_{3,3} = 0.5, \quad q_{1,3} = 0, \end{cases}$$

then we obtain

$$\overline{q}^3 = \begin{pmatrix} 0 \\ 0.33(3) \\ 0.5 \\ 0.5 \end{pmatrix}.$$

For $j = 4$, we have $q_{4,4} = 0.5$, $X_4 = \{2,\ 3,\ 4\}$, $N \setminus X_4 = \{1\}$. In this case, it is necessary to solve the system of linear equations

$$\overline{q}^4 = P\overline{q}^4; \quad q_{4,4} = 0.5; \quad q_{1,4} = 0.$$

If we solve the system of linear equations

$$\begin{cases} q_{1,4} = q_{1,4}; \\ q_{2,4} = 0.25q_{1,4} + 0.25q_{2,4} + 0.25q_{3,4} + 0.25q_{4,4}; \\ q_{3,4} = \qquad\qquad\qquad\qquad 0.25q_{3,4} + 0.25q_{4,4}; \\ q_{4,4} = \qquad\qquad\qquad\qquad 0.25q_{3,4} + 0.25q_{4,4}; \\ q_{4,4} = 0.5, \quad q_{1,4} = 0, \end{cases}$$

then we find

$$\overline{q}^4 = \begin{pmatrix} 0 \\ 0.33(3) \\ 0.5 \\ 0.5 \end{pmatrix}.$$

So, the limiting matrix is

$$Q = \begin{pmatrix} 1 & 0 & 0 & 0 \\ 0.33(3) & 0 & 0.33(3) & 0.33(3) \\ 0 & 0 & 0.5 & 0.5 \\ 0 & 0 & 0.5 & 0.5 \end{pmatrix}.$$

1.7.3 A Modified Algorithm to Find the Limiting Matrix

In the following, we propose a modification of the algorithm for calculating the limiting matrix from the previous section. The theoretical estimation of the number of elementary operations of the modified algorithm is also $O(n^3)$. However, we can see that in this algorithm, we can solve less than $|Y^+|$ systems of linear equations of the form

$$P\overline{q}^j = \overline{q}^j; \quad q_{j,j} = 1; \quad q_{i,j} = 0, \quad \forall i \in N \setminus I(X_j).$$

Moreover, we show that these systems can be simplified. We describe and characterize the proposed modification for calculating the limiting matrix Q, using the structure properties of the graph of the probability transitions $G_p = (X, E_p)$ and its strongly connected deadlock components $G^{i_r} = (X^{i_r}, E^{i_r})$, $r = 1, 2, \ldots, k'$.

Algorithm 1.21 Determining the Limiting Matrix for a Markov Multichain

The algorithm consists of two parts:

The first part determines the limiting probabilities $q_{x,y}$ for $x \in \bigcup_{r=1}^{k'} X^{i_r}$ and $y \in X$.

The second procedure calculates the limiting probabilities $q_{x,y}$ for $x \in X \setminus \bigcup_{r=1}^{k'} X^{i_r}$ and $y \in X$.

Procedure 1:

1. For each ergodic class X^{i_r}, we solve the system of linear equations

$$\pi^{i_r} = \pi^{i_r} P^{(i_r)}, \quad \sum_{y \in X^{i_r}} \pi_y^{i_r} = 1,$$

where $P^{(i_r)}$ is the matrix of probability transitions corresponding to the ergodic class X^{i_r}, i.e., $P^{(i_r)}$ is a submatrix of P, and π^{i_r} is a row vector with the components $\pi_y^{i_r}$ for $y \in X^{i_r}$. If $\pi_y^{i_r}$ are known, then $q_{x,y}$ for $x \in X^{i_r}$, and $y \in X$ can be calculated as follows:

Set $q_{x,y} = \pi_y^{i_r}$ if $x, y \in X^{i_r}$, and $q_{x,y} = 0$ if $x \in X^{i_r}$, $y \in X \setminus X^{i_r}$.

Procedure 2:

1. We construct an auxiliary directed acyclic graph $GA = (XA, EA)$, which is obtained from the graph $G_p = (X, E_p)$, by using the following transformations: We contract each set of vertices X^{i_r} into one vertex z^{i_r}, where X^{i_r} is a set of vertices of a strongly connected deadlock component $G^{i_r} = (X^{i_r}, E^{i_r})$.

   If the obtained graph contains parallel directed edges $e^1 = (x, z)$, $e^2 = (x, z)$, ..., $e^{m'} = (x, z)$ with the corresponding probabilities $p^1_{x,z}, p^2_{x,z}, \dots, p^{m'}_{x,z}$, then we replace these parallel directed edges with one directed edge $e = (x, z)$ with the probability $p_{x,z} = \sum_{i=1}^{m'} p^i_{x,z}$; after this transformation, we associate with each vertex z^i_r a directed edge $e = (z^r, z^r)$ with the probability $p'_{z^r,z^r} = 1$.

2. We fix the directed graph $GA = (XA, EA)$ obtained by the construction described in step 1, where $XA = \big(X \setminus (\bigcup_{r=1}^{k'} X^{i_r})\big) \cup Z^r$, $Z^r = (z^1, z^2, \dots, z^{k'})$. In addition, we fix a new probability matrix $P' = (p'_{x,y})$ that corresponds to this graph GA.

3. For each $x \in XA$ and every $z^i \in Z^r$, we find the probability $\pi'_x(z^i)$ of the system's transition from state x to state z^r. The probabilities $\pi'_x(z^i)$ can be found by solving the following k' systems of linear equations:

$$
\begin{array}{llllll}
P'\pi'(z^1) = \pi'(z^1), & \pi'_{z^1}(z^1) = 1, & \pi'_{z^2}(z^1) = 0, & \dots, & \pi'_{z^{k'}}(z^1) = 0; \\
P'\pi'(z^2) = \pi'(z^2), & \pi'_{z^1}(z^2) = 0, & \pi'_{z^2}(z^2) = 1, & \dots, & \pi'_{z^{k'}}(z^2) = 0; \\
\vdots & \vdots & \vdots & & \vdots \\
P'\pi'(z^{k'}) = \pi'(z^{k'}), & \pi'_{z^1}(z^{k'}) = 0, & \pi'_{z^2}(z^{k'}) = 0, & \dots, & \pi'_{z^{k'}}(z^{k'}) = 1,
\end{array}
$$

   where $\pi'(z^i)$, $i = 1, 2, \dots, k'$ are the column vectors with components $\pi'_x(z^i)$ for $x \in XA$. So, each vector $\pi'_x(z^i)$ gives the probabilities of the systems' transitions from the states $x \in XA$ to the ergodic classes X^i.

4. We put $q_{x,y} = 0$ for every $x, y \in X \setminus \bigcup_{r=1}^{k'} X^{i_r}$ and $q_{x,y} = \pi'_x(z^r)\pi^{i_r}_y$ for every $x \in X \setminus \bigcup_{r=1}^{r} X^{i_r}$ and $y \in X^{i_r}$, $X^{i_r} \subset X$. If $x \in X^{i_r}$ and $y \in X \setminus X^{i_r}$, then fix $q_{x,y} = 0$.

In the algorithm, $\pi(z^i)'$ can be treated as a vector where each component $\pi_x(z^i)'$ expresses the probability that the system will be in the positive recurrent class X^i after a large number of states' transitions of the system if it starts transitions in the state $x \in X$. Therefore, $q_{x,y} = \pi'_x(z^i)\pi^i_y$ if $y \in X^i$.

By comparing this algorithm with the previous one, we observe that here, we solve the system of linear equations

$$
P'\pi(z^r) = \pi'_{z^r}, \;\; \pi'_{z^r} = 1, \;\; \pi'_{z^i} = 0, \; i = 1, 2, \dots k' \;\; (i \neq r), \;\; r = 1, 2, \dots, k'
$$

instead of the system of equations

$$P\overline{q}^j = \overline{q}^j; \quad q_{j,j} = 1; \quad q_{i,j} = 0, \quad \forall i \in N \setminus I(X_j)$$

for $j = 1, 2, \ldots, k'$.

Theorem 1.22 *The algorithm correctly finds the limiting matrix Q, and the running time of the algorithm is $O(|X|^3)$.*

Proof The correctness of Procedure 1 of the algorithm follows from the definition of the ergodic Markov class (positive recurrent Markov chain). So, Procedure 1 finds the probabilities $q_{x,y}$ for $x \in \bigcup_{r=1}^{k'} X^{i_r}$ and $y \in X$. Let us show that Procedure 2 correctly finds the remaining elements $q_{x,y}$ of the matrix Q. Indeed, each vertex $x \in X \setminus \bigcup_{r=1}^{r} X^{i_r}$ in GA corresponds to transient states of the Markov chain, and therefore, we have $q_{x,y} = 0$ for every $x, y \in X \setminus \bigcup_{r=1}^{k'} X^{i_r}$. If $x \in X^{i_r}$, then the system cannot reach a state $y \in X \setminus X^{i_r}$, and therefore, for such two arbitrary states x, y, we have $q_{x,y} = 0$. Finally, we show that the algorithm correctly determines the limiting probability $q_{x,y}$ if $x \in X \setminus \bigcup_{r=1}^{r} X^{i_r}$ and $y \in X^{i_r}$. In this case, the limiting probability $q_{x,y}$ is equal to the limiting probability of the system to reach the ergodic class X^{i_r} multiplied by the limiting probability of the system to remain in the state $y \in X^{i_r}$, i.e., $q_{x,y} = \pi_x(z^r)\pi_y^{i_r}$. Here $\pi_y^{i_r}$ is the probability of the system to remain in the state $y \in X^{i_r}$, and $\pi_x(z^{i_r})$ is the limiting probability of the system to reach the absorbing state z_{i_r} in GA. According to the construction of the auxiliary graph GA, the value $\pi_x(z^{i_r})$ coincides with the limiting probability of the system to reach the ergodic class X^{i_r}. The correctness of this fact can be easily obtained from Lemma 1.15 and Theorem 1.20. According to Lemma 1.15, the probabilities $\pi_x(z^r)$ for $x \in X \setminus \bigcup_{r=1}^{k'} X^{i_r}$ can be found by solving the following system of linear equations:

$$P'\pi'(z^r) = \pi'(z^r),$$

$$\pi'_{z^1}(z^r) = 0, \ \pi'_{z^2}(z^r) = 0, \ \ldots, \ \pi'_{z^r}(z^r) = 1, \ \ldots, \ \pi'_{z^{k'}}(z^r) = 0,$$

which determine them correctly. So, the algorithm correctly finds the limiting matrix Q.

Now let us show that the running time of the algorithm is $O(n^3)$. We obtain this estimation from steps 3 and 4. In step 3, we solve $k' \leq n$ systems of linear equations $\pi^{i_r} = \pi^{i_p}P^{(i_r)}$, $\sum_{y \in X^{i_r}} \pi_y^{i_r} = 1$, where $\sum_{r=1}^{k'} |X_{i_r}| \leq n$. Therefore, here we determine the solution to these systems by using $O(n^3)$ elementary operations. In step 4, we also solve $k' \leq n$ systems of linear equations. Each of these systems contains not more than n variables.

All these systems have the same left part and therefore can be solved simultaneously by applying the Gaussian elimination. The simultaneous solution to these r systems with the same left part (by using the Gaussian elimination method) uses $O(n^3)$ elementary operations. □

As we have shown in the proof of the theorem, each component $\pi'_x(z^i)$ of the vector $\pi'(z^i)$ represents the probability that the system will occupy a state of the recurrent class X^i after a large number of states' transitions if the system starts transitions in the state $x \in X$, i.e., the vector $\pi'(z^i)$ gives the limiting probabilities $\pi'_x(z^i)$ for every $x \in X$ and the arbitrary recurrent class X^i. Therefore, $q_{x,y} = \pi'_x(z^i)\pi^i_y$ if $y \in X^i$.

It is easy to observe that if the subgraph $G' = \left(X \setminus \bigcup_{r=1}^{k'} X^{i_r}, E'\right)$ of GA induced by the subset of vertices $X \setminus \bigcup_{r=1}^{k'} X^{i_r}$ is an acyclic graph, then

$$\pi_x(z^i)' = P_x(z^i, 0 \le t(z^i) \le |XA|),$$

where $P_x(z^i, 0 \le t(z^i) \le |XA|$ is the probability of the system to reach z^i in GA from x, using $T(z)$ transitions such that $0 \le t(z^i) \le |XA|$. These probabilities can be calculated by using the algorithms from Sect. 1.7. Thus, in the case that the subgraph G' is acyclic, Procedure 2 can be modified by introducing the calculation of the vectors $\pi_x(z^i)$ based on the algorithms from Sect. 1.7.

Below is an example that illustrates the calculation procedure in Algorithm 1.21.

Example Consider the problem of determining the limiting matrix of the probabilities Q for the example from the previous subsection, i.e., the Markov chain is determined by the stochastic matrix of probabilities

$$P = \begin{pmatrix} 1 & 0 & 0 & 0 \\ 0.25 & 0.25 & 0.25 & 0.25 \\ 0 & 0 & 0.5 & 0.5 \\ 0 & 0 & 0.5 & 0.5 \end{pmatrix}.$$

The graph $G_p = (X, E_p)$ of the Markov process is represented in Fig. 1.6.

If we apply Procedure 1 of the algorithm, then we find the probabilities

$$\pi^1_1 = 1, \quad \pi^2_3 = 0.5, \quad \pi^2_4 = 0.5,$$

which represent the solutions to the system

$$\pi^1 = \pi^1 P^{(1)}, \quad \pi^1_1 = 1;$$
$$\pi^2 = \pi^2 P^{(2)}, \quad \pi^2_3 + \pi^2_4 = 1.$$

The first system represents the ergodicity condition for the recurrent class that corresponds to the deadlock component $X^1 = \{1\}$, and the second one represents the ergodicity condition that corresponds to the deadlock component $X^2 = \{3, 4\}$, i.e.,

$$P^{(1)} = (1); \quad P^{(2)} = \begin{pmatrix} 0.5 & 0.5 \\ 0.5 & 0.5 \end{pmatrix}.$$

In such a way, we determine

$$\begin{array}{llll} q_{1,1} = 1, & q_{1,2} = 0, & q_{1,3} = 0, & q_{1,4} = 0, \\ q_{3,3} = 1, & q_{3,4} = 0, & q_{3,1} = 0, & q_{3,2} = 0, \\ q_{4,4} = 0.5, & q_{4,3} = 0.5, & q_{4,1} = 0, & q_{4,2} = 0. \end{array}$$

After that, we apply Procedure 2. To apply this procedure, we construct the auxiliary graph $GA = (XA, EA)$. This graph is represented in Fig 1.7. Graph GA is obtained from G_p, where the components $X^{i_1} = \{1\}$ and X^{i_2} are contracted into vertices $Z^1 = 1'$ and $Z^2 = 3'$, respectively. The matrix P' for this graph is given by

$$P' = \begin{pmatrix} 1 & 0 & 0 \\ 0.25 & 0.25 & 0.5 \\ 0 & 0 & 1 \end{pmatrix}.$$

We solve the following two systems of linear equations:

$$\begin{array}{lll} P'\pi'(1') = \pi'(1'), & \pi'_{1'}(1') = 1, & \pi'_{3'}(3') = 0; \\ P'\pi'(3') = \pi'(3'), & \pi'_{3'}(1') = 0, & \pi'_{1'}(3') = 0, \end{array}$$

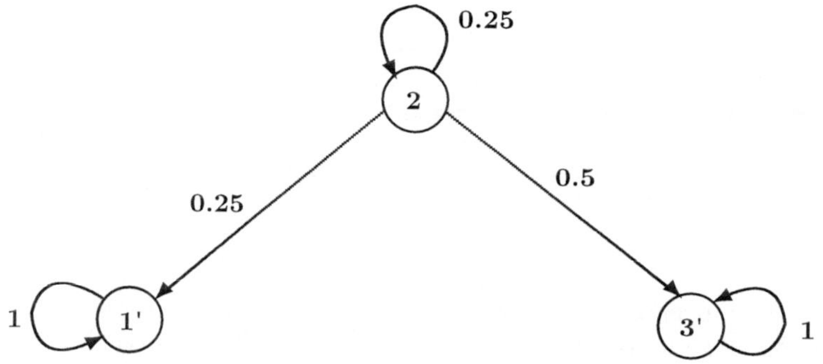

Fig. 1.7 Graph GA obtained from $G_p = (X, E_p)$

where

$$\pi'(1') = \begin{pmatrix} \pi'_{1'}(1') \\ \pi'_{2'}(1') \\ \pi'_{3'}(1') \end{pmatrix}, \quad \pi'(3') = \begin{pmatrix} \pi'_{1'}(3') \\ \pi'_{2'}(3') \\ \pi'_{3'}(3') \end{pmatrix}.$$

The first system of equations can be written in the following form:

$$\begin{cases} \pi'_{1'}(1') = \pi'_{1'}(1'), \\ 0.25\pi'_{1'}(1') + 0.25\pi'_{2'}(1') + 0.5\pi'_{3'}(1') = \pi'_{2'}(1'), \\ \qquad\qquad\qquad\qquad\qquad\qquad \pi'_{3'}(1') = \pi'_{3'}(1'), \\ \pi'_{1'}(1') = 1, \quad \pi'_{3'}(1') = 0 \end{cases}$$

and we obtain

$$\pi'_{1'}(1') = 1; \quad \pi'_{2'}(1') = 0.33(3); \quad \pi'_{3'}(1') = 0.$$

The second system of equations can be represented as follows:

$$\begin{cases} \pi'_{1'}(3') = \pi'_{1'}(3'), \\ 0.25\pi'_{1'}(3') + 0.25\pi'_{2'}(3') + 0.5\pi'_{3'}(3') = \pi'_{2'}(3'), \\ \qquad\qquad\qquad\qquad\qquad\qquad \pi'_{3'}(3') = \pi'_{3'}(3'), \\ \pi'_{3'}(3') = 1, \quad \pi'_{1'}(3') = 0 \end{cases}$$

and we obtain

$$\pi'_{3'}(3') = 1; \quad \pi'_{2'}(3') = 0.66(6); \quad \pi'_{1'}(3') = 0.$$

After that, we calculate

$$q_{2,1} = \pi'_2(1') \cdot \pi_1^1 = 0.33(3), \qquad q_{2,2} = 0,$$
$$q_{2,3} = \pi'_2(3') \cdot \pi_3^2 = 0.66(6) \cdot 0.5 = 0.33(3),$$
$$q_{2,4} = \pi'_2(3') \cdot \pi_4^2 = 0.66(6) \cdot 0.5 = 0.33(3).$$

Thus, we obtain the limiting matrix

$$Q = \begin{pmatrix} 1 & 0 & 0 & 0 \\ 0.33(3) & 0 & 0.33(3) & 0.33(3) \\ 0 & 0 & 0.5 & 0.5 \\ 0 & 0 & 0.5 & 0.5 \end{pmatrix}.$$

1.7.4 Calculation of the First Hitting Probability of a State

In this section, we consider Markov processes in which the system may stop transitions as soon as a given state $y \in X$ is reached. We call such processes *Markov processes with a stopping state*. We show that the dynamic programming algorithms from Sect. 1.7 can be used to determine the probability of first hitting the stopping state from an arbitrary starting state.

Let a Markov chain with the corresponding matrix of probability transitions P be given. Fix a state $y \in X$ and consider the problem of determining the probabilities π_i of first hitting the state y when the system starts transitioning in a state $x_i \in X$.

It is evident that if we have a Markov unichain and y as an absorbing state, then the probability of first hitting the state y from a given state $x_i \in X$ is equal to the limiting probability $q_{x,y}$ from x to y, where $q_{x,y} = 1$ for every $x \in X$. If we apply Algorithm 1.13 and if we take into account that $\tau \to \infty$, then we obtain the following system of linear equations:

$$\pi' = P\pi', \quad \pi'_{i_y} = 1,$$

where π' is the column vector with components π'_{i_x}, i.e.,

$$\pi' = \begin{pmatrix} \pi'_1 \\ \pi'_2 \\ \vdots \\ \pi'_n \end{pmatrix}.$$

So, this system for Markov unichains with an absorbing state always has a unique solution: $p_i = 1, i = 1, 2, \ldots, n$ and $q_{x_i,y} = p_i = 1, \forall x_i \in X$.

If we have an arbitrary Markov multichain that contains an absorbing state $y \in X$, then the limiting probabilities $q_{x_i,y}$ from $x_i \in X$ to y also coincide with the limiting probabilities π_i of first hitting the state $y \in X$. However, here these probabilities may be different from 1, and some of them may be equal to zero. The zero components of the vector π can easily be determined from the graph $G_p = (X, E_p)$ of this Markov process. Indeed, let X_y be the set of vertices $x \in X$ for which in G_p there exists a directed path from x to y, and we consider $I_y = \{i \in \{1, 2, \ldots, n\} \mid x_i \in X_y\}$. Then it is evident that $\pi_i = 0$ if and only if $i \in N \setminus I_y$, where $N = \{1, 2, \ldots, n\}$. Therefore, the probabilities π_i for $x_i \in X$ can be found by solving the following system of linear equations:

$$\pi' = P\pi', \quad \pi'_{i_y} = 1 \quad \text{and} \quad \pi_i = 0, \quad \forall i \in N \setminus I_y.$$

We obtain this system of equations from Algorithm 1.13 if $\tau \to \infty$.

In the general case for an arbitrary Markov chain, if y is not an absorbing state, the probability of first hitting the state y from $x \in X$ can be found in the following way:

We consider a new matrix of probability transitions P', which is obtained from P, by replacing the elements $p_{y,z}$ with the new elements $p'_{y,z}$, where $p'_{y,y} = 1$ and $p_{y,z} = 0, \forall z \in X \setminus \{y\}$. After that, we obtain a new Markov chain with a new matrix of probability transitions P', where y is an absorbing state, and we determine the limiting probabilities of first hitting the state y from $x \in X$, using the procedure described above. So, in order to determine the vector of first hitting probabilities, we have to solve the following system of linear equations:

$$\pi' = P'\pi'; \quad \pi'_{i_y} = 1, \quad \pi'_i = 0, \quad \forall i \in N \setminus I_y.$$

Thus, for an arbitrary starting state $x \in X$ and an arbitrary positive recurrent state y in a Markov unichain, the first hitting probability $\pi'_{x,y}$ is equal to 1, i.e., $\pi'_{x,y} = 1$, $\forall x \in X$.

Example Consider the problem of determining the probabilities of first hitting for the Markov process with the matrix of probability transitions

$$P = \begin{pmatrix} 0.3 & 0.3 & 0.4 & 0 \\ 0.5 & 0 & 0.3 & 0.2 \\ 0 & 0.6 & 0 & 0.4 \\ 0 & 0 & 0 & 1.0 \end{pmatrix}$$

and a given stopping state $y = 3$. To determine $\pi'_1, \pi'_2, \pi'_3, \pi'_4$, we form the matrix

$$P' = \begin{pmatrix} 0.3 & 0.3 & 0.4 & 0 \\ 0.5 & 0 & 0.3 & 0.2 \\ 0 & 0 & 1.0 & 0 \\ 0 & 0 & 0 & 1.0 \end{pmatrix}$$

and find the set $X_y = \{4\}$ for the corresponding graph $G'_p = (X, E'_p)$. This means that we have to solve the system of linear equations

$$\pi' = P'\pi', \quad \pi'_3 = 1, \quad \pi'_4 = 0.$$

The solution to this system is

$$\pi'_1 = 0.8909(09), \quad \pi'_2 = 0.745(45), \quad \pi'_3 = 1, \quad \pi'_4 = 0.$$

It is evident that the vector of the first hitting limiting probabilities π' for different fixed stopping states $y \in X$ may be different, i.e., it depends on the stopping state y.

Therefore, in the following, we denote this vector for a given stopping state $y \in X$ by $\pi'(y)$, where

$$\pi'(y) = \begin{pmatrix} \pi'_{1,j_y} \\ \pi'_{2,j_y} \\ \vdots \\ \pi'_{n,j_y} \end{pmatrix}.$$

If we find the vectors of first hitting probabilities $\pi'(y)$ for every stopping state $y \in X$, then we determine the matrix $\pi' = (\pi_{i,j})'$, where an arbitrary element $\pi'_{i,j}$ represents the limiting probability of first hitting state x_j if the system starts transitions in state x_i.

It easy to observe that the following relationship between the elements of the limiting matrix $Q = (q_{i,j})$ and the elements of the matrix of the first hitting probabilities π' in the Markov chain holds:

$$q_{i,j} = \pi'_{i,j} q_{j,j}, \quad i, j = 1, 2, \ldots, n.$$

Using this property and the property of the limiting matrix, we obtain the system of linear equations

$$\sum_{j=1}^{n} \pi'_{i,j} q_{j,j} = 1, \quad i = 1, 2, \ldots, n.$$

In the general case, the rank of this system may be less than n, and therefore, the values $q_{j,j}$ cannot be uniquely determined if the matrix π' is known. This system has a unique solution with respect to $q_{j,j}$ only in the case if in the Markov chain, each positive recurrent state is an absorbing state.

1.8 State-Time Probabilities for Non-stationary Markov Processes

We consider Markov processes, in which the probabilities of transitions of systems from one state to another depend on time. Such processes are called non-stationary Markov processes. In this case, the process is defined by a dynamic matrix $P(t) = (p_{x,y}(t))$ and a given starting state x_{i_0}, where the dynamic matrix $P(t)$ is assumed to be stochastic for every discrete moment of time $t = 1, 2, \ldots$. The state-time probabilities $P_{x_{i_0}}(x, t)$ for non-stationary Markov processes are defined and calculated in a similar way as stationary processes by using the recursive formula

$$P_{x_{i_0}}(x, \tau+1) = \sum_{y \in X} P_{x_{i_0}}(y, \tau) p_{y,x}(\tau), \quad \tau = 0, 1, 2, \ldots, t-1,$$

where $P_{x_{i_0}}(x_{i_0}, 0) = 1$ and $P_{x_{i_0}}(x_{i_0}, 0) = 0$ for $x \in X \setminus \{x_{i_0}\}$. It is evident that if $p_{x,y}(t)$ does not depend on time, then this formula becomes the formula from Sect. 1.7. The matrix form of this formula can be represented as follows:

$$\pi(\tau+1) = \pi(\tau)P(\tau), \quad \tau = 0, 1, 2, \ldots, t-1,$$

where $\pi(\tau) = (\pi_1(\tau), \pi_2(\tau), \ldots, \pi_n(\tau))$ is the vector with the components $\pi_i(\tau) = P_{x_{i_0}}(x_i, \tau)$.

At the starting moment of time $\tau = 0$, the vector $\pi(\tau)$ is given in the same way as for stationary processes, i.e., $\pi_{i_0}(0) = 1$ and $\pi_i(0) = 0$ for arbitrary $i \neq i_0$. If we apply this formula for a given starting vector $\pi(0)$ and $\tau = 0, 1, 2, \ldots, t-1$, then we obtain

$$\pi(t) = \pi(0)P(0)P(1)P(2) \cdots P(t-1).$$

So, an arbitrary element $\overline{p}_{x_i,x_j}(t)$ of the matrix $\overline{P}(t) = P(0)P(1)P(2) \cdots \cdots P(t-1)$ expresses the probability of the system $\mathbb{L}$ to reach the state x_j from x_i by using t units of time. Now let us show how to calculate the probability $P_{x_{i_0}}(y,\ t_1 \leq t(y) \leq t_2)$ in the case of non-stationary Markov processes. In the same way as for the stationary case, we consider the non-stationary Markov process with a given absorbing state $y \in X$. So, we assume that the dynamic matrix $P(t)$, which is stochastic for every $t = 0, 1, 2, \ldots$ and $p_{y,y}(t) = 1$ for arbitrary t, is given. Then the probabilities $P_x(y,\ 0 \leq t(y) \leq t)$ for $x \in X$ can be determined if we tabulate the values $P_x(y,\ t-\tau \leq t(y) \leq t)$, $\tau = 0, 1, 2, \ldots, t$, using the following recursive formula:

$$P_x(y,\ t-\tau-1 \leq t(y) \leq t) = \sum_{z \in X} p_{x,z}(t-\tau-1) P_z(y,\ t-\tau \leq t(y) \leq t),$$

where for $\tau = 0$ we fix

$$P_x(y,\ t \leq t(y) \leq t) = 0 \ \text{ if } \ x \neq y \ \text{ and } \ P_y(y,\ t \leq t(y) \leq t) = 1.$$

We can represent this recursive formula in the following matrix form:

$$\pi''(\tau+1) = P(t-\tau-1)\pi''(\tau), \quad \tau = 0, 1, 2, \ldots, t-1.$$

At the starting moment of time $t = 0$, the vector $\pi''(0)$ is given: All components are equal to zero except the component corresponding to the absorbing state, which is equal to one, i.e.,

$$\pi_i''(0) = \begin{cases} 0, & \text{if } x_i \neq y, \\ 1, & \text{if } x_i = y. \end{cases}$$

If we apply this formula to $\tau = 0, 1, 2, \ldots, t-1$, then we obtain

$$\pi''(t) = P(0)P(1)P(2)\cdots P(t-1)\pi''(0), \quad t = 1, 2, \ldots.$$

So, if we consider the matrix $\overline{P}(t) = P(0)P(1)P(2)\cdots P(t-1)$, then an arbitrary element $\overline{p}_{i,j_y}(t)$ of the column j_y in the matrix $\overline{P}(t)$ expresses the probability of the system $\mathbb{L}$ to reach the state y from x_i, using not more than t units of time, i.e., $\overline{p}_{i,j_y}(t) = P_{x_i}(y,\ 0 \le t(y) \le t)$. Here, the matrix $P(t)$ is a stochastic matrix for $t = 0, 1, 2, \ldots$, where $p_{y,y}(t) = 1$ for $t = 1, 2, \ldots$ and

$$\pi''(\tau) = \begin{pmatrix} \pi_1''(\tau) \\ \pi_2''(\tau) \\ \vdots \\ \pi_n''(\tau) \end{pmatrix}, \quad \tau = 0, 1, 2, \ldots$$

is the column vector, where an arbitrary component $\pi_i''(\tau)$ expresses the probability of the dynamical system to reach the state y from x_i, using not more than τ units of time if the system starts transitions in the state x at the moment of time $t - \tau$, i.e., $\pi_i''(\tau) = P_{x_i}(y,\ t - \tau \le t(y) \le t)$.

This means that in the case that y is an absorbing state, the probability $P_x(y,\ t_1 \le t(y) \le t_2)$ can be found in the following way:

(a) Find the matrices

$$\overline{P}(t_1 - 1) = P(0)P(1)P(2)\cdots P(t_1 - 2)$$

and

$$\overline{P}(t_2) = P(0)P(1)P(2)\cdots P(t_2 - 1).$$

(b) Calculate

$$\begin{aligned} P_x(y,\ t_1 \le t(y) \le t_2) &= P_x(y,\ 0 \le t(y) \le t_2) - P_x(y,\ 0 \le t(y) \le t_1 - 1) \\ &= \overline{p}_{i_x,j_y}(t_2) - \overline{p}_{i_x,j_y}(t_1 - 1), \end{aligned}$$

where $\overline{p}_{i_x,j_y}(t_1 - 1)$ and $\overline{p}_{i_x,j_y}(t_2)$ represent the corresponding elements of the matrices $\overline{P}(t_1 - 1)$ and $\overline{P}(t_2)$.

The results described above allow us to formulate algorithms for the calculation of the probabilities $P_x(y,\ 0 \le t(y) \le t)$ for an arbitrary non-stationary Markov process. Such algorithms can be obtained if in the general steps of Algorithms 1.12 and 1.13, we replace the matrix P with the matrix $P(t-\tau-1)$ and $\pi'(\tau)$ with $\pi''(\tau)$. Below we describe these algorithms. They can be derived in an analogous way like the algorithms from the previous subsection.

Algorithm 1.23 Calculation of the State-Time Probabilities of the System in Matrix Form (Non-stationary Case)

Preliminary step (Step 0): Fix the vector $\pi''(0) = (\pi_1''(0), \pi_2''(0), \ldots, \pi_n''(0))$, where $\pi_i''(0) = 0$ for $i \neq i_y$ and $\pi_{i_y}''(0) = 1$.

General step (Step $\tau+1,\ \tau \geq 0$): For a given τ, calculate

$$\pi''(\tau+1) = P(t-\tau-1)\pi''(\tau)$$

and then put

$$\pi_{i_y}''(\tau+1) = 1.$$

If $\tau < t-1$, then go to the next step, i.e., $\tau = \tau+1$; otherwise, stop.

Algorithm 1.24 Calculation of the State-Time Probabilities of the System with Known Probability of Its Remaining in the Final State (Non-stationary Case)

Preliminary step (Step 0): Fix the vector $\pi''(0) = (\pi_1''(0), \pi_2''(0), \ldots, \pi_n''(0))$, where $\pi_i''(0) = 0$ for $i \neq i_y$ and $\pi_{i_y}''(0) = 1$.

General step (Step $\tau+1,\ \tau \geq 0$): For a given τ, calculate

$$\pi''(\tau+1) = P(t-\tau-1)\pi''(\tau)$$

and then put

$$\pi_{i_y}''(\tau+1) = q(y).$$

If $\tau < t-1$, then go to the next step, i.e., $\tau = \tau+1$; otherwise, stop.

Note that Algorithm 1.24 finds the probabilities $P_x(y,\ 0 \leq t(y) \leq t)$ when the value $q(y)$ is given. We treat this value as the probability of the system to remain in the state y; for the case $q(y) = 1$, this algorithm coincides with the previous one.

To calculate the probability $P_{x_{i_0}}(x,\ t_1 \leq t(x) \leq t_2)$ for $x \in X$ in the case of non-stationary Markov processes, we also use the following auxiliary result:

Lemma 1.25 *Let a Markov process determined by the stochastic matrix of probabilities $P = (p_{x,y})$ and the starting state x_{i_0} be given.*

Then the following formula holds:

$$\begin{aligned} P_{x_{i_0}}(x,\ t_1 \le t(x) \le t_2) = P_{x_{i_0}}(x, t_1) + P_{x_{i_0}}^{t_1}(x, t_1+1) + \\ + P_{x_{i_0}}^{t_1,t_1+1}(x, t_1+2) + \cdots + \\ + P_{x_{i_0}}^{t_1,t_1+1,\ldots,t_2-1}(x, t_2), \end{aligned} \tag{1.64}$$

where $P_{x_{i_0}}^{t_1,t_1+1,\ldots,t_1+i-1}(x, t_1+i)$, $i = 1, 2, \ldots, t_2 - t_1$ *is the probability of the dynamical system to reach state* x *from* x_0 *by using* $t_1 + i$ *transitions such that it does not pass through* x *at the moments of time* $t_1, t_1+1, t_1+2, \ldots, t_1+i-1$.

Proof Taking into account that $P_{x_{i_0}}(x,\ t_1 \le t(x) \le t_1+i)$ expresses the probability of the system $\mathbb{L}$ to reach state x from x_0 at least at one of the moments of time $t_1, t_1+1, \ldots, t_1+i$, we can use the following recursive formula:

$$\begin{aligned} P_{x_{i_0}}(x,\ t_1 \le t(x) \le t_1+i) = P_{x_{i_0}}(x,\ t_1 \le t(x) \le t_1+i-1) + \\ + P_{x_{i_0}}^{t_1,t_1+1,\ldots,t_1+i-1}(x, t_1+i). \end{aligned} \tag{1.65}$$

Applying formula (1.65) $t_2 - t_1$ times to $i = 1, 2, \ldots, t_2 - t_1$, we obtain equality (1.64).

□

Note that formulae (1.64) and (1.65) cannot be used directly for the calculation of the probability $P_{x_0}(x,\ t_1 \le t(x) \le t_2)$. Nevertheless, we can see that such a representation of the probability $P_{x_{i_0}}(x,\ t_1 \le t(x) \le t_2)$ in the time-expanded network method allows us to formulate a suitable algorithm to calculate this probability and to develop new algorithms for solving the problems.

Corrolary 1.26 *If the state* x *of the dynamical system* $\mathbb{L}$ *in the graph of probability transitions* $G_p = (X, E_p)$ *corresponds to a deadlock vertex, then*

$$P_{x_{i_0}}(x,\ t_1 \le t(x) \le t_2) = \sum_{t=t_1}^{t_2} P_{x_{i_0}}(x, t). \tag{1.66}$$

Let X_f be a subset of X and assume that at the moment of time $t = 0$, the dynamical system $\mathbb{L}$ is in the state x_0. Denote by $P_{x_{i_0}}(X_f,\ t_1 \le t(X_f) \le t_2)$ the probability that at least one of the states $x \in X_f$ will be reached at the time moment $t(x)$ such that $t_1 \le t(x) \le t_2$. Then the following corollary holds:

Corrolary 1.27 *If the subset of states* $X_f \subset X$ *of the dynamical system* $\mathbb{L}$ *in the graph* $G_p = (X, E_p)$ *corresponds to the subset of deadlock vertices, then for the probability* $P_{x_{i_0}}(X_f,\ t_1 \le t(X_f) \le t_2)$, *the following formula holds:*

$$P_{x_{i_0}}(X_f,\ t_1 \le t(X_f) \le t_2) = \sum_{x \in X_f} \sum_{t=t_1}^{t_2} P_{x_{i_0}}(x, t). \tag{1.67}$$

It is easy to see that formula (1.67) generalizes formula (1.66).

1.9 Markov Processes with Rewards

Markov processes with rewards were introduced in [6, 9, 10, 71]. The main problems related to such processes and approaches for their solving can be found in [71, 152, 199]. In this section, we consider Markov processes with rewards and analyze some classical results that we use in Chaps. 2 and 3. For Markov processes with rewards, the probability matrix $P = (p_{x,y})$, a starting state, and the rewards matrix $R = (r_{x,y})$, where an element $r_{x,y}$ expresses the reward if the system makes a transition from a state $x \in X$ to a state $y \in X$, are given. Thus, in a Markov process with rewards, when the system makes transitions from one state to another, a sequence of rewards is obtained. Such a sequence of rewards is a random variable with a probability distribution induced by the probability relations of the Markov process and for which the *expected total reward* during T transitions of the system can be defined.

1.9.1 The Expected Total Reward

Consider a Markov process with the transition probability matrix $P = (p_{x,y})$, a given starting state, and the matrix $R = (r_{x,y})$, where an element $r_{x,y}$ of matrix R expresses the reward of the system's transition from state x to state y. This Markov process generates a sequence of rewards, which is a random variable induced by transition probability distributions in the states. The expected total reward during t transitions, when the system starts transitions in a given state $x \in X$, we denote by $\sigma_x(t)$. The values $\sigma_x(t)$ for $x \in X$ are defined and calculated based on the following recursive formula:

$$\sigma_x(\tau + 1) = \sum_{y \in X} p_{x,y}(r_{x,y} + \sigma_y(\tau)), \qquad \tau = 0, 1, 2, \ldots, t - 1, \tag{1.68}$$

where $\sigma_x(0) = 0$ for every $x \in X$. Formula (1.68) can be treated as follows: If the system makes a transition from state x to state y, then it earns the amount $r_{x,y}$ plus the amount it expects to earn during the next $\tau - 1$ transitions if the system starts transitions in state y at the moment of time $\tau = 0$. Taking into account that in state x, the system makes transitions randomly (with the probability distribution $p_{x,y}$), we obtain the values $r_{x,y} + \sigma_y(\tau)$, which should be weighted by the transition probabilities $p_{x,y}$. In the case of a non-stationary process, i.e., if the probabilities and the rewards are changing in time, the expected total cost of the dynamical system is defined and calculated in a similar way; in formula (1.68), we replace $p_{x,y}$ with $p_{x,y}(\tau)$ and $r_{x,y}$ with $r_{x,y}(\tau)$.

It is easy to observe that the formula for the expected total reward given above can be represented as follows:

$$\sigma_x(\tau+1) = \sum_{y\in X} p_{x,y} r_{x,y} + \sum_{y\in X} p_{x,y}\sigma_y(\tau), \quad \tau = 0, 1, 2, \dots, t-1.$$

If for an arbitrary state $x_i \in X$, in this formula we denote

$$r_{x_i} = \sum_{y\in X} p_{x_i,y} r_{x_i,y}, \quad \sigma_i(\tau) = \sigma_{x_i}(\tau), \quad i = 1, 2, \dots, n$$

and if we regard r_{x_i} and $\sigma_i(\tau)$ as the components of the corresponding vectors

$$\mathbf{r} = \begin{pmatrix} r_{x_1} \\ r_{x_2} \\ \vdots \\ r_{x_n} \end{pmatrix}, \quad \sigma(\tau) = \begin{pmatrix} \sigma_1(\tau) \\ \sigma_2(\tau) \\ \vdots \\ \sigma_n(\tau) \end{pmatrix},$$

then the formula for calculating the expected total reward can be written in the following matrix form:

$$\sigma(\tau+1) = \mathbf{r} + P\sigma(\tau), \quad \tau = 0, 1, 2, \dots, t-1. \tag{1.69}$$

The component $\mathbf{r}_i$ of the vector $\mathbf{r}$ may be interpreted as the reward to be expected in the next transition out of state x_i, and therefore, we call it the *expected immediate reward* in state x_i. An arbitrary component $\sigma_i(\tau)$ of the vector $\sigma(\tau)$ expresses the expected total reward of the system during τ transitions if the system starts transitions in state x_i.

Applying t times this formula and taking into account that $\sigma(0) = \mathbf{0}$, where $\mathbf{0}$ is a vector with zero components, we obtain

$$\sigma(t) = P^0\mathbf{r} + P^1\mathbf{r} + P^2\mathbf{r} + \cdots + P^{t-1}\mathbf{r},$$

where P^0 is the identity matrix, i.e., $P^0 = I$.

For an arbitrary state $x_i \in X$, we denote

$$\omega_i(t) = \frac{1}{t}\sigma_i(t), \quad t = 1, 2, \dots.$$

This value expresses the *expected average reward per transition* of the system during t state transitions if the system starts transitions in x_i. We call the vector $\omega(t)$ with the components $\omega_i(t)$, $i = 1, 2, \dots, n$ the *vector of average rewards* of the system if it starts transitions in x_i.

Let us consider an arbitrary discrete Markov process for which there exists the limit

$$\lim_{t\to\infty} P^t = Q.$$

Then there exists the limit

$$\lim_{t\to\infty} \omega(t) = \omega$$

and $\omega = Q\mu$.

This fact can be proved by using the following property: Let $Q(0), Q(1), Q(2), \dots, Q(t), \dots$ be an arbitrary sequence of real matrices for which there exists the limit

$$\lim_{t\to\infty} Q(t) = Q.$$

Then there exists the limit

$$\lim_{t\to\infty} \frac{\sum_{k=0}^{t-1} Q(k)}{t}$$

and this limit is equal to Q, i.e.,

$$\lim_{t\to\infty} \frac{\sum_{k=0}^{t-1} Q(k)}{t} = Q.$$

So, if we put $Q(t) = P^t$, then we obtain

$$\lim_{t\to\infty} \frac{1}{t} \sum_{k=0}^{t-1} P^k = Q.$$

This means that

$$\lim_{t\to\infty} \omega(t) = \lim_{t\to\infty} \frac{1}{t} \sum_{k=0}^{t-1} P^k \mathbf{r} = Q\mathbf{r}.$$

Using the results described above, we can determine the vector of expected average rewards $\omega(t)$ if $t \to \infty$ in the following way: We find the limiting matrix Q and the vector μ and then calculate $\omega = Q\mathbf{r}$. This means that in the case of a large number of states' transitions t, the vector of the expected total rewards $\sigma(t)$ can be approximated with the vector $tQ\mathbf{r}$, i.e., $\sigma(t) \approx tQ\mathbf{r}$.

We have proved the property mentioned above in the case if there exists the limit $\lim_{t\to\infty} P^t = Q$. In the general case, the existence of the limit $\lim_{t\to\infty} \omega(t) = \omega$ for an arbitrary matrix P can be proved by using the Cesàro limit

$$\lim_{t\to\infty} \frac{1}{t} \sum_{k=0}^{t-1} P^k = Q.$$

The result described above allows us to prove the following lemma:

Lemma 1.28 *For any arbitrary Markov process, the vector of limiting average rewards ω satisfies the following equations:*

$$\omega = P\omega \tag{1.70}$$

$$\omega = Q\omega. \tag{1.71}$$

Proof Using formula (1.69), we can write the following relation:

$$\frac{\sigma(t+1)}{t} = \frac{\mathbf{r}}{t} + P\frac{\sigma(t)}{t},$$

i.e.,

$$\frac{t+1}{t}\frac{\sigma(t+1)}{t+1} = \frac{\mathbf{r}}{t} + P\frac{\sigma(t)}{t}.$$

If in the last equation, we take the limit if $t \to \infty$, then we obtain the equality (1.70).

We prove formula (1.71), using (1.70). From (1.70) we obtain

$$\omega = P\omega, \quad \omega = P^2\omega, \quad \ldots, \quad \omega = P^t\omega.$$

This implies

$$\omega = \frac{1}{t} \sum_{\tau=0}^{t-1} P^{\tau}\omega.$$

If in this formula we take the limit for $t \to \infty$, then we obtain (1.71). □

In the following, we study the asymptotic behavior of the expected total reward and the average reward per transition in the Markov processes with transition costs using the z-transform. We can see that the z-transform allows us to formulate a more adequate asymptotic formula for determining the expected total reward and the average reward per transition for the dynamical system in Markov chains.

1.9.2 Asymptotic Behavior of the Expected Total Reward

Now we show that the z-transform can be used for the estimation of the expected total reward of the dynamical system in the case of a large number of transitions. To obtain an asymptotic formula for the expected total reward in Markov processes, we apply the z-transform to the equation

$$\sigma(t+1) = \mathbf{r} + P\sigma(t).$$

Based on the properties of the z-transform from Sect. 1.2, we get

$$z^{-1}(F_\sigma(z) - F_\sigma(0)) = \frac{1}{1-z}\mathbf{r} + PF_\sigma(z).$$

Through rearrangement, we obtain

$$F_\sigma(z) - \sigma(0) = \frac{z}{1-z}\mathbf{r} + zPF_\sigma(z),$$

$$(I - zP)F_\sigma(z) = \frac{z}{1-z}\mathbf{r} + \sigma(0)$$

or

$$F_\sigma(z) = \frac{z}{1-z}(I - zP)^{-1}\mathbf{r} + (I - zP)^{-1}\sigma(0). \tag{1.72}$$

For many practical problems, $\sigma(0)$ is identically zero and therefore, Eq. (1.72) reduces to

$$F_\sigma(z) = \frac{z}{1-z}(I - zP)^{-1}\mathbf{r}.$$

In general, $\sigma(0)$ may be different from zero.

In Sect. 1.2, we showed that the inverse transformation of $(I - zP)^{-1}$ has the form $Q + T(t)$, where Q is the limiting matrix and $T(t)$ is a sum of differential matrices with geometrically decreasing coefficients. This means that for $(I - zP)^{-1}$, the following relation can be written:

$$(I - zP)^{-1} = \frac{1}{1-z}Q + \mathbb{T}(z),$$

where $\mathbb{T}(z)$ is the z-transform of $T(t)$.

If we substitute this formula in (1.72), then we obtain

$$F_\sigma(z) = \frac{z}{(1-z)^2}Q\mathbf{r} + \frac{z}{1-z}\mathbb{T}(z)\mathbf{r} + \frac{1}{1-z}Q\sigma(0) + \mathbb{T}(z)\sigma(0). \tag{1.73}$$

To identify the components of $\sigma(t)$ by using the inverse transformation for $F_\sigma(z)$, we have to analyze each component of $F_\sigma(z)$. First, we observe that the term $zQ\mathbf{r}/(1-z)^2$ represents the ramp of the magnitude $Q\mathbf{r}$.

A more detailed analysis of the term $z\mathbb{T}(z)\mu/(1-z)$ allows us to conclude that it represents a step of magnitude $\mathbb{T}(1)\mathbf{r}$ plus geometric terms that tend to zero as t becomes very large. Furthermore, let us assume that all roots of the equation $\det(I - zP) = 0$ are different. Then $\mathbb{T}(z)$ can be expressed as follows:

$$\mathbb{T}(z) = \sum_i \frac{D_i}{1-\alpha_i z},$$

where D_i are the matrices that do not depend on z and α_i represent some constants, each of which is less than 1. Therefore, $z\mathbb{T}(z)/(1-z)$ can be represented in the following way:

$$\frac{z}{1-z}\mathbb{T}(z) = \frac{1}{1-z}\sum_i \frac{D_i}{1-\alpha_i} - \sum_i \frac{D_i}{(1-\alpha_i)(1-\alpha_i z)}.$$

If after that we take the inverse transformation, then we obtain

$$\sum_i \frac{D_i}{1-\alpha_i} - \sum_i \frac{D_i \alpha_i^t}{1-\alpha_i}.$$

Now it is evident that

$$\sum_i \frac{D_i}{1-\alpha_i} = \mathbb{T}(1).$$

In a similar way, we can show that the inverse transformation of $z\mathbb{T}(z)/(1-z)$ is $T(1)$ if the equation $\det(I - zP) = 0$ admits multiple roots.

So, the term $z\mathbb{T}(z)\mathbf{r}/(1-z)$ in (1.73) expresses the ramp of the magnitude $\mathbb{T}(1)\mathbf{r}$. The quantity $Q\sigma(0)/(1-z)$ is a step of magnitude $Q\sigma(0)$, and $\mathbb{T}(z)\sigma(0)$ represents geometric components that vanish if t is large.

Finally, if we take the inverse transformation for $F_\sigma(z)$, then we obtain the asymptotic representation of the expected total reward

$$\sigma(t) = tQ\mathbf{r} + \mathbb{T}(1)\mathbf{r} + Q\sigma(0) + \epsilon(t),$$

where $\epsilon(t) \to 0$ if $t \to \infty$. If we come back to the expression $\omega = Q\mathbf{r}$, then our formula can be written as follows:

$$\sigma(t) = t\omega + \varepsilon + \epsilon(t), \tag{1.74}$$

where

$$\varepsilon = \mathbb{T}(1)\mathbf{r} + Q\sigma(0)$$

and ω represents the limiting vector of the average rewards; a component ω_j of the vector ω expresses the average reward per transition of the system if it starts transitions in state x_j. The vector ω is called the *gain vector* of the Markov process with rewards, and vector ε is called the *bias vector*. As we have shown, $\mathbb{T}(1)$ can be determined using $O(n^4)$ elementary operations based on the approach described in Sects. 1.4 and 1.5.

Below we give an example that illustrates how to calculate the expected total cost in Markov processes, based on the formula above.

Example 3 Consider the Markov process with the corresponding matrices of probability transitions and reward transitions

$$P = \begin{pmatrix} 0.5 & 0.5 \\ 0.4 & 0.6 \end{pmatrix}; \quad R = \begin{pmatrix} 9 & 3 \\ 3 & -7 \end{pmatrix}.$$

We seek the estimation of the vector of the expected total rewards $\sigma(t)$ and the vector of average costs per transition in this Markov process.

First of all, we find the vector $\mathbf{r} = \begin{pmatrix} r_{x_1} \\ r_{x_2} \end{pmatrix}$, where $r_{x_1} = 0.5 \cdot 9 + 0.5 \cdot 3 = 6$, $r_{x_2} = 0.4 \cdot 3 - 0.6 \cdot 7 = -3$, i.e., $\mathbf{r} = \begin{pmatrix} 6 \\ -3 \end{pmatrix}$. If we take σ identically equal to zero, then we have

$$\sigma(t) = tQ\mathbf{r} + \mathbb{T}(1)\mathbf{r}. \tag{1.75}$$

In Sect. 1.2, it is shown that for our example, we have

$$(I - zP)^{-1} = \frac{1}{1-z}\begin{pmatrix} \frac{4}{9} & \frac{5}{9} \\ \frac{4}{9} & \frac{5}{9} \end{pmatrix} + \frac{1}{1 - \frac{1}{10^t}}\begin{pmatrix} \frac{5}{9} & -\frac{5}{9} \\ -\frac{4}{9} & \frac{4}{9} \end{pmatrix} = \frac{1}{1-z}Q + \mathbb{T}(z)$$

that implies

$$Q = \begin{pmatrix} \frac{4}{9} & \frac{5}{9} \\ \frac{4}{9} & \frac{5}{9} \end{pmatrix}; \quad \mathbb{T}(1) = \begin{pmatrix} \frac{50}{81} & -\frac{50}{81} \\ -\frac{40}{81} & \frac{40}{81} \end{pmatrix}.$$

If we introduce S, $\mathbb{T}(1)$ in (1.75) and take into account that $\mathbf{r} = \begin{pmatrix} 6 \\ -3 \end{pmatrix}$, then we obtain

$$\sigma(t) = t \begin{pmatrix} 1 \\ 1 \end{pmatrix} + \begin{pmatrix} \frac{50}{9} \\ \frac{40}{9} \end{pmatrix}.$$

So,

$$\sigma_1(t) = t + \frac{50}{9}, \quad \sigma_2(t) = t - \frac{40}{9}.$$

The vector of average reward per transition ω is determined as follows:

$$\omega = Q\mathbf{r} = \begin{pmatrix} 1 \\ 1 \end{pmatrix}.$$

A more detailed interpretation of this example can be found in [71].

1.9.3 The Expected Total Reward for Non-stationary Processes

Consider a Markov process with a given transition probability matrix P and a reward matrix R where the elements of these matrices depend on time, i.e., $p_{x,y} = p_{x,y}(t)$, $r_{x,y} = r_{x,y}(t)$ and $\sum_{y \in X} p_{x,y}(t) = 1$ for every $t = 0, 1, 2, \dots$.

So, we consider a non-stationary Markov process determined by the dynamic matrices $P(t) = (p_{x,y}(t))$ and $R(t) = (r_{x,y}(t))$. For an arbitrary $x \in X$, the expected total reward $\sigma_x(t)$ during t transitions of the system (if it starts transitions in x) can be calculated using the following recursive formula:

$$\sigma_x(\tau) = \sum_{y \in X} p_{x,y}(t-\tau)(r_{x,y}(t-\tau) + \sigma_y(\tau - 1)), \quad \tau = 1, 2, \dots, t,$$

starting with $\sigma_x(0) = 0$ for every $x \in X$. After t steps, we obtain $\sigma_x(t)$.

Note that in the recursive formula above, $\sigma_x(\tau)$ expresses the expected total reward during τ transitions of the system if it starts transitions in x at the moment of time $t - \tau$.

Based on this calculation procedure, we can show that the following formula for $\sigma(t)$ holds:

$$\sigma(t) = \mathbf{r}(0) + P(0)\mathbf{r}(1) + P(0)P(1)\mathbf{r}(2) + \dots$$
$$\dots + P(0)P(1)P(2)\cdots P(t-2)\mathbf{r}(t-1),$$

where

$$\mathbf{r}(\tau) = \begin{pmatrix} r_{x_1}(\tau) \\ r_{x_2}(\tau) \\ \vdots \\ r_{x_n}(\tau) \end{pmatrix}, \qquad \tau = 0, 1, 2, \ldots, t-1$$

are the column vectors with the components

$$r_{x_j}(\tau) = \sum_{y \in X} p_{x_j,y}(\tau) r_{x_j,y}(\tau), \qquad \tau = 0, 1, 2, \ldots, t-1; \qquad j = 1, 2, \ldots, n.$$

This formula can easily be proved by using the induction principle on the number of transitions t. The value $\overline{r}_j(\tau)$ in this formula expresses the immediate cost of the system in the state x at the moment of time τ.

1.9.4 The Variance of the Expected Total Reward

For a Markov process with probability matrix $P = (p_{x,y})$ and associated reward matrix $R = (r_{x,y})$, we can define the variance $D_{x_{i_0}}(t)$ of the total reward during t transitions for the dynamical system if it starts transitions in the state $x(0) = x_{i_0}$ at the moment of time $t = 0$ [51, 152]. For an arbitrary state $x \in X$ and an arbitrary t, we define and calculate the variance $D_x(t)$ during t transitions, using the following recursive formula:

$$D_x(\tau+1) = \sum_{y \in X} p_{x,y}((r_{x,y} - r_x)^2 + D_y(\tau)), \qquad \tau = 0, 1, 2, \ldots, t-1,$$

where

$$r_x = \sum_{w \in X} p_{x,w} r_{x,w}$$

and $D_x(0) = 0$ for every $x \in X$. This formula can be treated in the same way as the formula for the expected total reward. The variance for the non-stationary case is defined in a similar way. If we denote the column vector by $D(t)$ with components $D_x(t)$ and the column vector by μ^v with components

$$r_x^v = \sum_{y \in X} (r_{x,y} - r_x)^2,$$

then the formula above in the extended form can be expressed as follows:

$$D(t) = P^0\mathbf{r}^v + P^1\mathbf{r}^v + P^2\mathbf{r}^v + \cdots + P^{t-1}\mathbf{r}^v,$$

where $\mathbf{r}$ is the column vector with components $\overline{r}_x^v$ for $x \in X$.

In a similar way as for the average reward, it can be shown that for $D(t)$, there exists

$$\overline{D} = \lim_{t\to\infty} \frac{1}{t} D(t)$$

and $\overline{D} = Q\mathbf{r}^v$.

A component $\overline{D}_x$ of vector D can be treated as the limiting average variance for the dynamical system if it starts transitions in state x [152]. So, for the case of a large number of transitions, the variance vector $D(t)$ can be expressed as

$$D(t) = t\overline{D} + \varepsilon' + \epsilon'(t),$$

where

$$\varepsilon' = \mathbb{T}(1)\mathbf{r}^v + QD(0)$$

and $\epsilon'(t)$ tends to zero if $t \to \infty$, i.e., for a large t, the value $D(t)$ can be approximated with $t\overline{D}$, i.e., $D(t) \approx tQ\mathbf{r}^v$.

1.10 Markov Processes with Discounted Rewards

Consider a Markov Process with a given stochastic matrix of transition probabilities $P = (p_{x,y})$ and a matrix of transition rewards $R = (r_{x,y})$. Assume that the future rewards of states' transitions of the dynamical system from one state to another in the considered process are discounted according to a given *discount factor* (discount rate) γ, where $0 < \gamma < 1$. This means that at the moment of time t, the reward of the system's transition from a state $x \in X$ to a state $y \in X$ is $r_{x,y}(t) = \gamma^t r_{x,y}$. For such a process, the expected total discounted reward during t transitions (if the system starts transitions in a state $x \in X$) is defined and calculated on the basis of the following recursive formula:

$$\sigma_x^\gamma(\tau + 1) = \sum_{y\in X} p_{x,y}(r_{x,y} + \gamma\sigma_y^\gamma(\tau)), \quad \tau = 0, 1, 2, \ldots t - 1.$$

Using the vector notations from Sect. 1.9.1, this formula can be expressed in the matrix form

$$\sigma^{\gamma}(\tau+1)=\mathbf{r}+\gamma P\sigma^{\gamma}(\tau),\qquad \tau=0,1,2,\dots t-1,$$

where $\sigma(0)$ is the vector with zero components. This implies that the expected total discounted reward during t transitions can be calculated by applying the formula

$$\sigma^{\gamma}(t)=P^{0}\mathbf{r}+\gamma P^{1}\mathbf{r}+\gamma^{2}P^{2}\mathbf{r}+\cdots+\gamma^{t-1}P^{t-1}\mathbf{r},\tag{1.76}$$

where $\mathbf{r}$ is the vector with the components

$$r_{x_i}=\sum_{y\in X}p_{x_i,y}r_{x_i,y},\ i=1,2,\dots,n.$$

In the following, we can see that for Markov processes with discounted rewards, the limit $\lim_{t\to\infty}\sigma^{\gamma}(t)=\sigma^{\gamma}<\infty$ always exists (see [71, 194, 199]). Using this property, we can take the limit in the relation $\sigma^{\gamma}(t+1)=\mathbf{r}+\gamma P\sigma^{\gamma}(t)$ for $t\to\infty$, and we obtain the equation

$$\sigma^{\gamma}=\mathbf{r}+\gamma P\sigma^{\gamma}.$$

For $0\le\gamma<1$, this equation has a unique solution with respect to σ^{γ}. So, the vector of limiting expected total discounted rewards σ^{γ} can be found by solving the system of linear equations

$$(I-\gamma P)\sigma^{\gamma}=\mathbf{r},$$

where $\det(I-\gamma P)\neq 0$ for $0\le\gamma<1$.

To prove the properties mentioned above and to study the asymptotic behavior of the expected total discounted reward in Markov processes, we use the z-transform again.

If we apply the z-transform to the relation $\sigma^{\gamma}(t+1)=\mathbf{r}+\gamma P\sigma^{\gamma}(t)$, then we have

$$z^{-1}(F_{\sigma^{\gamma}}(z)-F_{\sigma^{\gamma}}(0))=\frac{1}{1-z}\mathbf{r}+\gamma PF_{\sigma^{\gamma}}(z).$$

After rearrangement, we obtain

$$F_{\sigma^{\gamma}}(z)=\frac{z}{1-z}(I-\gamma zP)^{-1}\mathbf{r}+(I-\gamma zP)^{-1}F_{\sigma^{\gamma}}(0).$$

In this case, the inverse matrix of $(I - \gamma z P)$ can be represented as follows:

$$(I - \gamma z P)^{-1} = \frac{1}{1 - \gamma z} Q + \mathbb{T}(\gamma z). \tag{1.77}$$

Thus,

$$F_{\sigma^\gamma}(z) = \frac{z}{(1 - z)(1 - \gamma z)} Q\mathbf{r} + \frac{z}{1 - z} \mathbb{T}(\gamma z)\mathbf{r} + \left(\frac{1}{1 - \gamma z} Q + \mathbb{T}(\gamma z) \right) F_{\sigma^\gamma}(0).$$

If we substitute here

$$\frac{z}{(1 - z)(1 - \gamma z)} = \frac{1}{1 - \gamma} \left(\frac{1}{1 - z} - \frac{1}{1 - \gamma z} \right),$$

then we can find the inverse z-transform of the component

$$\frac{z}{(1 - z)(1 - \gamma z)} Q\mathbf{r};$$

the pre-image of this component corresponds to

$$\frac{1}{1 - \gamma} Q\mathbf{r} - \frac{\gamma^t}{1 - \gamma} Q\mathbf{r}.$$

It is easy to observe that the inverse transform of $z\mathbb{T}(z)/(1 - z)$ is $T(\gamma)$. Indeed, if all roots of the characteristic polynomial $\det(I - zP) = 0$ are different, then

$$\mathbb{T}(\gamma z) = \sum_i \frac{D_i}{1 - \alpha_i \gamma z},$$

where D_i represent matrices that do not depend on z and α_i are the corresponding proper values. Therefore,

$$\frac{z}{1 - z} \mathbb{T}(z) = \frac{1}{1 - z} \sum_i \frac{D_i}{1 - \alpha_i \gamma z} - \sum_i \frac{D_i}{(1 - \alpha_i \gamma)(1 - \alpha_i \gamma z)}.$$

Based on this formula for $z\mathbb{T}(z)/(1 - z)$, we can find the inverse z-transform, which has the form

$$\sum_i \frac{D_i}{1 - \alpha_i \gamma} - \sum_i \frac{D_i (\alpha_i \gamma)^t}{1 - \alpha_i \gamma}.$$

This allows us to conclude that

$$\sum_i \frac{D_i}{1-\alpha_i \gamma} = T(\gamma).$$

In an analogous way, we can show that the inverse z-transform of $z\mathbb{T}(z)/(1-z)$ is $T(\gamma)$ if the equation $\det(I - zP) = 0$ admits multiple roots. Finally, we can also observe that all coefficients in the representation of $F_{\sigma^\gamma}(0)$ tend to zero for $t \to \infty$. Therefore, for a large t, the expected total discounted rewards can be expressed by the following asymptotic formula:

$$\sigma^\gamma(t) = \frac{1}{1-\gamma} Q\mathbf{r} + T(\gamma)\mathbf{r} + \epsilon(t),$$

where $\epsilon(t) \to 0$ for $t \to \infty$. If we take relation (1.77) into account, we obtain the following: For $\sigma(t)$, there exists $\lim_{t\to\infty} \sigma^\gamma(t) = \sigma$ and

$$\sigma^\gamma = (I - \gamma P)^{-1}\mathbf{r}, \tag{1.78}$$

where

$$(I - \gamma P)^{-1} = \frac{1}{1-\gamma} Q + T(\gamma).$$

So, if $0 < \gamma < 1$, then there exists $(I - \gamma P)^{-1}$ and the vector of limiting discounted rewards σ^γ can be calculated by using formula (1.77).

In general, formula (1.77) can be obtained from the recursive formula (1.76) for $t \to \infty$. Indeed, for $0 < \gamma < 1$, we have

$$\lim_{t\to\infty} \sigma^\gamma(t) = \sum_{\tau=0}^{\infty} (\gamma P)^\tau \mathbf{r}.$$

The matrix P is stochastic and $0 < \gamma < 1$. Therefore, all proper values are less than 1 and we obtain

$$\sum_{t=0}^{\infty} (\gamma P)^\tau = (I - \gamma P)^{-1}.$$

This means that $\lim_{t\to\infty} \sigma^\gamma(t) = \sigma^\gamma = (I - \gamma P)^{-1}\mathbf{r}$, i.e., formula (1.78) holds.

Example Let a Markov process with discounted rewards be given, where $\gamma = 0.8$,

$$P = \begin{pmatrix} 0.5 & 0.5 \\ 0.6 & 0.4 \end{pmatrix}; \quad R = \begin{pmatrix} 5 & 3 \\ 6 & -4 \end{pmatrix},$$

and consider the problem of determining $\sigma^{\gamma} = \begin{pmatrix} \sigma_1^{\gamma} \\ \sigma_2^{\gamma} \end{pmatrix}$. This vector can be found according to formula (1.78) or by solving the system of linear equations $(I - \gamma P)\sigma^{\gamma} = \mathbf{r}$, where $\mathbf{r} = \begin{pmatrix} r_{x_1} \\ r_{x_2} \end{pmatrix}$.

So, if we calculate $r_{x_1} = 5 \times 0.5 + 3 \times 0.5 = 4$; $r_{x_2} = 6 \times 0.6 - 4 \times 0.4 = 2$, then we find σ_1^{γ} and σ_2^{γ} by solving the system of linear equations

$$\begin{cases} 0.6\sigma_1^{\gamma} - 0.4\sigma_2^{\gamma} = 4, \\ -0.48\sigma_1^{\gamma} + 0.58\sigma_2^{\gamma} = 2. \end{cases}$$

The solution to this system is $\sigma_1^{\gamma} = 20$, $\sigma_2^{\gamma} = 20$.

1.11 Semi-Markov Processes with Rewards

So far, we have studied Markov processes in which the time between transitions is constant. Now will consider a more general class of stochastic discrete processes, where the transition time τ from one state $x \in X$ to another state $y \in Y$ is an integer random variable that takes values from a given interval $[1, t']$, i.e., $\tau \in \{1, 2, \ldots, t'\}$. Such processes are related to semi-Markov processes [136]. We consider a stochastic discrete process that is determined by a finite set of states X and a probability function

$$p : X \times X \times \{1, 2, \ldots, t'\} \rightarrow [0, 1]$$

that satisfies the condition

$$\sum_{y \in X} \sum_{\tau=1}^{t} p_{x,y,\tau} = 1, \quad \forall x \in X,$$

where $p_{x,y,\tau}$ expresses the probability of the system to pass from state x to state y, using τ units of time. Let $P_{x_{i_0}}(x, \bar{t})$ denote the probability of system $\mathbb{L}$ to reach state x at the moment of time $\bar{t}$ if it starts transitions in state x_{i_0} at the moment of time $t = 0$. It is easy to observe that this probability for the considered process can be calculated by using the following recursive formula:

$$P_{x_{i_0}}(x, t) = \sum_{y \in X} \sum_{\tau=1}^{t} p_{y,x,\tau} P_{x_{i_0}}(y, t - \tau), \quad t = 0, 1, 2, \ldots, \bar{t},$$

where $P_{x_{i_0}}(x_{i_0}, 0) = 1$ and $P_{x_{i_0}}(x, 0) = 0$ for $x \in X \setminus \{x_{i_0}\}$. Here we set $p_{x,y,\tau} = 0$ if $\tau > t'$. If in the considered process for arbitrary $x, y \in X$ the probabilities $p_{x,y,\tau}$ satisfy the condition

$$p_{x,y,\tau} = 0, \quad \tau = 2, 3, \ldots t', \tag{1.79}$$

then we obtain a Markov process, and the formula above for calculating $P_{x_{i_0}}(x, t)$ is transformed into formula (1.2).

For a semi-Markov process with transition rewards along the reward function

$$r : X \times X \times \{1, 2, \ldots, t'\} \mapsto \mathbb{R}$$

is given, which determines the values $r_{x,y,\tau}$ for every $x, y \in X$ and $\tau \in \{1, 2, \ldots, t\}$. Here $r_{x,y,\tau}$ is the reward of the system $\mathbb{L}$ to pass from state x to state y by using τ units of time. The expected total reward $\sigma_x(\bar{t})$ at the moment of time $\bar{t}$ if the system starts transitions in a state x at the moment of time $t = 0$ is a random variable induced by the reward and probability transition functions.

We define and calculate this value, using the following formula:

$$\sigma_x(t) = \sum_{y \in X} \sum_{\tau=1}^{t} p_{x,y,\tau}(r_{x,y,\tau} + \sigma_y(t - \tau)), \quad t = 1, 2 \ldots, \bar{t} - 1, \tag{1.80}$$

where $\sigma_x(0) = 0, \forall x \in X$. If the probabilities $p_{x,y,\tau}$ satisfy the condition (1.79), then formula (1.80) becomes formula (1.68) for the calculation of the expected total reward in Markov processes. Formula (1.80) can be written in the following form:

$$\sigma_x(t) = r_x^t + \sum_{y \in X} \sum_{\tau=1}^{t} p_{x,y,\tau} \sigma_y(t - \tau), \quad t = 1, 2, \ldots, \bar{t},$$

where

$$r_x^t = \sum_{y \in X} \sum_{\tau=1}^{t} p_{x,y,\tau} r_{x,y,\tau}$$

represents the immediate reward in the state $x \in X$. In an analogous way as for Markov processes, here we can introduce the discount factor γ, $0 < \gamma < 1$. If we assume that the reward at the next discrete moment of time is discounted with the rate γ, then we can define the expected total discounted reward $\sigma_x^{\gamma}(\bar{t})$ by using the following recursive formula:

$$\sigma_x^{\gamma}(t) = \sum_{y \in X} \sum_{\tau=1}^{t} p_{x,y,\tau}(r_{x,y,\tau} + \gamma^{\tau} \sigma_y(t - \tau)), \quad t = 0, 1, 2 \ldots, \bar{t}.$$

It is easy to observe that for an arbitrary semi-Markov process with transition rewards, an auxiliary Markov process with transition rewards can be constructed. We obtain this auxiliary Markov process if we represent a transition from state x to state y by using $\tau \in \{1, 2, \ldots, t'\}$ units of time in a semi-Markov process as a sequence of τ unit time transitions via $\tau - 1$ fictive intermediate states $x_1^\tau, x_2^\tau, \ldots, x_{\tau-1}^\tau$. So, we regard a transition from x to y by using τ units of time in the semi-Markov process as a sequence of transitions

$$x \to x_1^\tau, \quad x_1^\tau \to x_2^\tau, \quad \ldots, \quad x_{\tau-2}^\tau \to x_{\tau-1}^\tau, \quad x_{\tau-1}^\tau \to y$$

for an auxiliary Markov process, where the corresponding transition probabilities are defined as follows:

$$p_{x,x_1^\tau} = p_{x,y,\tau}; \quad p_{x_1^\tau,x_2^\tau} = p_{x_2^\tau,x_3} = \cdots = p_{x_{\tau-2}^\tau,x_{\tau-1}^\tau} = p_{x_{\tau-1}^\tau,y} = 1.$$

We define the transition rewards for the auxiliary Markov process with new fictive states as follows:

$$r_{x,x_1^\tau} = r_{x,y,\tau}; \quad r_{x_1^\tau,x_2^\tau} = r_{x_2^\tau,x_3^\tau} = \cdots = r_{x_{\tau-2}^\tau,x_{\tau-1}^\tau} = r_{x_{\tau-1}^\tau,y} = 0.$$

After the construction above, the problem of calculating the probabilities $P_x(y, t)$, and the expected total rewards in semi-Markov processes can be reduced to the problem of calculating the corresponding characteristics for the auxiliary Markov process. Moreover, the proposed approach allows us to determine the limiting state probabilities, the average reward per transition, and the expected total reward for semi-Markov processes with transition rewards.

1.12 Expected Total Reward for Processes with Stopping States

For Markov processes with rewards, the expected total reward may not exist if the system makes transitions indefinitely. In this section, we consider a class of Markov processes with rewards for which the expected total reward exists and can be efficiently calculated.

Let a Markov process determined by the matrix of transition probability $P = (p_{x,y})$ and the reward matrix $R = (r_{x,y})$ be given. In addition, we assume that for the Markov process, a stopping state $z \in X$, as it is defined in Sect. 1.7.4, is given. We consider the problem of determining the expected total reward for the dynamical system if it starts transitions in a state $x \in X$ and stops transitions in a given state z as soon as this state is reached. We show that if the stopping state $z \in X$ corresponds to a positive recurrent state of the unichain Markov process, then the expected total reward exists for an arbitrary starting state and it can be efficiently calculated.

We also show how to define and calculate the expected total reward in a process with a given stopping state for an arbitrary Markov chain with rewards.

First, let us consider our problem in the case of unichain processes if the stopping state coincides with an absorbing state $z \in X$. In this case, the expected total reward for the problem with a stopping state z can be calculated for an arbitrary starting state $x \in X$ considering $r_{z,z} = 0$, $\sigma_z = 0$ and applying the recursive formula (1.69). So, here we only have to fix $r_{z,z} = 0$ in the matrix R if $r_{z,z} \neq 0$ and then apply formula (1.69). Based on the results from Sect. 1.9.2, we may conclude that in the considered case, this iterative procedure is convergent. Thus, the expected total reward σ_x in the unichain Markov process exists for an arbitrary starting state $x \in X$. Therefore, if in the recursive formula (1.69) we take the limit when t tends to ∞, then we obtain the system of linear equations

$$\sigma = \mathbf{r} + P\sigma, \quad \sigma_{z,z} = 0,$$

which has a unique solution. The existence of a unique solution to this system of equations for the unichain Markov process with stopping state $z \in X$, where $r_{z,z} = 0$, can be proved if we represent $\sigma = \mathbf{r} + P\sigma$ in the following form:

$$(I - P)\sigma = \mathbf{r}. \tag{1.81}$$

The rank of the matrix $(I - P)$ is equal to $n - 1$ because the column vectors of the matrix P that correspond to the states $x \in X \setminus \{z\}$ are linearly independent. Therefore, if we add the condition $\sigma_z = 0$ to the system (1.81), then we obtain the system of linear equations

$$(I - P)\sigma = \mathbf{r}, \quad \sigma_z = 0, \tag{1.82}$$

which has a unique solution because $\mu_z = 0$. So, we can determine σ_x for every $x \in X$. The solution to the system (1.82) can be found using the well-known classical numerical method as well as the iterative procedure from [159, 172].

After replacing $\sigma_{z,z} = 0$ in $\sigma = \mathbf{r} + P\sigma$, we obtain the system for which the iterative procedure from [159, 172] can be efficiently used for an approximation.

Below there is an example for calculating the expected total reward in a Markov unichain with an absorbing stopping state.

Example 1 Let a unichain Markov process with absorbing stopping state $z = 4$ be given, where the probability and the reward matrices are defined as follows:

$$P = \begin{pmatrix} 0.3 & 0.3 & 0.4 & 0 \\ 0.5 & 0 & 0.3 & 0.2 \\ 0 & 0.6 & 0 & 0.4 \\ 0 & 0 & 0 & 1.0 \end{pmatrix}, \quad R = \begin{pmatrix} 2 & 1 & 2 & 0 \\ -3 & 0 & -1 & 1 \\ 0 & 1 & 0 & 2 \\ 0 & 0 & 0 & 0 \end{pmatrix}.$$

So, $X = \{1, 2, 3, 4\}$ and $z = 4$.

Consider the problem of determining the expected total rewards when the system starts transitions in the states $x \in X \setminus \{4\}$ and stops transitions in state $z = 4$, i.e., we have to calculate $\sigma_1, \sigma_2, \sigma_3, \sigma_4$, where $\sigma_4 = 0$.

We find

$$r_{x_1} = 0,3 \cdot 2 + 0.3 \cdot 1 + 0.4 \cdot 2 = 1.7,$$
$$r_{x_2} = 0.5 \cdot (-3) + 0.3 \cdot (-1) + 0.2 \cdot 1 = -1.6,$$
$$r_{x_3} = 0.6 \cdot 1 + 0.4 \cdot 2 = 1.4$$

and solve the system of linear equations (1.82), i.e., we determine the solution to the following system of linear equations:

$$\begin{cases} 0.7\sigma_1 - 0.3\sigma_2 - 0.4\sigma_3 = 1.7, \\ -0.5\sigma_1 + \sigma_2 - 0.3\sigma_3 - 0.2\sigma_4 = -1.6, \\ -0.6\sigma_2 + \sigma_3 - 0.4\sigma_4 = 1.4, \\ \sigma_4 = 0. \end{cases}$$

The solution to this system is

$$\sigma_1 = 4, \quad \sigma_2 = 1, \quad \sigma_3 = 2, \quad \sigma_4 = 0.$$

If the stopping state $z \in X$ corresponds to a positive recurrent state in the unichain Markov process, then the problem of determining the expected total reward can be calculated by modifying state z into an absorbing state. This means that the matrix P is changed into a new matrix P' which is obtained from P, where the elements $p_{z,y}$ that correspond to row z are changed as follows: $p_{z,z} = 1$, $p_{z,y} = 0$, $\forall y \; X \setminus \{z\}$. In such a way, we obtain a new unichain Markov process with a new probability transition matrix, where z is an absorbing state, and we can calculate the expected total reward for an arbitrary starting state $x \in X$.

For defining and calculating the expected total reward of first hitting a given state $z \in X$ of an arbitrary Markov process, it is necessary to introduce an additional hypothesis related to probability transitions because the first hitting probability $\pi'_{x,z}$ for some $x \in X$ may be less than one or equal to zero. It is evident that if $\pi'_{x,z} = 0$, then the expected total reward σ_x in the considered Markov process with stopping state z makes no sense. Therefore, we can consider the problem of calculating σ_x only for $x \in \overline{X} = X \setminus X^0$, where

$$X^0 = \{x \in X \mid \pi_{x,z} = 0\}.$$

This means that the expected total reward in the considered process with a given stopping state z can be calculated by using only the matrices $\overline{P} = (p_{x,y})$, $\overline{R} = (r_{x,y})$, generated by the elements $p_{x,y}$ and $r_{x,y}$ for $x, y \in \overline{X}$. Here, $\overline{P}$ is a submatrix of P, and the condition $\sum_{y \in \overline{X}} p_{x,y} = 1$ for some $x \in \overline{X}$ may fail to hold.

We obtain the condition $\sum_{y\in\overline{X}} p_{x,y} = 1, \forall x \in \overline{X}$ if we change the matrix $\overline{P}$ into a new matrix $\overline{P}' = (p'_{x,y})$, where

$$p'_{x,y} = \frac{1}{\sum\limits_{v\in\overline{X}} p_{x,v}} p_{x,y}, \quad \forall x \in \overline{X} \setminus \{z\}; \quad p_{z,z} = 1; \quad p_{z,x} = 0, \ \forall x \in \overline{X} \setminus \{z\}.$$

Here, the probabilities $p'_{x,y}$ can be treated as the conditional probability of the system's transition from state x to state y in the case if state z is reached, i.e., when $\pi'_{x,y} = 1$. Such a transformation of the probabilities allows us to estimate the expected total reward if the dynamical system stops transitions in state z. Therefore, in this case, for each state $x \in \overline{X}$, the corresponding probability transitions $p_{x,y}$ from x for $y \in \overline{X}$ and $p'_{x,y}$ from x for $y \in \overline{X}$ should be proportional. It is easy to observe that after the transformation mentioned above, the immediate reward $\overline{r}'_x$ and the probability of first hitting the absorbing state in the auxiliary problem satisfy the following properties:

- $\min\limits_{y\in\overline{X}} r'_{x,y} \leq \mu_x \leq \max\limits_{y\in\overline{X}} r'_{x,y}, \quad \forall x \in \overline{X}.$
- The probability of first hitting state z from arbitrary $x \in \overline{X}$ is equal to 1.

Thus, to calculate the expected total reward in a Markov process with stopping state z, we have to construct the auxiliary Markov process with probability matrix $\overline{P}$ and reward matrix $\overline{R}$, then determine the matrices $\overline{P}'$ and $\overline{R}'$, where $\overline{R}'$ is obtained from $\overline{R}$ by fixing $r_{z,x} = 0, \forall x \in \overline{X}$, and after that, calculate the expected total cost of first hitting the state z from an arbitrary state $x \in \overline{X}$. These values express the corresponding expected total rewards in the initial Markov process with the stopping state z.

Example 2 Consider the problem of determining the expected total rewards σ_x for the Markov process with the matrices P, R from Example 1 if $z = 3$.

For this example, we have $X^0 = \{4\}$ and $\overline{X} = \{1, 2, 3\}$ because $\pi'_{4,3} = 0$ and $\pi'_{x,3} \neq 0$ for $x = 1, 2, 3$. Thus, we obtain

$$\overline{P} = \begin{pmatrix} 0.3 & 0.3 & 0.4 \\ 0.5 & 0 & 0.3 \\ 0 & 0.6 & 0 \end{pmatrix}, \quad \overline{R} = \begin{pmatrix} 2 & 1 & 2 \\ -3 & 0 & -1 \\ 0 & 1 & 2 \end{pmatrix}.$$

Using the procedures mentioned above, we determine

$$\overline{P}' = \begin{pmatrix} 0.3 & 0.3 & 0.4 \\ 0.625 & 0 & 0.375 \\ 0 & 0 & 1 \end{pmatrix}, \quad \overline{R}' = \begin{pmatrix} 2 & 1 & 2 \\ -3 & 0 & -1 \\ 0 & 0 & 0 \end{pmatrix}.$$

For the auxiliary Markov process with the matrices $\overline{P}'$ and $\overline{R}'$, we solve the system of linear equations $(I - \overline{P}')\sigma' = \mathbf{r}'$, $\sigma'_z = 0$.

So, we calculate

$$r'_{x_1} = 0,3 \cdot 2 + 0.3 \cdot 1 + 0.4 \cdot 2 = 1.7,$$

$$r'_{x_2} = 0,625 \cdot (-3) + 0.375 \cdot (-1) = 2.25,$$

$$r'_{x_3} = 0$$

and solve the following system of linear equations:

$$\begin{cases} 0.7\sigma'_1 - 0.3\sigma'_2 - 0.4\sigma'_3 = 1.7, \\ -0.625\sigma'_1 + \sigma'_2 - 0.375\sigma'_3 = -2.25, \\ \sigma'_3 = 0. \end{cases}$$

The solution to this system is $\sigma'_1 = 1.041$, $\sigma'_2 = 1.5994$, $\sigma'_3 = 0$, and for the initial problem we can take $\sigma_1 = \sigma'_1$, $\sigma_2 = \sigma'_2$, $\sigma_3 = \sigma'_3$. These values express the expected total rewards from $x \in \{1, 2, 3\}$ to $z = 3$ in the case if the system stops transitions in $z = 3$.

The expected total reward for the discounted Markov processes with stopping state z can be determined by using the following system of linear equations:

$$(I - \gamma P)\sigma = \mathbf{r}, \quad \sigma_z = 0.$$

If $z \in X$ is an absorbing state for a unichain Markov process with $r_{z,z} = 0$, then this system has a unique solution for an arbitrary $\gamma \in (0, 1]$.

Chapter 2
Markov Decision Processes and Stochastic Control Problems on Networks

Abstract In this chapter, we study a class of problems for Markov decision process models with finite state and action spaces. We consider finite and infinite horizon models. For a finite horizon model, the problem with an expected total reward optimization criterion is considered, which can be efficiently solved by using the backward dynamic programming technique. For infinite horizon models, two basic problems are studied: the problem with an expected total discounted reward optimization criterion and the problem with an expected average reward optimization criterion. We present some classical results concerned with determining the optimal solutions to these problems and show how these results can be extended for a class of control problems on networks. The main attention is addressed to the linear programming approach for Markov decision processes and control problems on networks. Our emphasis is on formulating and studying the infinite horizon decision problems in terms of stationary strategies. We show that infinite horizon Markov decision problems with average and discounted optimization criteria can be formulated in terms of stationary strategies as classical mathematical programming problems with quasi-monotonic (quasi-convex and quasi-concave) object functions and linear constraints. In the following, we show that such quasi-monotonic programming models for infinite horizon decision problems are useful for studying stochastic games with average and discounted payoffs.

Keywords Markov decision processes · Control problems on networks · Iterative algorithms · Linear programming approach · Quasi-linear programming approach

2.1 Markov Decision Processes

A Markov decision process is a model for sequential decision making when the outcome is uncertain. In such a model, the stochastic process is under the partial control of an external observer, called *decision maker*. At each discrete moment of time, the state of the process is observed, and based on this observation, the decision maker selects an action that influences the future evolution of the process.

D. Lozovanu, S. W. Pickl, *Markov Decision Processes and Stochastic Positional Games*, International Series in Operations Research & Management Science 349,
https://doi.org/10.1007/978-3-031-40180-0_2

Depending on the action chosen in a state, the decision maker receives a reward at each time step.

In this section, we formulate the basic problems for such processes and present some classical results that we use in the following for studying a class of stochastic control problems on networks and some classes of stochastic games.

2.1.1 Model Formulation and Basic Problems

A *Markov decision process* consists of the following elements:

- a state space X (which we assume to be finite);
- a finite set $A(x)$ of actions in each state $x \in X$, where $A = \cup_{x \in X} A(x)$ refers to the action space;
- a step reward (or a step cost) $r_{x,a}$ for each state $x \in X$ and for an arbitrary action $a \in A(x)$;
- a transition probability function $p : X \times A \times X \to [0, 1]$ that gives the transition probabilities $p^a_{x,y}$ from an arbitrary $x \in X$ to each $y \in X$ for a fixed action $a \in A(x)$, where $\sum_{y \in X} p^a_{x,y} = 1, \ \forall x \in X, \ a \in A(x)$;
- a starting state $x_0 \in X$.

We refer these elements to a stationary model of a Markov decision process. For a non-stationary model, we assume that the rewards and the probability transitions depend on time, i.e., we have the reward functions $r_{x,a}(t)$ for $(x, a) \in X \times A$ and the *transition probability functions* $p^a_{x,y}(t)$ for $(x, y, a) \in X \times X \times A$, where $\sum_{y \in X} p^a_{x,y}(t) = 1, \forall (x, a) \in X \times A$ and $t = 0, 1, 2, \dots$.

The framework of a Markov decision process is the following: at time moment $t = 0$, a dynamical system is in the state x_0, where the decision maker selects an action $a \in A(x_0)$ and receives the reward $r_{x_0,a_0}(0)$. After that, the system passes randomly from x_0 to a state $x_1 \in X$ according to probability distribution $\{p^{a_0}_{x_0,y}(0)\}_{x \in X}$. Then at time moment $t = 1$, the decision maker observes the state x_1, selects an action $a_1 \in A(x_1)$, and receives the reward $r_{x_1,a_1}(1)$. After that, the system passes from x_1 randomly to a state $x_2 \in X$ according to probability distribution $\{p^{a_1}_{x_1,y}(1)\}_{x \in X}$ and so on. In general, at time moment t, $t = 0, 1, 2, \dots$, the decision maker observes *the history*

$$h_t = (x_0, a_0, x_1, a_1, \dots, x_{t-1}, a_{t-1}, x_t)$$

and selects an action $a_t \in A(x_t)$. Such a process has a planning horizon, which is the result of the time points at which the system can be controlled by the decision maker. This horizon may be finite, infinite, or of random length [77, 152]. We consider processes with finite and infinite time horizons and study problems in which the decision maker selects actions in the states in order to optimize one of the following criteria:

- expected total reward over a finite horizon;
- expected total discounted reward over an infinite horizon;
- expected average reward over an infinite horizon.

We specify these criteria for a Markov decision process, assuming that the decision maker applies a *policy (a strategy)* of selecting the actions in the states, based on the observed *histories*.

Let

$$H_t = \{h_t = (x_0, a_0, x_1, a_2, \dots, x_{t-1}, a_{t-1}, x_t)\}$$

be the set of possible histories up to time t and

$$H_\infty = \{h_\infty = (x_0, a_0, x_1, a_2, \dots) | (x_k, a_k) \in X \times A,\ k = 0, 1, 2, \dots\}.$$

A *policy (or a strategy)* is a sequence of decision rules $\pi = (\pi^0, \pi^1, \dots, \pi^t, \dots)$, where π^t is the decision rule at the moment of time t, $t = 0, 1, 2, \dots$. The *decision rule* π^t *at the moment of time* t is a function

$$\pi^t : H_t \times A \mapsto [0, 1]$$

that gives the probability $\pi^t_{h_t, a_t}$ of the action $a_t \in A(x_t)$ to be taken at the moment of time t for a given history $h_t \in H_t$, where

$$\pi^t_{h_t, a_t} \geq 0,\ \forall a \in A(x_t) \text{ and } \sum_{a_t \in A(x_t)} \pi^t_{h_t, a_t} = 1,\ \forall h_t \in H_t. \tag{2.1}$$

So, the decision rule π^t at the moment of time t may depend on the entire history of the system until time t, i.e., on the states at the time moments $0, 1, 2, \dots, t$ and the actions at the time moments $0, 1, 2, \dots, t-1$. The set of all policies we denote by $\mathcal{P}$. A policy $\pi = (\pi^0, \pi^1, \pi^2, \dots)$ is called *memoryless* if the decision rule π^t is independent of $(x_0, a_0, \dots, x_{t-1}, a_{t-1})$ for every $t = 0, 1, 2, \dots$. So, for a memoryless policy, the decision rule at the moment of time t depends only on the state x_t, and therefore, instead of the notation $\pi^t_{h_t, a_t}$, $\pi^t_{x_t, a_t}$ can be used. Memoryless policies are called *Markov policies (or Markov strategies)*. We denote the *set of all Markov policies* by $\mathcal{P}(M)$. A *memoryless policy* is said to be *stationary* if the decision rules are independent of time t, i.e., $\pi^0 = \pi^1 = \pi^2 = \dots$. This means that a stationary policy is determined by a non-negative function s on $X \times A$ such that $\sum_{a \in A} s_{x,a} = 1$ for every $x \in X$. The set of stationary policies we denote by $\mathcal{P}(S)$. If the decision rule of stationary policy s is nonrandomized, i.e., $s_{x,a}$ for $x \in X$, and $a \in A$ takes only values 0 and 1, then the policy is called *deterministic*. So, a deterministic policy can be identified with a function $\mathfrak{s}$ on X, where $\mathfrak{s}(x) = a_x$ means that the action $a_x \in A(x)$ is chosen in the state $x \in X$. We denote the set of deterministic policies by $\mathcal{P}(D)$.

For a given Markov policy $\pi = (\pi^0, \pi^1, \pi^2, \dots)$, the transition probability matrix $P(\pi^t)$ and the reward vector $\mathbf{r}(\pi^t)$ can be defined as follows:

$$\{P_{x,y}(\pi^t)\}_{x,y} = \sum_{a \in A(x)} p^a_{x,y}(t)\pi^t_{x,a}, \ \forall x \in X, \ y \in Y, \quad t = 0, 1, 2, \dots; \tag{2.2}$$

$$\{\mathbf{r}(\pi^t)\}_x = \sum_{a \in A(x)} r_{x,a}(t)\pi^t_{x,a}, \quad \forall x \in X, \quad t = 0, 1, 2, \dots. \tag{2.3}$$

Let θ be a distribution on X, where $\theta(x) = \theta_x$ is the probability that the system starts transitions in x and consider a policy π. Then according to the Ionescu-Tulcea theorem (see [16], page 140), there exists a unique probability measure $\mathbb{P}_{\theta,\pi}$ on H_∞ and an expectation $\mathbb{E}_{\theta,\pi}$ with respect to $\mathbb{P}_{\theta,\pi}$.

Thus, if we take the random variables ξ_t and η_t that denote the state and action at time t, $t = 0, 1, 2, \dots$, then according to the Ionescu-Tulcea theorem, there exists $\mathbb{P}_{\theta,\pi}\{\xi_t = x, \eta_t = a\}$ that represents the probability that at time t, the state of the system is x and the action is a. Similarly, there exists $\mathbb{P}_{\theta,\pi}\{\xi_t = x\}$ that represents the probability that at time t, the state of the system is x. Moreover, $\mathbb{P}_{\theta,\pi}\{\xi_t = x\} = \mathbb{P}_{\theta,\pi}\{\xi_t = x, \eta_t = a\}$. Additionally, based on the results of Sect. 1.8 and formulae (2.2), (2.3), the following formulae can be derived:

(1) $\mathbb{P}_{\theta,\pi}\{\xi_t = y, \eta_t = a\} = \sum_{x \in X} \theta_x \cdot \{P(\pi^0)P(\pi^1)\cdots P(\pi^{t-1})\}_{x,y} \cdot \pi^t_{y,a}, \forall (y, a) \in X \times A$, $t = 0, 1, 2, \dots$, where $P(\pi^0)P(\pi^1) \cdots P(\pi^{t-1})$ is defined as identity matrix I if $t = 0$;

(2) $\mathbb{E}_{\theta,\pi}\{r_{\xi_t,\eta_t}(t)\} = \sum_{x \in X} \theta_x \cdot \{P(\pi^0)P(\pi^1) \cdots P(\pi^{t-1})\}_{x,y} \cdot \mathbf{r}(\pi^t)\}_x$.

In [77], the following theorem is proved:

Theorem 2.1 *For any initial distribution θ on X, any sequence of policies $\pi_0, \pi_1, \pi_2, \dots$, and any sequence of non-negative real numbers $p_0, p_1, p_2, \dots$ satisfying $\sum_{k=0}^{\infty} p_k = 1$, there exists a Markov policy π_* such that*

$$\mathbb{P}_{\theta,\pi^*}\{\xi_t = y, \eta_t = a\} = \textstyle\sum_{k=0}^{\infty} p_k \cdot \mathbb{P}_{\theta,\pi_k}\{\xi_t = y, \eta_t = a\}, \\ \forall (y, a) \in X \times A, \ t = 0, 1, 2, \dots. \tag{2.4}$$

From this theorem, we obtain the following result:

Corollary 2.2 *For any starting state x and any policy π, there exists a Markov policy π_* such that*

$$\mathbb{P}_{x,\pi^*}\{\xi_t = y, \eta = a\} = \mathbb{P}_{x,\pi}\{\xi_t = y, \eta_t = a\}, \ \forall (y, a) \in X \times A, \ t = 0, 1, 2, \dots$$

and

$$\mathbb{E}_{x,\pi^*}\{r_{\xi_t,\eta_t}(t)\} = \mathbb{E}_{x,\Pi}\{r_{\xi_t,\eta_t}\},\ t = 0, 1, 2, \ldots.$$

2.1.2 Optimality Criteria for Markov Decision Processes

The main criteria for Markov decision processes with finite state and action spaces that we consider are the following:

- *expected total reward over a finite horizon*;
- *expected total discounted reward over an infinite horizon*;
- *expected average reward over an infinite horizon.*

We consider the expected total criterion for *finite horizon Markov decision processes* in case the rewards and the transition probability depend on time, i.e., for *non-stationary Markov decision processes*. For *infinite horizon Markov decision processes* with expected total reward and expected average reward criteria, we assume that the reward and the probability transition do not depend on time, i.e., we consider only stationary Markov decision processes. In general, for infinite horizon Markov decision processes, the expected total reward criterion (see [77, 151, 152]) can be used; however, such a criterion only makes sense for a special class of decision problems.

Expected Total Reward over a Finite Horizon

Consider a Markov decision process with a finite planning time period $[0, T]$. The *expected total reward* of a policy $\pi \in \mathcal{P}$ when the initial state of the system is $x \in X$ we denote by $\sigma_x^T(\pi)$ and define it as

$$\sigma_x^T(\pi) = \sum_{t=0}^{T} \mathbb{E}_{x,\pi}\{r_{\xi_t,\eta_t}(t)\} = \sum_{t=0}^{T} \sum_{y\in X} \sum_{a\in A(x)} \mathbb{P}_{x,\pi}\{\xi_t = y, \eta_t = a\} \cdot r_{y,a}(t).$$

Taking into account that here, the interchange between the summation and the expectation is allowed, then $\sigma_x^T(\pi)$ can also be defined as

$$\sigma_x^T(\pi) = \mathbb{E}_{x,\pi}\Big\{\sum_{t=0}^{T} r_{\xi_t,\eta_t}(t)\Big\}, \text{ for } x \in X.$$

Let

$$\sigma_x^T = \sup_{\pi\in\mathcal{P}} \sigma_x^T(\pi), \text{ for } x \in X, \tag{2.5}$$

where $\mathcal{P}$ is the *set of all policies*. In the vector notation (2.5), we can write

$$\sigma^T = \sup_{\pi \in \mathcal{P}} \sigma^T(\pi).$$

The vector σ^T is called the *value vector*. From Corollary 2.2 and formulae (2.2), (2.3), it follows that

$$\sigma^T = \sup_{\pi \in \mathcal{P}(M)} \sigma^T(\pi) \tag{2.6}$$

and

$$\sigma^T(\pi) = \sum_{i=0}^{T} P(\pi^0) P(\pi^1) P(\pi^2) \ldots P(\pi^{t-1}) \cdot r(\pi^t) \text{ for } \pi = (\pi^0, \pi^1, \ldots) \in \mathcal{P}(M), \tag{2.7}$$

where $\mathcal{P}(M)$ is the *set of all Markov policies*. A policy π^* is called an *optimal policy* if

$$\sigma^T(\pi^*) = \sigma^T. \tag{2.8}$$

So, for a finite horizon Markov decision process, there exists an optimal policy for which the supremum is attained, and it is attained simultaneously for all starting states. Moreover, we can see in the following that for a finite horizon Markov decision process, there exists an optimal Markov policy $\pi^* = (\pi^{*0}, \pi^{*1}, \pi^{*2}, \ldots, \pi^{*T})$, where π^{*T} is a deterministic decision rule for $t = 0, 1, 2, \ldots, T$.

Expected Total Discounted Reward over an Infinite Horizon
The *expected total discounted reward* for a Markov decision process with an initial state x and a policy $\pi \in \mathcal{P}$ we denote by $\sigma_x^\gamma(\pi)$ and define it as

$$\sigma_x^\gamma(\pi) = \sum_{t=0}^{\infty} \mathbb{E}_{x,\pi}\left\{\gamma^t \cdot r_{\xi_t,\eta_t}\right\},$$

where γ is a given *discount factor* that satisfies the condition $0 < \gamma < 1$. So,

$$\sigma_x^\gamma(\pi) = \sum_{t=0}^{\infty} \gamma^t \sum_{x \in X} \sum_{a \in A(x)} \mathbb{P}_{x,\pi}\{\xi_t = x, \eta_t = a\} \cdot r_{y,a}.$$

For any policy π and any initial state $x \in X$, the *expected total discounted reward* can also be defined as $\mathbb{E}_{x,\pi}\left\{\gamma^t \cdot r_{\xi_t,\eta_t}\right\}$. If the rewards $r_{x,a}, \forall x \in X, a \in A(x)$ are bounded and $M = \max_{x \in X, a \in A(x)}\{r_{x,a}\}$, then according to the theorem of dominated convergence [8], the following formula holds:

$$\mathbb{E}_{x,\pi}\Big\{\sum_{t=0}^{\infty}\gamma^t \cdot r_{\xi_t,\eta_t}\Big\} = \sum_{t=0}^{\infty}\mathbb{E}_{x,\pi}\big\{\gamma^t \cdot r_{\xi_t,\eta_t}\big\} = \sigma_x^{\gamma}(\pi).$$

So, the expected total discounted reward and the expected total discounted reward criteria are equivalent.

Let $\pi = (\pi^0, \pi^1, \pi^2, \dots)$ be a Markov policy. Then

$$\sigma^{\gamma}(\pi) = \sum_{t=0}^{\infty}\gamma^t \cdot P(\pi^0)P(\pi^1)P(\pi^2)\dots P(\pi^{t-1}) \cdot \mathbf{r}(\pi^t).$$

Therefore, for a stationary policy π, this formula holds

$$\sigma^{\gamma}(\pi) = \sum_{t=0}^{\infty}\gamma^t \cdot P^t(\pi) \cdot \mathbf{r}(\pi)$$

and the value vector can be defined as

$$\sigma^{\gamma} = \sup_{\pi\in\mathcal{P}} \sigma^{\gamma}(\pi). \tag{2.9}$$

A policy π^* is an *optimal policy* if

$$\sigma^{\gamma}(\pi^*) = \sigma^{\gamma}.$$

In the following, we can see that for a Markov decision process with an expected total γ-discounted reward criterion, there exists an optimal deterministic policy s^*, the supremum in (2.9) is attained, and the value vector σ^{γ} is the unique solution to the following equation:

$$v_x = \max_{a\in A(x)} \big\{r_{x,a} + \gamma \sum_{y\in X} p_{x,y}^a v_y\big\},\ x \in X, \tag{2.10}$$

that corresponds to the policy s^* satisfying the condition

$$r_{x,s^*} + \gamma\sum_{y\in X} p_{x,y}^{s^*}\sigma_x^{\gamma} = \max_{a\in A(x)} \big\{r_{x,a} + \gamma\sum_{y\in X} p_{x,y}^a\sigma_y^{\gamma}\big\},\ \forall a \in A(x), \forall x \in X. \tag{2.11}$$

Expected Average Reward over an Infinite Horizon
The *expected average reward over an infinite horizon* for a given policy $\pi \in \mathcal{P}$ and an initial state $x \in X$ can be defined using one of the following four evaluation measures:

1. Lower limit of the expected average reward:

$$\phi_x(\pi) = \lim_{T\to\infty}\inf \frac{1}{T+1}\sum_{t=0}^{T}\mathbb{E}_{x,\pi}\{r_{\xi_t,\eta_t}\} \text{ with } value\ vector\ \phi = \sup_{\pi\in\mathcal{P}}\phi(\pi).$$

2. Upper limit of the expected average reward:

$$\overline{\phi}_x(\pi) = \lim_{T\to\infty}\sup \frac{1}{T+1}\sum_{t=0}^{T}\mathbb{E}_{x,\pi}\{r_{\xi_t,\eta_t}\} \text{ with } value\ vector\ \overline{\phi} = \sup_{\pi\in\mathcal{P}}\overline{\phi}(\pi).$$

3. Expectation of the lower limit of the average reward:

$$\psi_x(\pi)=\mathbb{E}_{x,\pi}\Big\{\lim_{T\to\infty}\inf \frac{1}{T+1}\sum_{t=0}^{T}r_{\xi_t,\eta_t}\Big\} \text{ with } value\ vector\ \psi = \sup_{\pi\in\mathcal{P}}\psi(\pi).$$

4. Expectation of the upper limit of the average reward:

$$\overline{\psi}_x(\pi)=\mathbb{E}_{x,\pi}\Big\{\lim_{T\to\infty}\sup \frac{1}{T+1}\sum_{t=0}^{T}r_{\xi_t,\eta_t}\Big\} \text{ with } value\ vector\ \overline{\psi} = \sup_{\pi\in\mathcal{P}}\overline{\psi}(\pi).$$

In general, the relation between these four criteria is the following (see Kallenberg [77]):

$$\psi_x(\pi) \le \phi_x(\pi) \le \overline{\phi}_x(\pi)) \le \overline{\psi}_x(\pi),\ \forall x \in X \text{ and } \forall \pi \in \mathcal{P}.$$

However, in [8, 77]), it was shown that

$$\psi(\pi) = \phi_x(\pi) = \overline{\phi}(\pi)) = \overline{\psi}(\pi) \text{ for every stationary policy } \pi \in \mathcal{P}(S)$$

and that there exists a deterministic optimal policy that is optimal for all these four criteria. This means that in the case of a stationary policy, the considered four criteria are equivalent in the sense that an optimal deterministic policy for one criterion is also optimal for the other criteria.

Expected Total Reward over an Infinite Horizon
The expected total reward criterion can be extended for an infinite horizon Markov decision process if there exists $\sum_{t=0}^{T} \mathbb{E}_{x,\pi}\{r_{\xi_t,\eta_t}(t)\}$ and if $T \to \infty$. However, $\sum_{t=0}^{\infty} \mathbb{E}_{x,\pi}\{r_{\xi_t,\eta_t}(t)\}$ may not be well-defined. As shown in [77, 152], the expected total reward criterion over an infinite horizon can be applied for *substochastic models* if $\sum_{y\in X} p_{x,y}^{a} \leq 1, \forall x \in X$ and for some special cases when $\sum_{t=0}^{\infty} \mathbb{E}_{x,\pi}\{r_{\xi_t,\eta_t}(t)\}$ is well-defined. In [178, 179], the following total reward criterion over the infinite horizon was introduced:

$$\sigma_x(\pi) = \lim_{T\to\infty} \inf \frac{1}{T+1} \sum_{t=0}^{T} \sum_{\tau=0}^{t} \mathbb{E}_{x,\pi}\{r_{\xi_t,\eta_t}(\tau)\}.$$

Some extensions of Markov decision process models to semi-Markov decision models were considered in [201].

In the following, we study these problems for Markov decision processes:

- A *finite horizon Markov decision problem* for a finite horizon Markov decision process in which we seek a policy that provides the maximal expected total reward.
- A *discounted Markov decision problem* for an infinite horizon Markov decision process in which it is necessary to determine the policy with the maximal expected total discounted reward.
- An *average Markov decision problem* for an infinite horizon Markov decision process in which it is necessary to determine a policy with the maximal expected average reward.

2.2 Finite Horizon Markov Decision Problems

We present the optimality equations and a dynamic programming algorithm for determining the optimal policies for finite horizon Markov decision problems, when the rewards and the transition probabilities depend on time.

2.2.1 *Optimality Equations for Finite Horizon Problems*

Let us consider a non-stationary Markov decision problem with a finite planning period $[0, T]$, the rewards $r_x^a(t)$, and the transition probabilities $p_{x,y}^a(t)$ depending on time t, where $\sum_{y\in X} p_{x,y}^a(t) = 1, \ \forall x \in X, a \in A(x)$ and $t = 0, 1, 2, \ldots T$. An optimal policy for such a problem can be determined based on the following theorem:

Theorem 2.3 *Let u_x^t for $x \in X$ and $t = 0, 1, \dots, T$ represent the solution to the optimal Bellman equations:*

$$u_x^t = \max_{a \in A(x)} \{r_{x,a}(t) + \sum_{y \in X} p_{x,y}^a(t) u_y^{t+1}\}, \ \forall x \in X, \ t = T-1, T-2, \dots, 0, \tag{2.12}$$

$$u_x^T = \max_{a \in A(x)} \{r_{x,a}(T)\}, \ \forall\, x \in X. \tag{2.13}$$

A deterministic policy $\pi^* = (\mathfrak{s}^0, \mathfrak{s}^1 \dots, \mathfrak{s}^T)$ for which

$$\mathfrak{s}_x^t \in \text{argmin}\{r_{x,a} + \sum_{y \in X} p_{x,y}^a u_y^t\} \ \textit{for } x \in X \textit{ and } t = T-1, \ t-2, \dots, 0$$

is an optimal policy of a finite horizon Markov decision problem, and the vector u^0 represents the value vector of an expected total reward, i.e., $u^0 = \sigma^T$.

Proof We prove the theorem by induction on T. Let $\pi = (\pi^0, \pi^1, \dots, \pi^T)$ be an arbitrary policy. For $T = 0$, the theorem holds because we have

$$\sigma_x^T(\pi) = \sum_{x \in X} \sum_{a \in A(x)} \mathbb{P}\{\xi_0 = x, \eta_0 = a\} \cdot r_{x,a}(0) = \sum_{a \in A(x)} r_{x,a}(0) \pi_{x,a}^0$$

$$\leq \max_{a \in A(x)} \{r_{x,a}(0)\} = v_x^0, \ \forall x \in X.$$

Assume that the theorem holds for $T = 0, 1, 2, \dots, \tau$, and let us show that it holds for $T = \tau + 1$. Consider an arbitrary state $x \in X$. According to Corollary 2.2, there exists a Markov policy $\overline{\pi} = (\pi^0, \pi^1, \dots, \pi^{\tau+1})$ such that $\sigma_x^{\tau+1}(\overline{\pi}) = \sigma_x^{\tau+1}(\pi)$. If we consider the policy $\hat{\pi} = (\hat{\pi}^0, \hat{\pi}^1, \dots, \hat{\pi}^{\tau})$, where $\hat{\pi}_{y,a}^k = \overline{\pi}_{y,a}^{k+1}$ for all $(y, a) \in x \times A$ and $k = 0, 1, \dots, \tau$, then based on the induction assumption, we have $\sigma_y^{\tau} \leq u_y^1$ for all $y \in X$, because for a planning period $[0, \tau + 1]$, the vector u^1 is the same as u^0 for a planning period $[0, \tau]$.

Therefore,

$$\sigma_x^{\tau+1}(\pi) = \sigma_x^{\tau+1}(\overline{\pi}) = \sum_{a \in A(x)} \overline{\pi}_{x,a}^0 \cdot \{r_{x,a}(0) + \sum_{y \in X} p_{x,y}^a(0) \sigma_y^{\tau}(\hat{\pi})\}$$

$$\leq \sum_{a \in A(x)} \overline{\pi}_{x,a}^0 \cdot \{r_{x,a}(0) + \sum_{y \in X} p_{x,y}^a(0) u_y^1\} \leq \max_{a \in A(x} \{r_{x,a}(0) + \sum_{y \in X} p_{x,y}^a(0) u_y^1\} = u_x^0.$$

On the other hand, for vector u^0 and policy π^*, we have

$$u^0 = \mathbf{r}(\mathfrak{s}^0) + P(\mathfrak{s}^0)u^1 = \mathbf{r}(\mathfrak{s}^0) + P(\mathfrak{s}^0)\{\mathbf{r}(\mathfrak{s}^1) + P(\mathfrak{s}^1)u^2\}$$

$$= \cdots = \sum_{l=0}^{\tau+1}\{P(\mathfrak{s}^0)P(\mathfrak{s}^1)\cdots P(\mathfrak{s}^{l-1})\mathbf{r}(\mathfrak{s}^l)\} = \sigma^{\tau+1}(\pi^*),$$

i.e., $\sigma^{\tau+1}(\pi^*) = u^0 \geq \sigma^{\tau+1}(\pi)$. So, π^* is an optimal policy, and u^0 is the value vector. □

2.2.2 *The Backward Induction Algorithm*

Let $\mathbf{r}(\mathfrak{s}^t)$ and $P(\mathfrak{s}^t)$, respectively, be the reward vector and the transition matrix of a deterministic decision rule $\mathfrak{s}^t$ at the time moment t. Then based on Theorem 2.3, an optimal policy π^* and the value vector σ^T of the finite horizon Markov decision problem can be found by using the following algorithm:

1. Set $t = T$ and $\sigma = \mathbf{0}$.
2. Take $\mathfrak{s}^t$ such that

$$\{\mathbf{r}(\mathfrak{s}^t) + P(\mathfrak{s}^t)\sigma\}_x = \max_{a\in A(x)} \{r_{x,a}(t) + \sum_{y\in X} p^a_{x,y}(t)\sigma_y\} \text{ for all } x \in X,$$

and replace σ with $\mathbf{r}(\mathfrak{s}^t) + P(\mathfrak{s}^t)\sigma$.
3. If $t \neq 0$, then substitute $t-1$ with t and return to step 2; otherwise, fix $\pi^* = (\mathfrak{s}^0, \mathfrak{s}^1 \ldots, \mathfrak{s}^T)$ as the optimal stationary deterministic policy and $\sigma^T = \sigma$ as the optimal value vector.

2.3 Discounted Markov Decision Problems

In this section, we consider infinite horizon Markov decision problems with expected total discounted optimization criteria and show how to determine the optimal solutions to these problems. Additionally, we show that a discounted Markov decision problem can be formulated and studied in terms of stationary strategies as a quasi-monotonic programming problem with linear constraints.

2.3.1 The Optimality Equation and Algorithms

Recall that the vector of expected total discounted rewards $\sigma^\gamma(\pi)$ of a stationary policy π with given discount factor $\gamma \in (0, 1)$ is represented as

$$\sigma^\gamma(\pi) = \sum_{t=0}^{\infty} \gamma^t P^t(\pi)\mathbf{r}(\pi).$$

Taking into account that $\{\gamma P(\pi)\}^t \rightarrow 0$ for $t \rightarrow \infty$, we have

$$\sigma^\gamma(\pi) = (I - \gamma P(\pi))^{-1}\mathbf{r}(\pi). \tag{2.14}$$

The γ-discounted value vector σ^γ for a Markov decision processes is defined as

$$\sigma^\gamma = \sup_{\pi \in \mathcal{P}} \sigma^\gamma(\pi). \tag{2.15}$$

A policy π^* is an *optimal policy* if

$$\sigma^\gamma(\pi^*) = \sigma^\gamma.$$

As noted in the previous section, the optimal policy π for a discounted Markov decision problem exists. For such a policy, the supremum in (2.15) is attained, and the value vector σ^γ represents the unique solutions to the following *optimality equation for the discounted Markov decision problem*:

$$\nu_x = \max_{a \in A(x)} \{r_{x,a} + \gamma \sum_{y \in X} p^a_{x,y}\nu_y\}, \ x \in X. \tag{2.16}$$

This fact follows from the Banach fixed point theorem [66, 77, 152] because σ^γ represents a fixed point of mapping $U : \mathbb{R}^T \rightarrow \mathbb{R}^T$, defined by

$$(U\nu)_x = \max_{a \in A(x)} \{r_{x,a} + \gamma \sum_{y \in X} p^a_{x,y}\nu_y\}, \forall x \in X,$$

where U is a contraction mapping with contraction factor $\gamma \in (0, 1)$. Therefore, σ^γ can be computed by iterative procedures based on the mentioned fixed point theorem. Two basic algorithms for determining the solution to a discounted Markov decision process are known: the value iteration algorithm [10] and the policy iteration algorithms [71]. Some modifications and improvements of these algorithms were presented in [77, 152, 157, 193].

The Value Iteration Algorithm

1. Set $t = 0$, fix an arbitrary ν ($\|\nu\| < \infty$), and specify $\epsilon > 0$.

2. For each $x \in X$, find

$$\nu_x^{t+1} = \max_{a \in A(x)} \{r_{x,a} + \gamma \sum_{y \in X} p_{x,y}^a \nu_y^t\}.$$

3. If

$$\|\nu^{t+1} - \nu^t\| < \frac{\epsilon(1-\gamma)}{2\gamma},$$

then go to step **4**; otherwise, replace t with $t+1$ and return to step **2**.

4. For each $x \in X$, choose an action

$$a_x^* \in \underset{a \in A(x)}{\operatorname{argmax}} \{r_{x,a} + \gamma \sum_{y \in X} p_{x,y}^a \nu_y^{t+1}\}$$

and fix

$$\pi_{x,a}^{\epsilon} = \begin{cases} 1, & \text{if } a = a_x^*, \\ 0, & \text{if } a \neq a^*. \end{cases}$$

Then π^ϵ is an ϵ-optimal solution.

In [47, 150, 152, 161, 197] it was shown that the sequence ν^t, $t = 0, 1, 2, \ldots,$ obtained by the value iteration algorithm, converges in norm to $\sigma^{\gamma *}$.

The Policy Iteration Algorithm

1. Set $t = 0$ and fix a deterministic decision rule π^0.

2. Find a solution ν^t of the following equation:

$$(I - \gamma P(\pi^t))\nu = \mathbf{r}(\pi^t).$$

3. For each $x \in X$, choose an action

$$a_x^* \in \underset{a \in A(x)}{\operatorname{argmax}} \{r_{x,a} + \gamma \sum_{y \in X} p_{x,y}^a \nu_y^t\},$$

and find

$$\pi_{x,a}^{t+1} = \begin{cases} 1, & \text{if } a = a_x^*, \\ 0, & \text{if } a \neq a^*. \end{cases}$$

4. If $\pi^{t+1} = \pi^t$, then stop and fix $\pi^* = \pi^t$; otherwise, replace t with $t+1$ and return to step **2**.

In [77, 152] it was shown that the policy iteration algorithm in the case of a finite set of states and a finite set of actions finds the solution to the problem in a finite number of steps.

2.3.2 The Linear Programming Approach

Linear programming represents an important mathematical tool for studying and solving Markov decision problems [38, 69, 70, 76, 77, 152]. In this section, we show how to apply the linear programming approach to the discounted Markov decision problem.

If value vector σ^{γ} is the unique solution to equation (2.16), then σ^{γ} satisfies the following system of inequalities:

$$\nu_x \geq r_{x,a} + \gamma \sum_{y \in X} p^a_{x,y} \nu_y, \;\; \forall x \in X, \forall a \in A(x) \tag{2.17}$$

and σ^{γ} is the smallest vector that satisfies (2.17). This means that σ^{γ} represents the unique basic optimal solution to the following linear programming problem:
Minimize

$$\phi^{\gamma}_{\theta}(\nu) = \sum_{x \in X} \theta_x \nu_x$$

subject to

$$\nu_x \geq r_{x,a} + \gamma \sum_{y \in X} p^a_{x,y} \nu_y, \;\; \forall x \in X, \;\; \forall a \in A(x),$$

where $\theta_x, \; x \in X$ represent arbitrary positive values such that $\sum_{x \in X} \theta_x = 1$. Here, θ_y for $y \in X$ can be treated as the probabilities of choosing the starting state $y \in X$ in the decision problem. The optimal value of the object function of the problem is $\upsilon^{\gamma}_{\theta} = \phi_{\theta}(\upsilon^{\gamma})$. In the case $\theta_y = 1$ for $y = x_0$ and $\theta_y = 0$ for $y \in X \setminus \{x_0\}$, we obtain the linear programming model for the discounted Markov decision problem with starting state x_0. In general, the value vector υ^{γ} is the optimal solution to the linear programming problem for arbitrary $\theta_x > 0, \; \forall x \in X$ that gives the solution to the problem with an arbitrary starting state $x \in X$.

The arguments presented above prove the following theorem:

Theorem 2.4 *Let ν^*_x $(x \in X)$ be an optimal basic solution to the linear programming problem:*

Minimize

$$\phi_\theta^\gamma(\nu) = \sum_{x \in X} \theta_x \nu \tag{2.18}$$

subject to

$$\nu_x - \gamma \sum_{y \in X} p_{x,y}^a \nu_y \ge r_{x,a}, \qquad \forall x \in X,\ a \in A(x), \tag{2.19}$$

where $0 < \gamma < 1$. *Then* ν_x^* *for* $x \in X$ *represents the optimal total expected discounted cost for the problem with starting states* $x \in X$.

An optimal stationary deterministic policy can be found by fixing a map

$$s^* : x \to a \in A^*(x) \text{ for } x \in X,$$

where

$$A^*(x) = \left\{ a \in A(x) \mid \nu_x - \gamma \sum_{y \in X} p_{x,y}^a \nu_y = r_{x,a} \right\}, \forall x \in X \setminus \{z\}.$$

By solving the linear programming problem (2.18), (2.19), we determine the value vector υ^γ. If υ^γ is known, then we find an optimal policy π^* in the same way as in step 4 of the value iteration algorithm. The dual problem for the linear programming problem (2.18), (2.19) is the following:
Maximize

$$\varphi_\theta^\gamma(\boldsymbol{\alpha}) = \sum_{x \in X} \sum_{a \in A(x)} r_{x,a} \alpha_{x,a} \tag{2.20}$$

subject to

$$\begin{cases} \sum\limits_{a \in A(y)} \alpha_{y,a} - \gamma \sum\limits_{x \in X} \sum\limits_{a \in A(x)} p_{x,y}^a \cdot \alpha_{x,a} = \theta_y, \quad \forall y \in X; \\ \\ \alpha_{x,a} \ge 0, \qquad \forall x \in X,\ a \in A(x), \end{cases} \tag{2.21}$$

where θ_y for $y \in X$ can be treated in the same way as in the primal problem.

It is easy to observe that for an arbitrary feasible solution $\boldsymbol{\alpha}$ of problem (2.20), (2.21), this formula holds $\sum_{a \in A(x)} \alpha_{x,a} > 0,\ \forall x \in X$. Therefore, we can calculate the values

$$s_{x,a} = \frac{\alpha_{x,a}}{\sum\limits_{a \in A(x)} \alpha_{x,a}} \quad \text{for } x \in X,\ a \in A(x) \tag{2.22}$$

that satisfy the condition

$$\sum_{a\in A(x)} s_{x,a} = 1, \forall x \in X, a \in A(x). \tag{2.23}$$

These values for a feasible solution α to problem (2.20), (2.21) determine a stationary policy $\mathbf{s}$, where $s_{x,a}$ for $x \in X$, and $a \in A(x)$ expresses the corresponding probability of selecting the actions a in the states x when the starting state is chosen randomly according to the distribution $\{\theta_x\}$. This means that if the stationary policy $\mathbf{s}$ for the considered Markov decision problem is applied, then the expected total discounted reward is equal to $\varphi_\theta(\boldsymbol{\alpha})$. Additionally, if for given α we denote

$$q_x = \sum_{a\in A(x)} \alpha_{x,a} \text{ for } x \in X, \tag{2.24}$$

then q_x for $x \in X$ can be treated as the limiting probabilities in the states $x \in X$ in the Markov process induced by stationary policy $\mathbf{s}$ when the starting state is chosen randomly according to the distribution θ_x on X.

Thus, if we find an optimal solution α^* of the dual problem (2.20), (2.21), then we determine an optimal stationary policy $\mathbf{s}^*$ for the discounted Markov decision problem, where

$$s^*_{x,a} = \frac{\alpha^*_{x,a}}{\sum\limits_{a\in A(x)} \alpha^*_{x,a}} \quad \text{for } x \in X,\ a \in A.$$

Taking into account that a discounted Markov problem can be represented as a linear programming problem, we may conclude that for an optimal deterministic policy for a discounted Markov decision problem, there always exists a polynomial algorithm.

2.3.3 A Nonlinear Model for the Discounted Problem

From the dual linear programming problem (2.20), (2.21), the following nonlinear programming problem can be derived:
Maximize

$$\psi^\gamma_\theta(\mathbf{s}, \mathbf{q}) = \sum_{x\in X} \sum_{a\in A(x)} r_{x,a} s_{x,a} q_x \tag{2.25}$$

subject to

$$\begin{cases} q_y - \gamma \sum_{x\in X} \sum_{a\in A(x)} p^a_{x,y}\, s_{x,a} q_x = \theta_y, & \forall y \in X; \\ \sum_{a\in A(y)} s_{y,a} = 1, & \forall y \in X; \\ s_{x,a} \geq 0, \ \forall x \in X, \ \forall a \in A(x), \end{cases} \tag{2.26}$$

where θ_y are the same values as in problems (2.20), (2.21), and $s_{x,a}, q_x$ for $x \in X,\ a \in A(x)$ represent the variables that must be found. We obtain the problem (2.25), (2.26) if we take into account (2.22)–(2.24) in (2.20), (2.21). In this nonlinear problem, q_x for $x \in X$ can be treated as the limiting probabilities in the states when the starting state is chosen randomly according to the distribution function θ_x on X. One can easily check that if α is a feasible solution to problem (2.20), (2.21), then $s_{x,a},\ q_x$ for $x \in X, a \in A(x)$, determined according to (2.22), (2.24), and they represent a feasible solution to problem (2.25), (2.26). Conversely, if $s_{x,a}, q_x$ for $x \in X$, and $a \in A(x)$ is a feasible solution to problem (2.25), (2.26), then $\alpha_{x,a} = s_{x,a}q_x$, for $x \in X$, and $a \in A(x)$ represent a feasible solution to problem (2.20), (2.21).

So, there exists a bijective mapping between feasible solutions to problems (2.20), (2.21) and (2.25), (2.26) that preserve the values of the corresponding object functions. Therefore, an optimal solution to problem (2.25), (2.26) corresponds to an optimal solution to a discounted Markov decision problem. In general, the nonlinear programming model (2.25), (2.26) can be derived based on Eq. (2.14) if we represent it as follows:

$$(I - P(s))\sigma^{\gamma} = \mathbf{r}(s), \tag{2.27}$$

where σ^{γ} is the value vector that corresponds to a given stationary policy s. For a fixed s, (2.27) represents the following linear programming problem:
Minimize

$$\phi^{\gamma}_{\theta,s}(\nu) = \sum_{x\in X} \theta_x \nu_x \tag{2.28}$$

subject to

$$\nu_x - \gamma \sum_{y\in X} \sum_{a\in A(x)} p^a_{x,y} s_{x,a} \nu_y \geq \sum_{a\in A(x)} r_{x,a} s_{x,a} \ \forall x \in X, \tag{2.29}$$

where $\theta_x,\ x \in X$ represent arbitrary positive values such that $\sum_{x\in X} \theta_x = 1$ and are treated as the probabilities of choosing the starting state $y \in X$ in the decision

problem. If we consider the dual model for the linear programming problem (2.28), (2.29), then we obtain the following problem:
Maximize

$$\psi^{\gamma}_{\theta,s}(\mathbf{q}) = \sum_{x\in X}\sum_{a\in A(x)} r^a_x s_{x,a} q_x \tag{2.30}$$

subject to

$$\begin{cases} q_y - \gamma \sum\limits_{x\in X}\sum\limits_{a\in A(x)} p^a_{x,y}\, s_{x,a} q_x = \theta_y, & \forall y \in X; \\ \qquad q_x \geq 0, & \forall x \in X. \end{cases} \tag{2.31}$$

Here, q_x for $x \in X$ are positive and uniquely determined by the equations above, i.e., (2.31) can be replaced by the following system of equations:

$$q_y - \gamma \sum_{x\in X}\sum_{a\in A(x)} p^a_{x,y}\, s_{x,a} q_x = \theta_y, \qquad \forall y \in X. \tag{2.32}$$

So, if we vary s and maximize (2.30), (2.32) on the set S determined by

$$\begin{cases} \sum\limits_{a\in A(x)} s_{x,a} = 1, & \forall x \in X; \\ s_{x,a} \geq 0, & \forall x \in X,\ a \in A(x), \end{cases}$$

then we obtain the problem (2.25), (2.26). As we have shown, using the notation (2.22)–(2.24) in (2.25), (2.26), we obtain the linear programming problem (2.28), (2.29).

2.3.4 The Quasi-monotonic Programming Approach

In this section, based on the results from the previous sections, we show that the problem of determining the optimal solution to a discounted Markov decision problem can be represented as a quasi-monotonic programming problem with linear constraints. For this, we use the following theorem:

Theorem 2.5 *Let an average Markov decision problem be given and consider the function*

$$\psi^{\gamma}_{\theta}(\mathbf{s}) = \sum_{x\in X}\sum_{a\in A(x)} r^a_x s_{x,a}\, q_x, \tag{2.33}$$

where q_x for $x \in X$ satisfies the condition

$$q_y - \gamma \sum_{x\in X} \sum_{a\in A(x)} p^a_{x,y}\, s_{x,a} q_x = \theta_y, \qquad \forall y \in X. \tag{2.34}$$

Then on the set S of solutions to the system

$$\begin{cases} \sum\limits_{a\in A(x)} s_{x,a} = 1, \ \forall x \in X; \\ \qquad s_{x,a} \geq 0, \ \forall x \in X, \ a \in A(x), \end{cases} \tag{2.35}$$

the function $\psi^\gamma_\theta(\mathbf{s})$ depends only on $s_{x,a}$ for $x \in X$, $a \in A(x)$, and $\psi^\gamma_\theta(s)$ is quasi-monotonic on $\mathbf{S}$ (i.e., $\psi^\gamma_\theta(\mathbf{s})$ is quasi-convex and quasi-concave on $\mathbf{S}$ [23]).

Proof For an arbitrary $\mathbf{s} \in \mathbf{S}$, system (2.34) uniquely determines q_x for $x \in X$, and therefore, $\psi_\theta(\mathbf{s})$ is uniquely determined for an arbitrary $\mathbf{s} \in \mathbf{S}$, i.e., the first part of the theorem holds.

Now let us prove the second part of the theorem. We show that the function $\psi(\mathbf{s})$ is quasi-monotonic on $\mathbf{S}$. To prove this, it is sufficient to show that for an arbitrary $c \in \mathbf{R}^1$, the *sublevel set*

$$L^-_c(\psi^\gamma_\theta) = \{\mathbf{s} \in \mathbf{S} |\ \psi^\gamma_\theta(\mathbf{s}) \leq c\}$$

and the *superlevel set*

$$L^+_c(\psi^\gamma_\theta) = \{\mathbf{s} \in \mathbf{S} |\ \psi^\gamma_\theta(\mathbf{s}) \geq c\}$$

of function $\psi^\gamma_\theta(\mathbf{s})$ are convex. These sets can be obtained from the *sublevel set*

$$L^-_c(\varphi^\gamma_\theta) = \{\alpha |\ \varphi^\gamma_\theta(\alpha) \leq c\}$$

and the *superlevel set*

$$L^+_c(\varphi^\gamma_\theta) = \{\alpha |\ \varphi_\theta\gamma(\alpha) \geq c\}$$

of function $\varphi^\gamma_\theta(\alpha)$ for the linear programming problem (2.20), (2.21).

Denote by α^i, $i = \overline{1,k}$ the basic solutions to system (2.21). All feasible strategies of problem (2.20), (2.21) can be obtained as convex combinations of basic solutions α^i, $i = \overline{1,k}$. Each $\alpha^i \in \{1, 2, \ldots, k\}$ determines a stationary strategy

$$s^{(i)}_{x,a} = \frac{\alpha^i_{x,a}}{q^i_x}, \quad x \in X, \ a \in A(x) \tag{2.36}$$

for which $\psi_\theta^\gamma(s^{(i)}) = \varphi(\alpha^i)$, where

$$q_x^i = \sum_{a\in A(x)} \alpha_{x,a}^i, \quad \forall x \in X. \tag{2.37}$$

An arbitrary feasible solution α to system (2.21) determines a stationary strategy

$$s_{x,a} = \frac{\alpha_{x,a}}{q_x} \quad \text{for } x \in X,\ a \in A(x) \tag{2.38}$$

for which $\psi_\theta^\gamma(\mathbf{s}) = \varphi_\theta(\alpha)$, where $q_x = \sum_{a\in A(x)} \alpha_{x,a}$, $\forall x \in X$. Taking into account that α can be represented as $\alpha = \sum_{i=1}^k \lambda^i \alpha^i$, where $\sum_{i=1}^k \lambda^i = 1$, $\lambda^i \geq 0$, $i = \overline{1,k}$, we have $\varphi_\theta(\alpha) = \sum_{i=1}^k \varphi_\theta(\alpha^i)\lambda^i$, and we can consider

$$\alpha = \sum_{i=1}^k \lambda^i \alpha^i; \quad q = \sum_{i=1}^k \lambda^i q^i. \tag{2.39}$$

Using (2.36)–(2.39), we obtain

$$s_{x,a} = \frac{\alpha_{x,a}}{q_x} = \frac{\sum\limits_{i=1}^k \lambda^i \alpha_{x,a}^i}{q_x} = \frac{\sum\limits_{i=1}^k \lambda^i s_{x,a}^i q_x^i}{q_x} = \sum_{i=1}^k \frac{\lambda^i q_x^i}{q_x} s_{x,a}^i, \quad \forall x \in X_\alpha,\ a \in A(x)$$

and

$$q_x = \sum_{i=1}^k \lambda^i q_x^i, \quad \text{for } x \in X. \tag{2.40}$$

So,

$$s_{x,a} = \sum_{i=1}^k \frac{\lambda^i q_x^i}{q_x} s_{x,a}^{(i)} \quad \text{for } x \in X,\ a \in A(x), \tag{2.41}$$

where q_x for $x \in X$ are determined according to (2.40). The strategy s defined by (2.41) is a feasible strategy because $s_{x,a} \geq 0, \forall x \in X, a \in A(x)$ and $\sum_{a\in A(x)} s_{x,a} = 1$, $\forall x \in X$. Moreover, we can observe that $q_x = \sum_{i=1}^k \lambda^i q_x^i$, for $x \in X$ represents a solution to system (2.34) for the strategy s defined by (2.41). This can be verified by introducing (2.40) and (2.41) in (2.34); after such a substitution, all equations from (2.34) are transformed into identities.

For $\psi_\theta^\gamma(\mathbf{s})$, we have

$$\psi_\theta^\gamma(\mathbf{s}) = \sum_{x\in X}\sum_{a\in A(x)} r_{x,a}s_{x,a}q_x = \sum_{x\in X}\sum_{a\in A(x)} r_{x,a}\sum_{i=1}^{k}\left(\frac{\lambda^i q_x^i}{q_x}s_{x,a}^{(i)}\right)q_x =$$

$$\sum_{i=1}^{k}\left(\sum_{x\in X}\sum_{a\in A(x)} r_{x,a}s_{x,a}^{(i)}q_x^i\right)\lambda^i = \sum_{i=1}^{k}\psi_\theta(s^{(i)})\lambda^i,$$

i.e.,

$$\psi_\theta^\gamma(s) = \sum_{i=1}^{k}\psi_\theta^\gamma(s^{(i)})\lambda^i, \tag{2.42}$$

where s is the strategy that corresponds to α. This means that if strategies $s^{(1)}, s^{(2)}, \ldots, s^{(k)}$ correspond to the basic solutions $\alpha^1, \alpha^2, \ldots, \alpha^k$ to problem (2.20), (2.21), and $s \in \mathbf{S}$ corresponds to an arbitrary solution α that can be expressed as a convex combination of basic solutions to problem (2.20), (2.21) with the corresponding coefficients $\lambda^1, \lambda^2, \ldots, \lambda^k$, then we can express the strategy s and the corresponding value $\psi_\theta^\gamma(s)$ by (2.40)–(2.42).

Thus, an arbitrary strategy $s \in \mathbf{S}$ is determined according to (2.40), (2.41), where $\lambda^1, \lambda^2, \ldots, \lambda^k$ correspond to a solution to the following system:

$$\sum_{i=1}^{k}\lambda^i = 1;\ \ \lambda^i \geq 0,\ \ i = \overline{1,k}.$$

Consequently, the sublevel set $L_c^-(\psi_\theta^\gamma)$ of function $\psi_\theta^\gamma(s)$ represents the set of strategies s determined by (2.40), (2.41), where $\lambda^1, \lambda^2, \ldots, \lambda^k$ satisfy the condition

$$\begin{cases} \sum_{i=1}^{k}\psi_\theta^\gamma(s^i)\lambda^i \leq c; \\ \sum_{i=1}^{k}\lambda^i = 1;\ \ \lambda^i \geq 0,\ \ i = \overline{1,k}, \end{cases} \tag{2.43}$$

and the superlevel set $L_c^+(\psi_\theta^\gamma)$ of $\psi_\theta^\gamma(s)$ represents the set of strategies s determined by (2.40), (2.41), where $\lambda^1, \lambda^2, \ldots, \lambda^k$ satisfy the condition

$$\begin{cases} \sum_{i=1}^{k}\psi_\theta^\gamma(s^{(i)})\lambda^i \geq c; \\ \sum_{i=1}^{k}\lambda^i = 1;\ \ \lambda^i \geq 0,\ \ i = \overline{1,k}. \end{cases} \tag{2.44}$$

Let us show that $L_c^-(\psi_\theta^\gamma)$, $L_c^+(\psi_\theta^\gamma)$ are convex sets. We present the proof of the convexity of sublevel set $L_c^-(\psi_\theta^\gamma)$. The proof of the convexity of $L_c^+(\psi_\theta^\gamma)$ is similar to the proof of the convexity of $L_c^-(\psi_\theta^\gamma)$. Denote by Λ the set of solutions $(\lambda^1, \lambda^2, \dots, \lambda^k)$ of system (2.43). Then from (2.40), (2.41), (2.43), we have

$$L_c^-(\psi_\theta^\gamma) = \prod_{x\in X} \hat{S}_x,$$

where $\hat{S}_x$ represents the set of strategies

$$s_{x,a} = \frac{\sum_{i=1}^{k} \lambda^i q_x^i s_{x,a}^{(i)}}{\sum_{i=1}^{k} \lambda^i q_x^i}, \quad \text{for } a \in A(x)$$

in the state $x \in X$ determined by $(\lambda^1, \lambda^2, \dots, \lambda^k) \in \Lambda$. Here, $\sum_{i=1}^{k} \lambda^i q_x^i > 0$, and $s_{x,a}$ for a given $x \in X$ represents a linear-fractional function with respect to $\lambda^1, \lambda^2, \dots, \lambda^k$, defined on a convex set Λ_x, and $\hat{S}_x$ is the image of $s_{x,a}$ on Λ_x. Therefore, $\hat{S}_x$ is a convex set (see [23]). □

Corollary 2.6 *The function $\psi_\theta^\gamma(\mathbf{s})$ is continuous on the set of solutions of system* (2.35).

Thus, a discounted Markov decision problem can be formulated as the quasi-monotonic programming problem of how to maximize the quasi-monotonic object function (2.33), (2.34) on a polyhedron set $\mathbf{S}$ of solutions of system (2.35). By solving this problem, we determine a stationary policy s^* for the discounted Markov decision problem. In Chap. 3, we use this quasi-monotonic model for studying stochastic games with discounted payoffs.

2.4 Average Markov Decision Problems

This section presents the results concerned with determining the optimal policies for average Markov decision problems. First, we study the problem of determining the optimal policies for average Markov decision problems with a unichain property, and then we present the main results for the general case of the problem, i.e., for the case when policies may induce multichain processes. Additionally, a new approach based on quasi-monotonic programming for determining the optimal solution to these problems is proposed. In the following, we can see that such an approach can be used for studying stochastic control problems on networks and average stochastic games with average payoffs.

2.4.1 The Main Results for the Unichain Model

An average Markov decision problem with a unichain property is a problem for which the transition probability matrix induced by an arbitrary deterministic policy is unichain. We show that for such a problem, the optimality equation can be represented as follows:

$$\varepsilon_x + \omega = \max_{a \in A(x)} \left\{ r_{x,a} + \sum_{y \in X} p_{x,y}^a \varepsilon_y \right\}, \quad \forall x \in X. \tag{2.45}$$

To prove this, we first need to show how to estimate the average reward of a deterministic stationary policy π for the unichain problem.

Denote by ω^π the average reward per transition in the stationary Markov process with transition probability matrix $P(\pi)$ and reward vector $\mathbf{r}(\pi)$ induced by a stationary deterministic policy π.

According to the results from Sect. 1.9.2 (see formula (1.74)), we have

$$\sigma^\pi(t) = t\omega^\pi + \varepsilon^\pi + \epsilon^\pi(t).$$

Using this formula, we can write the following two equivalent equations:

$$\sigma^\pi(t) = t\omega^\pi + \varepsilon^\pi + \epsilon^\pi(t),$$
$$\sigma^\pi(t-1) = (t-1)\omega^\pi + \varepsilon^\pi + \epsilon^\pi(t-1),$$

where $\epsilon^\pi(t)$ and $\epsilon^\pi(t-1)$ tend to zero if t tends to infinity. If we introduce the expression of $\sigma^\pi(t)$ and $\sigma^\pi(t-1)$ into the recursive formula

$$\sigma^\pi(t) = \mathbf{r}(\pi) + P(\pi)\sigma^\pi(t-1),$$

then we obtain

$$t\omega^\pi + \varepsilon^\pi + \epsilon^\pi(t) = \mathbf{r}(\pi) + P(\pi)\big((t-1)\omega^\pi + \varepsilon^\pi + \epsilon^\pi(t-1)\big).$$

Through rearrangement, we get

$$\varepsilon^\pi + t\omega^\pi - (t-1)P(\pi)\omega^\pi = \mathbf{r}(\pi) + P(\pi)\varepsilon^\pi + P(\pi)\epsilon^\pi(t-1) - \epsilon^\pi(t).$$

Here,

$$\omega^{\pi} = P^{\pi}\omega^{\pi}. \tag{2.46}$$

In addition, for a Markov unichain, all components of the vector ω^{π} are the same, i.e., $\omega^{\pi}_{x_1} = \omega^{\pi}_{x_2} = \cdots = \omega^{\pi}_{x_n} = \omega$. So, if $t \to \infty$, then $\epsilon^{\pi}(t),\ \epsilon^{\pi}(t-1) \to 0$ and we obtain

$$\varepsilon^{\pi}_x + \omega = \{\mathbf{r}(\pi) + P(\pi)\varepsilon^{\pi}\}_x, \forall x \in X. \tag{2.47}$$

This is the system of equations for a unichain Markov process. It is well known that in the case of unichain processes, the rank of the matrix $(I - P)$ is equal to $n-1$ (see [137]). Based on this fact, it was shown in [137] that the system of Eqs. (2.47) has a unique solution once $\varepsilon^{\pi}_i = 0$ for some i has been set. This means that two different vectors $\bar{\varepsilon}^{\pi}$ and $\tilde{\varepsilon}^{\pi}$, which represent the solutions to this equation, differ only by some constant for each component. Therefore, the system of Eqs. (2.47) allows us to determine the average reward per transition in the unichain Markov process with rewards induced by a deterministic stationary policy π.

Equation (2.47) holds for an arbitrary deterministic stationary policy $\pi \in S$. But a deterministic stationary policy means that in each state $x \in X$, an action $a \in A(x)$ is chosen. Taking into account that we have a Markov decision process with finite state and action spaces, we can select actions a^*_x for $x \in X$ in such a way that the average reward per transition ω^* will be maximal, i.e.,

$$\omega^* = r_{x,a^*} + \sum_{y \in X} p^{a^*}_{x,y}\varepsilon_y - \varepsilon_x, \forall x \in X.$$

So, the equation

$$\omega^* = \max_{a \in A(x)} \{r_{x,a} + \sum_{y \in Y} p^{a}_{x,y}\varepsilon_y - \varepsilon_x\}, \ \forall x \in X$$

has a solution. This involves that (2.45) has solutions. Based on this, for the average Markov decision problem with a unichain property, the following algorithms can be proposed:

The Value Iteration Algorithm

1. Set $t = 0$, specify $\epsilon > 0$, and select an arbitrary initial vector ε^0.
2. For each $x \in X$, find

$$\varepsilon^{t+1}_x = \max_{a \in A(x)} \{r_{x,a} + \sum_{y \in X} p^{a}_{x,y}\varepsilon^{t}_y\}.$$

3. If

$$(\max_{x \in X}\{\varepsilon_x^{t+1}\} - \max_{x \in X}\{\varepsilon_x^t\}) < \epsilon,$$

then go to step 4; otherwise, change t to $t + 1$ and return to step 2.
4. For each $x \in X$, choose an action

$$a_x^* \in \operatorname*{argmax}_{a \in A(x)}\{r_{x,a} + \sum_{y \in X} p_{x,y}^a v_y^t\},$$

fix

$$\pi_{x,a}^{\epsilon} = \begin{cases} 1, & \text{if } a = a_x^*, \\ 0, & \text{if } a \neq a^*, \end{cases}$$

and stop.

The details of the convergence of this algorithm and the conditions under which the algorithm finds the solution in finite time can be found in [77, 152].

The Policy Iteration Algorithm

1. Set t=0 and select an arbitrary policy π^t.
2. Find a solution ω^t, ε_x^t for $x \in X$ of equation

$$\varepsilon_x^t + \omega^t = \{\mathbf{r}(\pi^t) + P(\pi^t)\varepsilon^t\}_x, \forall x \in X.$$

3. If

$$(\max_{x \in X}\{\varepsilon_x^{t+1}\} - \max_{x \in X}\{\varepsilon_x^t\}) < \epsilon,$$

then go to step 4; otherwise, change t to $t + 1$ and return to step 2.
4. For each $x \in X$, choose an action

$$a_x^* \in \operatorname*{argmax}_{a \in A(x)}\{r_{x,a} + \sum_{y \in X} p_{x,y}^a v_y^t\},$$

fix

$$\pi_{x,a}^{\epsilon} = \begin{cases} 1, & \text{if } a = a_x^*, \\ 0, & \text{if } a \neq a^*, \end{cases}$$

and stop.

2.4.2 Linear Programming for a Unichain Problem

If for optimality Eq. (2.45) we consider the following inequalities

$$\varepsilon_x + \omega \geq r_{x,a} + \sum_{y \in X} p_{x,y}^a \varepsilon_y, \quad \forall x \in X, \tag{2.48}$$

then we can observe that a solution to Eq. (2.45) represents a solution to a system of linear in Eqs. (2.48) with the smallest ω. So, a solution to the average Markov decision problem with a unichain property is a solution to system (2.45) for the smallest ω. This means that an average Markov decision problem with a unichain property can be formulated as the following linear programming problem:
Minimize

$$\phi(\varepsilon, \omega) = \omega \tag{2.49}$$

subject to

$$\varepsilon_x - \sum_{y \in X} p_{x,y}^a \varepsilon_y + \omega \geq r_{x,a}, \quad \forall x \in X, \quad \forall a \in A(x). \tag{2.50}$$

If an optimal basic solution ω^*, ε_x^* for $x \in X$ is known, then an optimal deterministic stationary policy π^* can be found according to step 4 of the value iteration algorithm.

The dual model for the linear programming problem (2.49), (2.50) is:
Maximize

$$\varphi(\alpha) = \sum_{x \in X} \sum_{a \in A(x)} r_{x,a}\, \alpha_{x,a} \tag{2.51}$$

subject to

$$\begin{cases} \sum\limits_{a \in A(y)} \alpha_{y,a} - \sum\limits_{x \in X} \sum\limits_{a \in A(x)} p_{x,y}^a\, \alpha_{x,a} = 0, \quad \forall y \in X; \\ \sum\limits_{x \in X} \sum\limits_{a \in A(x)} \alpha_{x,a} = 1; \\ \alpha_{x,a} \geq 0, \quad \forall x \in X, \ a \in A(x). \end{cases} \tag{2.52}$$

If α^* is an optimal basic solution to this problem, then an optimal stationary policy of a unichain problem can be found as

$$\pi^*_{x,a} = \begin{cases} \dfrac{\alpha^*_{x,a}}{\sum_{a \in A(x)} \alpha^*_{x,a}}, & \text{if } \sum_{a \in A(x)} \alpha^*_{x,a} \neq 0, \\ \text{arbitrary}, & \text{if } \sum_{a \in A(x)} \alpha_{x,a} = 0. \end{cases}$$

2.4.3 A Nonlinear Model for the Unichain Problem

In [100], it was shown that an average Markov decision problem with a unichain property can be formulated as the following optimization problem:
Maximize

$$\psi(s,q) = \sum_{x \in X} \sum_{a \in A(x)} r_{x,a} s_{x,a}\, q_x \tag{2.53}$$

subject to

$$\begin{cases} q_y - \sum\limits_{x \in X} \sum\limits_{a \in A(x)} p^a_{x,y} s_{x,a} q_x = 0, & \forall y \in X; \\ \sum\limits_{x \in X} q_x = 1; & \\ \sum\limits_{a \in A(x)} s_{x,a} = 1, & \forall x \in X; \\ s_{x,a} \geq 0, \ \forall x \in X, \ a \in A(x). & \end{cases} \tag{2.54}$$

The variables $s_{x,a}$ correspond to policies, where $s_{x,a}$ expresses the probability of selecting the action $a \in A(x)$ in the state $x \in X$, and q_x for $x \in X$ represents the corresponding limiting probabilities in the states $x \in X$ for the probability transition matrix $P^s = (p^s_{x,y})$ induced by s, i.e., $p^s_{x,y} = \sum_{a \in A(x)} p_{x,y} s_{x,a}$.

In this problem, the average reward $\psi(s,q)$ is maximized under the conditions (2.54) that determine the stationary policy in the unichain problem. An optimal solution (s^*, q^*) to problem (2.53), (2.54) with $s^*_{x,a} \in \{0,1\}$ corresponds to an optimal stationary strategy $s^* : X \to A$, where $a^* = s^*(x)$ for $x \in X$ if $s^*_{x,a} = 1$. Using the notations $\alpha_{x,a} = s_{x,a} q_x$ for $x \in X$, $a \in A(x)$, problem (2.53), (2.54) can be easily transformed into the following linear programming problem:

Maximize

$$\varphi(\alpha) = \sum_{x \in X} \sum_{a \in A(x)} r_{x,a}\, \alpha_{x,a} \tag{2.55}$$

subject to

$$\begin{cases} q_y - \sum\limits_{x \in X} \sum\limits_{a \in A(x)} p^a_{x,y} \alpha_{x,a} = 0, & \forall y \in X; \\ \sum\limits_{x \in X} q_x = 1; & \\ \sum\limits_{a \in A(x)} \alpha_{x,a} - q_x = 0, & \forall x \in X; \\ \alpha_{x,a} \geq 0, \ \ \forall x \in X, \ a \in A(x). & \end{cases} \tag{2.56}$$

This problem can be simplified by eliminating q_x from (2.56), and finally, after such an elimination, we obtain the linear programming (2.51), (2.52).

Based on the relationship mentioned above, between problem (2.51), (2.54) and problem (2.53), (2.54) in [100], the following result has been proved:

Lemma 2.7 *Let an average Markov decision problem be given, where an arbitrary stationary strategy* s *generates a Markov unichain, and consider the function*

$$\psi(s) = \sum_{x \in X} \sum_{a \in A(x)} r_{x,a} s_{x,a}\, q_x, \tag{2.57}$$

where q_x *for* $x \in X$ *are uniquely determined by the following unichain condition:*

$$\begin{cases} q_y - \sum\limits_{x \in X} \sum\limits_{a \in A(x)} p^a_{x,y} s_{x,a} q_x = 0, & \forall y \in X; \\ \sum\limits_{x \in X} q_x = 1. & \end{cases} \tag{2.58}$$

Then the function $\psi(s)$ *on the set* $\overline{S}$ *of solutions to the system*

$$\begin{cases} \sum\limits_{a \in A(x)} s_{x,a} = 1, & \forall x \in X; \\ s_{x,a} \geq 0, \ \ \forall x \in X, \ a \in A(x) & \end{cases}$$

depends only on $s_{x,a}$ *for* $x \in X$, $a \in A(x)$, *and* $\psi(s)$ *is quasi-monotonic on* $\overline{S}$ *(i.e.,* $\psi(s)$ *is quasi-concave and quasi-convex on* $\overline{S}$ *[23]).*

Moreover, $\psi(s) = \omega_x(s)$, $\forall x \in X$.

In Sect. 2.4.1, we present the proof of this lemma for the general case of a multichain average Markov decision problem. An important property for the average Markov decision problem with a unichain property that we use in the next chapter is the following:

Corollary 2.8 *The function* $\psi(s)$ *defined according to (2.57), (2.58) is continuous on* $\overline{S}$.

2.4.4 Optimality Equations for Multichain Processes

In Sect. 2.4.1, it was shown that the gain and bias for a Markov reward process with the transition matrix $P(\pi)$ and reward vector $\mathbf{r}(\pi)$ corresponding to a deterministic stationary policy π can be found by solving Eqs. (2.46), (2.47) if the matrix $P(\pi)$ is unichain. In the case of unichain processes, the components ω_x of the gain vector ω are the same, and therefore, for a solution $(\omega^*, \varepsilon^*)$ to Eq. (2.46), ω^* is uniquely determined. The evaluation of the gain and the bias in a multichain process can also be found based on Eqs. (2.46), (2.47); however, if $P(\pi)$ contains several closed irreducible recurrent classes $X_1, X_2, \ldots, X_p$ $(p > 1)$ and the class of transient states X_0, then the components of ω for different states $x \in X$ may be different, but for each recurrent class X_k, we obtain the solution to equations

$$\varepsilon_x + \omega_x^\pi = \{\mathbf{r}(\pi) + P(\pi)\varepsilon^\pi\}_x, \ \forall x \in X_k \tag{2.59}$$

by setting $\varepsilon_{x_k} = 0$ for some $x_k \in X_k$. Moreover, for each $x \in X \setminus X_0$, we have $\omega_x = P(\pi)\omega_x$. In this case, the rank of the matrix $(I - P(\pi))$ is equal to $n - p$. For $x \in X_0$, we can determine ω_x and ε_x by calculating $\omega_x = \{P(\pi)\omega\}_x$, $\varepsilon_x = \{P(\pi)\varepsilon\}_x$.

Based on the arguments above, we can see that for an arbitrary deterministic policy π of a multichain Markov decision process, Eq. (2.59) has a solution that satisfies the following condition:

$$\omega_x = \{P(\pi)\omega\}_x, \ \forall x \in X. \tag{2.60}$$

Thus, for an average Markov decision problem, in the general case, the following result holds:

Theorem 2.9 *For an arbitrary average Markov decision problem, the system of equations*

$$\varepsilon_x + \omega_x = \max_{a \in A(x)} \Big\{ r_{x,a} + \sum_{y \in X} p_{x,y}^a \varepsilon_y \Big\}, \ \forall x \in X \tag{2.61}$$

has a solution under the set of solutions to the system of equations

$$\omega_x = \max_{a \in A(x)} \Big\{ \sum_{y \in X} p_{x,y}^a \omega_y \Big\}, \quad \forall x \in X. \tag{2.62}$$

If ω_x^* *and* ε_x^* *for* $x \in X$ *represent a solution to Eqs. (2.61), (2.62), then* ω_x^* *for* $x \in X$ *represents the optimal average reward per transition when the process starts in the corresponding states* $x \in X$*. An optimal stationary strategy*

$$s^* : x \rightarrow a \in A(x) \ for\ x \in X$$

for the average Markov decision problem can be found by fixing a map $s^*(x) = a^* \in A(x)$ *such that*

$$a^* \in \operatorname*{argmax}_{a \in A(x)} \Big\{ \sum_{y \in X} p_{x,y}^a \omega_y^* \Big\}$$

and

$$a^* \in \operatorname*{argmax}_{a \in A(x)} \Big\{ r_{x,a} + \sum_{y \in X} p_{x,y}^a \varepsilon_y^* \Big\}.$$

Corollary 2.10 *For an arbitrary strategy* $s : X \rightarrow A$*, the system of linear equations*

$$\begin{cases} \varepsilon_x + \omega_x = r_{x,s(x)} + \sum\limits_{y \in X} p_{x,y}^{s(x)} \varepsilon_y, & \forall x \in X; \\ \omega_x = \sum\limits_{y \in X} p_{x,y}^{s(x)} \omega_y, & \forall x \in X \end{cases} \tag{2.63}$$

has a solution.

Based on Theorem 2.9 and Corollary 2.10, the following policy iteration algorithm for determining an optimal deterministic policy for an average Markov decision problem in the general case can be proposed.

The Policy Iteration Algorithm for a Multichain Problem
Preliminary step (Step 0): Fix an arbitrary deterministic stationary strategy

$$s^0 : x_i \rightarrow a \in A(x_i) \ \text{for} \ x_i \in X.$$

General step (Step k, $k \geq 1$*): Determine the matrix* $P(s^{k-1})$ *and vector* $\mathbf{r}(s^{k-1})$ *that correspond to strategy* s^{k-1}.

Solve the following system of equations with respect to ω *and* ε:

$$\begin{cases} (P(s^{k-1}) - I)\omega = 0; \\ \mathbf{r}(s^{k-1}) + (P(s^{k-1}) - I)\varepsilon - \omega = 0 \end{cases}$$

and determine $\omega^{s^{k-1}}$ and $\varepsilon^{s^{k-1}}$ (here, $\omega^{s^{k-1}}$ is uniquely determined and $\varepsilon^{s^{k-1}}$ is uniquely determined up to ε, satisfying $(P(s^{k-1}) - I)\varepsilon = 0)$.

After that, determine a deterministic strategy s^k such that

$$s^k \in \operatorname*{argmin}_s \left\{ P(s)\omega^{s^{k-1}} \right\}$$

and set $s^k = s^{k-1}$ if

$$s^{k-1} \in \operatorname*{argmin}_s \left\{ P(s)\omega^{s^{k-1}} \right\}.$$

If $s^k \neq s^{k-1}$, then go to the next step $k + 1$; otherwise, check if

$$s^k \in \operatorname*{argmin}_s \left\{ \mathbf{r}(s) + P(s)\varepsilon^{s^{k-1}} \right\}$$

and set $s^k = s^{k-1}$ if

$$s^{k-1} \in \operatorname*{argmin}_s \left\{ \mathbf{r}(s) + P(s)\varepsilon^{s^{k-1}} \right\}.$$

If $s^k = s^{k-1}$, then stop and set $s^* = s^{k-1}$; otherwise, go to the next step $k + 1$.

The details concerned with the convergence, computational complexity, and applications of policy iteration algorithms for average Markov decision problems can be found in [152, 162–164].

2.4.5 Linear Programming for Multichain Problems

From Theorem 2.9, we can draw the following conclusion: To determine a solution to the Markov decision problem, it is necessary to determine ω_x for $x \in X$ that satisfies (2.62) and for which there exists ε_x for $x \in X$ that satisfies (2.61). This is equivalent to the problem of determining the "minimal" vector ω with the components ω_x for $x \in X$ that satisfies the conditions

$$\varepsilon_x + \omega_x \geq r_{x,a} + \sum_{y \in X} p_{x,y}^a \varepsilon_y, \quad \forall x \in X, \ \forall a \in A(x);$$

$$\omega_x \geq \sum_{y \in X} p_{x,y}^a \omega_y, \quad \forall x \in X, \ \forall a \in A(x).$$

Thus, we have to minimize a positive linear combination of components of ω under the restrictions given above, i.e., we obtain the following linear programming problem:
Minimize

$$\phi(\varepsilon, \omega) = \sum_{x \in X} \theta_x \omega_x \tag{2.64}$$

subject to

$$\begin{cases} \varepsilon_x - \sum\limits_{y \in X} p^a_{x,y}\varepsilon_y + \omega_x \geq r_{x,a}, & \forall x \in X, \ \forall a \in A(x); \\ \omega_x - \sum\limits_{y \in X} p^a_{x,y}\omega_y \geq 0, & \forall x \in X, \ \forall a \in A(x), \end{cases} \tag{2.65}$$

where $\theta_x > 0$, $\forall x \in X$, and $\sum_{x \in X} \theta_x = 1$. So, if we find an optimal basic solution $(\omega^*, \varepsilon^*)$ to this linear programming problem, then ω^* represents the unique part of the solution to Eqs. (2.59), (2.60). This means that the following theorem holds:

Theorem 2.11 *Let ε^*, ω^* be an optimal basic solution to the linear programming problem (2.64), (2.65). Then ω^*_x for $x \in X$ represents the optimal average reward of the Markov decision problem when the transitions start in x, and an optimal deterministic stationary policy s^* can be found as follows: $s_{x,a^*} \in A^*_1(x) \cap A^*_2$ for $x \in X$, where*

$$A^*_1(x) = \left\{a* \in A(x) | a* = \operatorname*{argmin}_{a \in A(x)} \left\{\varepsilon^*_x - \sum_{y \in X} p^a_{x,y}\varepsilon^*_y + \omega^*_x - r_{x,a}\right\}\right\},$$

$$A^*_2(x) = \left\{a* \in A(x) | a* = \operatorname*{argmin}_{a \in A(x)} \left\{\sum_{y \in X} p_{x,y}\omega^*_y - \omega^*_x\right\}\right\}.$$

If we consider the dual model for (2.64), (2.65), then we obtain the following linear programming problem:
Maximize

$$\varphi(\alpha, \beta) = \sum_{x \in X} \sum_{a \in A(x)} r_{x,a}\alpha_{x,a} \tag{2.66}$$

subject to

$$\begin{cases} \sum_{a\in A(y)} \alpha_{y,a} - \sum_{x\in X}\sum_{a\in A(x)} p^a_{x,y}\,\alpha_{x,a} = 0, \ \ \forall y \in X; \\ \sum_{a\in A(y)} \alpha_{y,a} + \sum_{a\in A(y)} \beta_{y,a} - \sum_{x\in X}\sum_{a\in A(x)} p^a_{x,y}\beta_{x,a} = \theta_y, \ \ \forall y \in X; \\ \qquad \alpha_{x,a} \geq 0, \ \beta_{y,a} \geq 0, \ \ \forall x \in X, \ a \in A(x), \end{cases} \tag{2.67}$$

where $\theta_y > 0$, $\forall y \in X$, and $\sum_{y\in X}\theta_y = 1$.

This problem generalizes the unichain linear programming problem (2.51), (2.52) from Sect. 2.4.2. In (2.67), the restrictions

$$\sum_{a\in A(y)} \alpha_{y,a} + \sum_{a\in A(y)} \beta_{y,a} - \sum_{x\in X}\sum_{a\in A(x)} p^a_{x,y}\beta_{x,a} = \theta_y, \ \ \forall y \in X \tag{2.68}$$

with the condition $\sum_{y\in X}\theta_y = 1$ generalize the constraint

$$\sum_{x\in X}\sum_{a\in A(y)} \alpha_{y,a} = 1 \tag{2.69}$$

in the unichain model. It is easy to check that by summing (2.68) over y, we obtain the equality (2.69).

The relationship between feasible solutions to problem (2.66), (2.67) and stationary strategies in the average Markov decision problem is the following (see [77, 152]):

Lemma 2.12 *Let* $(\boldsymbol{\alpha}, \boldsymbol{\beta})$ *be a feasible solution to the linear programming problem (2.66), (2.67) and denote* $X_\alpha = \{x \in X \mid \sum_{a\in X}\alpha_{x,a} > 0\}$. *Then* (α, β) *possesses the properties that* $\sum_{a\in A(x)}\beta_{x,a} > 0$ *for* $x \in X \setminus X_\alpha$, *and a stationary strategy* $s_{x,a}$ *that corresponds to* (α, β) *is determined as*

$$s_{x,a} = \begin{cases} \dfrac{\alpha_{x,a}}{\sum\limits_{a\in A(x)} \alpha_{x,a}} & \text{if } x \in X_\alpha; \\ \dfrac{\beta_{x,a}}{\sum\limits_{a\in A(x)} \beta_{x,a}} & \text{if } x \in X \setminus X_\alpha, \end{cases} \tag{2.70}$$

where $s_{x,a}$ *expresses the probability of choosing the actions* $a \in A(x)$ *in the states* $x \in X$.

Proof If $x \in X_\alpha$, then $\sum_{a \in A(x)} \alpha_{x,a} > 0$; therefore, we can calculate $s_{x,a} = \frac{\alpha_{x,a}}{\sum_{a \in A(x)} \alpha_{x,a}}$ for $a \in A(x)$, where $s_{x,a} \geq 0$ and $\sum_{a \in A(x)} \alpha_{x,a} = 1$. If $x \notin X_\alpha$, then $\sum_{a \in A(x)} \alpha_{x,a} = 0$ and from (2.67), we have $\sum_{a \in A(x)} \beta_{x,a} > 0$; therefore, we can calculate $s_{x,a} = \frac{\beta_{x,a}}{\sum_{a \in A(x)} \beta_{x,a}}$, where $s_{x,a} \geq 0$ and $\sum_{a \in A(x)} \alpha_{x,a} = 1$. □

Based on this lemma, we obtain the following theorem:

Theorem 2.13 *Let α^*, β^* be an optimal basic solution to the linear programming problem (2.66), (2.67). Then an optimal stationary strategy (policy) s^* can be found as follows:*

$$s^*_{x,a} = \begin{cases} \dfrac{\alpha^*_{x,a}}{\sum_{a \in A(x)} \alpha^*_{x,a}} & \text{if } x \in X_{\alpha^*}; \\ \dfrac{\beta^*_{x,a}}{\sum_{a \in A(x)} \beta^*_{x,a}} & \text{if } x \in X \setminus X_{\alpha^*}. \end{cases} \tag{2.71}$$

Thus, an optimal stationary policy s^* for an average Markov decision problem can be found by using the following algorithms:

Algorithm 1

- Form the linear programming problem (2.64), (2.65) and find an optimal basic solution $(\varepsilon^*, \omega^*)$.
- For each $x \in X$, find

$$A_1^*(x) = \left\{a* \in A(x) | a* = \underset{a \in A(x)}{\operatorname{argmin}} \left\{\varepsilon_x^* - \sum_{y \in X} p_{x,y}^a \varepsilon_y^* + \omega_x^* - r_{x,a}\right\}\right\},$$

$$A_2^*(x) = \left\{a* \in A(x) | a* = \underset{a \in A(x)}{\operatorname{argmin}} \left\{\sum_{y \in X} p_{x,y} \omega_y^* - \omega_x^*\right\}\right\}$$

and fix an arbitrary map $s^* : x \to a \in A_1^*(x) \cap A_2^*(x)$ for $x \in X$.

Algorithm 2

- Form the linear programming problem (2.66), (2.67) and find an optimal basic solution (α^*, β^*).
- Find $s^*_{x,a}$ for $x \in X$ and $a \in A(x)$ according to (2.165).

Remark 2.14 Algorithm 1 determines an optimal deterministic policy for the average Markov decision problem if ε^*, ω^* is a basic optimal solution to problem (2.64), (2.65). Algorithm 2 determines a stationary policy; however, the policy may not be deterministic if (α^*, β^*) is a basic optimal solution to problem (2.66), (2.67).

Remark 2.15 Problem (2.66), (2.67) can also be considered for the case when $\theta_x = 0$ for some $x \in X$. In particular, if $\theta_x = 0, \forall x \in X \setminus \{x_0\}$ and $\theta_{x_0} = 1$, then this problem is transformed into the model with fixed starting state x_0. In this case, for a feasible solution (α, β), the subset $X \setminus X_\alpha$ may contain states for which $\sum_{a \in A(x)} \beta_{x,a} = 0$. In such states, (2.165) cannot be used for determining $s_{x,a}$. Formula (2.165) can be used for determining the strategies $s_{x,a}$ in the states $x \in X$ for which either $\sum_{a \in A(x)} \alpha_{x,a} > 0$ or $\sum_{a \in A(x)} \beta_{x,a} > 0$, and these strategies determine the value of the object function in the decision problem. In the states $x \in X_0$, where

$$X_0 = \{x \in X | \sum_{a \in A(x)} \alpha_{x,a} = 0, \sum_{a \in A(x)} \beta_{x,a} = 0\},$$

the strategies of selecting the actions may be arbitrary because they do not affect the value of the objective function.

So, the discounted and average Markov decision problems can be represented as linear programming problems. This means that for these problems, there exist polynomial time algorithms to determine their solutions. In [77, 152], it was noted that the value and policy iteration algorithms for discounted Markov decision problems and unichain average Markov decision problems efficiently work and allow us to determine the solution to the problems of large dimensions.

The average Markov decision problems can also be solved efficiently by using iterative algorithms; however, the multichain problems are more difficult from a computational point of view than the unichain problems for which the iterative and linear programming algorithms work more efficiently. In [77], it was shown that the problem of determining whether or not a Markov decision problem is unichain or multichain is an NP-complete problem.

2.4.6 A Nonlinear Model for the Multichain Problem

We show that an average Markov decision problem in terms of stationary strategies can be formulated as follows:

Maximize

$$\psi(\mathbf{s}, \mathbf{q}, \mathbf{w}) = \sum_{x \in X} \sum_{a \in A(x)} r_{x,a} q_x \tag{2.72}$$

subject to

$$\begin{cases} q_y - \sum\limits_{x\in X}\sum\limits_{a\in A(x)} p^a_{x,y}\, s_{x,a} q_x = 0, & \forall y \in X; \\ q_y + w_y - \sum\limits_{x\in X}\sum\limits_{a\in A(x)} p^a_{x,y} s_{x,a} w_x = \theta_y, & \forall y \in X; \\ \sum\limits_{a\in A(y)} s_{y,a} = 1, & \forall y \in X; \\ s_{x,a} \geq 0,\ \forall x \in X,\ \forall a \in A(x); \quad w_x \geq 0,\ \forall x \in X, \end{cases} \tag{2.73}$$

where θ_y are the same values as in problem (2.66), (2.67) and $s_{x,a},\ q_x,\ w_x$ for $x \in X, a \in A(x)$ represent the variables that must be found.

Theorem 2.16 *Optimization problem (2.72), (2.73) determines the optimal stationary strategies of the multichain average Markov decision problem.*

Proof Indeed, if we assume that each action set $A(x), x \in X$ contains a single action a', then system (2.67) is transformed into the following system of equations:

$$\begin{cases} q_y - \sum\limits_{x\in X} p_{x,y} q_x = 0, & \forall y \in X; \\ q_y + w_y - \sum\limits_{x\in X} p_{x,y} w_x = \theta_y, & \forall y \in X, \end{cases}$$

with conditions $q_y, w_y \geq 0$ for $y \in X$, where $q_y = \alpha_{y,a'},\ w_y = \beta_{y,a'},\ \forall y \in X$ and $p_{x,y} = p^{a'}_{x,y},\ \forall x, y \in X$. This system uniquely determines q_x for $x \in X$ and determines w_x for $x \in X$ up to an additive constant in each recurrent class of $P = (p_{x,y})$ (see [152]). Here, q_x represents the limiting probability in state x when transitions start in states $y \in X$ with probabilities θ_y, and therefore, the condition $q_x \geq 0$ for $x \in X$ can be released.

Note that w_x may be negative for some states; however, the additive constants in the corresponding recurrent classes can always be chosen so that w_x becomes non-negative. In general, we can observe that in (2.73), the condition $w_x \geq 0$ for $x \in X$ can be released and this does not influence the value of the objective function of the problem. In the case $|A(x)| = 1,\ \forall x \in X$, the average reward is determined as $\psi = \sum_{x\in X} r_x q_x$, where $r_x = r^a_x, \forall x \in X$.

If the action sets $A(x),\ x \in X$ may contain more than one action, then for a given stationary strategy $\mathbf{s} \in \mathbf{S}$ of selecting the actions in the states, we can find the average reward $\psi(s)$ in a similar way as above by considering the probability matrix $P^{\mathbf{s}} = (p^{\mathbf{s}}_{x,y})$, where

$$p^{\mathbf{s}}_{x,y} = \sum_{a\in A(x)} p^a_{x,y} s_{x,a} \tag{2.74}$$

expresses the probability transition from a state $x \in X$ to a state $y \in X$ when the strategy $\mathbf{s}$ of selecting the actions in the states is applied. This means that we have to solve the following system of equations:

$$\begin{cases} q_y - \sum_{x\in X} p^s_{x,y} q_x = 0, & \forall y \in X; \\ q_y + w_y - \sum_{x\in X} p^s_{x,y} w_x = \theta_y, & \forall y \in X, \end{cases}$$

where q_x is the limiting probability in the states $x \in X$ when the process starts transitions in the states $x \in X$ with probabilities θ_x. If in this system we take (2.74) into account, then this system can be written as follows:

$$\begin{cases} q_y - \sum_{x\in X} \sum_{a\in A(x)} p^a_{x,y}\, s_{x,a} q_x = 0, & \forall y \in X; \\ q_y + w_y - \sum_{x\in X} \sum_{a\in A(x)} p^a_{x,y} s_{x,a} w_x = \theta_y, & \forall y \in X. \end{cases} \tag{2.75}$$

An arbitrary solution $(\mathbf{q}, \mathbf{w})$ of the system of Eq. (2.75) uniquely determines q_y for $y \in X$ that allows us to determine the average reward per transition

$$\psi(\mathbf{s}) = \sum_{x\in X} \sum_{a\in X} r^a_x s_{x,a} q_x \tag{2.76}$$

when the stationary strategy $\mathbf{s}$ is applied and the initial state is chosen according to probability distribution θ on X. If we seek an optimal stationary strategy, then we should add to (2.75) the conditions

$$\sum_{a\in A(x)} s_{x,a} = 1, \ \forall x \in X; \ s_{x,a} \geq 0, \ \forall x \in X, a \in A(x) \tag{2.77}$$

and maximize (2.76) under the constraints (2.75), (2.77). In such a way, we obtain problem (2.72), (2.73) without conditions $w_x \geq 0$ for $x \in X$.

As we have noted, the conditions $w_x \geq 0$ for $x \in X$ do not influence the values of the objective function (2.72), and therefore, we can preserve such conditions that show the relationship of the problem (2.72), (2.73) with problem (2.66), (2.67). □

The relationship between feasible solutions to problem (2.66), (2.67) and feasible solutions to problem (2.72), (2.73) can be established based on the following lemma:

Lemma 2.17 *Let* $(\mathbf{s}, \mathbf{q}, \mathbf{w})$ *be a feasible solution to problem (2.72), (2.73). Then*

$$\alpha_{x,a} = s_{x,a} q_x, \ \ \beta_{x,a} = s_{x,a} w_x, \ \ \forall x \in X, a \in A(x) \tag{2.78}$$

represent a feasible solution $(\boldsymbol{\alpha}, \boldsymbol{\beta})$ *to problem (2.66), (2.67), and* $\psi(\mathbf{s}, \mathbf{q}, \mathbf{w}) = \varphi(\boldsymbol{\alpha}, \boldsymbol{\beta})$. *If* $(\boldsymbol{\alpha}, \boldsymbol{\beta})$ *is a feasible solution to problem (2.66), (2.67), then a feasible solution* $(\mathbf{s}, \mathbf{q}, \mathbf{w})$ *to problem (2.72), (2.73) can be determined as follows:*

$$s_{x,a} = \begin{cases} \dfrac{\alpha_{x,a}}{\sum\limits_{a \in A(x)} \alpha_{x,a}} & \text{for } x \in X_\alpha,\ a \in A(x); \\ \dfrac{\beta_{x,a}}{\sum\limits_{a \in A(x)} \beta_{x,a}} & \text{for } x \in X \setminus X_\alpha,\ a \in A(x); \end{cases} \quad (2.79)$$

$$q_x = \sum_{a \in A(x)} \alpha_{x,a}, \quad w_x = \sum_{a \in A(x)} \beta_{x,a} \ \text{for } x \in X.$$

Proof Assume that $(\mathbf{s}, \mathbf{q}, \mathbf{w})$ is a feasible solution to problem (2.72), (2.73), and $(\boldsymbol{\alpha}, \boldsymbol{\beta})$ is determined according to (2.78). Then by introducing (2.78) to (2.66), (2.67), we can observe that (2.67) is transformed into (2.73) and $\psi(\mathbf{s}, \mathbf{q}, \mathbf{w}) = \varphi(\boldsymbol{\alpha}, \boldsymbol{\beta})$, i.e., $(\boldsymbol{\alpha}, \boldsymbol{\beta})$ is a feasible solution to problem (2.66), (2.67). The second part of the lemma follows directly from the properties of feasible solutions to problems (2.66), (2.67) and (2.72), (2.73). □

Note that a pure stationary strategy s of problem (2.72), (2.73) corresponds to a basic solution $(\boldsymbol{\alpha}, \boldsymbol{\beta})$ of problem (2.66), (2.67) for which (2.79) holds; however, system (2.67) may contain basic solutions for which stationary strategies determined by (2.79) do not correspond to pure stationary strategies. Moreover, two different feasible solutions to problem (2.66), (2.67) may generate the same stationary strategy through (2.79). Such solutions to system (2.67) are considered *equivalent solutions* to the decision problem.

Corollary 2.18 *If* $(\alpha^i, \beta^i),\ i = \overline{1,k}$ *represent the basic solutions to system (2.67), then the set of solutions*

$$M = \left\{(\alpha, \beta) |\ (\alpha, \beta) = \sum_{i=1}^{k} \lambda^i (\alpha^i, \beta^i),\ \sum_{i=1}^{k} \lambda^i = 1,\ \lambda^i > 0,\ i = \overline{1,k}\right\}$$

determines all feasible stationary strategies of problem (2.72), (2.73) through (2.79).

An arbitrary solution (α, β) *to system (2.67) can be represented as follows:* $\alpha = \sum_{i=1}^{k} \lambda^i \alpha^i$, *where* $\sum_{i=1}^{k} \lambda^i = 1;\ \lambda^i \geq 0,\ i = \overline{1,k}$, *and* β *represents a solution to the system*

$$\begin{cases} \sum\limits_{a \in A(y)} \beta_{x,a} - \sum\limits_{z \in X} \sum\limits_{a \in A(z)} p^a_{z,x} \beta_{z,a} = \theta_x - \sum\limits_{a \in A(x)} \alpha_{x,a}, \ \forall x \in X; \\ \beta_{y,a} \geq 0, \ \forall x \in X,\ a \in A(x). \end{cases}$$

If (α, β) *is a feasible solution to problem (2.66), (2.67), and* $(\alpha, \beta) \notin M$, *then there exists a solution* $(\alpha', \beta') \in M$ *that is equivalent to* (α, β) *and* $\varphi(\alpha, \beta) = \varphi(\alpha', \beta')$.

2.4.7 A Quasi-monotonic Programming Approach

Based on the results from the previous section, we show now that an average Markov decision problem in terms of stationary strategies can be represented as a quasi-monotonic programming problem.

Theorem 2.19 *Let an average Markov decision problem be given and consider the function*

$$\psi(\mathbf{s}) = \sum_{x\in X}\sum_{a\in A(x)} r_{x,a}s_{x,a}\,q_x, \tag{2.80}$$

where q_x for $x \in X$ satisfy the condition

$$\begin{cases} q_y - \sum\limits_{x\in X}\sum\limits_{a\in A(x)} p_{x,y}^a\, s_{x,a}q_x = 0, & \forall y \in X;\\ q_y + w_y - \sum\limits_{x\in X}\sum\limits_{a\in A(x)} p_{x,y}^a s_{x,a}w_x = \theta_y, & \forall y \in X. \end{cases} \tag{2.81}$$

Then on the set S of solutions to the system

$$\begin{cases} \sum\limits_{a\in A(x)} s_{x,a} = 1, & \forall x \in X;\\ s_{x,a} \geq 0, & \forall x \in X,\ a \in A(x), \end{cases} \tag{2.82}$$

the function $\psi(\mathbf{s})$ depends only on $s_{x,a}$ for $x \in X$, $a \in A(x)$, and $\psi(s)$ is quasi-monotonic on $\mathbf{S}$ *(i.e., $\psi(\mathbf{s})$ is quasi-convex and quasi-concave on* $\mathbf{S}$*).*

Proof For an arbitrary $\mathbf{s} \in \mathbf{S}$, system (2.81) uniquely determines q_x for $x \in X$ and determines w_x for $x \in X$ up to a constant in each recurrent class of $P^{\mathbf{s}} = (p_{x,y}^{\mathbf{s}})$, where $p_{x,y}^{\mathbf{s}} = \sum_{a\in A(x)} p_{x,y}^a s_{x,a}$, $\forall x, y \in X$. This means that $\psi(\mathbf{s})$ is uniquely determined for an arbitrary $\mathbf{s} \in \mathbf{S}$, i.e., the first part of the theorem holds.

Now let us prove the second part of the theorem.

Assume that $\theta_x > 0$, $\forall x \in X$, where $\sum_{x\in X}\theta_x = 1$ and consider arbitrary two strategies $\mathbf{s}', \mathbf{s}'' \in \mathbf{S}$ for which $\mathbf{s}' \neq \mathbf{s}''$. Then according to Lemma 2.17, there exist feasible solutions (α', β') and (α'', β'') of linear programming problem (2.66), (2.67) for which

$$\psi(\mathbf{s}') = \varphi(\alpha', \beta'),\quad \psi(\mathbf{s}'') = \varphi(\alpha'', \beta''), \tag{2.83}$$

where

$$\alpha'_{x,a} = s'_{x,a}q'_x, \quad \alpha''_{x,y} = s''_{x,a}q''_x, \quad \forall x \in X, \ a \in A(x);$$

$$\beta'_{x,a} = s'_{x,a}w'_x, \quad \beta''_{x,y} = s''_{x,a}q''_x, \quad \forall x \in X, \ a \in A(x);$$

$$q'_x = \sum_{a \in A(x)} \alpha'_{x,a} \quad w'_{x,a} = \sum_{a \in A(x)} \beta'_{x,a}, \quad \forall x \in X;$$

$$q''_x = \sum_{a \in A(x)} \alpha''_{x,a} \quad w''_{x,a} = \sum_{a \in A(x)} \beta''_{x,a}, \quad \forall x \in X.$$

The function $\varphi(\alpha, \beta))$ is linear, and therefore, for an arbitrary feasible solution $(\overline{\alpha}, \overline{\beta})$ of problem (2.66), (2.67), the following formula holds:

$$\varphi(\overline{\alpha}, \overline{\beta}) = t\varphi(\alpha', \beta') + (1-t)\varphi(\alpha'', \beta'') \tag{2.84}$$

if $0 \le t \le 1$ and $(\overline{\alpha}, \overline{\beta}) = t(\alpha', \beta') + (1-t)(\alpha'', \beta'')$.

Note that $(\overline{\alpha}, \overline{\beta})$ corresponds to a stationary strategy $\overline{\mathrm{s}}$ for which

$$\psi(\overline{\mathrm{s}}) = \varphi(\overline{\alpha}, \overline{\beta}), \tag{2.85}$$

where

$$\overline{s}_{x,a} = \begin{cases} \dfrac{\overline{\alpha}_{x,a}}{\overline{q}_x} & \text{if } x \in X_{\overline{\alpha}}; \\ \dfrac{\beta_{x,a}}{\overline{w}_x} & \text{if } x \in X \setminus X_{\overline{\alpha}}. \end{cases} \tag{2.86}$$

Here, $X_{\overline{\alpha}} = \{x \in X | \ \sum_{a \in A(x)} \overline{\alpha}_{x,a} > 0\}$ is the set of recurrent states induced by $P^{\overline{s}} = (p^{\overline{s}}_{x,y})$, where $p^{\overline{s}}_{x,y}$ are calculated according to (2.74) for $s = \overline{s}$ and

$$\overline{q}_x = tq'_x + (1-t)q'', \quad \overline{w}_x = tw'_x + (1-t)w''_x, \quad \forall x \in X.$$

We can see that $X_{\overline{\alpha}} = X_{\alpha'} \cup X_{\alpha''}$, where $X_{\alpha'} = \{x \in X | \ \sum_{a \in A(x)} \alpha'_{x,a} > 0\}$ and $X_{\alpha''} = \{x \in X | \ \sum_{a \in A(x)} \alpha''_{x,a} > 0\}$. The value

$$\psi(\overline{\mathrm{s}}) = \sum_{x \in X} \sum_{a \in A(x)} r^a_x \overline{s}_{x,a} \overline{q}_x$$

is determined by r_x^a, $\overline{s}_{x,a}$ and $\overline{q}_x$ in recurrent states $x \in X_{\overline{\alpha}}$, and it is equal to $\varphi(\overline{\alpha}, \overline{\beta})$. If we use (2.86), then for $x \in X_{\overline{\alpha}}$ and $a \in A(x)$, we have

$$\overline{s}_{x,a} = \frac{t\alpha'_{x,a} + (1-t)\alpha''_{x,a}}{tq'_x + (1-t)q''_x} = \frac{ts'_{x,a}q'_x + (1-t)s''_{x,a}q''_x}{tq'_x + (1-t)q''_x} =$$

$$= \frac{tq'_x}{tq'_x + (1-t)q''_x}s'_{x,a} + \frac{(1-t)q''_x}{tq'_x + (1-t)q''_x}s''_{x,a}$$

and for $x \in X \setminus X_{\overline{\alpha}}$ and $a \in A(x)$, we have

$$\overline{s}_{x,a} = \frac{t\beta'_{x,a} + (1-t)\beta''_{x,a}}{tw'_x + (1-t)w''_x} = \frac{ts'_{x,a}w'_x + (1-t)s''_{x,a}w''_x}{tw'_x + (1-t)w''_x} =$$

$$= \frac{tw'_x}{tw'_x + (1-t)w''_x}s'_{x,a} + \frac{(1-t)w''_x}{tw'_x + (1-t)w''_x}s''_{x,a}.$$

So, we obtain

$$\overline{s}_{x,a} = t_x s'_{x,a} + (1-t_x)s''_{x,a}, \quad \forall a \in A(x), \tag{2.87}$$

where

$$t_x = \begin{cases} \dfrac{tq'_x}{tq'_x + (1-t)q''_x} & \text{if } x \in X_{\overline{\alpha}}; \\[2ex] \dfrac{tw'_x}{tw'_x + (1-t)w''_x} & \text{if } x \in X \setminus X_{\overline{\alpha}} \end{cases} \tag{2.88}$$

and from (2.83)–(2.85), we have

$$\psi(\overline{\mathbf{s}}) = t\psi(\mathbf{s}') + (1-t)\psi(\mathbf{s}''). \tag{2.89}$$

This means that if we consider the set of strategies

$$S(\mathbf{s}', \mathbf{s}'') = \{\overline{\mathbf{s}} \mid \overline{\mathbf{s}}_{x,a} = t_x s'_{x,a} + (1-t_x)s''_{x,a}, \quad \forall x \in X, a \in A(x)\},$$

then for an arbitrary $\overline{\mathbf{s}} \in S(\mathbf{s}', \mathbf{s}'')$, we have

$$\min\{\psi(\mathbf{s}'), \psi(\mathbf{s}'')\} \le \psi(\overline{\mathbf{s}}) \le \max\{\psi(\mathbf{s}'), \psi(\mathbf{s}'')\}, \tag{2.90}$$

i.e., $\psi(\mathbf{s})$ is monotonic on $S(\mathbf{s}', \mathbf{s}'')$. Moreover, using (2.87)–(2.90), we can see that $\bar{\mathbf{s}}$ possesses the properties

$$\lim_{t\to 1} \bar{s}_{x,a} = s'_{x,a},\ \forall x \in X, a \in A(x); \quad \lim_{t\to 0} \bar{s}_{x,a} = s''_{x,a},\ \forall x \in X, a \in A(x) \tag{2.91}$$

and

$$\lim_{t\to 1} \psi(\bar{\mathbf{s}}) = \psi(s'); \quad \lim_{t\to 0} \psi(\bar{\mathbf{s}}) = \psi(\mathbf{s}'').$$

In the following, we show that the function $\psi(\mathbf{s})$ is quasi-monotonic on $\mathbf{S}$. To prove this, it is sufficient to show that for an arbitrary $c \in R$, the *sublevel set*

$$L_c^-(\psi) = \{\mathbf{s} \in \mathbf{S} |\ \psi(\mathbf{s}) \le c\}$$

and the *superlevel set*

$$L_c^+(\psi) = \{\mathbf{s} \in \mathbf{S} |\ \psi(\mathbf{s}) \ge c\}$$

of function $\psi(\mathbf{s})$ are convex. These sets can be obtained from the *sublevel set* $L_c^-(\varphi) = \{(\alpha, \beta) |\ \varphi(\alpha, \beta) \le c\}$ and the *superlevel set* $L_c^+(\varphi) = \{(\alpha, \beta) |\ \varphi(\alpha, \beta) \ge c\}$ of function $\varphi(\alpha, \beta)$ for the linear programming problem (2.66), (2.67), using (2.79).

Denote by (α^i, β^i), $i = \overline{1, k}$ the basic solutions to system (2.67).

According to Corollary 2.18, all feasible strategies of problem (2.66), (2.67) can be obtained through (2.79), using the basic solutions (α^i, β^i), $i = \overline{1, k}$. Each (α^i, β^i), $i = \overline{1, k}$ determines a stationary strategy

$$s^i_{x,a} = \begin{cases} \dfrac{\alpha^i_{x,a}}{q^i_x}, & \text{for } x \in X_{\alpha^i},\ a \in A(x); \\[2mm] \dfrac{\beta^i_{x,a}}{w^i_x}, & \text{for } x \in X \setminus X_{\alpha^i},\ a \in A(x) \end{cases} \tag{2.92}$$

for which $\psi(s^i) = \varphi(\alpha^i, \beta^i)$, where

$$X_{\alpha^i} = \{x \in X |\ \sum_{a\in A(x)} \alpha^i_{x,a} > 0\},\ q^i_x = \sum_{a\in A(x)} \alpha^i_{x,a},\ w^i_x = \sum_{a\in A(x)} \beta^i_{x,a},\ \forall x \in X. \tag{2.93}$$

An arbitrary feasible solution (α, β) to system (2.67) determines a stationary strategy

$$s_{x,a} = \begin{cases} \dfrac{\alpha_{x,a}}{q_x}, & \text{for } x \in X_\alpha,\ a \in A(x); \\ \dfrac{\beta_{x,a}}{w_x}, & \text{for } x \in X \setminus X_\alpha,\ a \in A(x), \end{cases} \tag{2.94}$$

for which $\psi(\mathbf{s}) = \varphi(\alpha, \beta)$, where

$$X_\alpha = \{x \in X | \sum_{a \in A(x)} \alpha_{x,a} > 0\}, \quad q_x = \sum_{a \in A(x)} \alpha_{x,a}, \quad w_x = \sum_{a \in A(x)} \beta_{x,a}, \ \forall x \in X.$$

Taking into account that (α, β) can be represented as

$$(\alpha, \beta) = \sum_{i=1}^{k} \lambda^i (\alpha^i, \beta^i), \text{ where } \sum_{i=1}^{k} \lambda^i = 1, \ \lambda^i \geq 0, \ i = \overline{1, k}, \tag{2.95}$$

we have $\varphi(\alpha, \beta) = \sum_{i=1}^{k} \varphi(\alpha^i, \beta^i)\lambda^i$, and we can consider

$$X_\alpha = \bigcup_{i=1}^{k} X_{\alpha^i}; \quad \alpha = \sum_{i=1}^{k} \lambda^i \alpha^i; \quad q = \sum_{i=1}^{k} \lambda^i q^i; \quad w = \sum_{i=1}^{k} \lambda^i w^i. \tag{2.96}$$

Using (2.92)–(2.96), we obtain

$$s_{x,a} = \frac{\alpha_{x,a}}{q_x} = \frac{\sum\limits_{i=1}^{k} \lambda^i \alpha^k_{x,a}}{q_x} = \frac{\sum\limits_{i=1}^{k} \lambda^i s^i_{x,a} q^i_x}{q_x} = \sum_{i=1}^{k} \frac{\lambda^i q^i_x}{q_x} s^i_{x,a}, \ \forall x \in X_\alpha, \ a \in A(x);$$

$$s_{x,a} = \frac{\beta_{x,a}}{w_x} = \frac{\sum\limits_{i=1}^{k} \lambda^i \beta^k_{x,a}}{w_x} = \frac{\sum\limits_{i=1}^{k} \lambda^i s^i_{x,a} w^i_x}{w_x} = \sum_{i=1}^{k} \frac{\lambda^i w^i_x}{w_x} s^i_{x,a}, \ \forall x \in X \setminus X_\alpha, a \in A(x)$$

and

$$q_x = \sum_{i=1}^{k} \lambda^i q^i_x, \qquad w_x = \sum_{i=1}^{k} \lambda^i w^i_x \ \text{ for } \ x \in X. \tag{2.97}$$

So,

$$
s_{x,a} = \begin{cases} \sum_{i=1}^{k} \frac{\lambda^i q_x^i}{q_x} s_{x,a}^i & \text{if } q_x > 0; \\ \sum_{i=1}^{k} \frac{\lambda^i w_x^i}{w_x} s_{x,a}^i & \text{if } q_x = 0, \end{cases} \tag{2.98}
$$

where q_x and w_x are determined according to (2.97).

We can see that if $\lambda^i, s^i, q^i,\ i = \overline{1,k}$ are given, then the strategy s defined by (2.98) is a feasible strategy because $s_{x,a} \geq 0, \forall x \in X, a \in A(x)$ and $\sum_{a \in A(x)} s_{x,a} = 1,\ \forall x \in X$. Moreover, we can observe that $q_x = \sum_{i=1}^{k} \lambda^i q_x^i,\ w_x = \sum_{i=1}^{k} \lambda^i w_x^i$ for $x \in X$ represent a solution to system (2.81) for the strategy s defined by (2.98). This can be verified by introducing (2.97) and (2.98) into (2.81); after such a substitution, all equations from (2.81) are transformed into identities. For $\psi(\mathbf{s})$, we have

$$
\psi(\mathbf{s}) = \sum_{x \in X} \sum_{a \in A(x)} r_{x,a} s_{x,a} q_x = \sum_{x \in X_\alpha} \sum_{a \in A(x)} r_{x,a} \sum_{i=1}^{k} \left(\frac{\lambda^i q_x^i}{q_x} s_{x,a}^i \right) q_x =
$$

$$
\sum_{i=1}^{k} \Bigg(\sum_{x \in X_{\alpha^i}} \sum_{a \in A(x)} r_{x,a} s_{x,a}^i q_x^i \Bigg) \lambda^i = \sum_{i=1}^{k} \psi(s^i) \lambda^i,
$$

i.e.,

$$
\psi(\mathbf{s}) = \sum_{i=1}^{k} \psi(s^i) \lambda^i, \tag{2.99}
$$

where $\mathbf{s}$ is the strategy that corresponds to (α, β).

Thus, assuming that the strategies $\mathbf{s}^1, \mathbf{s}^2, \ldots, \mathbf{s}^k$ correspond to basic solutions $(\alpha^1, \beta^1), (\alpha^2, \beta^2), \ldots, (\alpha^k, \beta^k)$ to problem (2.66), (2.67), and $\mathbf{s} \in \mathbf{S}$ corresponds to an arbitrary solution (α, β) to this problem, which can be expressed as a convex combination of basic solutions to problem (2.66), (2.67) with the corresponding coefficients $\lambda^1, \lambda^2, \ldots, \lambda^k$, we can express the strategy s and the corresponding value $\psi(\mathbf{s})$ by (2.97)–(2.99). In general, the representation (2.97)–(2.99) of strategy $\mathbf{s}$ and of the value $\psi(\mathbf{s})$ is valid for an arbitrary finite set of strategies from $\mathbf{S}$ if (α, β) can be represented as a convex combination of the finite number of feasible solutions $(\alpha^1, \beta^1), (\alpha^2, \beta^2), \ldots, (\alpha^k, \beta^k)$ that correspond to $\mathbf{s}^1, \mathbf{s}^2, \ldots, \mathbf{s}^k$; in the case $k = 2$ from (2.97)–(2.99), we obtain (2.87)–(2.89). It is evident that for a feasible strategy $\mathbf{s} \in \mathbf{S}$, the representation (2.97), (2.98) may not be unique, i.e., two different vectors $\overline{\Lambda} = (\overline{\lambda}^1, \overline{\lambda}^2, \ldots, \overline{\lambda}^k)$ and $\overline{\overline{\Lambda}} = \overline{\overline{\lambda}}^1, \overline{\overline{\lambda}}^2, \ldots, \overline{\overline{\lambda}}^k$ can determine the same strategy $\mathbf{s}$

via (2.97), (2.98). In the following, we assume that $\mathbf{s}^1, \mathbf{s}^2, \ldots, \mathbf{s}^k$ represent the system of linearly independent basic solutions to system (2.82), i.e., each $\mathbf{s}^i \in \mathbf{S}$ corresponds to a pure stationary strategy.

Thus, an arbitrary strategy $\mathbf{s} \in \mathbf{S}$ is determined according to (2.97), (2.98), where $\lambda^1, \lambda^2, \ldots, \lambda^k$ correspond to a solution to the following system:

$$\sum_{i=1}^{k} \lambda^i = 1; \ \ \lambda^i \geq 0, \ \ i = \overline{1,k}.$$

Consequently, the sublevel set $L_c^-(\psi)$ of function $\psi(s)$ represents the set of strategies s determined by (2.97), (2.98), where $\lambda^1, \lambda^2, \ldots, \lambda^k$ satisfy the condition

$$\begin{cases} \sum\limits_{i=1}^{k} \psi(s^i)\lambda^i \leq c; \\ \sum\limits_{i=1}^{k} \lambda^i = 1; \quad \lambda^i \geq 0, \ \ i = \overline{1,k}, \end{cases} \tag{2.100}$$

and the superlevel set $L_c^+(\psi)$ of $\psi(s)$ represents the set of strategies s determined by (2.97), (2.98), where $\lambda^1, \lambda^2, \ldots, \lambda^k$ satisfy the condition

$$\begin{cases} \sum\limits_{i=1}^{k} \psi(s^i)\lambda^i \geq c; \\ \sum\limits_{i=1}^{k} \lambda^i = 1; \quad \lambda^i \geq 0, \ \ i = \overline{1,k}. \end{cases} \tag{2.101}$$

The level set $L_c(\psi) = \{s \in \overline{S} | \ \psi(s) = c\}$ of function $\psi(s)$ represents the set of strategies s determined by (2.97), (2.98), where $\lambda^1, \lambda^2, \ldots, \lambda^k$ satisfy the condition

$$\begin{cases} \sum\limits_{i=1}^{k} \psi(s^i)\lambda^i = c; \\ \sum\limits_{i=1}^{k} \lambda^i = 1; \quad \lambda^i \geq 0, \ \ i = \overline{1,k}. \end{cases} \tag{2.102}$$

Let us show that $L_c^-(\psi)$, $L_c^+(\psi)$, $L_c(\psi)$ are convex sets. We present the proof of convexity of sublevel set $L_c^-(\psi)$. The proof of convexity of $L_c^+(\psi)$ and $L_c(\psi)$ is similar to the proof of convexity of $L_c^-(\psi)$.

Denote by Λ the set of solutions $(\lambda^1, \lambda^2, \ldots, \lambda^k)$ of system (2.100). Then from (2.97), (2.98), (2.100), we have

$$L_c^-(\psi) = \prod_{x \in X} \hat{S}_x,$$

where $\hat{S}_x$ represents the set of strategies

$$s_{x,a} = \begin{cases} \dfrac{\sum\limits_{i=1}^{k} \lambda^i q_x^i s_{x,a}^i}{\sum\limits_{i=1}^{k} \lambda^i q_x^i} & \text{if } \sum_{i=1}^{k} \lambda^i q_x^i > 0, \\ \dfrac{\sum\limits_{i=1}^{k} \lambda^i w_x^i s_{x,a}^i}{\sum\limits_{i=1}^{k} \lambda^i w_x^i} & \text{if } \sum_{i=1}^{k} \lambda^i q_x^i = 0 \end{cases} \qquad a \in A(x)$$

in the state $x \in X$ determined by $(\lambda^1, \lambda^2, \dots, \lambda^k) \in \Lambda$.

For an arbitrary $x \in X$, the set Λ can be represented as follows: $\Lambda = \Lambda_x^+ \cup \Lambda_x^0$, where

$$\Lambda_x^+ = \{(\lambda^1, \lambda^2, \dots, \lambda^k) \in \Lambda | \sum_{i=1}^{k} \lambda^i q_x^i > 0\},$$

$$\Lambda_x^0 = \{(\lambda^1, \lambda^2, \dots, \lambda^k) \in \Lambda | \sum_{i=1}^{k} \lambda^i q_x^i = 0\}$$

and $\sum_{i=1}^{k} \lambda^i w_x^i > 0$ if $\sum_{i=1}^{k} \lambda^i q_x^i = 0$.

Therefore, $\hat{S}_x$ can be expressed as follows: $\hat{S}_x = \hat{S}_x^+ \cup \hat{S}_x^0$, where $\hat{S}_x^+$ represents the set of strategies

$$s_{x,a} = \frac{\sum\limits_{i=1}^{k} \lambda^i q_x^i s_{x,a}^i}{\sum\limits_{i=1}^{k} \lambda^i q_x^i}, \quad \text{for } a \in A(x) \tag{2.103}$$

in the state $x \in X$ determined by $(\lambda^1, \lambda^2, \dots, \lambda^k) \in \Lambda_x^+$, and $\hat{S}_x^0$ represents the set of strategies

$$s_{x,a} = \frac{\sum\limits_{i=1}^{k} \lambda^i w_x^i s_{x,a}^i}{\sum\limits_{i=1}^{k} \lambda^i w_x^i}, \quad \text{for } a \in A(x) \tag{2.104}$$

in the state $x \in X$ determined by $(\lambda^1, \lambda^2, \dots, \lambda^k) \in \Lambda_x^0$.

Thus, if we analyze (2.103), then we observe that $s_{x,a}$ for a given $x \in X$ represents a linear-fractional function with respect to $\lambda^1, \lambda^2, \ldots, \lambda^k$ defined on convex set Λ_x^+ and $\hat{S}_x^+$ is the image of $s_{x,a}$ on Λ_x^+. Therefore, $\hat{S}_x^+$ is a convex set. If we analyze (2.104), then we can observe that $s_{x,a}$ for given $x \in X$ represents a linear-fractional function with respect to $\lambda^1, \lambda^2, \ldots, \lambda^k$ on the convex set Λ_x^0 and $\hat{S}_x^0$ is the image of $s_{x,a}$ on Λ_x^0.

Therefore, $\hat{S}_x^0$ is a convex set (see [23]). Additionally, we can observe that $\Lambda_x^+ \cap \Lambda_x^0 = \emptyset$, and in the case $\Lambda_x^+, \Lambda_x^0, \neq \emptyset$, the set Λ_x^0 represents the limit inferior to Λ_x^+. Using this property and taking into account (2.91), we can conclude that each strategy $s_x \in \hat{S}_x^0$ can be regarded as the limit of a sequence of strategies $\{s_x^t\}$ from $\hat{S}_x^+$. Therefore, we can see that $\hat{S}_x = \hat{S}_x^+ \cup \hat{S}_x^0$ is a convex set. This involves the convexity of the sublevel set $L_c^-(\psi)$. In an analogous way, using (2.101) and (2.102), we can show that the superlevel set $L_c^+(\psi)$ and the level set $L_c(\psi)$ are convex sets. This means that the function $\psi(\mathbf{s})$ is quasi-monotonic on $\mathbf{S}$. So, if $\theta_x > 0, \forall x \in X$ and $\sum_{x \in X} \theta_x = 1$, then the theorem holds.

If $\theta_x = 0$ for some $x \in X$, then the set $X \setminus X_\alpha$ may contain states for which $\sum_{a \in A(x)} \alpha_{x,a} = 0$ and $\sum_{a \in A(x)} \beta_{x,a} = 0$ (see Remark 2.15 and Lemma 2.17). In this case, X can be represented as follows: $X = (X \setminus X_0) \cup X_0$, where $X_0 = \{x \in X| \sum_{a \in A(x)} \alpha_{x,a} = 0;\ \sum_{a \in A(x)} \beta_{x,a} = 0\}$. For $x \in X \setminus X_0$, the convexity of $\hat{S}_x$ can be proved in the same way as for the case $\theta_x > 0, \forall x \in X$. If $X_0 \neq \emptyset$, then for $x \in X_0$, we have $\hat{S}_x = S_x$, and the convexity of $\hat{S}_x$ is evident. So, the theorem holds. □

Remark 2.20 For a multichain average Markov decision problem, the function $\psi(\mathbf{s})$ on the set of solutions to system (2.82) may not be continuous.

2.5 Stochastic Discrete Control Problems on Networks

We consider a class of stochastic control problems on networks that generalizes the deterministic discrete control problems with a finite set of states from [9, 11, 12, 15, 21, 88]. This class of problems we formulate and study by applying the results from previous sections for Markov decision processes with a finite state and action spaces.

2.5.1 Deterministic Discrete Optimal Control Problems

Let a discrete dynamical system $\mathbb{L}$ with a finite set of states $X \subset \mathbb{R}^n$ be given, where at every time step $t = 0, 1, 2, \ldots$, the state of the system $\mathbb{L}$ is $x(t) \in X$. At the starting moment of time $t = 0$, the state of the dynamical system $\mathbb{L}$ is $x(0) = x_0$.

Assume that the dynamics of the system $\mathbb{L}$ are described by the system of difference equations

$$x(t+1) = g_t(x(t), u(t)), \quad t = 0, 1, 2, \ldots, \tag{2.105}$$

where

$$x(0) = x_0, \tag{2.106}$$

and

$$u(t) = (u_1(t), u_2(t), \ldots, u_m(t)) \in \mathbb{R}^m$$

represents the *vector of the control parameters* (see [12, 21, 89, 191]). For any time step t and an arbitrary state $x(t) \in X$, the feasible set $U_t(x(t))$ of the vector $u(t)$ of the control parameters is given, i.e.,

$$u(t) \in U_t(x(t)), \quad t = 0, 1, 2, \ldots. \tag{2.107}$$

We assume that in (2.105), the vector functions

$$g_t(x(t), u(t)) = (g_t^1(x(t), u(t)),\ g_t^2(x(t), u(t)), \ldots,\ g_t^n(x(t), u(t)))$$

are uniquely determined by $x(t)$ and $u(t)$ at every time step $t = 0, 1, 2, \ldots$. So, $x(t+1)$ is uniquely determined by $x(t)$ and $u(t)$.

Additionally, we assume that at each moment of time t, the cost

$$c_t(x(t), x(t+1)) = c_t(x(t),\ g_t(x(t), u(t)))$$

of the system's transition from state $x(t)$ to state $x(t+1)$ is known.

Let

$$x_0 = x(0), x(1), x(2), \ldots, x(t), \ldots$$

be a trajectory, generated by given vectors of the control parameters

$$u(0), u(1), \ldots, u(t-1), \ldots.$$

Then after a fixed number of transitions τ of the dynamical system, we can calculate the *integral time cost (total cost)*, which we denote by $F_{x_0}^{\tau}(u(t))$, i.e.,

$$F_{x_0}^{\tau}(u(t)) = \sum_{t=0}^{\tau-1} c_t(x(t),\ g_t(x(t), u(t))). \tag{2.108}$$

In [12, 21], the following discrete optimal control problem with a finite time horizon is considered: Find for given τ the vectors of control parameters

$$u(0), u(1), u(2), \ldots, u(\tau - 1),$$

which satisfy the conditions (2.105)–(2.107) and minimize the functional (2.108). The solution to this optimal control problem can be found by using the *dynamic programming algorithms* from [12, 112].

We mainly focus our attention on the *infinite horizon control problems.*

It is well-known that infinite horizon decision problems can be used for studying and solving problems with a finite time horizon in the case of a large sequence of decisions because it is often easier to solve the infinite horizon problem and to use the solution to this in order to obtain a solution to the finite horizon problem with a large number of solutions.

Thus, we assume that in the control problem above, τ is not bounded, i.e., $\tau \to \infty$. It is evident that if $\tau \to \infty$, then the integral time cost

$$\lim_{\tau\to\infty} \sum_{t=0}^{\tau-1} c_t(x(t),\ g_t(x(t), u(t)))$$

for a given control may not exist. Therefore, in this case, we study the asymptotic behavior of the integral time cost $F^{\tau}_{x_0}(u(t))$ using a trajectory determined by a feasible or optimal control. To estimate this value, we apply the concept from [11, 12], i.e., for a fixed control u, if τ is too large, we estimate $F^{\tau}_{x_0}(u(t))$ asymptotically, using the function $\phi_u(\tau) = K\varphi(\tau)$ such that

$$\lim_{\tau\to\infty} \frac{1}{\varphi(\tau)} \sum_{t=0}^{\tau-1} c_t(x(t),\ g_t(x(t), u(t))) = K, \tag{2.109}$$

where K is a constant.

So, in control problems with an infinite time horizon, we seek a control u^* with a suitable limiting function $\phi_{u^*}(\tau)$. Based on the asymptotic approach mentioned above, we may conclude that for a given control, if τ is too large, the value $F^{\tau}_{x_0}(u(t))$ can be approximated by $K\varphi(\tau)$. Moreover, we can see that for the stationary case of the control model with the costs that do not depend on time, the function $\phi_u(\tau)$ is linear. This means that $\varphi(\tau) = \tau$ and $F^{\tau}_{x_0}(u(t))$ for a large τ can be approximated by $\phi_u(\tau) = K\tau$.

In the following, we study only stationary control problems. For such problems, the vector functions g_t and the feasible sets $U_t(x(t))$ do not depend on time, i.e., $g_t(x, u) = g(x, u)$ and $U_t(x) = U(x)$, $\forall x \in X$, $t = 0, 1, 2, \ldots$. Moreover, the control at every discrete moment of time depends only on the state $x \in X$, and the cost of the system's transition from the state $x \in X$ to the state $y \in Y$ does not depend on time, i.e., $c_t(x(t), x(t+1)) = c(x, y)$, $\forall x, y \in X$ and every $t = 0, 1, 2, \ldots$ if $x = x(t)$, $y = x(t+1)$.

Thus, for the considered stationary control problems, the integral time cost by a trajectory during τ transitions can be asymptotically expressed as $F^{\tau}_{x_0}(u(t)) = K\varphi(\tau)$, where $\varphi(\tau) = \tau$. In this case, for the dynamical system $\mathbb{L}$, the constant K in (2.109) expresses the *average cost per transition* along a trajectory determined by the control $u(t)$. Therefore, for the infinite horizon optimal control problem, the objective function which has to be minimized is defined as follows:

$$F_{x_0}(u(t)) = \lim_{\tau \to \infty} \frac{1}{\tau} \sum_{t=0}^{\tau-1} c(x(t),\ g(x(t), u(t))). \tag{2.110}$$

In [11], it was shown that for the stationary case of the problem, the optimal control u^* does depend on neither time nor the starting state, and it can be found in the set of stationary controls.

Another class of control problems with an infinite time horizon, which is widely used for practical problems, is characterized by a discounting objective cost function [17]:

$$\widehat{F}_{x_0}(u(t)) = \sum_{t=0}^{\infty} \gamma^t c_t(x(t),\ g_t(x(t), u(t))). \tag{2.111}$$

Here, γ is a *discount factor* that satisfies the condition $0 < \gamma < 1$, and $\widehat{F}_{x_0}(u(t))$ is called the *total discounted cost*. In a control problem with such an optimization criterion, the control that minimizes the functional (2.111) is sought.

In [41, 152, 182, 199], it was shown that if $0 < \gamma < 1$ and the costs $c_t(x(t), g_t(x(t), u(t)))$ are bounded, then for the stationary case of the control problem with a discounted objective optimization criterion, the optimal stationary control exists.

The problems formulated above correspond to deterministic models in which the decision maker is able to fix the vector of control parameters $u(t)$ from a given feasible set $U_t(x(t))$ in each dynamical state $x(t)$; the states $x(t) \in X$ in these models are called *controllable states*.

2.5.2 *Stochastic Discrete Optimal Control Problems*

We consider the control models in which the dynamical system in the control process may admit dynamical states $x(t)$, where the corresponding vector of control parameters $u(t)$ is changed in a random way, according to given distribution functions

$$p : U_t(x(t)) \to [0, 1], \quad \sum_{i=1}^{k(x(t))} p\big(u^i_{x(t)}\big) = 1 \tag{2.112}$$

on the corresponding dynamical feasible set $U_t(x(t))$. Here, $k(x(t)) = |U_t(x(t))|$, i.e., we consider the control models with finite feasible sets.

We regard each dynamical state $x(t)$ of the system in the considered control problem as a position (x, t), and we assume that the set of positions

$$Z = \{(x, t) = x(t) \mid x(t) \in X,\ t = 0, 1, 2, \dots\}$$

is divided into two subsets

$$Z = Z^C \cup Z^N, \quad Z^C \cap Z^N = \emptyset$$

such that Z^C corresponds to the set of *controllable states* and Z^N corresponds to the set of *uncontrollable states*. This means that for the stochastic control problems, we have the following behavior of the dynamics in the control process: If the starting state $x(0)$ belongs to the set of controllable states Z^C, then the decision maker fixes the vector of control parameters $u(0)$ from the feasible set $U_0(x(0))$, and we obtain the next state $x(1)$; if the state $x(0)$ belongs to the set Z^N, then the system passes to the next state $x(1)$ in a random way. If at the moment of time $t = 1$ the state $x(1)$ belongs to the set of controllable states Z^C, then the decision maker fixes the vector of control parameters $u(1)$ from $U_1(x(1))$, and we obtain the next state $x(2)$; if $x(1)$ belongs to the set of uncontrollable states Z^N, then the system indefinitely passes to the next state $x(2)$ in a random way and so on.

It is evident that for a fixed control, the average cost per transition and the discounted total cost in this process represent the random variables induced by the distribution functions on feasible sets in the uncontrollable states and the control in the controllable states.

To define the *expected average cost per transition* and *expected total discounted cost* in the considered stochastic control problems for a fixed control, we apply the concept of Markov decision processes in the following way.

Let $u'(t) \in U_t(x(t))$ be the given feasible vectors in the controllable states $x(t) \in Z^C$. Then we may assume that we have the following distribution functions:

$$p : U_t(x(t)) \to \{0, 1\} \text{ for } x(t) \in Z^C,$$

where $p(u'(t)) = 1$ and $p(u(t)) = 0$, $\forall u(t) \in U_t(x(t)) \setminus \{u'(t)\}$.

These distribution functions in the controllable states, together with the distribution functions (2.112) in the uncontrollable states, determine a Markov process. For this Markov process with transition probabilities $p_{z,v}$ and transition costs $c_{z,v}$ for $(z, v) \in Z \times Z$, we can determine the expected average cost and the expected total discounted cost, which we denote by $F_{x_0}(u(t))$ and $\widehat{F}_{x_0}(u(t))$, respectively. In such a way, we obtain the corresponding optimization problems in which we seek the controls that minimize the expected average cost and the total discounted cost.

Thus, we use the combined concept of deterministic and stochastic control models from [56, 114–123, 129–132, 135, 146, 147] and develop algorithms to determine optimal strategies of the considered problems. We mainly study the stationary versions of the control problems with a finite set of states for the dynamical system and describe algorithms based on the results for Markov decision processes.

2.6 Average Stochastic Control Problems on Networks

In this section, we study the stochastic discrete optimal control problem with an average cost criterion and show how to determine the optimal stationary control when the dynamics of the system are described by a directed graph of the states' transitions. We show that stochastic control problems on networks are tightly connected with Markov decision problems and are equivalent in the case of finite state and action spaces.

2.6.1 *Problem Formulation*

Let a discrete dynamical system $\mathbb{L}$ with a finite set of states X be given, where $|X| = n$. At every discrete moment of time $t = 0, 1, 2, \ldots$, the state of $\mathbb{L}$ is $x(t) \in X$. The dynamics of the system are described by a directed *graph of the states' transitions* $G = (X, E)$, where the set of vertices X corresponds to the set of states of the dynamical system, and an arbitrarily directed edge $e = (x, y) \in E$ expresses the possibility of the system $\mathbb{L}$ to pass from state $x = x(t)$ to state $y = x(t+1)$ at every discrete moment of time t. So, a directed edge $e = (x, y)$ in G corresponds to a stationary control of the system in the state $x \in X$, which provides a transition from $x = x(t)$ to $y = x(t+1)$ for every discrete moment of time t. We assume that graph G does not contain deadlock vertices, i.e., for each x, there exists at least one leaving directed edge $e = (x, y) \in E$. In addition, we assume that with each edge $e = (x, y) \in E$, a quantity c_e is associated, which expresses the cost (or the reward [71]) of the system $\mathbb{L}$ to pass from state $x = x(t)$ to state $y = x(t)$ for every $t = 0, 1, 2, \ldots$. The cost c_e for an arbitrary edge $e = (x, y)$ is denoted by $c_{x,y}$. A sequence of directed edges $E' = \{e_0, e_1, e_2, \ldots, e_t, \ldots\}$, where $e_t = (x(t), x(t+1))$, $t = 0, 1, 2, \ldots$ determines a control of the dynamical system with a fixed starting state $x_0 = x(0)$ in G. An arbitrary control in G generates a trajectory $x_0 = x(0), x(1), x(2), \ldots$ for which the average cost per transition can be defined in the following way:

$$f(E') = \lim_{t\to\infty} \frac{1}{t} \sum_{\tau=0}^{t-1} c_{e_\tau}.$$

In [11], it was shown that this value exists and $|f_{x_0}(E')| \leq \max_{e\in E'} |c_e|$. Moreover, in [11] it was also shown that if G is strongly connected, then for an arbitrary fixed starting state $x_0 = x(0)$, there exists the optimal control $E^* = \{e_0^*, e_1^*, e_2^* \dots\}$ for which

$$f(E^*) = \min_{E'} \lim_{t\to\infty} \frac{1}{t} \sum_{\tau=0}^{t-1} c_{e_\tau},$$

and this optimal control does neither depend on the starting state nor on time. Therefore, the optimal control for this problem can be found in the set of stationary strategies $\mathbb{S}$. A *stationary strategy* in G is defined as a map:

$$s : x \to y \in X(x) \quad \text{for} \quad x \in X,$$

where $X(x) = \{y \in X \mid e = (x, y) \in E\}$.

Let s be a stationary strategy. Denote by $G_s = (X, E_s)$ the subgraph of G generated by the edges of the form $e = (x, s(x))$ for $x \in X$. Then it is easy to observe that in G_s, there exists a unique directed cycle C_s, which can be reached from x_0 through the directed edges from E_s.

Moreover, we can see that the mean cost of this cycle is equal to the average cost per transition of the dynamical system using the trajectory generated by the stationary strategy s. Thus, if G is a strongly connected directed graph, then the problem of determining the optimal control on G is equivalent to the problem of finding in G the cycle C_G^* for which

$$\frac{\sum\limits_{e\in E(C_G^*)} c_e}{n(C_G^*)} = \min_{C_G} \frac{\sum\limits_{e\in E(C_G} c_e}{n(C_G)},$$

where $E(C_G)$ is the set of directed edges of the directed cycle C_G in G that can be reached from a starting vertex and $n(C_G)$ is the number of its edges. If the cycle C_G^* is known, then the optimal control for a given arbitrary starting state $x_0 = x(0)$ in G can be found in the following way: We fix the transitions through the directed edges of the graph in order to reach a vertex of the directed cycle C_G^*, and then we preserve transitions through the directed edges of this cycle. Polynomial and strongly polynomial time algorithms for determining the optimal average cost cycles in a weighted directed graph and the optimal stationary strategies for control problems on networks have already been proposed in [28, 79, 97, 112, 156]. In the following, we consider the stochastic version of the problem formulated above.

We assume that the set of states X of the dynamical system may admit states in which the system $\mathbb{L}$ makes transitions to the next state in a random way, according to a given distribution function of probabilities on the set of possible transitions from these states. So, the set of states X is divided into two subsets X_C and X_N ($X = X_C \cup X_N$, $X_C \cap X_N = \emptyset$), where X_C represents the set of states $x \in X$ in which the transitions of the system to the next state y can be controlled by the decision maker at every discrete moment of time t and X_N represents the set of states $x \in X$ in which the decision maker is not able to control the transition because the system passes to the next state y randomly. Thus, for each $x \in X_N$, a probability distribution function $p_{x,y}$ on the set of possible transitions (x, y) from x to $y \in X(x)$ is given, i.e.,

$$\sum_{y \in X(x)} p_{x,y} = 1, \quad \forall x \in X_N; \quad p_{x,y} \geq 0, \quad \forall y \in X(x). \tag{2.113}$$

Here, $p_{x,y}$ expresses the probability of the system's transition from state x to state y for every discrete moment of time t. Note that the condition $p_{x,y} = 0$ for a directed edge $e = (x, y) \in E$ is equivalent to the condition in which G does not contain this edge. In the same way as for the deterministic problem, here we assume that with each directed edge $e = (x, y) \in E$, a cost c_e is associated. We call the graph G with the properties mentioned above *decision network* and denote it by (G, X_C, X_N, c, p, x_0). So, this network is determined by the directed graph G with a fixed starting state x_0; the subsets X_C, X_N; the cost function $c : E \to \mathbb{R}$; and the probability function $p : E_N \to [0, 1]$ on the subset of the edges $E_N = \{e = (x, y) \in E \,|\, x \in X_N,\ y \in X\}$, where p satisfies the condition (2.113). If the control problem is considered for an arbitrary starting state, then we denote the network by (G, X_C, X_N, c, p).

We define a stationary control for the problem on networks as a map:

$$s : x \to y \in X(x) \quad \text{for} \quad x \in X_C. \tag{2.114}$$

Let s be an arbitrary stationary control. Then we can determine the graph $G_s = (X, E_s \cup E_N)$, where $E_s = \{e = (x, y) \in E \,|\, x \in X_C,\ y = s(x)\}$, $E_N = \{e = (x, y) \,|\, x \in X_N,\ y \in X\}$. This graph corresponds to a Markov process with the probability matrix $P(s) = (p^s_{x,y})$, where

$$p^s_{x,y} = \begin{cases} p_{x,y}, & \text{if} \quad x \in X_N \quad \text{and} \quad y = X; \\ 1, & \text{if} \quad x \in X_C \quad \text{and} \quad y = s(x); \\ 0, & \text{if} \quad x \in X_C \quad \text{and} \quad y \neq s(x). \end{cases} \tag{2.115}$$

In the considered Markov process for an arbitrary state $x \in X_C$, the transition $(x, s(x))$ from the states $x \in X_C$ to the states $y = s(x) \in X$ is made with the probability $p_{x,s(x)} = 1$ if the stationary control s is applied. For this Markov process, we can determine the average cost per transition $\omega^s_{x_i}$ for an arbitrary fixed starting state $x_i \in X$ as defined in Sect. 1.9.2. As we have shown, the vector ω^s can be calculated according to the formula $\omega^s = Q^s\mu^s$, where Q^s is the limiting matrix of the Markov process generated by the stationary policy s and μ^s is the corresponding vector of the immediate costs with components $\mu^s_x = \sum_{y\in X(x)} p^s_{x,y}c^s_{x,y}$. This means that for an arbitrary stationary policy s and arbitrary $x \in X$, we can calculate the average cost per transition ω^s_x, i.e., we can find

$$f_x(s) = \omega^s_x.$$

The control problem on the network (G, X_C, X_N, c, p, x_0) consists of finding a stationary policy s^* for which

$$f_{x_0}(s^*) = \min_s f_{x_0}(s).$$

2.6.2 Algorithms for Solving Average Control Problems

An average control problem on the network (G, X_C, X_N, c, p) can be represented as an average Markov decision problem consisting of the following elements:

- the set of states $X = X_C \cup X_N$;
- the set of actions $A = \cup_{x\in X}A(x)$, where $A(x)$ is the set of actions in $x \in X$;
  if $x \in X_N$, then $A(x)$ consists of a single action a_x, i.e., $A(x) = \{a_x\}$;
  if $x \in X_C$, then $A(x) = \{a_e | e \in E(x)\}$ contains several actions, where each a_e corresponds to a directed edge from $E(x) = \{e = (x, y) | e \in E\}$, i.e., $|A(x)| = |E(x)|$, $\forall x \in X_N$;
- The transition probabilities $p^a_{x,y}$ for $x, y \in X$ and $a \in A$ are defined as follows:

$$p^a_{x,y} = \begin{cases} p_{x,y} & \text{if } x \in X_N,\ a = a_x; \\ 1 & \text{if } x \in X_C,\ a = a_{(x,y)}; \\ 0 & \text{if } x \in X_C,\ a \neq a_{(x,y)}. \end{cases} \tag{2.116}$$

- The rewards r^a_x for $x \in X$ and $a \in A(x)$ we define as follows:
  if $x \in X_N$, then we set $r^a_x = -\sum_{y\in X} p_{x,y}c_{x,y}$ for $a = a_x$; if $x \in X_C$, then we set $r^a_x = -c_{x,y}$ for $a = a_{(x,y)}$ and $r^a_x = 0$ for $a \neq a_{(x,y)}$.

So, the average control problem on the network (G, X_C, X_N, c, p) represents an average Markov decision problem with the set of states $X = X_C \cup X_N$, the set of actions $A = \cup_{x\in X}A(x)$, the set of rewards $\mathbf{r} = \{r^a_x, |x \in X, a \in A(x)$, and the probability distributions $\{p^a_{x,y}\}_{y\in X}$ for $x \in X$ and $a \in A(x)$. In general, this Markov decision problem can be modified by directly considering the immediate costs μ^a_x involved instead of the rewards r^a_x, where $\mu^a_x = \sum_{y\in X} p_{x,y}c_{x,y}$ if $x \in X_N, a = a_x$; $\mu^a_x = c_{x,y}$ if $x \in X_C, a = a_{(x,y)}$; and $\mu^a_x = 0$ if $x \in X_C$, $a \neq a_{(x,y)}$. After such a modification, we obtain an average Markov decision problem in which we have to find a deterministic stationary policy that minimizes the average cost per transition. Therefore, the algorithms from Sect. 2.4 can be used to determine the optimal stationary controls for the problem on the network (G, X_C, X_N, c, p).

2.6.3 Linear Programming for Unichain Control Problems

A stochastic control problem with a unichain property is the problem for which the stochastic matrix $P(s)$ of an arbitrary stationary control s is unichain. If the matrix $P(s)$ for a given stationary control s is multichain, then we can say that the control problem is multichain. Based on the results from Sect. 2.4.2 and the reduction procedure of an average control problem to an auxiliary average Markov decision problem presented above, we can formulate the following linear programming model for a unichain control problem on the network (G, X_C, X_N, c, p):
Maximize

$$\phi(\varepsilon, \omega) = \omega \tag{2.117}$$

subject to

$$\begin{cases} \varepsilon_x - \varepsilon_y + \omega \leq c_{x,y}, \quad \forall x \in X_C, \; y \in X(x); \\ \varepsilon_x - \sum\limits_{y \in X} p_{x,y}\varepsilon_y + \omega \leq \mu_x, \quad \forall x \in X_N, \end{cases} \tag{2.118}$$

where $\mu_x = \sum_{y \in X} p_{x,y}c_{x,y}, \; \forall x \in X_N$.

The dual model for this problem is the following:
Minimize

$$\varphi(\alpha, q) = \sum_{x \in X_C} \sum_{y \in X(x)} c_{x,y}\alpha_{x,y} + \sum_{x \in X_N} \mu_x q_x \tag{2.119}$$

subject to

$$\begin{cases} \sum\limits_{x \in X_C^-(y)} \alpha_{x,y} - \sum\limits_{x \in X(y)} \alpha_{y,x} + \sum\limits_{x \in X_N} p_{x,y}q_x = 0, \quad \forall y \in X_C; \\ \sum\limits_{x \in X_C^-(y)} \alpha_{x,y} - q_y + \sum\limits_{x \in X_N} p_{x,y}q_x = 0, \quad \forall y \in X_N; \\ \sum\limits_{x \in X_C} \sum\limits_{y \in X(x)} \alpha_{x,y} + \sum\limits_{x \in X_N} q_x = 1; \\ \alpha_{x,y} \geq 0, \; \forall x \in X_C, \; y \in X; \; q_x \geq 0, \; \forall x \in X_N. \end{cases} \tag{2.120}$$

According to the results from Sect. 2.4, a feasible solution α, q of problem (2.119) determines a stationary control $s_{x,y}$ in the controllable states $x \in X_c$, and $q_x = \sum_{y \in X} \alpha_{x,y}$ represents the limiting probabilities in the Markov process induced by s. So, system (2.120) can also be expressed as follows:

$$\begin{cases} \sum\limits_{x \in X_C^-(y)} \alpha_{x,y} + \sum\limits_{x \in X_N} p_{x,y} q_x = q_y, \quad \forall y \in X_C; \\ \sum\limits_{x \in X_C} q_x + \sum\limits_{z \in X_N} q_z = 1; \\ \sum\limits_{y \in X(x)} \alpha_{x,y} = q_x, \quad \forall x \in X_C; \\ \alpha_{x,y} \geq 0, \ \forall x \in X_C, \ y \in X; \ q_x \geq 0, \ \forall x \in X, \end{cases} \tag{2.121}$$

where $X_C^-(y) = \{x \in X_C | (x, y) \in E\}$.

Theorem 2.21 *Let α^*, q^* be an optimal basic solution to the linear programming problem (2.119), (2.120). Then an optimal stationary control s^* on controllable states X_C of the unichain control problem on the network (G, X_C, X_N, c, p) can be found as follows: If $q_x = \sum_{y \in X(s)} \alpha_{x,y} > 0$ for $x \in X_C$, then*

$$s^*_{x,y} = \begin{cases} 1, & if \ \alpha^*_{x,y} > 0; \\ 0, & if \ \alpha_{x,y} = 0. \end{cases} \tag{2.122}$$

If $q_x * = 0$ for $x \in X_C$, then we fix $s^*_{x,y} = 1$ for an arbitrary $(x, y) \in E(x)$ and set $s_{x,z} = 0$ for the rest of the directed edges $(x, z) \in E(x) \setminus \{(x, y)\}$.

Proof If α^*, q^* is a basic optimal solution to problem (2.119), (2.119), then for an arbitrary $x \in X_C$, no more than one directed edge $e = (x, y)$ from $E(x)$ will satisfy the condition $\alpha^*_{x,y} > 0$. So, if $q^*_x > 0$, then we can calculate

$$s^*_{x,y} = \begin{cases} \frac{\alpha_{x,y}}{q_x}, & if \ \alpha^*_{x,y} > 0; \\ 0, & if \ \alpha_{x,y} = 0, \end{cases}$$

i.e., $s^*_{x,y}$ for $y \in X(x)$ are determined according to (2.122). If $q_x * = 0$ for $x \in X_C$, then we can fix $s^*_{x,y} = 1$ for an arbitrary $(x, y) \in E(x)$ by setting $s_{x,z} = 0$ for the rest of the directed edges $(x, z) \in E(x) \setminus \{(x, y)\}$. □

Thus, if we find an optimal basic solution (α^*, q^*) to the linear programming problem (2.119), (2.120) (or to problem (2.119), (2.121)), then by setting $s^*_{x,y} = 1$ for $x \in X_C$, where $\alpha^*_{x,y}$ corresponds to a basic variable, then we determine an optimal stationary control. Note that the conditions $q_y \geq 0$, $\forall x \in X_N$ in the unichain linear programming problem (2.119), (2.120) are redundant, because q_x for $x \in X$ are uniquely determined and non-negative.

Therefore, the constraints (2.118) in the problem (2.117), (2.118) can be replaced by the following constraints:

$$\begin{cases} \varepsilon_x - \varepsilon_y + \omega \le c_{x,y}, \ \forall x \in X_C, \ y \in X(x); \\ \varepsilon_x - \sum_{z\in X} p_{x,z}\varepsilon_z + \omega = \mu_x, \ \forall x \in X_N. \end{cases} \tag{2.123}$$

In general, an optimal stationary control for an average stochastic control problem is more simple to determine by using the linear programming problem *(2.117), (2.118)*. An optimal stationary control for an average control problem on networks can be found based on the following theorem:

Theorem 2.22 *An arbitrary optimal basic solution* ε_x^* $(x \in X)$, ω^* *of the problem* (2.117), (2.118) *for a unichain control model on the network* (G, X_C, X_N, c, p) *possesses the following property:*

(1) $\min_{y\in X(x)} \{c_{x,y} + \varepsilon_y^* - \varepsilon_x^* - \omega^*\} = 0, \ \forall x \in X_C.$

(2) $\mu_x + \sum_{z\in X(x)} p_{x,z}\varepsilon_z^* - \varepsilon_x^* - \omega^* = 0, \ \forall x \in X_N.$

(3) *A stationary strategy* $s^* : X_C \to X$ *is optimal if and only if* $(x, s^*(x)) \in E_C^*$, $\forall x \in X_C$, *where*

$$E_C^* = \{e = (x, y) \in E_C \mid c_{x,y} + \varepsilon_y^* - \varepsilon_x^* - \omega^* = 0\}.$$

The value ω^* *is equal to the optimal average cost in the unichain control problem on the network* (G, X_C, X_N, c, p).

Proof The properties (1) and (2) of the theorem represent the optimality conditions for the linear programming problems (2.117), (2.118). If α^*, q^* is a basic solution to problem (2.119), (2.120), where $\alpha_{x,y}^* = s_{x,y}^* q_x^*$, $q_x^* = \sum_{y\in X} \alpha_{x,y}$, then we can take $s_{x,y}^* = 1$ for $(x, y) \in E_C$ that satisfy the conditions (1) and $s_{x,y}^* = 0$ in the other case. This means that $s^* : X_C \to X$ for which $(x, s^*(x)) \in E_C^*$, $\forall x \in X_C$. □

Corollary 2.23 *Let* $G = (X, E)$ *be a strongly connected directed graph with* $X_N = \emptyset$ *and let* $\alpha_{x,y}^*$, $(x, y) \in E$ *be the basic optimal solution to the linear programming problem:*
Minimize

$$\varphi(\alpha) = \sum_{x\in X_C} \sum_{y\in X(x)} c_{x,y}\alpha_{x,y} \tag{2.124}$$

subject to

$$\begin{cases} \sum\limits_{x\in X^-(y)} \alpha_{x,y} - \sum\limits_{z\in X(y)} \alpha_{y,z} = 0, \quad \forall y \in X; \\ \sum\limits_{x\in X}\sum\limits_{y\in X(x)} \alpha_{x,y} = 1; \\ \alpha_{x,y} \geq 0, \ \forall (x,y) \in E, \end{cases} \tag{2.125}$$

where $X^-(y) = \{x \in X | (x,y) \in E\}$.

Then the subgraph $G' = (X', E')$ *generated by the directed edges* $(x,y) \in E$ *with* $\alpha^*_{x,y} > 0$ *has the structure of a directed cycle, and an optimal stationary strategy* s^* *for the control problem on G with a given starting state* x_0 *can be found as follows:*

- *Fix a simple directed path that connects* x_0 *with the directed cycle* G' *and find the set of edges* E'' *of this directed path.*
- *Fix the stationary strategy* s^**, where* $s^*_{x,y} = 1$ *if* $(x,y) \in E' \cup E''$*; otherwise, put* $s^*_{x,y} = 0$.

Based on the results above, we can determine a solution for an average control problem by using the following linear programming algorithms:

Algorithm 1

- Form the linear programming problem (2.117), (2.118) and determine an optimal basic solution $(\varepsilon^*, \omega^*)$.
- For given solution $(\varepsilon^*, \omega^*)$, find s^* that satisfies the conditions (1)–(3) of Theorem 2.22.

Algorithm 2

- Form the linear programming problem (2.117), (2.118) and determine an optimal basic solution (α^*, q^*).
- For each $x \in X_C$ with $q^*_x > 0$, fix $s^*_{x,y} = 1$ if $\alpha^*_{x,y} > 0$ and set $s^*_{x,y} = 0$ if $\alpha^*_{x,y} = 0$; for each $x \in X_C$ with $q_x = 0$, select an arbitrarily directed edge $(x,y) \in E(x)$ and set $s^*_{x,y} = 1$; for the rest $(x,z) \in E(x) \setminus \{(x,y)\}$, set $s^*_{x,z} = 0$.

If $X_N = \emptyset$, then we have the deterministic control problem, and its solution can be found by using the following algorithm:

Algorithm 3

- Formulate the linear programming problem (2.124), (2.125), and find a basic optimal solution $\alpha^*_{x,y}$ and the corresponding directed graph $G' = (X', E')$ that has the structure of a directed cycle.
- Fix a simple directed path that connects x_0 to the directed cycle G' and find the set of edges E'' of this directed path.

Fig. 2.1 Network for the unichain control problem

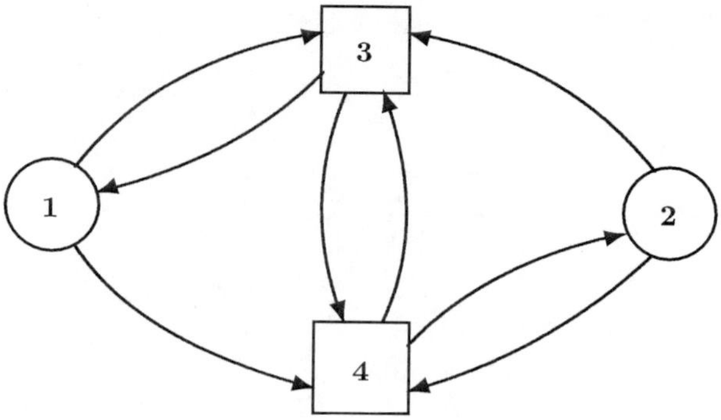

– Fix a stationary strategy s^*, where $s^*_{x,y} = 1$ if $(x, y) \in E' \cup E''$; otherwise, put $s^*_{x,y} = 0$.

Below we present an example that illustrates the linear programming algorithms.

Example Consider a stochastic control problem for which the network is represented in Fig. 2.1, i.e.,

$$G = (X, E), \quad X = \{1, 2, 3, 4\}, \quad X_C = \{1, 2\}, \quad X_N = \{3, 4\},$$
$$E = \{(1, 3), (1, 4), (2, 3), (2, 4), (3, 1), (3, 4), (4, 2), (4, 3)\}.$$

The transition cost for directed edges from E and the transition probabilities for directed edges originating in the vertices 3 and 4 are given by

$$\begin{array}{llll} c_{1,3} = 1, & c_{2,3} = 3, & c_{3,1} = 2, & c_{4,2} = 1, \\ c_{1,4} = 2, & c_{2,4} = 1, & c_{3,4} = 4, & c_{4,3} = 3, \\ p_{3,1} = 0.5, & p_{3,4} = 0.5, & p_{4,2} = 0.5, & p_{4,3} = 0.5\,. \end{array}$$

We seek the optimal stationary strategy s^* that provides the solution to the problem for an arbitrary starting state $x \in X$. This average control problem is unichain because each stationary control s generates a Markov unichain. Therefore, we can determine the optimal stationary control by solving the linear programming problem (2.117), (2.118) or the linear programming (2.119), (2.121).

For this example, we have

$$\mu_3 = p_{3,1}c_{3,1} + p_{3,4}c_{3,4} = 0.5 \cdot 2 + 0.5 \cdot 4 = 3,$$
$$\mu_4 = p_{4,2}c_{4,2} + p_{4,3}c_{4,3} = 0.5 \cdot 1 + 0.5 \cdot 3 = 2.$$

If we use the linear programming model (2.117), (2.118), then we obtain the following problem:

Maximize

$$\phi(\varepsilon, \omega) = \omega$$

subject to

$$\begin{cases} \varepsilon_1 - \varepsilon_3 + \omega \le 1, \\ \varepsilon_1 - \varepsilon_4 + \omega \le 3, \\ \varepsilon_2 - \varepsilon_3 + \omega \le 3, \\ \varepsilon_2 - \varepsilon_4 + \omega \le 1, \\ \varepsilon_3 - (05\varepsilon_1 + 0,5\varepsilon 4) + \omega \le 3, \\ \varepsilon_4 - (05\varepsilon_2 + 0,5\varepsilon 3) + \omega \le 2. \end{cases}$$

An optimal basic solution to this problem is

$$\varepsilon_1^* = 0, \ \varepsilon_2^* = -1, \varepsilon_3^* = 1, \ \varepsilon_4 = 0, \ \omega^* = 2$$

and $E_C^* = \{e = (x, y) \in E_C \mid c_{x,y} + \varepsilon_y^* - \varepsilon_x^* - \omega^* = 0\}$. So, $E_C^* = \{(1, 3), (2, 4)\}$ and the optimal stationary control is $s^* : 1 \to 3, \ 2 \to 4$.

If we consider the dual problem (2.119), (2.120), then we obtain the following problem:
Minimize

$$\varphi(\alpha, q) = \alpha_{1,3} + 2\alpha_{1,4} + 3\alpha_{2,3} + \alpha_{2,4} + 3q_3 + 2q_4$$

subject to

$$\begin{cases} 0.5q_3 = q_1, \\ 0.5q_4 = q_2, \\ \alpha_{1,3} + \alpha_{2,3} + 0.5q_4 = q_3, \\ \alpha_{1,4} + \alpha_{2,4} + 0.5q_3 = q_4, \\ \alpha_{1,3} + \alpha_{1,4} = q_1, \\ \alpha_{2,3} + \alpha_{2,4} = q_2, \\ q_1 + q_2 + q_3 + q_4 = 1, \\ q_i \ge 0, \ i = 1, 2, 3, 4; \quad \alpha_{i,j} \ge 0, \ i, j = 1, 2, 3, 4. \end{cases}$$

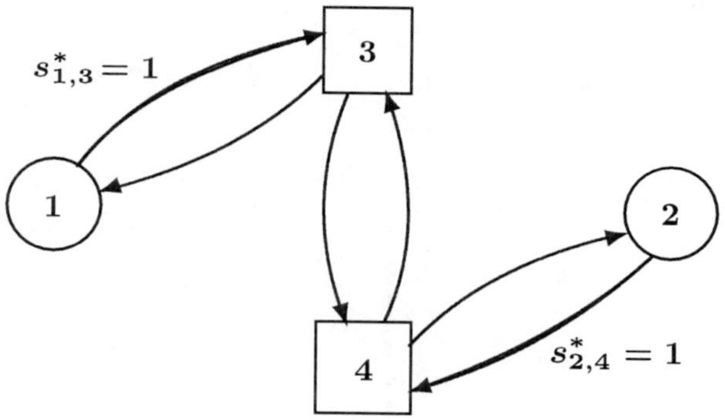

Fig. 2.2 Network corresponding to optimal strategies

It is easy to check that the optimal solution to this problem is

$$\alpha^*_{1,4} = 0, \quad \alpha^*_{2,3} = 0, \quad \alpha^*_{1,3} = \frac{1}{6}, \quad \alpha^*_{2,4} = \frac{1}{6},$$
$$q^*_1 = \frac{1}{6}, \quad q^*_2 = \frac{1}{6}, \quad q^*_3 = \frac{2}{6}, \quad q^*_4 = \frac{2}{6} \quad \text{and} \quad \varphi(\alpha^*, q^*) = 2.$$

So, $s^*_{1,4} = 0,\ s^*_{2,3} = 0,\ s^*_{1,3} = 1,\ s^*_{2,4} = 1$.

In Fig. 2.2, a network is presented which corresponds to an optimal stationary strategy $s^*_{1,3} = 1, s^*_{2,4} = 1$.

Remark 2.24 The linear programming problem (2.119), (2.120) can be considered on an arbitrary decision network (G, X_C, X_N, c, p). In this case, a basic optimal solution α^*, q^* determines a stationary control

$$s^*_{x,y} = \begin{cases} 1, & \text{if } \alpha^*_{x,y} > 0; \\ 0, & \text{if } \alpha^*_{x,y} = 0 \end{cases}$$

and a subset $X^* = \{x \in X \mid q_{x^*} > 0\}$, where s^* on X^* induces a separate irreducible Markov chain, and s^* provides the optimal average cost per transition for the problem with a starting state $x_0 \in X^*$. This means that the linear programming determines an optimal control for the problem when the starting state is $x \in X^*$. In general, if $x_0 \notin X^*$, then the unichain model does not find the optimal stationary control.

From Theorem 2.22, we can draw the following conclusions: For an arbitrary unichain control problem, there exists a function $\varepsilon^* : X \to \mathbb{R}$ and a value ω^* that satisfy the conditions

$$(1)\ \overline{c}_{x,y} = c_{x,y} + \varepsilon^*_y - \varepsilon^*_x - \omega^* \geq 0, \quad \forall x \in X_C,\ \forall y \in X(x);$$

$$(2)\ \min_{y \in X} \overline{c}_{x,y} = 0, \quad \forall x \in X_C;$$

$$(3)\ \overline{\mu}_x = \mu_x + \sum_{y \in X} p_{x,y}\varepsilon^*_x - \varepsilon^*_x - \omega^*_x = 0, \quad \forall x \in X_N.$$

If in the decision network (G, X_C, X_N, c, p) we change the cost function c by $\overline{c}$, then we obtain a new control problem on the network $(G, X_C, X_N, \overline{c}, p)$. Such a transformation of the cost function in the control problem does not change the optimal stationary strategies. In the new control problem, the cost function $\overline{c}$ satisfies the conditions $\min_{y \in X(x)} \overline{c}_{x,y} = 0,\ \forall x \in X_C$ and $\overline{\mu}_x = 0,\ \forall x \in X_N$. For this problem, the optimal average cost $\overline{\omega}^*_x$ for every $x \in X$ is equal to zero, and an optimal stationary strategy can be found by fixing an arbitrary map s^* such that $(x, s^*(x)) \in E^*_C$, where $E^*_C = \{(x, y) \in E_C \mid \overline{c}_{x,y} = 0\}$.

We call the cost function $\overline{c}_{x,y} = c_{x,y} + \varepsilon^*_y - \varepsilon^*_x - \omega^*,\ (x, y) \in E$ a *potential transformation* induced by the *potential function* $\varepsilon^* : X \to \mathbb{R}$ and the values ω^*_x for $x \in X$. Furthermore, we call the new problem with the cost function $\overline{c}$ a *control problem in canonical form*.

2.6.4 Optimality Equations for an Average Control Problem

In Sect. 2.6.2, it was shown that an average control problem on the network represents a special case of an average Markov decision problem. Therefore, if we specify Theorem 2.9 for an average control problem on the network (G, X_C, X_N, c, p), then we obtain the following result:

Theorem 2.25 *Let a control problem on the network (G, X_C, X_N, c, p) with immediate costs $\mu_x = \sum_{y \in X(x)} p_{x,y} c_{c,y}$ in uncontrollable states $x \in X_N$ be given. Then the system of equations*

$$\begin{cases} \varepsilon_x + \omega_x = \min_{y \in X}\{c_{x,y} + \varepsilon_y\}, & \forall x \in X_C; \\ \varepsilon_x + \omega_x = \mu_x + \sum_{y \in X} p_{x,y}\varepsilon_y, & \forall x \in X_N \end{cases} \tag{2.126}$$

has solutions with respect to ε_x *for* $x \in X$ *under the set of solutions to the system of equations*

$$\begin{cases} \omega_x = \min\limits_{y \in X(x)} \omega_y, & \forall x \in X_C; \\ \omega_x = \sum\limits_{y \in X(x)} p_{x,y}\omega_x, & \forall x \in X_N \end{cases} \tag{2.127}$$

with respect to ω_x *for* $x \in X$. *If* ε_x^*, ω_x^* $(x \in X)$ *is the solution to these equations, then* ω_x^* *for* $x \in X$ *represent the optimal average costs for the corresponding control problems with starting states* $x \in X$.

Corollary 2.26 *For an arbitrary average control problem on the network* (G, X_C, X_N, c, p), *the following system of inequalities:*

$$\begin{cases} \varepsilon_x - \varepsilon_y + \omega_x \le c_{x,y}, & \forall x \in X_C,\ \forall y \in X(x); \\ \varepsilon_x - \sum\limits_{y \in X(x)} p_{x,y}\varepsilon_y + \omega_x \le \mu_x, & \forall x \in X_N; \\ \omega_x - \omega_y \le 0, & \forall x \in X_C,\ \forall y \in X(x); \\ \omega_x - \sum\limits_{y \in X(x)} p_{x,y}\omega_y \le 0, & \forall x \in X_N; \end{cases} \tag{2.128}$$

has solutions. Moreover, if $\varepsilon_x^*(x \in X)$, $\omega_x^*(x \in X)$ *is a solution to the system of Eqs. (2.128) that satisfies the conditions of Theorem 2.25, then vector* ω^* *with components* ω_x^* *for* $x \in X$ *is the "maximal" vector that satisfies (2.128).*

2.6.5 Linear Programming for Multichain Control Problems

According to Corollary 2.26, in order to determine an optimal solution to an average control problem on the network (G, X_C, X_N, c, p), in the general case, it is necessary to determine a solution to system (2.128) for which the value vector ω is "maximal." Therefore, in order to determine such a vector, it is necessary to solve the following linear programming problem:
Maximize

$$\phi(\varepsilon, \omega) = \sum_{x \in X} \theta_x \omega_x \tag{2.129}$$

subject to

$$\begin{cases} \varepsilon_x - \varepsilon_y + \omega_x \le c_{x,y}, & \forall x \in X_C,\ \forall y \in X(x); \\ \varepsilon_x - \sum\limits_{y \in X(x)} p_{x,y}\varepsilon_y + \omega_x \le \mu_x, & \forall x \in X_N; \\ \omega_x - \omega_y \le 0, & \forall x \in X_C,\ \forall y \in X(x); \\ \omega_x - \sum\limits_{y \in X(x)} p_{x,y}\omega_y \le 0, & \forall x \in X_N, \end{cases} \tag{2.130}$$

where $\mu_x = \sum_{y \in X} p_{x,y}c_{x,y}$, $\forall x \in X_N$, and θ_x for $x \in X$ represent arbitrary positive values that satisfy the condition $\sum_{x \in X} \theta_x = 1$. In general, the values θ_x for $x \in X$ may be arbitrary positive values; however, in the following, we take these positive values such that $\sum_{x \in X} \theta_x = 1$, because we can treat θ_x as the probability of chosen x as a stating state in the control problem. We can see that in (2.130), the inequalities $\varepsilon_x - \sum_{y \in X}(x) p_{x,y}\varepsilon_y + \omega \le \mu_x$ and $\omega_x - \sum_{y \in X} p_{x,y}\omega_x$ for $x \in X_N$ can be changed by equalities, i.e., the constraints (2.130) in the problem (2.129), (2.130) can be replaced by the constraints

$$\begin{cases} \varepsilon_x - \varepsilon_y + \omega_x \le c_{x,y}, & \forall x \in X_C,\ \forall y \in X(x); \\ \varepsilon_x - \sum\limits_{y \in X(x)} p_{x,y}\varepsilon_y + \omega_x = \mu_x, & \forall x \in X_N; \\ \omega_x - \omega_y \le 0, & \forall x \in X_C,\ \forall y \in X(x); \\ \omega_x - \sum\limits_{y \in X(x)} p_{x,y}\omega_y = 0, & \forall x \in X_N. \end{cases} \tag{2.131}$$

This means that the optimal solutions to the problems (2.129), (2.130) and (2.129), (2.131) are the same.

Note that in the linear programming model (2.129), (2.130) in the case of the unichain control problem, the restrictions $\omega_x - \omega_y \le 0$ for $x \in X_C$, $y \in X(x)$ and $\omega_x - \sum_{y \in X(x)} p_{x,y}\omega_y \le 0$ for $x \in X_N$ become redundant in (2.130) because $\omega_x = \omega_y = \omega$, $\forall x, y \in X$, i.e., this model generalizes the linear programming model (2.117), (2.118).

The dual problem for the linear programming problem (2.129), (2.130) is the following:
Minimize

$$\varphi(\alpha, \beta, \lambda, q) = \sum_{x \in X_C} \sum_{y \in X(x)} c_{x,y}\alpha_{x,y} + \sum_{z \in X_N} \mu_z q_z \tag{2.132}$$

subject to

$$\begin{cases} \sum_{x \in X_C^-(y)} \alpha_{x,y} - \sum_{x \in X(y)} \alpha_{y,x} + \sum_{x \in X_N} p_{x,y} q_x = 0, \quad \forall y \in X_C; \\ \sum_{x \in X_C^-(y)} \alpha_{x,y} - q_y + \sum_{x \in X_N} p_{x,y} q_x = 0, \quad \forall y \in X_N; \\ \sum_{y \in X(x)} \alpha_{x,y} + \sum_{y \in X} \beta_{x,y} - \sum_{y \in X_C^-(x)} \beta_{y,x} - \sum_{y \in X_N^-(x)} p_{y,x} \lambda_y = \theta_x, \forall x \in X_C; \\ \quad q_x + \lambda_x - \sum_{y \in X_N^-(x)} p_{y,x} \lambda_x = \theta_x, \ \forall x \in X_N; \\ \quad \alpha_{x,y}, \beta_{x,y} \geq 0, \ \forall x \in X_C, \ y \in X(x); \ q_x, \lambda_x \geq 0, \ \forall x \in X_N, \end{cases} \tag{2.133}$$

where $X_C^-(y) = \{x \in X_C | (x, y) \in E\}$.

Theorem 2.27 *Let $\alpha^*, \beta^*, q^*, \lambda*$ be an optimal basic solution to the linear programming problem (2.132), (2.133). Then an optimal deterministic stationary control s^* on X_C for the average control problem on the network (G, X_C, X_N, c, p) can be found as follows:*

$$s_{x,y}^* = \begin{cases} 1 & \text{if } \alpha_{x,y}^* > 0 \text{ or } \beta_{x,y}^* > 0; \\ 0 & \text{if } \alpha_{x,y}^* = 0 \text{ and } \beta_{x,y}^* = 0. \end{cases} \tag{2.134}$$

Proof If $\alpha^*, \beta^*, q^*, \lambda*$ is an optimal solution to problem (2.132), (2.133), then according to Lemma 2.12, an optimal stationary control s^* on X_C can be found as follows:

$$s_{x,y} = \begin{cases} \dfrac{\alpha_{x,y}}{\sum_{y \in X(x)} \alpha_{x,y}} & \text{if } x \in X_C^\alpha; \\ \dfrac{\beta_{x,a}}{\sum_{y \in X(x)} \beta_{x,y}} & \text{if } x \in X_C \setminus X_N^\alpha, \end{cases} \tag{2.135}$$

where $X_C^\alpha = \{x \in X_N | \sum_{y \in X} \alpha_{x,y} > 0\}$. If $\alpha^*, \beta^*, q^*, \lambda*$ is a basic optimal solution, then for $x \in X_C$, there exists only one directed edge $(x, y) \in E(x)$ for which $\alpha_{x,y}^* > 0$ or $\beta_{x,y}^* > 0$ hold. Therefore, we obtain an optimal deterministic stationary control s^* that can be found according to (2.135). □

Theorem 2.28 *Let $\varepsilon_x^*, \omega_x^*$ for $x \in X$ be an optimal basic solution to linear programming problem (2.129), (2.130) and let*

$$s^* : x \to y \in X(x) \ for \ x \in X_C$$

be a map such that $(x, s^*(x)) \in E_{\overline{c}}^*(x) \cap E_{\omega^*}^*(x), \forall x \in X_C$, where

$$E_{\overline{c}}^*(x) = \Big\{(x, y) \in E_C \mid y \in \underset{z \in X(x)}{\operatorname{argmin}}\{c_{x,y} + \varepsilon_y^* - \varepsilon_x^* - \omega_x^*\}\Big\}, \quad \forall x \in X_C,$$

$$E_{\omega^*}^*(x) = \Big\{(x, y) \in E_C \mid y \in \underset{z \in X(x)}{\operatorname{argmin}}\{\omega_z^*\}\Big\}, \quad \forall x \in X_C.$$

Then s^* is an optimal stationary deterministic control for the average control problem on the network (G, X_C, X_N, c, p).

Proof Let $\varepsilon_x^*, \omega_x^*$ for $x \in X$ be an optimal basic solution to problem (2.129), (2.130) and $\alpha^*, \beta^*, q^*, \lambda*$ be an optimal basic solution to problem (2.132), (2.133). Then in the case $\alpha_{x,y}^* > 0$ or $\beta_x^* > 0$, either $q_x^* > 0$ or $\lambda_x > 0$, and according to the duality theory of linear programming for the optimal basic solution $\varepsilon_x^*, \omega_x^*, \in X$, we have $c_{x,y} + \varepsilon_y^* - \varepsilon_x^* - \omega_x^* = 0$ and $\omega_x^* - \omega_y^*$. So, if s^* is an optimal stationary control, then $(x, s^*(x)) \in E_{\overline{c}}^*(x) \cap E_{\omega^*}^*(x), \forall x \in X_C$. □

Thus, an optimal stationary control for a multichain average stochastic control problem on the network (G, X_C, X_N, c, p) can be found by using the following algorithms:

Algorithm 1

(1) Formulate the linear programming problem (2.132), (2.133) and find an optimal basic $\alpha^*, \ \beta^*, q^*, \lambda^*$.
(2) Find an optimal stationary control s^* according to (2.135).

Algorithm 2

(1) Formulate the linear programming problem (2.129), (2.130) and determine an optimal basic solution $\varepsilon^*, \ \omega^*$.
(2) Define the potential transformation

$$\overline{c}_{x,y} = c_{x,y} + \varepsilon_y^* - \varepsilon_x^* - \omega_x^*, \quad \forall (x, y) \in E.$$

(3) Determine the sets

$$E_{\overline{c}}^*(x) = \Big\{(x, y) \in E_C \mid y \in \underset{z \in X(x)}{\operatorname{argmin}}\{\overline{c}_{x,z}\}\Big\}, \quad \forall x \in X_C;$$

$$E_{\omega^*}^*(x) = \Big\{(x, y) \in E_C \mid y \in \underset{z \in X(x)}{\operatorname{argmin}}\{\omega_z^*\}\Big\}, \quad \forall x \in X_C.$$

(4) Fix a strategy control $s^* : X_C \to X$ on X_C, where

$$(x, s^*(x)) \in E_{\overline{c}^*}(x) \cap E^*_{\omega^*}(x), \quad \forall x \in X_C.$$

In the following, based on Algorithm 2, we show how to use the potential transformation for a detailed analysis of the solutions to the average control problem on the network (G, X_C, X_N, c, p).

Theorem 2.29 *For an arbitrary decision network* (G, X_C, X_N, c, p)*, there exists a potential transformation*

$$\overline{c}_{x,y} = c_{x,y} + \varepsilon^*_y - \varepsilon^*_x - h^*_x \ \textit{for}\ e = (x, y) \in E \tag{2.136}$$

of the cost function $c : R \to R^1$ *that satisfies the following conditions:*

$$(1)\ \overline{c}_{x,y} = c_{x,y} + \varepsilon^*_y - \varepsilon^*_x - h^*_x \geq 0, \quad \forall x \in X_C,\ y \in X(x);$$

$$(2)\ \min_{y \in X} \overline{c}_{x,y} = 0, \quad \forall x \in X_C;$$

$$(3)\ \overline{\mu}_x = \mu_x + \sum_{y \in X} p_{x,y}\varepsilon^*_y - \varepsilon^*_x - h^*_x = 0, \quad \forall x \in X_N;$$

$$(4)\ h^*_x = \min_{y \in X(x)} h^*_y, \quad \forall x \in X_C;$$

$$(5)\ h^*_x = \sum_{y \in X} p_{x,y} h^*_y, \quad \forall x \in X_N;$$

$$6)\ E^*_{h^*}(x) \cap E^*_{\overline{c}}(x) \neq \emptyset, \ \forall x \in X_C, \ \textit{where}$$

$$E^*_{h^*}(x) = \left\{ (x, y) \in E_C \mid y \in \underset{z \in X(x)}{\operatorname{argmin}}\{h_z\} \right\}, \ x \in X_C$$

and

$$E^*_{\overline{c}}(x) = \left\{ (x, y) \in E_C \mid y \in \underset{z \in X(x)}{\operatorname{argmin}}\{\overline{c}_{x,z}\} \right\}, \ x \in X_C.$$

The values ε^*_x *for* $x \in X$ *correspond to a basic solution to the system of linear equations*

$$\begin{cases} c_{x,y} + \varepsilon_y - \varepsilon_x - h^*_x = 0, \ \ \forall x \in X_C, \ (x, y) \in E^*_{h^*}(x); \\ \mu_x + \sum\limits_{y \in X} p_{x,y}\varepsilon_y - \varepsilon_x - h^*_x = 0, \ \ \forall x \in X_N \end{cases} \tag{2.137}$$

and determine the decision network in canonical form $(G, X_C, X_N, \overline{c}, p)$ *for the control problem on the network* (G, X_C, X_N, c, p), *where* $\overline{c}_{x,y} = c_{x,y} + \varepsilon^*_y - \varepsilon^*_x - h^*_x$, $\forall x \in X,\ y \in X(x)$ *of cost function* c *that satisfy conditions (1)–(6) of the theorem.*

The values h^*_x *for* $x \in X$ *coincide with the corresponding optimal average costs* ω^*_x *for* $x \in X$, *and an optimal stationary strategy for the control problem on the network can be found by fixing an arbitrary map* $s^* : X_C \to X$ *such that* $(x, s^*(x)) \in E^*_{h^*}(x) \cap E^*_{\overline{c}}(x),\ \forall x \in X_C$.

Proof Consider an average control problem on the network (G, X_C, X_N, c, p). Then an optimal stationary control of this problem can be found based on Theorem 2.28 by using an optimal basic solution $\varepsilon^*_x, \omega^*$ for $x \in X$ of the linear programming problem (2.129), (2.130). Let us show that for the decision network (G, X_C, X_N, c, p), there exists the potential transformation

$$\overline{c}_{x,y} = c_{x,y} + \varepsilon^*_y - \varepsilon^*_x - h^*_x, \quad \forall x \in X,\ y \in X(x)$$

that satisfies conditions (1)–(6) of the theorem.

It is easy to check that $\overline{\varepsilon}^*_x = \varepsilon^*_x$ and $\overline{\omega}^*_x = \omega^*_x - h^*_x = 0$ for $x \in X$ represent an optimal basic solution to the linear programming problem:
Maximize

$$\phi(\varepsilon, \omega) = \sum_{x \in X} \theta_x \omega_x \tag{2.138}$$

subject to

$$\begin{cases} \varepsilon_x - \varepsilon_y + \omega_x \le \overline{c}_{x,y}, & \forall x \in X_C,\ \forall y \in X(x); \\ \varepsilon_x - \sum\limits_{y \in X(x)} p_{x,y}\varepsilon_y + \omega_x \le \overline{\mu}_x, & \forall x \in X_N; \\ \omega_x - \omega_y \le 0, & \forall x \in X_C,\ \forall y \in X(x); \\ \omega_x - \sum\limits_{y \in X(x)} p_{x,y}\omega_y \le 0, & \forall x \in X_N. \end{cases} \tag{2.139}$$

This linear programming problem corresponds to the optimal control problem on the networks $(G, X_C, X_N, \overline{c}, p)$ with the cost function $\overline{c}$ on E, where $\overline{c}_{x,y} \ge 0, \forall (x, y) \in E$. Therefore, the optimal value of the object function of problem (2.138), (2.139) is equal to zero. Moreover, we can observe that if for the decision network there exists the potential transformation 2.136 that satisfies conditions (1)–(6) of the theorem, then $\omega^*_x = h^*_x, \forall x \in X$. □

If the decision network in canonical form is known, then the optimal stationary strategy for the stochastic multichain control problem can be found in a similar way as for the unichain case of the problem.

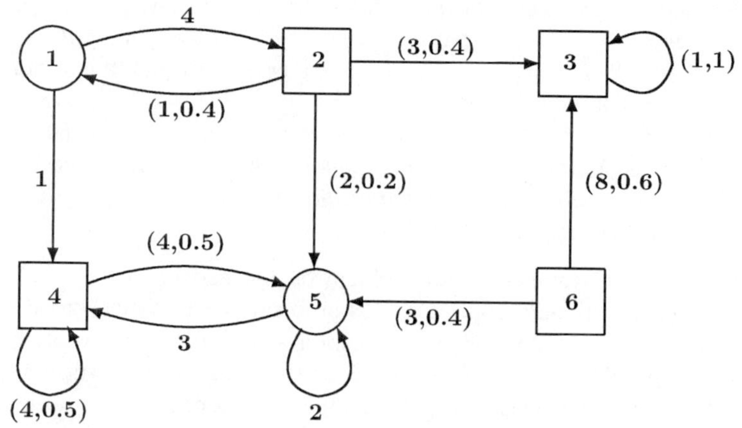

Fig. 2.3 Structure of graph $G = (X, E)$

We fix a strategy $s^* : X_C \to X$ such that $(x, s^*(x)) \in E^*_{\overline{c}}$. Moreover, the potential transformation $\overline{c}$ that satisfies the conditions (1)–(6) gives the values of the optimal average costs $\omega^*_x = h_x$ in the states $x \in X$ for a multichain control problem on the network (G, X_C, X_N, c, p).

Below we illustrate Algorithm 2 in an example and show how to use the potential transformation for the analyses of the optimal solutions to the control problem.

Example Consider the stochastic control problem on network (G, X_1, X_2, c, p) with the structure of graph $G = (X, E)$ given in Fig. 2.3.

In this graph, the vertices are represented by circles and squares. The vertices represented by circles correspond to the controllable states of the dynamical system, and the vertices represented by squares correspond to uncontrollable states.

So, $X = \{1, 2, 3, 4, 5, 6\}$; $X_C = \{1, 5\}$; $X_N = \{2, 3, 4, 6\}$;

$$E = \{(1,2),\ (1,4),\ (2,1),\ (2,3),\ (2,5),\ (3,3),\ (4,4),\ (4,5),\ (5,5),\ (5,4),\ (6,3),\ (6,5)\};$$

$$E_C = \{(1,2),\ (1,4),\ (5,5),\ (5,4)\};$$

$$E_N = \{(2,1),\ (2,3),\ (2,5),\ (3,3),\ (4,4),\ (4,5),\ (6,3),\ (6,5)\}.$$

The values of the cost function $c : E \to \mathbb{R}$ and of the transition probability function $p : E \to \mathbb{R}$ are written next to the edges in the graph. For the edges $e = (x, y) \in E_N$, these values are written in parentheses, where the first quantity expresses the cost, and the second one represents the probability transition from state x to state y. For the edges $e = (x, y) \in E_C$, only the costs are given, which are written also next to the edges. Thus, for this example, we obtain

$$c_{1,2} = 4, \quad c_{1,4} = 1, \quad c_{2,1} = 1, \quad c_{2,3} = 3, \quad c_{2,5} = 2, \quad c_{3,3} = 1,$$

$$c_{4,4} = 4, \quad c_{4,5} = 4, \quad c_{5,5} = 2, \quad c_{5,4} = 3, \quad c_{6,3} = 8, \quad c_{6,5} = 3;$$

$$p_{2,1} = 0.4, \quad p_{2,3} = 0.4, \quad p_{2,5} = 0.2, \quad p_{3,3} = 1, \quad p_{4,4} = 0.5,$$

$$p_{4,5} = 0.5, \quad p_{6,3} = 0.6, \quad p_{6,5} = 0.4.$$

After applying Algorithm 2, we solve the linear programming problem:
Maximize

$$\phi(\varepsilon, \omega) = \theta_1\omega_1 + \theta_2\omega_2 + \theta_3\omega_3 + \theta_4\omega_4 + \theta_5\omega_5 + \theta_6\omega_6$$

subject to

$$\begin{cases} \varepsilon_1 - \varepsilon_2 + \omega_1 \leq c_{1,2}; \\ \varepsilon_1 - \varepsilon_4 + \omega_1 \leq c_{1,4}; \\ \varepsilon_5 - \varepsilon_4 + \omega_5 \leq c_{5,4}; \\ \varepsilon_5 - \varepsilon_5 + \omega_5 \leq c_{5,5}; \\ \varepsilon_2 - (p_{2,1}\varepsilon_1 + p_{2,3}\varepsilon_3 + p_{2,5}\varepsilon_5) + \omega_2 \leq \mu_2; \\ \varepsilon_3 - p_{3,3}\varepsilon_3 + \omega_3 \leq \mu_3; \\ \varepsilon_4 - (p_{4,4}\varepsilon_4 + p_{4,5}\varepsilon_5) + \omega_4 \leq \mu_4; \\ \varepsilon_6 - (p_{6,3}\varepsilon_3 + p_{6,5}\varepsilon_5) + \omega_6 \leq \mu_6; \\ \omega_1 - \omega_2 \leq 0, \quad \omega_1 - \omega_4 \leq 0; \\ \omega_5 - \omega_5 \leq 0, \quad \omega_5 - \omega_4 \leq 0; \\ \omega_2 - (p_{2,1}\omega_1 + p_{2,3}\omega_3 + p_{2,5}\omega_5) \leq 0; \\ \omega_3 - p_{3,3}\omega_3 \leq 0; \\ \omega_4 - (p_{4,4}\omega_4 + p_{4,5}\omega_5) \leq 0; \\ \omega_6 - (p_{6,3}\omega_3 + p_{6,5}\omega_5) \leq 0. \end{cases}$$

Here

$$\theta_1 = \theta_2 = \theta_3 = \theta_4 = \theta_5 = \theta_6 = \frac{1}{6}$$

and

$$\mu_2 = 2,\ \mu_3 = 1,\ \mu_4 = 4,\ \mu_6 = 6.$$

If we introduce these data into the linear programming model above, then we obtain the problem:
Maximize

$$\phi(\varepsilon, \omega) = \frac{1}{6}\omega_1 + \frac{1}{6}\omega_2 + \frac{1}{6}\omega_3 + \frac{1}{6}\omega_4 + \frac{1}{6}\omega_5 + \frac{1}{6}\omega_6$$

subject to

$$\begin{cases} \varepsilon_1 - \varepsilon_2 + \omega_1 \le 4; \\ \varepsilon_1 - \varepsilon_4 + \omega_1 \le 1; \\ \varepsilon_5 - \varepsilon_4 + \omega_5 \le 3; \\ \qquad\qquad \omega_5 \le 2; \\ \varepsilon_2 - 0.4\varepsilon_1 - 0.4\varepsilon_3 - 0.2\varepsilon_5 + \omega_2 \le 2; \\ \qquad\qquad \omega_3 \le 1; \\ \varepsilon_4 - 0.5\varepsilon_4 - 0.5\varepsilon_5 + \omega_4 \le 4; \\ \varepsilon_6 - 0.6\varepsilon_3 - 0.4\varepsilon_5 + \omega_6 \le 6; \\ \omega_1 - \omega_2 \le 0, \quad \omega_1 - \omega_4 \le 0, \quad \omega_5 - \omega_4 \le 0; \\ \omega_2 - 0.4\omega_1 - 0.4\omega_3 - 0.2\omega_5 \le 0; \\ \omega_4 - 0.5\omega_4 - 0.5\omega_5 \le 0; \\ \omega_6 - 0.6\omega_3 - 0.4\omega_5 \le 0. \end{cases}$$

The optimal solution to this problem that satisfies the conditions of Theorem 2.29 is

$$\varepsilon_1^* = 0,\ \varepsilon_2^* = -\frac{8}{3},\ \varepsilon_3^* = -\frac{25}{3},\ \varepsilon_4^* = 4,\ \varepsilon_5^* = 0,\ \varepsilon_6^* = -\frac{2}{5};$$
$$\omega_1^* = \frac{4}{3},\ \omega_2^* = \frac{4}{3},\ \omega_3^* = 1,\ \omega_4^* = 2,\ \omega_5^* = 2,\ \omega_6^* = \frac{7}{5}.$$

If we determine the potential transformation

$$\overline{c}_{x,y} = c_{x,y} + \varepsilon_y^* - \varepsilon_x^* - \omega_x^*, \quad \forall (x, y) \in E,$$

then we obtain

$$\overline{c}_{1,2}=0,\ \ \overline{c}_{1,4}=\frac{11}{3},\ \ \overline{c}_{2,1}=\frac{7}{3},\ \ \overline{c}_{2,3}=-4,\ \ \overline{c}_{2,5}=\frac{10}{3},$$

$$\overline{c}_{3,3}=0,\ \ \overline{c}_{4,5}=-2,\ \ \overline{c}_{4,4}=2,\ \ \overline{c}_{5,4}=5,\ \ \overline{c}_{5,5}=0,\ \ \overline{c}_{6,3}=-\frac{4}{3},\ \ \overline{c}_{6,5}=2;$$

$$\overline{\mu}_2=0,\ \ \overline{\mu}_3=0,\ \ \overline{\mu}_4=0,\ \ \overline{\mu}_6=0.$$

The network $(G, X_C, X_N, \overline{c}, p)$ in canonical form is represented by Fig. 2.4. This network satisfies the conditions:

$$(1)\ \min\{\overline{c}_{1,1},\overline{c}_{1,4}\}=0,\ \min\{\overline{c}_{5,5},\overline{c}_{5,4}\}=0;$$

$$(2)\ \overline{\mu}_2=0,\ \overline{\mu}_3=0,\ \overline{\mu}_4=0,\ \overline{\mu}_6=0.$$

For a given optimal solution $\varepsilon_x^*, \omega_x^*$ for $x \in X$, we have $E^*_{h^*}(1)=E_{\overline{c}}(1)=\{(1,2)\}$ and $E^*_{h^*}(5)=E_{\overline{c}}(5)=\{(5,5)\}$.

Therefore, if we fix $s^*(1)=2$; $s^*(5)=5$, then we obtain the optimal stationary strategy $s^*: 1 \to 2;\ 5 \to 5$. The corresponding network induced by the optimal stationary strategy s^* is represented by Fig. 2.5.

Another optimal solution to the linear programming problem for this example is

$$\varepsilon_1^*=0,\ \varepsilon_2^*=-\frac{8}{3},\ \ \varepsilon_3^*=-\frac{13}{2},\ \varepsilon_4^*=\frac{1}{3},\ \varepsilon_5^*=-\frac{11}{3},\ \varepsilon_6^*=-\frac{7}{15};$$

$$\omega_1^*=\frac{4}{3},\ \ \omega_2^*=\frac{4}{3},\ \ \omega_3^*=1,\ \ \omega_4^*=2,\ \ \omega_5^*=2,\ \ \omega_6^*=\frac{7}{5}.$$

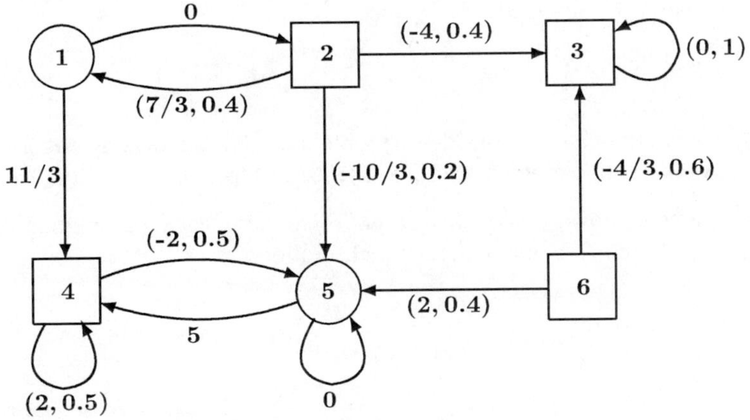

Fig. 2.4 Network $(G, X_C, X_N, \overline{c}, p)$ in canonical form

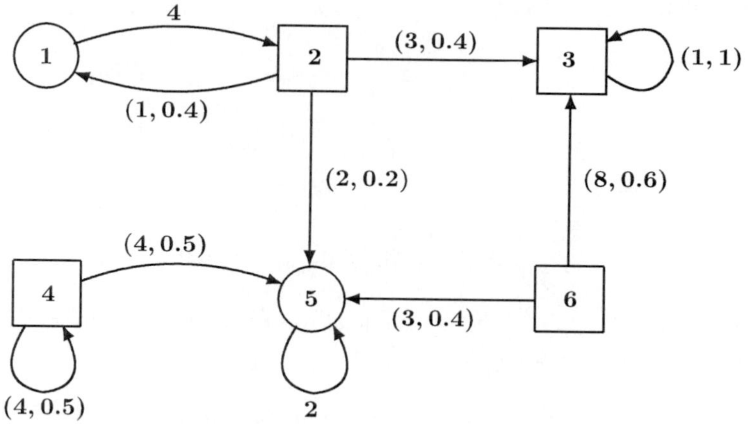

Fig. 2.5 Network determining optimal strategies

If we calculate $\overline{c}_{x,y}$ and $\overline{\mu}_x$ that correspond to this optimal solution, then we obtain

$$\overline{c}_{1,2} = 0,\ \overline{c}_{1,4} = 0,\ \overline{c}_{5,4} = 3,\ \overline{c}_{5,5} = 0,\ \overline{\mu}_2 = 0,\ \overline{\mu}_3 = 0,\ \overline{\mu}_4 = 0,\ \overline{\mu}_6 = 0.$$

It is easy to observe that in this case, $E^*_{\overline{c}}(x) \neq E^*_{h^*}(x)$ for $x = 1$. However, we can determine the optimal solution $s^*(1) = 2$, $s^*(5) = 5$ if we fix the strategy s^* such that $(x, s^*(x)) \in E^*_{\overline{c}}(x) \cap E^*_{h^*}(x)$ for $x = 1$ and $x = 2$, i.e., we obtain the same optimal stationary strategy as in the previous case.

Remark 2.30 If it is necessary for a multichain control problem to determine the optimal stationary strategy s^* only for a fixed starting state x_0, then it is sufficient to solve the linear programming problem:
Maximize

$$\phi(\varepsilon, \omega) = \omega_{x_0} \tag{2.140}$$

subject to (2.129).
The optimal strategy for the considered problem can be found using Algorithm 2 if in step 1) we exchange problem (2.129), (2.130) with problem (2.130), (2.140).

If in the example above, we fix $x_0 = 1$ and solve the linear programming problem (2.130), (2.140), then we obtain the optimal solution ε^*, ω^*, where

$$\varepsilon_1^* = 0,\ \ \varepsilon_2^* = -\frac{8}{3},\ \ \varepsilon_3^* = -\frac{25}{3},\ \ \varepsilon_4^* = 4,\ \ \varepsilon_5^* = 0;$$

$$\omega_1^* = \frac{4}{3},\ \ \omega_2^* = \frac{4}{3},\ \ \omega_3^* = 1,\ \ \omega_4^* = 2,\ \ \omega_5^* = 2,$$

and ε_6^*, ω_6^* are arbitrary values that satisfy the conditions

$$\varepsilon_3^* - 0.6\varepsilon_3^* - 0.4\varepsilon_5^* + \omega_6^* \le 6, \quad \omega_6^* - 0.6\omega_3^* - 0.4\omega_5^* \le 0.$$

Here, ε_6^* may differ from $-2/5$ and ω_6^* may differ from $7/5$. In this case, we obtain the same optimal strategy $s^* : 1 \to 2;\ 5 \to 5$, but we do not obtain ε_6^* and ω_6^*. If we solve the problem (2.130), (2.140) for $x_0 = 6$, then we obtain

$$\varepsilon_6^* = -\frac{25}{3},\ \omega_6^* = \frac{7}{5},\ \varepsilon_3^* = 0,\ \omega_3^* = 1,\ \varepsilon_5^* = 0,\ \omega_5^* = 2.$$

The remaining variables may be arbitrary.

Lemma 2.31 *Let $\varepsilon : X \to R^1$ be an arbitrary real function on X and h be an arbitrary real value. If the cost function c in the average control problem on the network (G, X_C, X_N, c, p) is changed to a cost function $\overline{c}$, where $\overline{c}_{x,y} = c_{x,y} + \varepsilon_y - \varepsilon_x - h$ for $(x, y) \in E$, then the optimal stationary control for the average control problem on the network $(G, X_C, X_N, \overline{c}, p)$ is the same as for the problem on the network (G, X_C, X_N, c, p), i.e., the optimal solutions to an average control problem on the network (G, X_C, X_N, c, p) are invariant with respect to the potential transformation $\overline{c}_{x,y} = c_{x,y} + \varepsilon_y - \varepsilon_x - h, \forall (x, y) \in E$.*

2.6.6 An Iterative Algorithm Based on a Unichain Model

For a multichain control problem, the optimal average costs in different states may be different. Therefore, the set of states X can be divided into several subsets $X_1, X_2, \dots, X_k$ such that each subset $X_i,\ i \in \{1, 2, \dots, k\}$ contains the states with the same optimal average costs and there are no states from different subsets with the same optimal average costs.

Let ω^i be the corresponding optimal average cost of the states $x \in X_i$, $i = 1, 2, \dots, k$ and assume that $\omega^1 < \omega^2 < \cdots < \omega^k$. In this section, we show that the average costs ω^i and the corresponding subsets X_i can be found successively by solving kn linear programming problems like (2.117), (2.118).

In the first step of the algorithm, we solve the linear programming problem:

Maximize

$$\phi(\varepsilon, h) = h \tag{2.141}$$

subject to

$$\begin{cases} \varepsilon_x - \varepsilon_y + h \le c_{x,y}, \quad \forall x \in X_C,\ y \in X(x); \\ \varepsilon_x - \sum_{z \in X} p_{x,z}\varepsilon_z + h \le \mu_x, \quad \forall x \in X_N. \end{cases} \tag{2.142}$$

Let ε_x^1 $(x \in X)$, h^1 be an optimal solution to this problem on the network (G, X_C, X_N, c, p). Then this solution satisfies the conditions:

(1) $c_{x,y}^1 = c_{x,y} + \varepsilon_y^1 - \varepsilon_x^1 - h^1 \geq 0, \quad \forall x \in X_C,\ y \in X(x).$

(2) $\mu_x^1 = \mu_x + \sum_{y \in X(x)} p_{x,y}\varepsilon_y^1 - \varepsilon_x^1 - h^1 \geq 0, \quad \forall x \in X_N.$

(3) There exists a non-empty subset X_1 from X, where
$\min_{y \in X(x)} c_{x,y}^1 = \min_{y \in X_1(x)} c_{x,y}^1 = 0,\ \forall x \in X_1 \cap X_C;$
$\mu_x^1 = 0,\ \forall x \in X_1 \cap X_N,$
and X_1 is a maximal subset in X with such a property.

If on the network (G, X_C, X_N, c, p) we make the potential transformation

$$c_{x,y}^1 = c_{x,y} + \varepsilon_y^1 - \varepsilon_x^1 - h^1, \qquad \forall x \in X,\ y \in X(x),$$

then we obtain the network (G, X_C, X_N, c^1, p) with a new cost function c^1 on E. According to Lemma 2.31, the optimal stationary strategies of the control problem on this network are the same as the optimal stationary strategies on the network (G, X_C, X_N, c, p). Moreover, here we have

$$\overline{\omega}_x^1 = \omega_x - h^1, \quad \forall x \in X,$$

where ω_x for $x \in X$ represents the corresponding optimal average costs of the states $x \in X$ in the primal problem and $\overline{\omega}_x^1$ are the optimal average costs of the states in the control problem on the network with transformation potential function c^1.

Thus, after the first step of the algorithm, we obtain the subset X_1, the value of the optimal average cost $\omega^1 = h^1$ for the states $x \in X_1$, the function $\varepsilon^1 : X \to \mathbb{R}$, and the network (G, X_C, X_N, c^1, p) with a new cost function c^1, where the optimal average costs $\overline{\omega}_x^1$ in the problem with the new network satisfy the condition:

$$\overline{\omega}_x^1 = 0,\ \forall x \in X_1; \quad \overline{\omega}_x^1 = \omega_x - h^1 > 0,\ \forall x \in X \setminus X_1.$$

At the second step of the algorithm, we solve the linear programming problem: *Minimize* the objective function (2.141)
subject to

$$\begin{cases} \varepsilon_x - \varepsilon_y + h \leq c_{x,y}^1, & \forall x \in X_C \setminus X_1,\ y \in X(x); \\ \varepsilon_x - \sum_{z \in X} p_{x,z}\varepsilon_z + h \leq \mu_x^1, & \forall x \in X_N \setminus X_1; \\ \varepsilon_x - \varepsilon_y \leq c_{x,y}^1, & \forall x \in X_1 \cap X_C,\ y \in X(x); \\ \varepsilon_x - \sum_{z \in X)} p_{x,z}\varepsilon_z \leq \mu_x^1, & \forall x \in X_1 \cap X_N. \end{cases} \tag{2.143}$$

This system is obtained from (2.130) by changing $c_{x,y}$ and μ_x to $c^1_{x,y}$ and μ^1_x, respectively, and by taking $h = 0$ for the inequalities that correspond to the states $x \in X_1 \cup X$.

Let ε^2_x ($x \in X_1$), h^2 be an optimal solution to this problem on the network (G, X_C, X_N, c^1, p). Then this solution satisfies the conditions:

(1) $c^2_{x,y} = c_{x,y} + \varepsilon^2_y - \varepsilon^2_x - h^2 \geq 0, \qquad \forall x \in X_C,\ y \in X(x).$

(2) $\mu^2_x = \mu_x + \sum\limits_{y \in X(x)} p_{x,y}\varepsilon^2_y - \varepsilon^2_x - h^2 \geq 0, \qquad \forall x \in X_N.$

(3) There exists a non-empty subset X_2 from X, where
$\min\limits_{y \in X(x)} c^2_{x,y} = \min\limits_{y \in X_2(x)} c^2_{x,y} = 0,\ \forall x \in X_2 \cap X_C;$
$\mu^2_x = 0,\ \forall x \in X^2 \cap X_N,$
and X_2 is a maximal subset in X with such a property.

After that, we make the potential transformation

$$c^2_{x,y} = c^1_{x,y} + \varepsilon^2_y - \varepsilon^2_x - h^2, \qquad \forall x \in X,\ y \in X(x)$$

on the network (G, X_C, X_N, c^1, p), and we obtain the network (G, X_C, X_N, c^2, p) with a new cost function c^2 on E. According to Lemma 2.31, the optimal stationary strategies of the control problem on this network are the same as the optimal stationary strategies on the network (G, X_C, X_N, c^1, p). Moreover, here we have

$$\overline{\omega}^2_x = \overline{\omega}^1_x - h^2, \quad \forall x \in X \setminus X_1,$$

where $\overline{\omega}^1_x$ for $x \in X \setminus X_1$ represent the corresponding optimal average costs of the states in the problem before the potential transformation is made and $\overline{\omega}^2_x$ are the optimal average costs of the states $x \in X \setminus X_1$ in the control problem after the potential transformation is made.

Thus, after the second step of the algorithm, we obtain the subset X_2, the value of the optimal average cost h^2 for the states $x \in X_2$, the function $\varepsilon^2 : X \to \mathbb{R}$, and the network (G, X_C, X_N, c^2, p) with a new cost function c^2, where for the optimal average costs $\overline{\omega}^2_x$ in the problem, we may set

$$\overline{\omega}^2_x = 0,\ \forall x \in X_1 \cup X_2; \quad \overline{\omega}^2_x = \overline{\omega}^1_x - h^2 > 0,\ \forall x \in X \setminus (X_1 \cup X_2).$$

In the next step of the algorithm, we solve the linear programming problem:

Minimize the objective function (2.141)
subject to

$$\begin{cases} \varepsilon_x - \varepsilon_y + h \le c^2_{x,y}, & \forall x \in X_C \setminus (X_1 \cup X_2),\ y \in X(x); \\ \varepsilon_x - \sum\limits_{z\in X} p_{x,z}\varepsilon_z + h \le \mu^2_x, & \forall x \in X_C \setminus (X_1 \cup X_2); \\ \varepsilon_x - \varepsilon_y \le c^2_{x,y}, & \forall x \in (X_1 \cup X_2) \cap X_C; \\ \varepsilon_x - \sum\limits_{z\in X} p_{x,z}\varepsilon_z \le \mu^2_x, & \forall x \in (X_1 \cup X_2) \cap X_N. \end{cases} \tag{2.144}$$

This system is obtained from (2.143) by changing $c^1_{x,y}$ and μ^1_x to $c^2_{x,y}$ and μ^2_x, respectively, and by taking $h = 0$ in the inequalities that correspond to the states $x \in (X_1 \cup X_2) \cap X$.

After a finite number of steps, we obtain the subsets

$$X_1, X_2, \ldots, X_k \ \ (X = X_1 \cup X_2 \cup \cdots \cup X_k),$$

the potential functions $\varepsilon^i : X \to \mathbb{R},\ i = 1, 2, \ldots, k$, and the values $h^1, h^2, \ldots, h^k$, where

$$\omega^i = \sum_{j=1}^{i} h^j, \ \ j = 1, 2, \ldots, k.$$

If we find $\varepsilon^*_x = \sum_{i=1}^{k} \varepsilon^i_x$ and fix $\omega^*_x = \omega^{i^*}$ for $x \in X_i$, then we determine the potential transformation

$$\overline{c}_{x,y} = c_{x,y} + \varepsilon^*_y - \varepsilon^*_x - \omega^*_x,\ \forall x \in X,\ y \in X(x),$$

that satisfies the conditions (1)–(6) of Theorem 2.29. This means that we determine the network $(G, X_C, X_N, \overline{c}, p)$ and the optimal stationary strategy s^*.

Example Consider the stochastic control problem on the network with the data from the example given in the previous section. The network is represented by Fig. 2.3, where $X = X_C \cup X_N$, $X_C = \{1, 5\}$, $X_N = \{2, 3, 4, 6\}$, and the costs and transition probabilities are written again along the edges.

We apply the algorithm described above. In the first step of the algorithm, we solve the linear programming problem:

Minimize

$$\phi(\varepsilon, h) = h$$

subject to

$$\begin{cases} \varepsilon_1 - \varepsilon_2 + h \le 4; \\ \varepsilon_1 - \varepsilon_4 + h \le 1; \\ \varepsilon_5 - \varepsilon_4 + h \le 3; \\ \varepsilon_5 - \varepsilon_5 + h \le 2; \\ \varepsilon_2 - 0.4\varepsilon_1 - 0.4\varepsilon_3 - 0.2\varepsilon_5 + h \le 2; \\ \varepsilon_3 - \varepsilon_3 + h \le 1; \\ \varepsilon_4 - 0.5\varepsilon_4 - 0.5\varepsilon_5 + h \le 4; \\ \varepsilon_6 - 0.6\varepsilon_3 - 0.4\varepsilon_5 + h \le 6. \end{cases}$$

An optimal solution to this problem is $h^1 = 1$, $\varepsilon_1^1 = 0$, $\varepsilon_2^1 = 0$, $\varepsilon_3^1 = 0$, $\varepsilon_4^1 = 0$, $\varepsilon_5^1 = 0$, $\varepsilon_6^1 = 0 = 1$.

We calculate $c_{x,y}^1$ and μ_x^1, using the formula

$$c_{x,y}^1 = c_{x,y} + \varepsilon_y^1 - \varepsilon_x^1 - h^1, \ \forall x \in X_1, \ y \in X(x);$$

$$\mu_x^1 = \mu_x + \sum_{y \in X(x)} p_{x,y} c_{x,y} - \varepsilon_x^1 - h^1, \ x \in X_2$$

and determine $c_{1,2}^1 = 3$, $c_{1,4}^1 = 0$, $c_{5,4}^1 = 2$, $c_{5,5}^1 = 1$; $\mu_2^1 = 1$, $\mu_3^1 = 0$, $\mu_4^1 = 3$, $\mu_6^1 = 5$. After the first step of the algorithm, we obtain

$$X_1 = \{3\}; \ h^1 = 1; \ \varepsilon_1^1 = 0, \ \varepsilon_2^1 = 0, \ \varepsilon_3^1 = 0, \ \varepsilon_4^1 = 0, \ \varepsilon_5^1 = 0, \ \varepsilon_6^1 = 0.$$

In the second step of the algorithm, we solve the linear programming problem:

Minimize

$$\overline{\psi}'(\varepsilon, h) = h$$

subject to

$$\begin{cases} \varepsilon_1 - \varepsilon_2 + h \le 3; \\ \varepsilon_1 - \varepsilon_4 + h \le 0; \\ \varepsilon_5 - \varepsilon_4 + h \le 2; \\ \varepsilon_5 - \varepsilon_5 + h \le 1; \\ \varepsilon_3 - \varepsilon_3 \le 0; \\ \varepsilon_2 - 0.4\varepsilon_1 - 0.4\varepsilon_3 - 0.2\varepsilon_5 + h \le 1; \\ \varepsilon_4 - 0.5\varepsilon_4 - 0.5\varepsilon_5 + h \le 3; \\ \varepsilon_6 - 0.6\varepsilon_3 - 0.4\varepsilon_5 + h \le 5. \end{cases}$$

An optimal solution to this problem is

$$h^2 = \frac{1}{3},\ \varepsilon_1^2 = 0,\ \varepsilon_2^2 = -\frac{8}{3},\ \varepsilon_3^2 = -\frac{25}{3},\ \varepsilon_4^2 = 4,\ \varepsilon_5^2 = 0,\ \varepsilon_6^2 = -\frac{2}{5}.$$

We calculate $c_{x,y}^2$ and μ_x^2, using formula

$$c_{x,y}^2 = c_{x,y}^1 + \varepsilon_y^2 - \varepsilon_x^2 - h^2;\ \ \mu_x^2 = \mu_x^1 + \sum_{z \in X} p_{x,y}\varepsilon_z - h^2$$

and find

$$c_{1,2}^2 = 0,\ c_{1,4}^2 = \frac{2}{3},\ c_{5,4}^2 = \frac{17}{3},\ c_{5,5}^2 = \frac{2}{3},\ \mu_2^2 = 0,\ \mu_3^2 = 0,\ \mu_4^2 = \frac{2}{3},\ \mu_6^2 = \frac{1}{15}.$$

After the second step of the algorithm, we obtain $X_2 = \{1, 2\}$;

$$h^2 = \frac{1}{3},\ \ \varepsilon_1^2 = 0,\ \ \varepsilon_2^2 = -\frac{8}{3},\ \ \varepsilon_3^2 = -\frac{25}{3},\ \ \varepsilon_4^2 = 4,\ \ \varepsilon_5^2 = 0,\ \ \varepsilon_6^2 = -\frac{2}{5}.$$

In the third step of the algorithm, we solve the linear programming problem:

Minimize

$$\phi(\varepsilon, h) = h$$

subject to

$$\begin{cases} \varepsilon_1 - \varepsilon_2 \leq 0; \\ \varepsilon_1 - \varepsilon_4 \leq \frac{11}{3}; \\ \varepsilon_5 - \varepsilon_4 + h \leq \frac{17}{3}; \\ \varepsilon_3 - \varepsilon_3 \leq 0; \\ \varepsilon_5 - \varepsilon_5 + h \leq \frac{2}{3}; \\ \varepsilon_2 - 0.4\varepsilon_1 - 0.4\varepsilon_3 - 0.2\varepsilon_5 \leq \frac{2}{3}; \\ \varepsilon_4 - 0.5\varepsilon_4 - 0.5\varepsilon_5 + h \leq \frac{2}{3}; \\ \varepsilon_6 - 0.6\varepsilon_3 - 0.4\varepsilon_5 + h \leq \frac{1}{15}. \end{cases}$$

An optimal solution to this problem is

$$h^3 = \frac{1}{15}, \; \varepsilon_1^3 = 0, \; \varepsilon_2^3 = 0, \; \varepsilon_3^3 = 0, \; \varepsilon_4^3 = 0, \; \varepsilon_5^3 = 0, \; \varepsilon_6 = 0.$$

Using this solution, we find

$$c_{1,2}^3 = 0, \; c_{1,4}^4 = \frac{11}{3}, \; c_{5,5}^3 = \frac{3}{5}, \; c_{5,4}^3 = \frac{26}{5}, \; \mu_2^3 = 0, \; \mu_3^3 = 0, \; \mu_4^3 = \frac{3}{5}, \; \mu_6^3 = 0.$$

After this step, we obtain

$$X_3 = \{6\}; \; h^3 = \frac{1}{15}, \; \varepsilon_1^3 = 0, \; \varepsilon_2^3 = 0, \; \varepsilon_3^3 = 0, \; \varepsilon_4^3 = 0, \; \varepsilon_5^3 = 0, \; \varepsilon_6 = 0.$$

In the fourth step of the algorithm, we solve the linear programming problem:

Minimize

$$\overline{\psi}'(\varepsilon, h) = h$$

subject to

$$\begin{cases} \varepsilon_1 - \varepsilon_2 \le 0; \\ \varepsilon_1 - \varepsilon_4 \le \frac{11}{3}; \\ \varepsilon_5 - \varepsilon_4 + h \le \frac{28}{5}; \\ \varepsilon_3 - \varepsilon_3 \le 0; \\ \varepsilon_5 - \varepsilon_5 + h \le \frac{3}{5}; \\ \varepsilon_2 - 0.4\varepsilon_1 - 0.4\varepsilon_3 - 0.2\varepsilon_5 \le \frac{2}{3}; \\ \varepsilon_4 - 0.5\varepsilon_4 - 0.5\varepsilon_5 + h \le \frac{3}{5}; \\ \varepsilon_6 - 0.6\varepsilon_3 - 0.4\varepsilon_5 \le 0. \end{cases}$$

An optimal solution to this system is $h^4 = 3/5$, $\varepsilon_1^3 = 0$, $\varepsilon_2^3 = 0$, $\varepsilon_3^3 = 0$, $\varepsilon_4^3 = 0$, $\varepsilon_5^3 = 0$, $\varepsilon_6 = 0$. Using this solution, we find $c_{1,2}^4 = 0$, $c_{2,4}^4 = 11/3$, $c_{5,4}^4 = 5$, $c_{5,5}^4 = 0$, $\mu_2^4 = 0$, $\mu_3^4 = 0$, $\mu_4^4 = 0$, $\mu_6^4 = 0$. After this step, we obtain $X_4 = \{4, 5\}$ and $h^4 = 3/5$. Thus, finally we have $X = X_1 \cup X_2 \cup X_3 \cup X_4$, where

$$X_1 = \{3\},\ X_2 = \{1, 2\},\ X_3 = \{6\},\ X_4 = \{4, 5\},$$

and

$$\omega^1 = h^1,\ \omega^2 = h^1 + h^2,\ \omega 3 = h^1 + h^2 + h^3,\ \omega^4 = h^1 + h^2 + h^3 + h^4,$$

i.e.,

$$\omega^1 = 1,\ \omega^2 = \frac{4}{3},\ \omega^3 = \frac{7}{5},\ \omega^4 = 2.$$

In addition, we can find

$$\begin{aligned} \varepsilon_1^* &= \varepsilon_1^1 + \varepsilon_1^2 + \varepsilon_1^3 + \varepsilon_1^4 = 0; \quad & \varepsilon_2^* &= \varepsilon_2^1 + \varepsilon_2^2 + \varepsilon_2^3 + \varepsilon_2^4 = -\frac{8}{3}; \\ \varepsilon_3^* &= \varepsilon_3^1 + \varepsilon_3^2 + \varepsilon_3^3 + \varepsilon_3^4 = -\frac{25}{3}; \quad & \varepsilon_4^* &= \varepsilon_4^1 + \varepsilon_4^2 + \varepsilon_4^3 + \varepsilon_4^4 = 4; \\ \varepsilon_5^* &= \varepsilon_5^1 + \varepsilon_5^2 + \varepsilon_5^3 + \varepsilon_5^4 = 0; \quad & \varepsilon_6^* &= \varepsilon_6^1 + \varepsilon_6^2 + \varepsilon_6^3 + \varepsilon_6^4 = -\frac{2}{5}. \end{aligned}$$

If we make the potential transformation of the cost function c for ω^* and ε^* found above, then we obtain the network in canonical form $(G, X_C, X_N, \overline{c}, p)$, represented by Fig. 2.4, which provides the optimal stationary strategies.

2.6.7 Markov Decision Problems vs. Control on Networks

In Sect. 2.6.2, it was shown that an average control problem on the network (G, X_C, X_N, c, p) can be represented as an average Markov decision problem. Here, we show that an arbitrary average Markov decision problem can be reduced to an average stochastic control problem on the network with special structure and such a reduction procedure can be made in polynomial time. So, we show that these problems are equivalents in the context of computational complexity theory. To prove this, we show how to construct the decision network (G', X'_C, X'_N, p', c') for an average control problem determined by the tuple $(X, A, \mathbf{r}, p)$, where X is the set of states of the decision problem, $A = \cup_{x\in X} A(x)$ is the set of actions, $\mathbf{r} = \{r_{x,a} | x \in X, a \in A(x)\}$ is the set of rewards, and p is the probability function of the Markov decision problem.

For the considered Markov decision problem, we formulate an average stochastic control problem on the network (G', X'_C, X'_N, p', c'), where $G' = (X', E')$, X'_C, X'_N, p' and c' are defined as follows:

The set of vertices X'_C consists of the same number of states as the set of states of the average Markov decision problem. The set of uncontrollable states X'_N of the auxiliary average control problems corresponds to the set of action $A = \cup_{x\in X} A(x)$, i.e., $|X_N| = |A|$.

So,

$$X'_C = \{x' = x \mid x \in X\}; \quad X'_N = \{(x, a) \mid x \in X, a \in A(x)\}.$$

The set of directed edges E' we represent as a couple of two disjoint subsets $E' = E'_C \cup E'_N$, where E'_C is the set of outgoing edges from $x' \in X'_C$ and E'_N is the set of outgoing edges from $x^a \in X'_N$. The sets E'_C and E'_N are defined as follows:

$$E'_C = \{(x, (x, a)) \mid x \in X'_C; \ (x, a) \in X'_N, \ a \in A\};$$
$$E'_N = \{((x, a), y) \mid (x, a) \in X'_N, \ y \in X'_C, \ p^a_{x,y} > 0, \ a \in A\}.$$

On the set of directed edges E', we define the cost function $c' : E' \to \mathbb{R}$, where

$$c'_{e'} = 0, \ \ \forall e' = (x, (x, a)) \in E'_C;$$

$$c'_{e'} = 2\, r_{x,a} \quad \text{for} \quad e' = ((x, a), y) \in E'_N \quad (x, y \in X, \ a \in A).$$

On E'_N we define the transition probability function $p' : E'_N \to [0, 1]$, where $p'_{e'} = p^a_{x,y}$ for $e' = ((x, a), y) \in E'_N$. The directed edges $e' = (x, (x, a))$ originate in controllable states of G'.

It is easy to observe that between the set of stationary strategies $\mathbb{S}$ in the Markov decision process and the set of strategies $\mathbb{S}'$ in the control problem on the network $(G', X'_C, X'_N, c', p', x_0)$, there exists a bijective mapping that preserves the average cost per transition. Therefore, if we find the optimal stationary strategy for the control problem on the network, then we can determine the optimal stationary strategy in the Markov decision process.

The network constructed above gives a graphical interpretation of the Markov decision process via the structure of graph G, where the actions and all possible transitions for an arbitrary fixed action are represented by arcs and nodes. A more simple graphical interpretation of the Markov decision process may be given using the graph of probability transitions $G_p = (X, E_p)$, which is induced by the probability function $p : X \times X \times A \to [0, 1]$. This graph may contain parallel directed edges, where each directed edge corresponds to an action. The set of vertices X corresponds to the set of states, and the set of edges E_p consists of $|A|$ subsets $E_p^1, E_p^2, \ldots, E_p^{|A|}$ $\left(E_p = \bigcup_{i=1}^{|A|} E_p^i\right)$, where $E_p^i = \{e^{a_i} = (x, y)^{a_i} \mid p_{x,y}^{a_i} > 0\}$, $i = 1, 2, \ldots, |A(x)|$.

An example of how to construct the graph $G_p = (X, E_p)$ and how to determine the solution to the Markov decision problem using the reduction procedure for an auxiliary control problem on the network is given below.

Example Consider a Markov decision process $(X, A, p, \mathbf{r})$, where $X = \{1, 2\}$, $A = 1, 2$, and the possible values of the corresponding probability and cost functions $p : X \times X \times A \to [0, 1]$, $\mathbf{r} : X \times A \to \mathbb{R}$ are defined as follows:

$$
\begin{aligned}
p_{1,1}^{a_1} &= 0.7, \quad p_{1,2}^{a_1} = 0.3, \quad p_{2,1}^{a_1} = 0.6, \quad p_{2,2}^{a_1} = 0.4, \\
p_{1,1}^{a_2} &= 0.4, \quad p_{1,2}^{a_2} = 0.6, \quad p_{2,1}^{a_2} = 0.5, \quad p_{2,2}^{a_2} = 0.5, \\
r_1^{a_1} &= 0.7, \quad r_1^{a_2} = 2.4, \quad r_2^{a_1} = 0.8, \quad r_2^{a_2} = -0.5.
\end{aligned}
$$

We consider the problem of finding the optimal stationary strategy for the corresponding Markov decision problem with minimal average costs and an arbitrary fixed starting state.

The data concerned with the actions in the considered Markov decision problem can be represented in a suitable form using the probability matrices

$$
P^{a_1} = \begin{pmatrix} 0.7 & 0.3 \\ 0.6 & 0.4 \end{pmatrix}, \quad P^{a_2} = \begin{pmatrix} 0.4 & 0.6 \\ 0.5 & 0.5 \end{pmatrix}
$$

and the reward vectors

$$
\mathbf{r}(a_1) = \begin{pmatrix} 0.7 \\ 0.8 \end{pmatrix}, \quad \mathbf{r}(a_2) = \begin{pmatrix} 2.4 \\ -0.5 \end{pmatrix}.
$$

In Fig. 2.6, this Markov process is represented by the multigraph $G_p = (X, E_p)$ with the set of vertices $X = \{1, 2\}$.

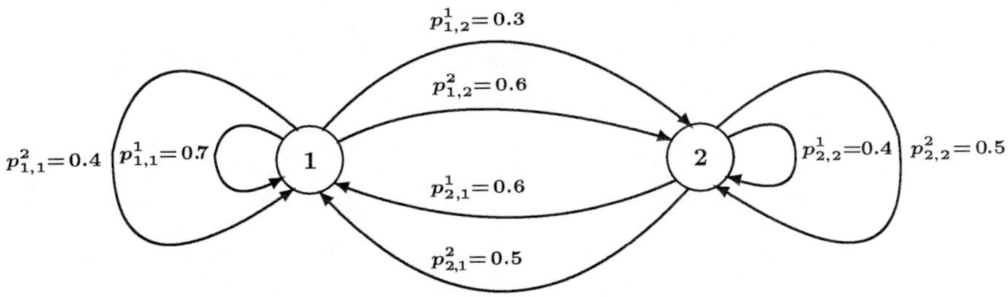

Fig. 2.6 Graph of the Markov decision problem

The set of directed edges E_p contains parallel directed edges that correspond to probability transitions from one state to another for different actions. We call this graph *multigraph of the Markov decision process.*

In Fig. 2.7, graph $G' = (X', E')$ is represented. In G', the sets X'_C, X'_N, E'_C, E'_N are defined as follows:

$$X'_C = \{1, 2\},\ X'_N = X^1 \cup X^2 = \{(1, 1), (1, 2), (2, 1), (2, 2)\},$$

where

$$X^1 = \{(1, 1), (1, 2)\},\ X^2 = \{(2, 1), (2, 2)\}$$

and

$$\begin{aligned} E'_C &= \{(1, (1, 1)),\ (1, (1, 2)),\ (2, (2, 1)),\ (2, (2, 2))\}, \\ E'_N &= \{((1, 1), 1),\ ((1, 1), 2),\ ((2, 1), 1),\ ((2, 2), 1), \\ &\qquad ((1, 2), 1),\ ((1, 2), 2),\ ((2, 1), 2),\ ((2, 2), 2)\}. \end{aligned}$$

The probabilities $p'_e = p'_{(x,a),y} = p^a_{x,y}$ for directed edges $((x, a), y) \in E'_N$ are written along the edges in Fig. 2.7, and the costs of the directed edges from E' are defined in the following way:

$$c'_{1,(1,1)} = c'_{1,(1,2)} = 0, \quad c'_{2,(2,1)} = c'_{2,(2,2)} = 0,$$

$$\begin{aligned} c'_{(1,1),1} = c'_{(1,1),2} &= 1.4, \quad c'_{(1,2),1} = c'_{(1,2),2} = 2.4, \\ c'_{(2,1),1} = c'_{(2,1),2} &= 0.8, \quad c'_{(2,2),1} = c'_{(2,2),2} = -1. \end{aligned}$$

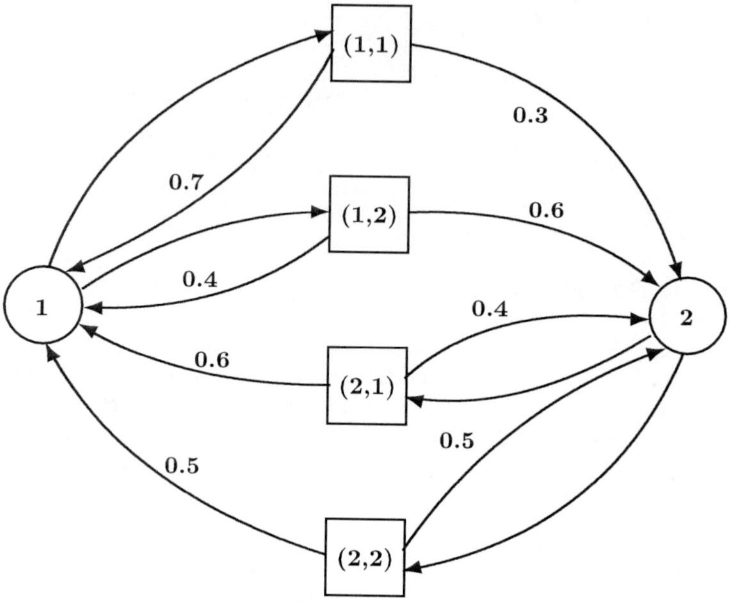

Fig. 2.7 Graph $G' = (X', E')$ of the control problem

The set of possible stationary strategies for this Markov decision process consists of four strategies, i.e., $\mathbb{S} = \{s^1, s^2, s^3, s^4\}$, where

$$\begin{aligned} s^1 &: 1 \to a_1, \quad 2 \to a_1; \\ s^2 &: 1 \to a_1, \quad 2 \to a_2; \\ s^3 &: 1 \to a_2, \quad 2 \to a_1; \\ s^4 &: 1 \to a_2, \quad 2 \to a_2. \end{aligned}$$

A fixed strategy s in the Markov decision process generates a simple Markov process determined by $P(s)$ and $\mathbf{r}(s)$. For example, if we fix strategy s_2, then we obtain a simple Markov process with transition costs generated by the following matrices $P(s_2)$ and $\mathbf{r}(s_2)$:

$$P(s_2) = \begin{pmatrix} 0.7 & 0.3 \\ 0.5 & 0.5 \end{pmatrix}, \quad \mathbf{r}(s) = \begin{pmatrix} 0.7 \\ -0.5 \end{pmatrix}.$$

It is easy to check that this Markov process is ergodic, and the limiting matrix of this process is

$$Q^{s_2} = \begin{pmatrix} \frac{5}{8} & \frac{3}{8} \\ \frac{5}{8} & \frac{3}{8} \end{pmatrix}.$$

In such a way, we determine $f_1(s_2) = f_2(s_2) = 1/4$. Analogously, it can be calculated by $f_1(s_1) = f_2(s_1) = 22/30$, $f_1(s_3) = f_2(s_3) = 16/10$ and $f_1(s_4) = f_2(4) = 9/11$. We can see that the optimal stationary strategy for the Markov decision problem with a minimal average cost criterion is s^2.

This strategy can be found by solving the following linear programming problem (2.132), (2.132) on the auxiliary network (G', X'_C, X'_N, p', c'):

Minimize

$$\varphi(\alpha, q) = 1.4q_{1,1} + 4.8q_{1,2} + 1.6q_{2,1} - q_{2,2}$$

subject to

$$\begin{cases} 0.7q_{1,1} + 0.4q_{1,2} + 0.6q_{2,1} + 0.5q_{2,2} = q_1, \\ 0.3q_{1,1} + 0.6q_{1,2} + 0.4q_{2,1} + 0.5q_{2,2} = q_2, \\ \alpha_{1,(1,1)} = q_{1,1}, \\ \alpha_{1,(1,2)} = q_{1,2}, \\ \alpha_{2,(2,1)} = q_{2,1}, \\ \alpha_{2,(2,2)} = q_{2,2}, \\ \alpha_{1,(1,1)} + \alpha_{1,(1,2)} = q_1, \\ \alpha_{2,(2,1)} + \alpha_{2,(2,2)} = q_2, \\ q_{1,1} + q_{1,2} + q_{2,1} + q_{2,2} + q_1 + q_2 = 1, \\ \alpha_{1,(1,1)}, \alpha_{1,(1,2)}, \alpha_{2,(2,1)}, \alpha_{2,(2,2)} \geq 0, \\ q_{1,1}, q_{1,2}, q_{2,1}, q_{2,2}, q_1, q_2 \geq 0. \end{cases}$$

The optimal solution to this problem is

$$q_1^* = \frac{5}{16}, \quad q_2^* = \frac{3}{16}, \quad q_{1,1}^* = \frac{5}{16}, \quad q_{2,2}^* = \frac{3}{16}, \quad q_{1,2}^* = 0, \quad q_{2,1}^* = 0,$$

$$\alpha_{1,(1,1)}^* = \frac{5}{16}, \quad \alpha_{2,(2,2)}^* = \frac{3}{16}, \quad \alpha_{1,(1,2)}^* = 0, \quad \alpha_{2,(2,1)}^* = 0,$$

and the optimal value of the objective function is $\overline{\psi}(\alpha^*, q^*) = 1/4$.

The optimal strategy s^* on G' we can find using Theorem 2.21, i.e., we fix

$$s_{1,(1,1)}^* = 1, \quad s_{1,(1,2)}^* = 0, \quad s_{2,(2,1)}^* = 0, \quad s_{2,(2,2)}^* = 1.$$

This means that the optimal stationary strategy for the Markov decision problem is

$$s^* : 1 \to a_1, \quad 2 \to a_2$$

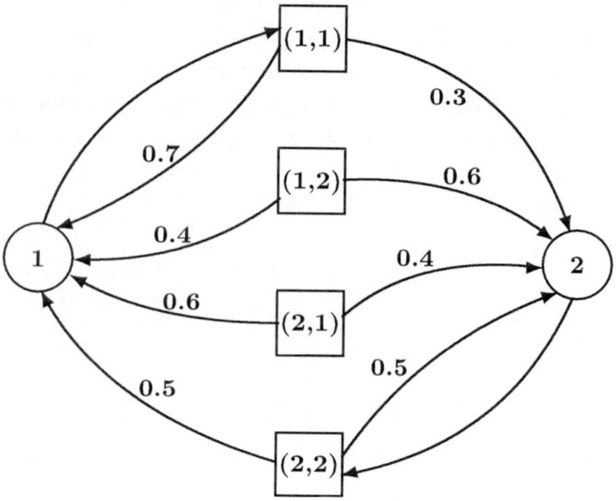

Fig. 2.8 Graph induced by the optimal strategy of the control problem

and the average cost per transaction is $f_1(s^*) = f_2(s^*) = 1/4$.

The auxiliary graph with distinguished optimal strategies in the controllable states $x_1 = 1$ and $x_2 = 2$ is represented in Fig. 2.8. The unique outgoing directed edge $(1, (1, 1))$ from vertex 1 that ends in vertex $(1, 1)$ corresponds to the optimal strategy $1 \to a_1$ in the state $x = 1$, and the unique outgoing directed edge $(2, (2, 2))$ from vertex 2 that ends in vertex $(2, 2)$ corresponds to the optimal strategy $2 \to a_2$ in the state $x = 2$.

In a similar way, a discounted Markov decision problem determined by a tuple $(X, A, \mathbf{r}, p)$ and a given discount factor γ can be transformed into a discounted control problem on the network (G', X'_C, X'_N, p', c') with a discount factor $\gamma' = \gamma^{\frac{1}{2}}$, where G', X'_C, X_N, p' are defined in the same way as for the network in the average case of a control problem, and the cost function c' on E' is defined as follows: $c' : E' \to \mathbb{R}$, where

$$c'_{e'} = 0, \quad \forall e' = (x, (x, a)) \in E'_C;$$

$$c'_{e'} = r_x^a \quad \text{for} \quad e' = ((x, a), y) \in E'_N \quad (x, y \in X,\ a \in A).$$

2.7 Discounted Control Problems on Networks

In this section, we study a class of stochastic control problems on networks with discounted costs. Based on the results from Sect. 2.3, we develop algorithms to determine the optimal stationary controls for the considered class of problems.

2.7.1 Problem Formulation

Let a discrete-time system $\mathbb{L}$ with a finite set of states X be given and assume that the dynamics of the system are described by a directed graph of the states' transitions $G = (X, E)$ with the vertex set X and edge set E. Thus, an arbitrarily directed edge $e = (x, y) \in E$ expresses the possibility of the system to pass from state $x = x(t)$ to state $y = x(t)$ at every discrete moment of time $t = 0, 1, 2, \ldots$. On an edge set E, a cost function $c : E \to \mathbb{R}$ is defined that indicates a cost c_e for each directed edge $e = (x, y) \in E$ if the system makes a transition from state $x = x(t)$ to state $y = x(t + 1)$ for every $t = 0, 1, 2, \ldots$. We define the stationary control for the system $\mathbb{L}$ in G as a map

$$s : x \to y \in X(x) \quad \text{for} \quad x \in X,$$

where $X(x) = \{y \in X \mid (x, y) \in E\}$.

Let s be an arbitrary stationary control. Then the set of edges of the form $(x, s(x))$ in G generates a subgraph $G_s = (X, E_s)$, where each vertex $x \in X$ contains one leaving directed edge. So, if the starting state $x_0 = x(0)$ is fixed, then the system makes transitions from one state to another through the corresponding directed edges $e_0^s, e_1^s, e_2^s, \ldots, e_t^s, \ldots$, where $e_t^s = (x(t), x(t + 1))$, $t = 0, 1, 2, \ldots$. This sequence of directed edges generates a trajectory $x_0 = x(0), x(1), x(2), \ldots$, which leads to a unique directed cycle. For an arbitrary stationary strategy s and a fixed starting state x_0, the expected total discounted cost $\sigma_{x_0}^{\gamma}(s)$ is defined as follows:

$$\sigma_{x_0}^{\gamma}(s) = \sum_{t=0}^{\infty} \gamma^t c_{e_t^s},$$

where γ, $0 < \gamma < 1$ is a given discount factor.

Based on the results from [71, 124, 152], it is easy to show that for an arbitrary stationary strategy s, there exists $\sigma_{x_0}^{\gamma}(s)$. If we denote by $\sigma^{\gamma}(s)$ the column vector with components $\sigma_x^{\gamma}(s)$ for $x \in X$, then $\sigma_{x_0}^{\gamma}(s)$ can be found by solving the system of linear equations

$$(I - \gamma P^s)\sigma^{\gamma}(s) = c^s, \tag{2.145}$$

where c^s is the vector with corresponding components $c_{x,s(x)}$ for $x \in X$, I is the identity matrix and P^s the matrix with elements $p_{x,y}^s$ for $x, y \in X$, defined as follows:

$$p_{x,y}^s = \begin{cases} 1, & \text{if} \ \ y = s(x); \\ 0, & \text{if} \ \ y \neq s(x). \end{cases}$$

It is well-known that for $0 < \gamma < 1$, the rank of the matrix $I - \gamma P^s$ is equal to $|X|$ and the system (2.145) has solutions for arbitrary c^s (see [152, 199]). Thus, we can determine $\sigma_{x_0}^{\gamma}(s^*)$ for an arbitrary starting state x_0.

In the considered deterministic discounted control problem on G, we seek a stationary control s^* such that

$$\sigma_{x_0}^{\gamma}(s^*) = \min_{s} \sigma_{x_0}^{\gamma}(s).$$

We formulate and study this problem in a more general case considering its *stochastic version*. We assume that the dynamical system may admit states in which the vector of control parameters is changed randomly. So, the set of states X is divided into two subsets $X = X_C \cup X_N$, $X_C \cap X_N = \emptyset$, where X_C represents the set of states in which the decision maker is able to control the dynamical system and where X_N represents the set of states in which the dynamical system makes random transitions to the next state. This means that for every $x \in X$ on the set of feasible transitions $E(x)$, the distribution function $p : E(x) \to \mathbb{R}$ is defined such that $\sum_{e \in E(x)} p_e = 1$, $p_e \geq 0$, $\forall e \in E(x)$, and the transitions from the states $x \in X_N$ to the next states are made randomly according to these distribution functions. Here, in a similar way as for the deterministic problem, we assume that with each directed edge $e = (x, y) \in E$, a cost c_e of the system's transition from state $x = x(t)$ to state $y = x(t + 1)$ for $t = 0, 1, 2, \ldots$ is associated. In addition, we assume that the discount factor γ, $0 < \gamma < 1$ and the starting state x_0 are given. We define a stationary control on G as a map

$$s : x \to y \in X(x) \quad \text{for} \quad x \in X_C.$$

Let s be an arbitrary stationary strategy. We define the graph $G_s = (X, E_s \cup E_N)$, where $E_s = \{e = (x, y) \in E \,|\, x \in X_C,\ y = s(x)\}$, $E_N = \{e = (x, y) \,|\, x \in X_N,\ y \in X\}$. This graph corresponds to a Markov process with the probability matrix $P^s = (p^s_{x,y})$, where

$$p^s_{x,y} = \begin{cases} p_{x,y} & \text{if} \quad x \in X_N \quad \text{and} \quad y = X; \\ 1 & \text{if} \quad x \in X_C \quad \text{and} \quad y = s(x); \\ 0 & \text{if} \quad x \in X_C \quad \text{and} \quad y \neq s(x). \end{cases}$$

For this Markov process with associated costs c_e, $e \in E$, we can define the expected total discounted cost $\sigma_{x_0}^{\gamma}(s)$ in a similar way as we defined the expected total discounted reward in Chap. 1. We consider the problem of determining the strategy s^* for which

$$\sigma_{x_0}^{\gamma}(s^*) = \min_{s} \sigma_{x_0}^{\gamma}(s).$$

Without loss of generality, we may consider that G has the property that an arbitrary vertex in G is reachable from x_0; otherwise, we can delete all vertices that cannot be reached from x_0.

2.7.2 Optimality Equations and Algorithms

In a similar way as in previous sections, a discounted control problem on the network $(G, X_C, X_N, c, p, \gamma)$ can be represented as a discounted Markov decision problem with the set of states $X = X_C \cup X_N$, the set of actions $A = \cup_{x \in X} A(x)$, the set of rewards $\mathbf{r} = \{r_{x,a} | x \in X, a \in A(x)\}$, the probability function p defined in Sect. 2.6.2, and a given discount factor γ. Based on such a representation of control problems and the optimality Eqs. (2.16) from Sect. 2.3.1, we can formulate the following result:

Theorem 2.32 *For an arbitrary discounted control problem on the network $(G, X_C, X_N, c, p, \gamma)$, the following system of equations:*

$$\begin{cases} \nu_x = \min\limits_{y \in X(x)} \{c_{x,y} + \gamma \nu_y\} & \forall x \in X_C, \\ \nu_x = \mu_x + \gamma \sum_{y \in X(x)} p_{x,y} \nu_y & \forall x \in X_N \end{cases} \tag{2.146}$$

with respect to ν_x, for $x \in X$ has a unique solution ν_x^* for $x \in X$, where ν_x^* represents the value of the expected total discounted cost for the control problem with fixed starting state $x \in X$.

From this theorem, we obtain the following result:

Corollary 2.33 *The value vector υ^γ is a maximal vector that satisfies the following system of inequalities:*

$$\begin{cases} \nu_x - \gamma \nu_y \le c_{x,y} & \forall x \in X_C, \ y \in X(x); \\ \nu_x - \gamma \sum\limits_{y \in X(x)} p_{x,y} \nu_y = \mu_x & \forall x \in X_N. \end{cases} \tag{2.147}$$

Based on this corollary, we can formulate the discounted control problem on the network $(G, X_C, X_N, c, p, \gamma)$ as the following linear programming problem:

$$\max \phi^\gamma(\nu) = \sum_{x \in X} \theta_x \nu_x \tag{2.148}$$

subject to (2.147),

where θ_x for $x \in X$ are arbitrary positive values; in particular, we can take $\theta_x = 1, \forall x \in X$. If $\theta_x > 0\ \forall x \in X$ and $\sum_{x \in X} \theta_x = 1$, then we can treat θ_x for $x \in X$ as the probability of choosing the starting state x in the control problem.

Note that the inequalities $\nu_x - \gamma \sum_{y \in X(x)} p_{x,y}\nu_y \leq \mu_x$ for $x \in X_N$ in (2.147) can be replaced by equalities, i.e., the system (2.147) can be replaced by the following system:

$$\begin{cases} \nu_x - \gamma \nu_y \leq c_{x,y} & \forall x \in X_C,\ y \in X(x); \\ \nu_x - \gamma \sum\limits_{y \in X(x)} p_{x,y}\nu_y = \mu_x & \forall x \in X_N. \end{cases} \tag{2.149}$$

We can observe that after such a transformation, the optimal solutions to problems (2.148), (2.149) and (2.147), (2.148) are the same.

If we dualize the linear programming problem (2.147), (2.148), then we obtain the following linear programming problem:
Minimize

$$\varphi^{\gamma}(\alpha, \beta) = \sum_{x \in X_C} \sum_{y \in X(x)} c_{x,y}\alpha_{x,y} + \sum_{x \in X_N} \mu_x \beta_x \tag{2.150}$$

subject to

$$\begin{cases} \sum\limits_{x \in X(y)} \alpha_{y,x} - \gamma \sum\limits_{x \in X_C^-(y)} \alpha_{x,y} - \gamma \sum\limits_{x \in X_N^-(y)} p_{x,y}\beta_x = \theta_x, \quad y \in X_C; \\ \beta_y - \gamma \sum\limits_{x \in X_C^-(y)} \alpha_{x,y} - \gamma \sum\limits_{x \in X_N^-(y)} p_{x,y}\beta_x = \theta_x, \quad y \in X_N; \\ \alpha_{x,y} \geq 0, \quad \forall x \in X_C,\ y \in X(x). \end{cases} \tag{2.151}$$

Theorem 2.34 *Let (α^*, β^*) be an optimal basic solution to the linear programming problem (2.159), (2.160). Then an optimal stationary control s^* of the discounted control problem on the network $(G, X_C, X_N, c, p, \gamma)$ can be found as follows: For each $x \in X_C$, set*

$$s^*_{x,y} = \begin{cases} 1 & \text{if } \alpha^*_{x,y} > 0; \\ 0 & \text{if } \alpha^*_{x,y} = 0. \end{cases} \tag{2.152}$$

Proof If (α^*, β^*) is an optimal basic solution to the linear programming (2.159), (2.160), then

$$s^*_{x,y} = \frac{\alpha^*_{x,y}}{\sum\limits_{y \in X(x)} \alpha_{x,y}}, \quad \forall x \in X_C, y \in X(x).$$

Taking into account that the basic solution (α^*, β^*) possesses the property that in each state $x \in X_C$, the formula $\alpha^*_{x,y} = 0$ holds only for one directed edge $(x, y) \in E(x)$, then $s^*_{x,y}$ are determined according to (2.152). □

Theorem 2.35 *For an arbitrary discounted control problem on the network $(G, X_C, X_N, c, p, \gamma)$, the linear programming problem (2.148), (2.149) has a basic optimal solution v^*_x for $x \in X$, that satisfies the following conditions:*

(1) $\overline{c}_{x,y} = c_{x,y} + \gamma v^*_y - v^*_x \geq 0, \ \forall x \in X_C, \, y \in X(x)$;

(2) $\min_{y \in X(x)} \{\overline{c}_{x,y}\} = 0, \ \forall x \in X_C$;

(3) $\overline{\mu}_x = \mu_x + \gamma \sum_{y \in X(z)} p_{x,y} v^*_y - v^*_x = 0, \ \forall x \in X_N$.

An arbitrary map $s^* : X_C \to X$ such that $(x, s^*(x)) \in E^*(x), \ \forall x \in X_C$, where $E^*(x) = \{(x, y) \mid y \in X(x), \overline{c}_{x,y} = 0\}$ represents an optimal stationary control for the discounted control problem on the network $(G, X_C, X_N, c, p, \gamma)$.

The proof of this theorem is similar to the proof of Theorem 2.28.

So, based on Theorems 2.34, 2.35, we can use the linear programming models (2.148), (2.149) and (2.159), (2.160) to determine the optimal solution to the discounted control problems on networks. Additionally, the iteration algorithms from Sect. 2.3.1 for such control problems can also be used.

2.8 Decision Problems with Stopping States

In Sect. 2.3, the problem of determining the optimal stationary policies for discounted Markov decision processes with a discount factor γ, where $0 < \gamma < 1$, was studied. If $\gamma = 1$, then the expected total reward (cost) for such a problem may not exist. Here, we study a class of unichain decision problems for which γ may be equal to 1, and the expected total reward (cost) exists. Moreover, we can see that for some problems, γ may be an arbitrary positive value. For the considered problems, we show how to determine the optimal stationary policies based on linear programming and iterative algorithms from Sect. 2.3. We assume that for the dynamical system in the considered problems, there exists a state in which transitions stop as soon as this state is reached [122] and formulate conditions for the existence of optimal policies with $\gamma > 0$.

2.8.1 Problem Formulation and Main Results

We consider the problem of determining the optimal stationary policies in a discounted Markov decision process with a finite set of states X, a finite set of actions $A = \cup_{x \in X} A(x)$, the probability function $p : A \times X \times X \to \mathbb{R}^+$ that satisfies the condition $\sum_{y\in X} p_{x,y}^a = 1$, $\forall x \in X, a \in A(x)$, the rewards r_x^a, $\forall x \in X, a \in A(x)$, and the discount factor γ, which is an arbitrary positive value. In the considered problem, we assume that the Markov process possesses the unichain property, and the transitions in the process stop in a given state $z \in X$ as soon as this state is reached. The mentioned conditions for the considered problem are equivalent to the conditions that any arbitrary stationary policy induces a Markov chain with the absorbing state z, where $p_{z,z} = 1$ and $r_{z,a} = 0$ for $a \in A(x)$. If such a property holds for a discounted Markov decision problem, then the expected total reward exists, and the optimal stationary policies can be found using the linear programming models and algorithms from Sect. 2.3. By specifying the results from Sect. 2.3 for the considered problem with a stopping state, we obtain the following results:

Theorem 2.36 *Let v_x^* $(x \in X)$ be an optimal basic solution to the linear programming problem:*
Minimize

$$\phi^\gamma(v) = \sum_{x\in X\setminus\{z\}} \theta_x v_x \tag{2.153}$$

subject to

$$\begin{cases} v_x - \gamma \sum_{y\in X} p_{x,y}^a v_y \geq r_{x,a}, & \forall x \in X \setminus \{z\},\ a \in A(x), \\ v_z = 0. \end{cases} \tag{2.154}$$

Then v_x^* for $x \in X$ represents the optimal total expected discounted cost for the problem with starting states $x \in X$. An optimal stationary strategy can be found by fixing a map

$$s^* : x \to a \in A^*(x) \text{ for } x \in X \setminus \{z\},$$

where

$$A^*(x) = \Big\{a \in A(x) \mid v_x - \gamma \sum_{y\in X} p_{x,y}^a v_y = r_{x,a}\Big\},\ \forall x \in X \setminus \{z\}.$$

If we consider the dual problem for (2.153), (2.154), then we obtain the following problem:
Maximize

$$\varphi^\gamma(\alpha) = \sum_{x\in X\setminus\{z\}} \sum_{a\in A(x)} r_{x,a}\alpha_{x,a} \tag{2.155}$$

subject to

$$\begin{cases} \sum_{a\in A(y)} \alpha_{y,a} - \gamma \sum_{x\in X\setminus\{z\}} \sum_{a\in A(x)} p^a_{x,y}, \alpha_{x,a} = \theta_x, & \forall y \in X \setminus \{z\}; \\ \alpha_{x,a} \geq 0, \ \forall x \in X \setminus \{z\}, \ a \in A(x). \end{cases} \tag{2.156}$$

Using this dual linear programming problem and Theorem 2.36, we obtain the following result:

Theorem 2.37 *Let* $\alpha^*_{x,a}$ $(x \in X \setminus \{z\},\ a \in A)$ *be a basic optimal solution to the linear programming problem (2.155), (2.156). Then the optimal stationary policy* s^* *for the discounted Markov decision problem with an arbitrary starting state* $x \in X \setminus \{z\}$ *and a given stopping state* z *is determined as follows:*

$$s^*_{x,a} = \begin{cases} 1, & \text{if } \alpha^*_{x,a} \neq 0; \\ 0, & \text{if } \alpha^*_{x,a} = 0. \end{cases}$$

Remark 2.38 Theorems 2.36, 2.37 are also valid for any arbitrary Markov decision problem with $\gamma \geq 1$ if the rewards are negative and there exists a policy that induces a Markov unichain with an absorbing state z.

2.8.2 *Optimal Control on Networks with Stopping States*

Consider a discounted control problem for the network $(G, X_C, X_N, c, p, \gamma)$ with a given stopping state $z \in X$, where $0 < \gamma \leq 1$.

Theorem 2.39 *If the control problem on the network* $(G, X_C, X_N, c, p, \gamma)$ *possesses the property that an arbitrary stationary control* s^* *induces a Markov unichain, then for the discounted control problem with a given stopping state* z*, there exists an optimal stationary control that can be found by solving the following linear programming problem:*
Maximize

$$\phi^\gamma(\nu) = \sum_{x\in X\setminus\{x_0\}} \theta_x \nu_x \tag{2.157}$$

subject to

$$\begin{cases} \nu_x - \gamma\, \nu_y \leq c_{x,y}, \ \forall x \in X_C \setminus \{z\}, \ y \in X(x); \\ \nu_x - \gamma \sum_{y\in X} p_{x,y}\nu_y = \mu_x, \quad \forall x \in X_N \setminus \{z\}, \\ \nu_z = 0. \end{cases} \tag{2.158}$$

If ν_x^* for $x \in X$ is an optimal basic solution to this problem, then there exists a map

$$s^* : x \to y \in X(x) \text{ for } x \in X_C \setminus \{z\}$$

such that $(x, s^*(x)) \in E^*(x)$, $E^*(x) = \{(x, y) \in E(x) | \nu_x^* - \gamma\nu_y^* - c_{x,y} = 0\}$ represents an optimal stationary control for the problem on the network $(G, X_C, X_N, c, p, \gamma)$.

The proof of the theorem follows from Theorem 2.36.

Corollary 2.40 *Let $(G, X_C, X_N, c, p, \gamma)$ be a network that satisfies the conditions of Theorem 2.39. Then there exist the values ν_x^* for $x \in X$ that satisfy the conditions:*

(1) $c_{x,y} + \gamma\,\nu_y^* - \nu_x^* \geq 0,\ \forall x \in X_C \setminus \{x\},\ y \in X(x);$
(2) $\min\limits_{y \in X(x)} (c_{x,y} + \gamma\,\nu_y^* - \nu_x^*) = 0,\ \forall x \in X_C \setminus \{z\};$
(3) $\mu_x + \gamma \sum\limits_{y \in X} p_{x,y}\nu_y^* - \nu_x^* = 0, \quad \forall x \in X_N \setminus \{z\};$
(4) $\nu_z^* = 0.$

A map $s^* : x \to y \in X(x)$ for $x \in X_C \setminus \{z\}$ such that $(x, s^*(x)) \in E^*(x)$, where $E^*(x) = \{(x, y) \in E(x) | \nu_x^* - \gamma\nu_y^* - c_{x,y} = 0\}$ is an optimal stationary control for the control problem on the network.

Remark 2.41 The control problem on the network with a given stopping state z in the case $X_N = 0$, $\gamma = 1$ becomes the problem of determining in G the minimum cost paths from $x \in X$ to z. If G is an acyclic graph with a sink vertex, then the problem has a solution for an arbitrary $\gamma > 0$.

If for the linear programming problem (2.157), (2.158) we consider the dual problem, then we obtain the following result:

Theorem 2.42 *Assume that in G an arbitrary stationary strategy $s : x \to X(x)$ for $x \in X_C$ generates a subgraph $G_s = (X, E_s \cup E_N)$, where the vertex z can be reached from any arbitrary $x \in X \setminus \{z\}$. Then the following linear programming problem has a solution:*
Minimize

$$\varphi^\gamma(\alpha, \beta) = \sum_{x \in X_C} \sum_{y \in X(x)} c_{x,y}\,\alpha_{x,y} + \sum_{x \in X_N} \mu_x \beta_x \tag{2.159}$$

subject to

$$\begin{cases} \sum\limits_{x \in X(y)} \alpha_{y,x} - \gamma \sum\limits_{x \in X_C^-(y)} \alpha_{x,y} - \gamma \sum\limits_{x \in X_N^-(y)} p_{x,y}\, \beta_x = \theta_x, \quad y \in X_C \setminus \{z\}; \\ \beta_y - \gamma \sum\limits_{x \in X_C^-(y)} \alpha_{x,y} - \gamma \sum\limits_{x \in X_N^-(y)} p_{x,y}\, \beta_x = \theta_x, \quad y \in X_N \setminus \{z\}; \\ \alpha_{x,y} \geq 0, \ \forall x \in X_C, \ y \in X(x). \end{cases} \tag{2.160}$$

If α^*, β^* *is an arbitrary basic solution to the problem (2.159), (2.160), then the optimal stationary strategy* s^* *for the discounted control problem with a stopping state* z *can be found by fixing* $s^*_{x,y} = 1$ *for* $x \in X_C, \ y \in X(x)$ *if* $\alpha^*_{x,y} > 0$ *and* $s_{x,y} = 0^*$ *in the other case.*

Remark 2.43 In Theorems 2.39, 2.159, the condition that each stationary policy induces a Markov chain with the absorbing state $z \in X$ can be replaced by the condition that there exists at least a policy with such a property. In this case, the theorems remain valid for the positive costs $c_e, \forall e \in E$.

Based on the results above, we may conclude that the stochastic control problems on networks with a stopping state can be regarded as finite horizon problems because the stopping state is reached in finite time. These problems in the case $\gamma = 1$ represent the control problems with expected total costs and can be solved by using linear programming. In case we fix the number of transitions when the final stopping state is reached, then we can apply the dynamic programming algorithms like the algorithms from [97, 120].

2.9 Deterministic Control Problems on Networks

The stochastic control problems on networks with average and discounted optimization criteria from Sects. 2.6, 2.7 in the case $X_N = \emptyset$ are deterministic problems, and therefore, the proposed algorithms from these sections can be adapted to the deterministic case of control problems considering $X_N = \emptyset$. In this section, we consider a class of finite horizon deterministic control problems with a cost function on edges depending on time and develop algorithms for solving them.

2.9.1 Dynamic Programming for Finite Horizon Problems

We consider a deterministic control problem when the dynamics of the system are described by a directed graph $G = (X, E)$, where the vertices $x \in X$ correspond to the states of the dynamical system $\mathbb{L}$ and an arbitrarily directed edge $e =$

$(x, y) \in E$ signifies the possibility of the system's transition from state $x = x(t)$ to state $y = x(t+1)$ at every discrete moment of time $t = 0, 1, 2, \dots$. So, the set $E(x) = \{e = (x, y) | (x, y) \in E\}$ corresponds to feasible control parameters that determine the next possible state $y = x(t+1)$. Additionally, we assume that $X_N = \emptyset$, and the cost function $c_e(t)$ that depends on t is associated with each directed edge $e = (x, y) \in E$ on the network. This means that if the system makes a transition from state $x = x(t)$ to state $y = x(t+1)$, then the cost is $c_{x,y}(t)$. Thus, in this case, the control problem in G is formulated in the following way.

For a given time moment $\bar{t}$ and fixed starting and stopping states $x_0, x_f \in X$, it is necessary to determine in G a sequence of the system's transitions $(x(0), x(1))$, $(x(1), x(2)), \dots, (x(\bar{t}-1), x(\bar{t}))$, which transfers the system $\mathbb{L}$ from a starting state $x_0 = x(0)$ to a stopping state $x_f = x(\bar{t})$ such that the total cost

$$F_{x_0 x_f}(\bar{t}) = \sum_{t=0}^{\bar{t}-1} c_{(x(t), x(t+1))}(t)$$

of the system's transitions by a trajectory

$$x_0 = x(0), x(1), x(2), \dots, x(\bar{t}) = x_f$$

is minimal, where $(x(t), x(t+1)) \in E,\ t = 0, 1, 2, \dots, \bar{t}-1$.

We describe the dynamic programming algorithm to solve this problem. Denote by

$$F^*_{x_0, x_f}(\bar{t}) = \min_{x_0 = x(0), x(1), \dots, x(\bar{t}) = x_f} \sum_{t=0}^{\bar{t}-1} c_{(x(t), x(t+1))}(t)$$

the minimal total cost of the system's transition from x_0 to x_f with $\bar{t}$ stages, where $F^*_{x_0, x_f}(0) = 0$ in the case $x_0 = x_f$ and $F^*_{x_0 x_f}(\bar{t}) = \infty$ if x_f cannot be reached from x_0 by using $\bar{t}$ transitions.

If we introduce the values $F^*_{x_0 x(t)}(t)$ for $t = 0, 1, 2, \dots, \bar{t}-1$, then it is easy to observe that for $F^*_{x_0 x(t)}(t)$, the following recursive formula can be obtained:

$$F^*_{x_0 x(t)}(t) = \min_{x(t-1) \in X^-_G(x(t))} \left\{ F^*_{x_0, x(t-1)}(t-1) + c_{(x(t-1), x(t))}(t-1) \right\},$$

where

$$F^*_{x_0, x(0)}(0) = 0$$

and

$$X_G^-(y) = \{x \in X \mid e = (x, y) \in E\}.$$

Based on this recursive formula, we can tabulate the values $F^*_{x_0 x(t)}(t)$, $t = 1, 2, \ldots, \bar{t}$ for every $x(t) \in X$. These values and the solution to the problem can be found using $O(|X|^2 \bar{t})$ elementary operations (here, we do not take into account the number of operations for calculating the values of the functions $c_e(t)$ for a given t).

The tabulation process should be organized in such a way that for every vertex $x = x(t)$ at a given time moment t, not only the cost $F^*_{x_0 x(t)}(t)$ but also the state $x^*(t-1)$ at the previous time moments are determined for which

$$\begin{aligned} F^*_{x_0, x(t)}(t) &= F^*_{x_0 x^*(t-1)} + c_{(x^*(t-1), x(t))}(t-1) \\ &= \min_{x(t-1) \in X_G^-(x(t))} \{F^*_{x_0, x(t-1)} + c_{(x(t-1), x(t))}(t-1)\}. \end{aligned}$$

So, if with each x at the time moments $t = 0, 1, 2, \ldots, \bar{t}$ we associate the labels $(t, x(t), F^*_{x_0 x(t)}, x^*(t-1))$, then the corresponding table allows us to find the optimal trajectory successively, starting from the final position $x_f = x^*(\bar{t}), x^*(\bar{t}-1), \ldots, x^*(1), x^*(0) = x_0$. In the example given below, all possible labels for every x and every t are represented in Table 2.1.

This problem can be extended to the case when the final state x_f should be reached at the time moment $t(x_f)$ from a given interval $[\bar{t}_1, \bar{t}_2]$. If $\bar{t}_1 \neq \bar{t}_2$, then the problem can be reduced to $\bar{t}_2 - \bar{t}_1 + 1$ problems with $\bar{t} = \bar{t}_1, \bar{t} = \bar{t}_1 + 1, \bar{t} = \bar{t}_1 + 2, \ldots, \bar{t} = \bar{t}_2$; by comparing the minimal total costs of these problems, we find the best one and $t(x_f)$.

An important case of the considered problem is if $\bar{t}_1 = 0$ and $\bar{t}_2 = \infty$. The solution to the problem in case the network may contain directed cycles only makes sense for positive and non-decreasing cost functions $c_e(t)$ on the edges $e \in E$. Obviously, in this case, we obtain $0 \le t(x_f) \le |X|$, and the problem can be solved in time $O(|X|^3)$ in the case of a free number of stages.

The proposed dynamic programming algorithm for the considered control problem can be extended to nonlinear dynamic minimum cost flow problems on networks. Algorithms based on *time-expanded network methods* for such a class of problems were described in [103, 104, 112, 131, 132].

Example Let the dynamic network determined by the graph $G = (X, E)$ represented in Fig. 2.9 be given. The cost functions are the following:

$$c_{(0,1)}(t) = c_{(0,3)}(t) = c_{(2,5)}(t) = 1;$$

$$c_{(2,3)}(t) = c_{(3,1)}(t) = 2t; \quad c_{(3,4)}(t) = 2t + 2;$$

$$c_{(1,2)}(t) = c_{(2,4)}(t) = c_{(1,5)}(t) = t; \quad c_{(4,5)}(t) = 2t + 1.$$

Table 2.1 The results of iterative procedure for the example from Fig. 2.9

t	x, F^*	0	1	2	3	4	5
0	$F^*_{x_0,x(0)}$	0	∞	∞	∞	∞	∞
	$x^*(0-1)$	–	–	–	–	–	–
1	$F^*_{x_0,x(1)}$	∞	1	∞	1	∞	∞
	$x^*(0)$	–	0*	–	0	–	–
2	$F^*_{x_0,x(2)}$	∞	3	2	∞	5	2
	$x^*(1)$	–	3	1*	–	3	1
3	$F^*_{x_0,x(3)}$	∞	∞	5	6	4	3
	$x^*(2)$	–	–	1	2*	2	2
4	$F^*_{x_0,x(4)}$	∞	12	∞	11	8	6
	$x^*(3)$	–	3*	–	2	2	2
5	$F^*_{x_0,x(5)}$	∞	19	16	∞	21	16
	$x^*(4)$	–	3	1	–	3	1*

Fig. 2.9 Structure of the dynamic network

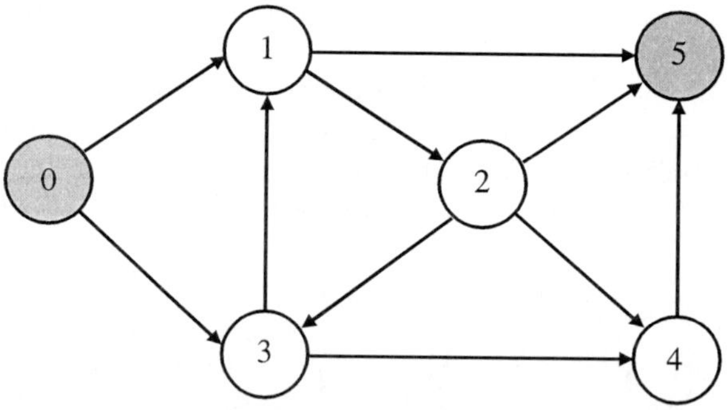

We consider the problem of finding a trajectory in G from $x(0) = x_0 = 0$ to $x_f = 5$, where $T = 5$. Using the recursive formula described above, we get Table 2.1 with values $F^*_{x_0 x(t)}(t)$ and $x^*(t-1)$. Then starting from the final state $x_f = 5$, we find the optimal trajectory

$$5^* \leftarrow 1^* \leftarrow 3^* \leftarrow 2^* \leftarrow 1^* \leftarrow 0^*$$

with total cost $F_{x_0,x(5)}(5) = 16$.

2.9.2 *Optimal Paths in Networks with Rated Costs*

In this section, we formulate and study optimal path problems on networks that extend the minimum cost path problems in the weighted directed graphs.

Let $G = (X, E)$ be a finite directed graph with vertex set X, $|X| = n$ and edge set E, where with each directed edge $e = (u, v) \in E$, a positive cost c_e

is associated. Assume that for two given vertices x, y, there exists a directed path $P(x, y) = \{x = x_0, e_0, x_1, e_1, x_2, e_2, \ldots, x_k = y\}$ from x to y. For this directed path, we define the total rated cost

$$C(x_0, x_k) = \sum_{t=0}^{k-1} \gamma^t c_{e_t},$$

where γ has a positive value. So, in this path, the costs c_{e_t} of the directed edges $e \in E$ at the moment of time t are rated by γ^t, and the cost of the directed edge e at the moment of time t in the directed path $P(x, y)$ when we pass through e is $\lambda^t c_e$. In G we seek the paths from x to y with minimal total rated cost.

We consider two basic problems:

1. to find in G a directed path $P^*(x, y)$ with minimal total cost in the case of a fixed number of transitions from x to y;
2. to find in G a directed path $P^*(x, y)$ with minimal total costs in the case of a free number of transitions from x to y.

If $\gamma = 1$, then these problems become the shortest path problems known in weighted directed graphs that can be efficiently solved by using the well-known combinatorial algorithms. In case $\gamma \neq 1$, we have the optimal path problems for graphs $G = (X, E)$ with cost functions $c_e(t) = \gamma^t c_e$ on edges $e \in E$ that depend on time. If the number k of the edges for the optimal path is fixed, then we can apply the dynamic programming algorithm from previous sections or the time-expanded network method from [4, 59, 86, 104, 112]. These algorithms determine the solution to the problem using $O(|x|^3 k)$ elementary operations. We show that the considered problems can be efficiently solved by using algorithms based on linear programming.

Algorithms for the Problem with a Free Number of Transitions

First, we consider the optimal path problem without restrictions on the number of transitions. We show that this problem can be efficiently solved based on linear programming. The linear programming model we use is the following:
Minimize

$$\phi(\alpha) = \sum_{e \in E} c_e \alpha_e \tag{2.161}$$

subject to

$$\begin{cases} \sum\limits_{e \in E^-(u)} \alpha_e - \gamma \sum\limits_{e \in E^+(u)} \alpha_e = 1, \quad u = x; \\ \sum\limits_{e \in E^-(u)} \alpha_e - \gamma \sum\limits_{e \in E^+(u)} \alpha_e = 0, \quad \forall u \in X \setminus \{x, y\}; \\ \alpha_e \geq 0, \ \forall e \in E, \end{cases} \tag{2.162}$$

where $E^-(u)$ is the set of directed edges that originate in the vertex $u \in X$ and $E^+(u)$ is the set of directed edges that enter u.

The following theorem holds:

Theorem 2.44 *If $\gamma \geq 1$ and in G there exists a directed path $P(x, y)$ from a given starting vertex x to a given final vertex y, then for positive costs c_e of edges $e \in E$, the linear programming problem (2.161), (2.162) has solutions. If α_e^* for $e \in E$ represents an optimal basic solution to this problem, then the set of directed edges $E^* = \{e \in E | \alpha_e^* > 0\}$ determines an optimal directed path from x to y.*

Proof Assume that $\gamma \geq 1$ and in G there exists at least a directed path $P(x, y) = \{x = x_0, e_0, x_1, e_1, x_2, e_2, \ldots, x_k = y\}$ from x to y. Denote by $E_P = \{e_0, e_1, e_2, \ldots, e_{k-1}\}$ the set of edges of directed path $P(x, y)$. Then it is easy to check that

$$\alpha_e = \begin{cases} \gamma^t, & if \;\; e = e_t \in E_P; \\ 0, & if \;\; e \in E \setminus E_P \end{cases} \tag{2.163}$$

represents a solution to system (2.162). Moreover, we can see that if the directed path $P(x, y)$ does not contain directed cycles, then the solution determined according to (2.163) corresponds to a basic solution to system (2.162). So, if in G there exists a directed path from x to y, then the set of solutions to system (2.163) is not empty. Taking into account that the costs c_e, $e \in E$ are non-negative, we obtain the optimal value of objective function (2.161) which is bounded, i.e., the linear programming problem (2.161), (2.162) has solutions.

Now let us prove that an arbitrary basic solution to system (2.162) corresponds to a simple directed path $P(x, y)$ from x to y. Let $\alpha = (\alpha_{e_1}, \alpha_{e_2}, \ldots, \alpha_{e_m})$ be a feasible solution to problem (2.161), (2.162) and denote $E_\alpha = \{e \in E | \alpha_e > 0\}$. Then it is easy to observe that the set of directed edges $E_\alpha \subseteq E$ in G induces a subgraph $G_\alpha = (X_\alpha, E_\alpha)$ in which vertex x is a source and y is a sink vertex. Indeed, if this is not the case, then we can determine the subset of vertices X'_α from X_α that can be reached in G_α from x, and X'_α does not contain vertex y. In G_α we can select the subgraph $G'_\alpha = (X'_\alpha, E'_\alpha)$ induced by the subset of vertices X'_α, and we can calculate

$$S = \sum_{u \in X'_\alpha} \sum_{e \in E^-(u)} \alpha_e,$$

where $E'^-(u) = \{e = (v, u) \in E' | v \in X'_\alpha\}$. It is easy to observe that the value S can also be expressed as follows:

$$S = \sum_{u \in X'_\alpha} \sum_{e \in E^+(u)} \alpha_e,$$

where $E'^{+}(u) = \{e = (u, v) \in E' | v \in X'_\alpha\}$. If we sum up the equalities from (2.163) that correspond to $u \in X'_\alpha$, then we obtain

$$\sum_{u \in X'_\alpha} \sum_{e \in E^-(u)} \alpha_e - \gamma \sum_{u \in X'_{alpha}} \sum_{e \in E^+(u)} \alpha_e = 1,$$

which involves $(1-\gamma)S = 1$. However, this cannot take place because $\gamma \geq 1$ and $S \geq 0$, i.e., we obtain a contradiction. So, if $\gamma \geq 1$, then in G_α there exists at least one directed path from x to y. Taking into account that an arbitrary vertex u in G_α contains at least an entering edge $e = (v, u)$ and at least an outgoing directed edge $e = (u, w)$, we may conclude that G_α has a structure of a directed graph, where x is a source and y is a sink.

Thus, to prove that a basic solution $\alpha = (\alpha_{e_1}, \alpha_{e_2}, \ldots, \alpha_{e_m})$ corresponds to a directed graph G_α that has a structure of a simple directed path from x to y, it is sufficient to show that G_α has a structure of a directed acyclic graph and G does not contain parallel directed paths $P'(u, w)$, $P"(u, w)$ from a vertex $u \in X_\alpha$ to $w \in X_\alpha$. We can prove the first part of the mentioned property as follows: If α is a basic solution and G_α contains a directed cycle, then there exists a directed path $P(x, y) = \{x = x_0, e_0, x_1, e_1, x_2, e_2, \ldots, x_r, e_r, \ldots x_k = y\}$ from x to y that contains a directed cycle $\{x_r, e_r, x_{r+1}, e_{r+1}, \ldots, x_{r+s-1}, e_{r+s-1}, x_r\}$ with the set of edges $E^0 = \{e_r, e_{r+1}, \ldots, e_{r+s-1}\}$. If we denote the set of edges of the directed path $P^1(x, x_r) = \{x = x_0, e_0, x_1, e_1, x_2, e_2, \ldots, x_r\}$ from x to x_r by $E^1 = \{e_0, e_1, e_2, \ldots, e_{r-1}\}$, and we denote the set of edges of the directed path $P^2(x_r, y) = \{x_r = x_{r+s}, e_{r+s}, x_{r+s+1}, e_{r+s+1}, \ldots, x_k = y\}$ from $x_r = x_{r+s}$ to $x_k = y$ by $E^2 = \{e_{r+s}, e_{r+s+1}, \ldots, e_{k-1}\}$, then for a small positive θ, we can construct the following feasible solution:

$$\alpha'_e = \begin{cases} \alpha_e, & \forall e \in E_\alpha \setminus (E^0 \cup E^2); \\ \alpha_{e_{r+i}} - \gamma^i \theta, & i = 0, 1, \ldots, s-1; \\ \alpha_{e_{r+s+i}} - \gamma^{s+i}\theta + \gamma^i \theta, & i = 0, 1, \ldots, k-r-s-1. \end{cases}$$

Here, θ can be chosen in such a way that $\alpha'_e = 0$ at least for an edge $e \in E^0 \cup E^2$. So, the number of non-zero components of the solution $\alpha' = (\alpha'_{e_1}, \alpha'_{e_2}, \ldots, \alpha'_{e_m})$ is less than the number of non-zero components of α.

Now let us show that for a basic solution, graph G_α cannot contain parallel directed paths from vertex x_r to a vertex $w \in X_\alpha$.

We prove this again by contradiction. We assume that in G_α, we have two directed paths $P'(x_r, w) = \{x_r, e'_r, x'_{r+1}, \ldots, e'_k, x'_k = w\}$ and $P''(x_r, w) = \{x_r, e_{r+1}, x''_{r+1}, \ldots, e''_l, x''_l = w\}$ from x to w with the corresponding edge sets $E' = \{e'_r, e'_{r+1}, \ldots, e'_k\}$ and $E'' = \{e''_r, e''_{r+1}, \ldots, e''_k\}$.

Then for a small positive θ, we can construct the following solution:

$$\alpha'_e = \begin{cases} \alpha_e, & if\ \ e \in E_\alpha \setminus (E' \cup E''); \\ \alpha_{e'_{r+i}} - \gamma^i\theta, & if e = e'_{r+i} \in E',\ i = 0, 1, \ldots, k - r; \\ \alpha_{e''_{r+i}} + \gamma^i\theta, & if e = e''_{r+i} \in E''.i = 0, 1, \ldots, l - r. \end{cases}$$

Here, we can choose θ in such a way that $\alpha'_{e_l} = 0$ at least for an edge $e_l \in E' \cup E''$, i.e., we obtain the number of non-zero components of the solution α' and it is less than the number of non-zero components of α. Thus, if α is a basic solution, then the corresponding graph G_α has a structure of a simple directed path from x to y. This means that if $\alpha^* = (\alpha^*_{e_1}, \alpha^*_{e_2}, \ldots, \alpha^*_{e_m})$ is an optimal basic solution $\alpha^* = (\alpha^*_{e_1}, \alpha^*_{e_2}, \ldots, \alpha^*_{e_m})$ to problem (2.161), (2.162), then the set of directed edges $E^* = \{e \in E | \alpha^*_e > 0\}$ determines an optimal directed path from x to y. □

Corollary 2.45 *If $\gamma \geq 1$ and vertex y is reachable in G from x, then for an arbitrary basic solution α to system (2.162), the corresponding graph G_α has a structure of a directed path from x to y.*

Corollary 2.46 *Assume that $0 < \gamma < 1$ and graph G contains directed cycles. Then for a basic solution α to system (2.162), either the corresponding graph G_α has a structure of a directed path from x to y, or this graph does not contain directed paths from x to y; in the second case, G_α contains a unique directed cycle that can be reached from x by using a unique directed path that connects vertex x with this cycle. Moreover, if G_α does not contain directed paths from x to y, then it consists of a set of vertices and edges $\{x = x_0, e_0, x_1 e_1, x_2, e_2, \ldots, x_r, e_r, x_{r+1}, e_{r+1}, \ldots, x_{r+s-1}, e_{r+s-1}, x_r\}$ with a unique directed cycle $\{x_r, e_r, x_{r+1}, e_{r+1}, \ldots, x_{r+s-1}, e_{r+s-1}, x_r\}$, where the non-zero components α_e of α can be expressed as follows:*

$$\alpha_e = \begin{cases} \gamma^t, & if e = e_t,\ t = 0, 1, \ldots, r - 1; \\ \gamma^{r+i}/(1 - \gamma^s), & if e = e_{r+i},\ i = 0, 1, \ldots, s - 1. \end{cases} \tag{2.164}$$

Remark 2.47 If $0 < \gamma < 1$, then the linear programming problem (2.161), (2.162) may have an optimal basic solution α^* for which graph G_{α^*} does not contain the directed path from x to y. This corresponds to the case when in G the optimal path from x to y does not exist.

Now we show that the linear programming model (2.161), (2.162) can be extended for the problem of determining the optimal paths from every $x \in X \setminus \{y\}$ to y. We can see that if $\lambda \geq 1$, then there exists the tree of optimal paths from every $x \in X \setminus \{y\}$ to y, and this tree of optimal paths can be found on the basis of the following theorem:

Theorem 2.48 *Assume that $\lambda \geq 1$ and in G for an arbitrary $u \in X \setminus \{y\}$ there exists at least a directed path $P(u, y)$ from u to y. Additionally, we assume that the costs c_e of edges $e \in E$ are positive. Then the following linear programming problem has solutions:*
Minimize

$$\phi(\alpha) = \sum_{e \in E} c_e \alpha_e \tag{2.165}$$

subject to

$$\begin{cases} \sum\limits_{e \in E^-(u)} \alpha_e - \gamma \sum\limits_{e \in E^+(u)} \alpha_e = 1, \quad \forall u \in X \setminus \{y\}, \\ \alpha_e \geq 0, \ \forall e \in E. \end{cases} \tag{2.166}$$

Moreover, if $\alpha^* = (\alpha^*_{e_1}, \alpha^*_{e_2}, \ldots, \alpha^*_{e_m})$ is an optimal basic solution to problem (2.165), (2.166), then the set of directed edges $E^* = \{e \in E | \alpha^*_e > 0\}$ determines a tree of the optimal directed path G_{α^*} from every $u \in X \setminus \{y\}$ to y.

Proof Let $\alpha = (\alpha_{e_1}, \alpha_{e_2}, \ldots, \alpha_{e_m})$ be a feasible solution to problem (2.165), (2.166) and consider the set of directed edges $E_\alpha = \{e \in E | \alpha_e > 0\}$ that corresponds to this solution. Then in the graph $G_\alpha = (X, E_\alpha)$ induced by the set of edges E_α, the vertex y is attainable from every $x \in X$. An arbitrary basic solution α to system (2.166) corresponds to a graph G_α which has a structure of a directed tree with a sink vertex y. Moreover, the optimal value of the objective function of the problem is bounded. Therefore, if we find an optimal basic solution α^* to the problem (2.165), (2.166), then we determine the corresponding tree of optimal paths G_{α^*}. □

If the graph $G = (X, E)$ does not contain directed cycles, then Theorem 2.44 and Theorem 2.48 can be extended for an arbitrary positive γ, i.e., in this case, the following theorem holds:

Theorem 2.49 *If $G = (X, E)$ has a structure of a directed acyclic graph with sink vertex y, then for an arbitrary $\gamma \geq 0$ and arbitrary costs $c_e, e \in E$, there exists the solution to the linear programming problem (2.161), (2.162). Moreover, if α^* is an optimal basic solution to this problem, then the set of directed edges $E^* = \{e \in E | \alpha^*_e > 0\}$ determines an optimal directed path from x to y.*

Proof The proof of this theorem is similar to the proof of Theorem 2.48. In this case, the set of edges E_α for a basic solution to problem (2.165), (2.166) induces graph $G_\alpha = (X_\alpha, E_\alpha)$ that has a structure of a directed tree with the sink vertex y. Therefore, the set of edges E_{α^*} for an optimal basic solution to problem (2.165), (2.166) corresponds to a directed tree $G_{\alpha^*} = (X_{\alpha^*}, E_{\alpha^*})$ of optimal paths from every $u \in X$ to the sink vertex y. □

As we have shown (see Corollary 2.46 and Remark 2.47), if $0 < \gamma < 1$ and graph $G = (X, E)$ contains directed cycles, then the linear programming problem (2.161), (2.162) may not find the optimal path from x to y, even in the case with positive costs $c_e, \forall e \in E$ because such an optimal path in G may not exist. Below we illustrate an example of the problem with $\gamma = 1/2$ and the network represented in Fig. 2.10. In the considered network, the vertices are represented by cycles and edges by arcs. Inside the cycles, the numbers of the vertices are written, and along the arcs, you can find the values α_e^* that correspond to the optimal solution to the problem with $x = 4,\ y = 1$ and $c_{(4,2)} = 1,\ c_{(2,1)} = 10,\ c_{(2,3)} = 1,\ c_{(3,2)} = 1$. The optimal basic solution to the linear programming (2.161), (2.162) for the considered example is $\alpha^*_{(4,2)} = 1,\ \alpha^*_{(2,1)} = 0,\ \alpha^*_{(2,3)} = 2/3,\ \alpha^*_{(3,2)} = 1/3$, and graph G_{α^*} is induced by the set of edges $\{(4, 2), (2, 3), (3, 2)\}$. Here, we can see that the values $\alpha^*_{(4,2)} = 1,\ \alpha^*_{(2,3)} = 2/3,\ \alpha^*_{(3,2)} = 1/3$ satisfy the condition (2.164). The corresponding graph G_{α^*} does not contain the directed path from vertex 4 to 1, i.e., the optimal path from vertex 4 to 1 does not exist.

In Fig. 2.11, the optimal solution to problem (2.161), (2.162) is represented by $x = 4,\ y = 1$ and $c_{(4,2)} = 1,\ c_{(2,1)} = 1,\ c_{(2,3)} = 2,\ c_{(3,2)} = 2$.

In this case, the optimal basic solution to problem (2.161), (2.162) is $\alpha^*_{(4,2)} = 1,\ \alpha^*_{(2,1)} = 1/2,\ \alpha^*_{(2,3)} = 0,\ \alpha^*_{(3,2)} = 0$. The corresponding non-zero components of this solution generate the subgraph $G_{\alpha^*} = (X_\alpha, E_{\alpha^*})$ in G, where $E_{\alpha^*} = \{(4, 2), (2, 1)\}$. The set of edges E_{α^*} generates a unique directed path from vertex 4 to 1, i.e., in the considered case, there exists the optimal path from vertex 4 to 1.

If for problem (2.165), (2.166) we consider the dual problem, then based on duality theorems of linear programming, we can prove the following result:

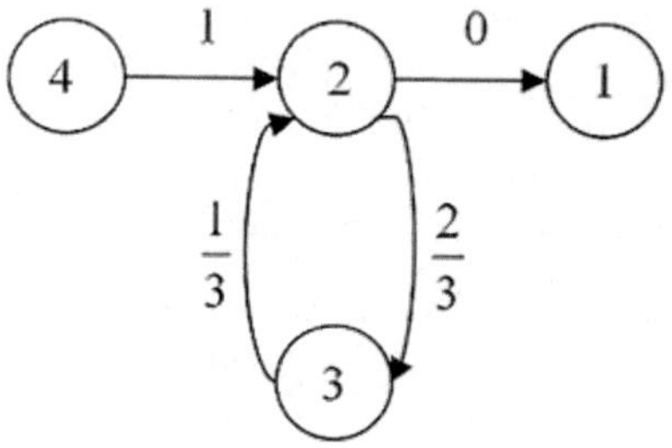

Fig. 2.10 Solution for case 1

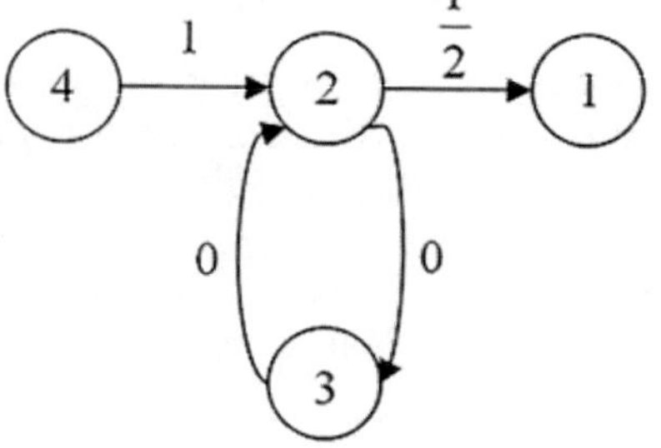

Fig. 2.11 Solution for case 2

Theorem 2.50 *Assume that $\gamma \geq 1$, and the costs $c_e, \in E$ are strictly positive. Let $\beta_u^*, \forall u \in X$ be a solution to the following linear programming problem:*
Maximize

$$\psi(\beta) = \sum_{x \in X \setminus \{x\}} \beta_x \tag{2.167}$$

subject to

$$\beta_u - \gamma \beta_v \leq c_{u,v}, \forall (u, v) \in E^0, \tag{2.168}$$

where

$$E^0 = \{e = (u, v) \in E | u \in X \setminus \{y\}, v \in X.$$

If β_u^* $u \in X$ is an optimal basic solution to problem (2.167), (2.168), then an arbitrary tree $T = (X, E'_{\beta^*})$ with sink vertex y of graph $G_{\beta^*} = (X, E_{\beta^*})$ induced by the set of directed edges

$$E_{\beta^*} = \{e = (x, y) \in E | \beta_u^* - \gamma \beta_v^* = c_{u,v}\}$$

represents the tree of optimal paths from $x \in X \setminus \{y\}$ to y. An optimal basic solution to problem (2.167), (2.168) can be found starting with $\beta_v^* = 0$ for $v = y$ and $\beta_u^* = \infty$ for $u \in X \setminus \{y\}$, and then repeat $|X| - 1$ times the following calculation procedure: Replace β_u^* for $u \in X \setminus \{y\}$ with $\beta_u^* = \min_{v \in X(u)} \left\{\gamma \beta_v^* + c_{u,v}\right\}$, where $X(v) = \{u \in X | (u, v) \in E\}$.

Proof Assume that α_e^*, $e \in E$ and β_u^*, $u \in X$ represent the optimal solutions to the primal linear programming problem (2.165), (2.166) and the dual linear programming problem (2.167), (2.168), respectively. Then, according to dual theorems of linear programming, these solutions satisfy the following condition:

$$\alpha_{u,v}^* (\beta_u^* - \gamma \beta_v^* - c_{u,v}) = 0 \quad \forall (u, v) \in E^0. \tag{2.169}$$

So, if α_e^*, $e \in E$ is an optimal basic solution, then $\beta_u^* - \gamma \beta_v^* - c_{u,v} = 0$ for an arbitrary $e = (u, v) \in E_{\alpha^*}$. Taking into account that the corresponding graph G_{α^*} for an optimal basic solution α^* has the structure of a directed tree with sink vertex y, then we obtain this tree, coinciding with the tree of optimal paths T_{β^*} that determines the solution β_u^*, $u \in X$ to the problem (2.167), (2.168).

Now let us prove that the procedure for calculating the values β_x^* correctly determines the optimal solution to the dual problem. Indeed, if in G the vertex y is attainable from each $v \in X$, then the rank of system (2.168) is equal to $|X| - 1$. This means that for an arbitrary optimal basic solution, not more than $|X| - 1$ components may be different from zero. Therefore, we can take $\beta_y^* = 0$. After that, taking condition (2.169) into account, we can find β_u^* for $u \in X \setminus \{y\}$, using the calculation procedure from the theorem starting with $\beta_v^* = 0$ for $v = y$ and $\beta_u^* = \infty$ for $u \in X \setminus \{y\}$. □

Thus, based on Theorem 2.50, we can find the tree of optimal paths in G for the problem with a free number of transitions as follows:

We determine the values β_u^* for $u \in X$, using the following steps:

Preliminary Step (Step 0) Fix $\beta_y^* = 0$ and $\beta_u^* = \infty$ for $u \in X \setminus \{y\}$.

General Step (Step $k(k > 0)$) For every $u \in X \setminus \{y\}$, replace the value

β_u^* with $\beta_u^* = \min_{v \in X(u)} \left\{\gamma\beta_v^* + c_{u,v}\right\}$, where $X(v) = \{u \in X | (u, v) \in E\}$.
If $k < |X| - 1$, then go to the next step; otherwise, stop.

If β_u^* for $u \in X$ are known, then we determine the set of directed edges E_{β^*} and the corresponding directed graph $G_{\beta^*} = (X, E_{\beta^*})$. After that, we find a directed tree $T_{\beta^*} = (X, E'_{\beta^*})$ in G_{β^*}. Then T_{β^*} represents the tree of optimal paths from $x \in X$ to y.

It can be easily observed that the proposed algorithm allows us to solve the considered problems in the general case with the same complexity as the problem with $\gamma = 1$, i.e., this algorithm in the case $\gamma \geq 1$ extends the algorithm for the shortest path problems (see [29, 59]).

Algorithms for the Problem with a Fixed Number of Transitions The optimal path problem with a fixed number of transitions from a starting vertex to a final one can be formulated and studied using the following linear programming model:
Minimize

$$\phi_{x,y}(\alpha) = \sum_{e \in E} c_e \alpha_e \tag{2.170}$$

subject to

$$\begin{cases} \sum\limits_{e \in E^-(u)} \alpha_e - \gamma \sum\limits_{e \in E^+(u)} \alpha_e = 1, & u = x; \\ \sum\limits_{e \in E^-(u)} \alpha_e - \gamma \sum\limits_{e \in E^+(u)} \alpha_e = 0, & \forall u \in X \setminus \{x, y\}; \\ \sum\limits_{e \in E^-(u)} \alpha_e - \gamma \sum\limits_{e \in E^+(u)} \alpha_e = -\gamma^{k-1}, & u = y; \\ \alpha_e \geq 0, \ \forall e \in E. \end{cases} \tag{2.171}$$

This model is valid for an arbitrary $\gamma > 0 (\gamma \neq 1)$. If we solve the linear programming (2.170), (2.171), then we find an optimal solution α^* that determines the optimal value of the objective function and the corresponding graph G_{α^*}. However, such an approach to solve this problem does not allow us to determine the order of the edges from G_{α^*} that form the optimal path $P(x, y)$ with a fixed number of transitions from x to y. The algorithms based on linear programming in this case determine only the optimal cost of the optimal path and the corresponding graph G_{α^*} in polynomial time. In order to determine the optimal path $P(x, y)$ with a given number of transitions K from x to y, it is necessary to solve the sequence of $K|X - 1|$ linear programming problems (2.170) (2.171) with a fixed starting vertex for $k = 1, 2, \dots, K$ and for an arbitrary final vertex $y \in X \setminus \{x\}$. For each of these problems, we determine the optimal value $\phi_{x,y}(\alpha^{k^*})$ and the corresponding graph G_{α^k}. After that, starting from a final vertex y, we find the optimal path $P(x, y)$ as follows: we fix a directed edge $e^{K-1} = (u^{K-1}, u^K = y)$ for which $\phi_{x,y}(\alpha^{K^*}) = \phi_{x,u^{K-1}}(\alpha^{K-1^*} + \gamma^{K-1} c_{e^{K-1}})$ and then find a directed edge $e^{K-2} = (u^{K-2}, u^K = y)$ for which $\phi_{x,u^{K-1}}(\alpha^{K-1^*}) = \phi_{x,u^{K-2}}(\alpha^{K-1^*} + \gamma^{K-2} c_{e^{K-2}})$ and so on. In such a way, we find the vertices $x = u^o, u^1, \dots, u^k = y$ of the path $P(x, y)$. More useful algorithms to solve the problem with a fixed number of transitions on the edges of the network are represented by the dynamic programming algorithms and the time-expanded network method from [4, 86, 104, 112]. For the application of these algorithms, it is sufficient to consider the network with cost functions $c_e(t) = \gamma^t c_e$ on edges $e \in E$.

2.9.3 Control Problems with Varying Time of Transitions

So far, in the control problems with average and discounted optimization cost criteria, we have considered that the time between transitions in the control process is constant and it is equal to 1.

Now we generalize these problems by assuming that the time of the system's transition from one state to another in the decision process varies and it may be different from 1. Such a problem statement may be useful for studying and solving the decision models in the case of semi-Markov processes. In this section, we show that the deterministic problem with a varying time of the states' transitions can be reduced to the problem with a fixed unit time of system transitions from one state to another. First, we formulate the control problem with an average cost optimization criterion when the transition time between the states is not constant. Let the dynamical system $\mathbb{L}$ with a finite set of states $X \subseteq \mathbb{R}^n$ be given, where at every discrete moment of time $t = 0, 1, 2, \dots,$ the state of $\mathbb{L}$ is $x(t) \in X$. Assume that the control of the system $\mathbb{L}$ at each time moment $t = 0, 1, 2, \dots$ for an arbitrary state $x(t)$ is realized by using the vector of control parameters $u(t) \in \mathbb{R}^m$ for which a feasible set $U_t(x(t))$ is given, i.e., $u(t) \in U_t(x(t))$. For arbitrary t and $x(t)$ on $U_t(x(t))$, an integer function is defined

$$\tau_{x(t)} : U_t(x(t)) \to \mathbb{N},$$

which represents an integer value $\tau_{x(t)}(u(t))$ for each control $u(t) \in U_t(x(t))$. This value expresses the time of a system's transition from state $x(t)$ to state $x\big(t + \tau_{x(t)}(u(t))\big)$ if the control $u(t) \in U_t(x(t))$ has been applied at the moment t for a given state $x(t)$.

The dynamics of the system $\mathbb{L}$ are described by the following system of difference equations:

$$\begin{cases} t_{j+1} &= t_j + \tau_{x(t_j)}\left(u(t_j)\right); \\ x(t_{j+1}) &= g_{t_j}\left(x(t_j), u(t_j)\right); \\ & u(t_j) \in U_{t_j}\left(x(t_j)\right); \\ & j = 0, 1, 2, \dots, \end{cases}$$

where

$$x(t_0) = 0, \quad t_0 = 0$$

is a given starting state of the dynamical system $\mathbb{L}$. Here, we suppose that the functions g_t and $\tau_{x(t)}$ are known and t_{j+1} and $x(t_{j+1})$ are uniquely determined by $x(t_j)$ and $u(t_j)$ at each step j. Let $u(t_j),\ j = 0, 1, 2, \dots$ be a control that generates the trajectory $x(0), x(t_1), x(t_2), \dots, x(t_k), \dots$. For this control, we define the mean integral time cost by a trajectory

$$F_{x_0}\left(u(t)\right) = \lim_{k\to\infty} \frac{\sum_{j=1}^{k-1} c_{t_j}\left(x(t_j),\ g_{t_j}\left(x(t_j),\ u(t_j)\right)\right)}{\sum_{j=0}^{k-1} \tau_{x(t_j)}(u(t_j))},$$

where $c_{t_j}\left(x(t_j),\ g_{t_j}\left(x(t_j),\ u(t_j)\right)\right) = c_{t_j}\left(x(t_j),\ x(t_{j+1})\right)$ represents the cost of system $\mathbb{L}$ to pass from state $x(t_j)$ to state $x(t_{j+1})$ at stage $[j, j+1]$.

We consider the problem of finding the time moments

$$t = 0,\ t_1,\ t_2,\ \dots,\ t_{k-1}, \dots$$

and the vectors of control parameters

$$u(0),\ u(t_1),\ u(t_2), \dots, u(t_{k-1}), \dots$$

which satisfy the conditions mentioned above and minimize the functional $F_{x_0}\left(u(t)\right)$.

In the case of $\tau_{x(t)}(u(t)) = 1$, for every t and $x(t)$, this problem becomes a control problem with a unit time of the states' transitions. The problem of determining the stationary control with a unit time of the states' transitions was studied in [11, 79, 97, 106, 156]. In the mentioned papers, it is assumed that $U_t(x(t))$, g_t and c_t do not depend on t, i.e., $g_t = g$, $c_t = c$ and $U_t(x) = U(x)$ for

$t = 0, 1, 2, \ldots$. Richard Bellman showed in [11] that for the stationary case of the problem with a unit time of the states' transitions, there exists an optimal stationary control $u^*(0), u^*(1), u^*(2), \ldots, u^*(t), \ldots$, such that

$$\lim_{k\to\infty} \frac{\sum_{t=0}^{k-1} c\big(x(t), g\left(x(t), u^*(t)\right)\big)}{k}$$

$$= \inf_{u(t)} \lim_{k\to\infty} \frac{\sum_{t=0}^{k-1} c\big(x(t), g\left(x(t), u(t)\right)\big)}{k} = \lambda < \infty.$$

Furthermore, in [97, 156] it was shown that the stationary case of the problem can be reduced to the problem of finding the optimal mean cost cycle in a graph of the states' transitions of a dynamical system. Based on these results in [28, 79, 106, 156], polynomial-time algorithms for finding the optimal stationary control are proposed. This variant of the problem can be solved by using the linear programming problem (2.124), (2.125) from Sect. 2.6.1 (see Corollary 2.23).

Below we extend the results mentioned above to the general stationary case of the problem with arbitrary transition-time functions τ_x. We show that this problem can be formulated as the problem of determining the optimal mean cost cycles in the graph of the states' transitions of the dynamical system for an arbitrary transition-time function on the edges.

For the discounted control problem with a varying time of the states' transitions, the dynamics are determined in the same way as for the problem above, but the objective function, which has to be minimized, is defined as follows:

$$\widehat{F}_{x_0}(u(t)) = \sum_{j=0}^{\infty} \gamma^{t_j} c\big(x(t_j), g\left(x(t_j), u(t_j)\right)\big),$$

where γ, $0 < \gamma < 1$, and it is a given discounted factor.

We consider the stationary case of the deterministic control problem, i.e., when g_t, c_t, $U_t(x(t))$, $u(t)$ do not depend on t and the transition function $\tau_{x(t)}$ depends only on the state x and on the control u_x in the state x. So, $g_t = g$, $c_t = c$, $U_t(x) = U(x)$, $\tau_{x(t)} = \tau(x, u_x)$ for $u(t) = u_x \in U(x)$, $\forall x \in X$, $t = 0, 1, 2, \ldots$.

In this case, it is convenient to study the problem on a network where the dynamics of the system are described by the graph of the states' transitions $G = (X, E)$. An arbitrary vertex x of G corresponds to a state $x \in X$, and an arbitrarily directed edge $e = (x, y) \in E$ expresses the possibility of the system $\mathbb{L}$ to pass from state $x(t)$ to state $x(t + \tau_e)$, where τ_e is the time of the system's transition from state x to state y through the edge $e = (x, y)$. So, on the edge set E, the function $\tau : E \to \mathbb{R}^+$ is defined, which associates a positive number τ_e with each edge, meaning that if the system $\mathbb{L}$ is in the state $x = x(t)$ at the moment of time t, then the system can reach the state y at the moment of time $t + \tau_e$ if it passes through the edge $e = (x, y)$, i.e., $y = x(t + \tau_e)$. In addition, on the edge set E, the cost function $c : E \to \mathbb{R}$ is defined, which associates with each edge the cost c_e of the system's transition from state $x = x(t)$ to state $y = x(t + \tau_e)$ for an arbitrary discrete moment of time t. So, finally, with each edge $e = (x, y) \in E$, the cost c_e and the transition time τ_e from x to y are associated. In G, an arbitrary edge $e = (x, y)$ corresponds to a control in the initial problem, and the set of edges $E(x) = \{e = (x, y) \mid (x, y) \in E\}$ originating in the vertex x corresponds to the feasible set $U(x)$ of the vectors of control parameters in the state x. The transition-time function τ in G is induced by the transition-time function τ_x for the stationary control problem. It is easy to observe that the infinite horizon control problem with a varying time of the states' transitions of the system in G can be regarded as the problem of finding in G the minimal mean cost cycle C_G^* that can be reached from the vertex x_0, where the vertex x_0 corresponds to the starting state $x_0 = x(0)$ of the dynamical system $\mathbb{L}$. Indeed, a stationary control in G corresponds to a fixed transition from a vertex $x \in X$ to another vertex $y \in X$ through a directed edge $e = (x, y)$ in G. Such a strategy of the states' transitions of the dynamical system in G generates a trajectory that leads to a directed cycle C_G with the set of edges $E(C_G)$. Therefore, the considered stationary control problem in G is reduced to the problem of finding the minimal mean cost cycle that can be reached from x_0, where in G, with each directed edge $e = (x, y) \in E$, the cost c_e and the transition time τ_e of the system's transition from state $x = x(t)$ to state $y = x(t + \tau_e)$ are associated. If the minimal mean cost cycle C_G^* in G is known, then the stationary optimal control for our problem can be found in the following way: In G, we fix an arbitrary simple directed path $P(x_0, x_k)$ with the set of edges $E(P(x_0, x_k))$, which connects the vertex x_0 with the cycle C_G^*.

After that, for an arbitrary state $x \in X$, we choose a stationary control that corresponds to a unique directed edge $e = (x, y) \in E(P(x_0, x_k)) \cup E(C^*)$. For such a stationary control, the following equality holds:

$$\inf_{u(t)} \lim_{k\to\infty} \frac{\sum\limits_{j=0}^{k-1} c\big(x(t_0), g\left(x(t_j), u(t_j)\right)\big)}{\sum\limits_{j=0}^{k-1} \tau_x(u(t_j))} = \frac{\sum\limits_{e\in E(\mathbb{C}^*)} c_e}{\sum\limits_{e\in E(\mathbb{C}^*)} \tau_e}.$$

Note that the condition $U(x) \neq \emptyset, \quad \forall x \in X$ for the stationary case of the control problem means that in G, each vertex x contains at least one leaving directed edge $e = (x, y)$. We assume that in G, every vertex $x \in X$ is attainable from x_0; otherwise, we can delete vertices from X for which there are no directed paths $P(x_0, x)$ from x_0 to x. Moreover, without loss of generality, we may consider that G is a strongly connected graph.

Then the problem of finding the optimal stationary control for the problem from Sect. 2.4 can be formulated as a combinatorial optimization problem in G in which it is necessary to find a directed cycle C^*_G such that

$$\frac{\sum\limits_{e\in E(C^*_G)} c_e}{\sum\limits_{e\in E(C^*_G)} \tau_e} = \min_{C_G} \frac{\sum\limits_{e\in E(C_G)} c_e}{\sum\limits_{e\in E(C_G)} \tau_e}.$$

The problem of determining the minimal mean cost cycle in a double-weighted directed graph was studied in [29, 79, 91, 156]. In the cited works, algorithms based on linear programming and parametrical methods are proposed. For the problem with a unit time of the states' transitions in [79], a strongly polynomial time algorithm was proposed. In the following, we describe an approach that is based on linear programming. We can see that such an approach may be used for solving a more general class of problems, for example, for the multi-criterion version of minimal mean cost cycle problems [115].

We consider the following linear programming problem:
Minimize

$$z = \sum_{e\in E} c_e\alpha_e \tag{2.172}$$

subject to

$$\begin{cases} \sum\limits_{e\in E^+(x)} \alpha_e - \sum\limits_{e\in E^-(x)} \alpha_e = 0, \ \ \forall x \in X; \\ \sum\limits_{e\in E} \tau_e\alpha_e = 1; \\ \alpha_e \geq 0, \ \ \forall e \in E, \end{cases} \tag{2.173}$$

where $E^+(x) = \{e = (x, y) \in E \mid y \in X\}$, $E^-(x) = \{e = (y, x) \in E \mid y \in X\}$.

The following lemma holds:

Lemma 2.51 *Let* $\alpha = (\alpha_{e_1}, \alpha_{e_2}, \dots, \alpha_{e_m})$ *be a feasible solution to the system* (2.173) *and* $G_\alpha = (X_\alpha, E_\alpha)$ *be the subgraph of* G*, generated by the set of edges* $E_\alpha = \{e_i \in E \mid \alpha_{e_i} > 0\}$. *Then an arbitrary extreme point* $\alpha^0 = (\alpha^0_{e_1}, \alpha^0_{e_2}, \dots, \alpha^0_{e_m})$ *of the polyhedron set determined by* (2.173) *corresponds to a subgraph* $G_{\alpha^0} = (X_{\alpha^0}, E_{\alpha^0})$*, which has the structure of a simple directed cycle and vice versa, i.e., if* $G_{\alpha^0} = (X_{\alpha^0}, E_{\alpha^0})$ *is a simple directed cycle in* G*, then the solution*

$\alpha^0 = \left(\alpha^0_{e_1}, \alpha^0_{e_2}, \ldots, \alpha^0_{e_m}\right)$ *with*

$$\alpha^o_{e_i} = \begin{cases} \dfrac{1}{\sum\limits_{e \in E_{\alpha^0}} \tau_e}, & \text{if } e_i \in E_{\alpha^0}; \\ 0, & \text{if } e_i \notin E_{\alpha^0} \end{cases}$$

corresponds to an extreme point of the set of solutions (2.173).

Proof Let $\alpha = \left(\alpha_{e_1}, \alpha_{e_2}, \ldots, \alpha_{e_m}\right)$ be an arbitrary feasible solution to the system (2.173). Then it is easy to observe that $G_\alpha = (X_\alpha, E_\alpha)$ contains at least one directed cycle. Indeed, for an arbitrary $x \in X_\alpha$, there exist at least one leaving edge $e' = (x, y) \in E_\alpha$ and at least one entering edge $e'' = (z, x) \in E_\alpha$; otherwise, α does not satisfy condition (2.173). Let us show that if G_α is not a simple directed cycle, then α does not represent an extreme point of the set of solutions to the system (2.173). If G_α does not have the structure of a simple directed cycle, then it contains a simple directed cycle C with the set of edges $E(C_G) \subset E_\alpha$, i.e., $m' = |E(C_G)| < m$. Without loss of generality, we may consider that $E(\mathbb{C}) = \{e_1, e_2, \ldots, e_{m'}\}$. Fix an arbitrary value θ such that $0 < \theta < \min_{e_i \in E(\mathbb{C})} \alpha_{e_i}$ and consider the following two solutions:

$$\alpha^1 = \frac{1}{1 - \theta \sum\limits_{i=1}^{m} \tau_{e_i}} \left(\alpha_{e_1} - \theta,\ \alpha_{e_2} - \theta,\ \ldots,\ \alpha_{e_{m'}} - \theta,\ \alpha_{e_{m'+1}},\ \ldots,\ \alpha_{e_m}\right);$$

$$\alpha^2 = \frac{1}{\theta \sum\limits_{i=1}^{m} \tau_{e_i}} \big(\underbrace{\theta, \theta, \ldots, \theta}_{m'}, 0, 0, \ldots, 0\big).$$

It is easy to check that α^1 and α^2 satisfy the condition (2.173), i.e., α^1 and α^2 are feasible solutions to the problem (2.172), (2.173). If we choose θ such that $0 < \theta \sum_{i=1}^{m'} \tau_{e_i} < 1$, then we can see that α can be represented as a convex combination of feasible solutions α^1 and α^2, i.e.,

$$\alpha = \left(1 - \theta \sum_{i=1}^{m'} \tau_e\right)\alpha^1 + \left(\theta \sum_{i=1}^{m'} \tau_e\right)\alpha^2. \tag{2.174}$$

So, α is not an extreme point in the set of solutions (2.173). If G_α represents a simple directed cycle, then the representation (2.174) is not possible, i.e., the second part of Lemma 2.51 holds. □

Using Lemma 2.51, we can prove the following result:

Theorem 2.52 *The optimal basic solution* $\alpha^* = (\alpha^*_{e_1}, \alpha^*_{e_2}, \ldots, \alpha^*_{e_m})$ *to problem* (2.172), (2.173) *corresponds to a minimal mean cycle* $C^*_G = G_{\alpha^*}$ *in* G*, i.e.,*

$$\alpha^*(e_i) = \begin{cases} \dfrac{1}{\sum\limits_{e\in E(\mathbb{C}^*)} \tau_e}, & \text{if } e \in E(\mathbb{C}^*); \\ 0, & \text{if } e \notin E(\mathbb{C}^*), \end{cases}$$

where $E(C^*_G)$ *is the set of edges of a directed cycle* C^*_G.

Proof According to Lemma 2.51, an arbitrary extreme point α^0 of the set of solutions to system (2.173) corresponds in G to the subgraph $G_{\alpha^0} = (X^0_\alpha, E^0_\alpha)$, which has the structure of a directed cycle. Taking into account that the optimal solution to problem (2.172), (2.173) is attained at an extreme point, we obtain the proof of the theorem. □

The linear programming problem (2.172), (2.173) allows us to find the minimal mean cycle in the graph G with positive values $\tau_e = \tau_{x,y}$ for $e = (x, y) \in E$. More efficient algorithms for solving the problem can be obtained using the dual problem (2.172), (2.173).

Theorem 2.53 *If* G *is a strongly connected directed graph, then there exists a function* $\varepsilon : X \to \mathbb{R}$ *and the value* λ *such that:*

(a) $\varepsilon_y - \varepsilon_x + c_{x,y} \geq \tau_{x,y} \cdot \lambda, \ \ \forall (x, y) \in E;$
(b) $\min\limits_{y\in O^-(x)} \{\varepsilon_y - \varepsilon_x + c_{x,y} - \tau_{x,y}\lambda\} = 0, \ \ \forall x \in X;$
(c) *an arbitrary cycle* $\mathbb{C}^*$ *of the subgraph* $G^0 = (X, E^0)$ *of* G*, generated by edges* $(x, y) \in E$ *for which* $\varepsilon_y - \varepsilon_x + c_{x,y} - \tau_{x,y} \cdot \lambda = 0$*, determines a minimal mean cycle in* G.

Proof We consider the dual problem for (2.172), (2.173):
Maximize

$$W = \lambda$$

subject to

$$\varepsilon_x - \varepsilon_y + \tau_{x,y}\lambda \leq c_{x,y}, \ \ \forall (x, y) \in E.$$

If p is the optimal value of the problem, then by using the duality properties of the solution to the problem, we obtain (a), (b), and (c). □

Based on the results described above, we can draw the following conclusions:

1. If $\lambda = 0$, then the values ε_x, $x \in X$ can be treated as the cost of minimal paths from vertices $x \in X$ to a vertex x_f that belongs to the minimal mean

cycle C_G^* (with $\lambda = 0$) in graph G with given costs c_e of edges $e \in E$. So, if x_f is known, then cycle C_G^* can be found in the following way: We construct the tree of minimum cost directed paths from $x \in X$ to x_f and determine the values ε_x, $\forall x \in X$. Then in G, we make a transformation of the costs $c'_{x,y} = \varepsilon_y - \varepsilon_x + c_{x,y}$ for $(x, y) \in E$ and find the subgraph $G^0 = (X, E^0)$ generated by the edges (x, y) with $c'_{x,y} = 0$. After that, we fix in G^0 a cycle C^* with zero cost of the edges. If the vertex x_f is not known, then we have to construct the tree of minimum cost paths with respect to each $x_f \in X$. So, in this case, with respect to each tree, we find the subgraph $G^0 = (X, E^0)$. Then at least for one of these subgraphs, we find a cycle C_G^* with zero cost ($c'_{x,y} = 0$) of the edges.

2. If $\lambda \neq 0$ and λ is known, then the minimal mean cost cycle C^* can be found in the following way: In G we change the costs $c_{x,y}$ of edges $(x, y) \in E$ for $c_{x,y} - \tau_{x,y}\lambda$ and, after that, solve the problem with the new costs according to point 1.
3. If $\lambda \neq 0$ and it is not known, then we find it using the bisection method on the segment $[h_0^1, h_0^2]$, where $h^1 = \min_{e\in E} c_e$, $h^2 = \max_{e\in E} c_e$. At each step k of the method, we find the midpoint $\lambda_k = \left(h_k^1 + h_k^2\right)/2$ of the segment $[h_k^1, h_k^2]$ and check if in G, with the cost $c_{x,y}^k - \tau_{x,y}\lambda_k$, there exists the cycle with a negative cost. If at a given step there exists the cycle with a negative cost, then we fix $h_{k+1}^1 = h_k^1, h_{k+1}^2 = \lambda_k$; otherwise, we put $h_{k+1}^1 = \lambda_k, h_{k+1}^1 = h_k^1$. In such a way, we find λ with a given precision. After that, the exact value of λ can be found from λ_k, using a special roundoff procedure from [85].

The algorithm described above allows us to determine the solution to the problem in the case if $\tau_e \geq 0, \forall e \in E$. In general, this problem can be considered for arbitrary τ_e and c_e. In this case, we may use the following fractional linear programming problem:
Minimize

$$z = \frac{\sum\limits_{e\in E} c_e\alpha_e}{\sum\limits_{e\in E} \tau_e\alpha_e} \tag{2.175}$$

subject to

$$\begin{cases} \sum\limits_{e\in E^+(x)} \alpha_e - \sum\limits_{e\in E^-(x)} \alpha_e = 0, \ \forall x \in X; \\ \sum\limits_{e\in E} \alpha_e = 1; \\ \qquad \alpha_e \geq 0, \ e \in E, \end{cases} \tag{2.176}$$

where $E^-(x) = \{e = (y, x) \in E \mid y \in X\}$; $E^+(x) = \{e = (x, y) \in E \mid y \in X\}$.

Of course, this model is valid if on the set of solutions to system (2.176), we have $\sum_{e\in E} \tau_e\alpha_e \neq 0$. Similarly, as in the linear programming problem, here we can show that an arbitrary optimal basic solution to the problem (2.175), (2.176) corresponds to an optimal mean directed cycle in G.

Let $\alpha = \left(\alpha_{e_1}, \alpha_{e_2}, \ldots, \alpha_{e_{|E|}}\right)$ be an arbitrary feasible solution to system (2.176) and denote by $G_\alpha = (X_\alpha, E_\alpha)$ the subgraph of G generated by the set of edges $E_\alpha = \{e \in E \,|\, \alpha_e > 0\}$. In [106], it was shown that an arbitrary extreme point $\alpha^0 = \left(\alpha^0_{e_1}, \alpha^0_{e_2}, \ldots, \alpha^0_{e_{|E|}}\right)$ of the set of solutions to system (2.176) corresponds to a subgraph $G_{\alpha^0} = (X_{\alpha^0}, E_{\alpha^0})$ that has the structure of an elementary directed cycle. Taking into account that for problem (2.175), (2.176), there exists an optimal solution $\alpha^* = \left(\alpha^*_{e_1}, \alpha^*_{e_2}, \ldots, \alpha^*_{e_{|E|}}\right)$ corresponding to an extreme point of the set of solutions (2.176), and we obtain

$$\max z = \frac{\sum\limits_{e\in E_{\alpha^*}} c_e\alpha^*_e}{\sum\limits_{e\in E_{\alpha^*}} \tau_e\alpha^*_e};$$

the set of edges E_{α^*} generates a directed cycle G_{α^*} for which $\alpha^*_e = 1/|E_{\alpha^*}|$, $\forall e \in E_{\alpha^*}$. Therefore,

$$\max z = \frac{\sum\limits_{e\in E_{\alpha^*}} c_e}{\sum\limits_{e\in E_{\alpha^*}} \tau_e}.$$

So, an optimal solution to problem (2.175), (2.176) corresponds to the minimal mean cost cycle in the directed graph of the states' transitions of the dynamical system.

This means that the fractional linear programming problem (2.175), (2.176) can be used to determine the optimal solution to the problem in the general case.

2.9.4 Reduction of the Problem in the Case of Unite Time of State Transitions

As we have shown, the deterministic control problem with an average cost criterion on the network can be solved for an arbitrary transition-time function, using a linear programming problem (2.172), (2.173) or the linear fractional programming problem (2.175), (2.176). For the discounted control problem with varying time of states' transitions, a similar linear programming model cannot be derived.

However, both problems can be reduced to the corresponding cases of the problems with a unit time of the states' transitions of the system.

Below we describe a general scheme of how to reduce the control problems with varying time of the states' transitions in the case with a unit time of the states' transition of the system. We show that our problems can be reduced to the case with a unit time of the states' transitions on an auxiliary graph $G' = (X', E')$, which is obtained from $G = (X, E)$, using a special construction. This means that after such a reduction, we can apply the linear programming approach described in Sect. 2.6.

Graph $G' = (X', E')$ with unit transitions on directed edges $e' \in E'$ is obtained from G, where each directed edge $e = (x, y) \in E$ with corresponding transition time τ_e is replaced with a sequence of directed edges

$$e'_1 = (x, x^e_1),\ e'_2 = (x^e_1, x^e_2),\ \dots,\ e'_{\tau_e} = (x^e_{\tau_e - 1}, y).$$

This means that we represent a transition from a state $x = x(t)$ at the moment of time t to the state $y = x(t + \tau_e)$ at the moment of time $t + \tau_e$ in G and in G' as the transition of a dynamical system from the state $x = x(t)$ at the time moment t to $y = x(t + \tau_e)$ if the system makes transitions through a new fictive intermediate set of states $x'_1, x'_2, \dots, x'_{\tau-1}$ at the corresponding discrete moments of time

$$t + 1,\ t + 2,\ \dots,\ t + \tau_e - 1.$$

The graphical interpretation of this construction is represented in Figs. 2.12 and 2.13. In Fig. 2.12, it is represented as an arbitrarily directed edge $e = (x, e)$ with the corresponding transition time τ_e in G. In Fig. 2.13, it is represented as the sequence of directed edges e'_i and the intermediate states $x_1, x_2, \dots, x_{\tau_e - 1}$ in G' that correspond to a directed edge $e = (x, y)$ in G.

So, the set of vertices X' of the graph G' consists of the set of states X and the set of intermediate states $XE = \{x^e_i \mid e \in E,\ i = 1, 2, \dots, \tau_e\}$, i.e., $X' = X \cup XE$.

Fig. 2.12 Edge $e = (x, y)$ with associated transition time τ_e

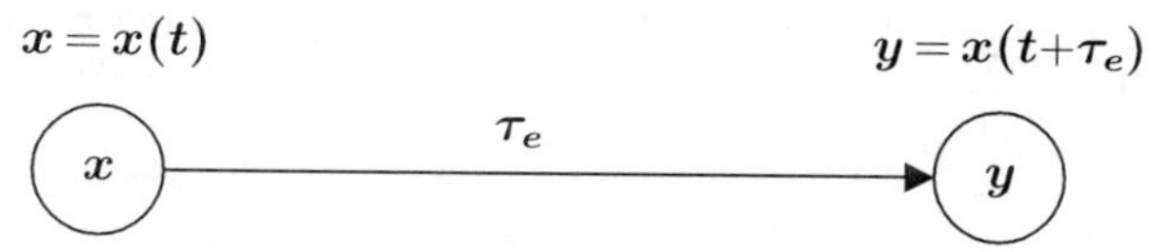

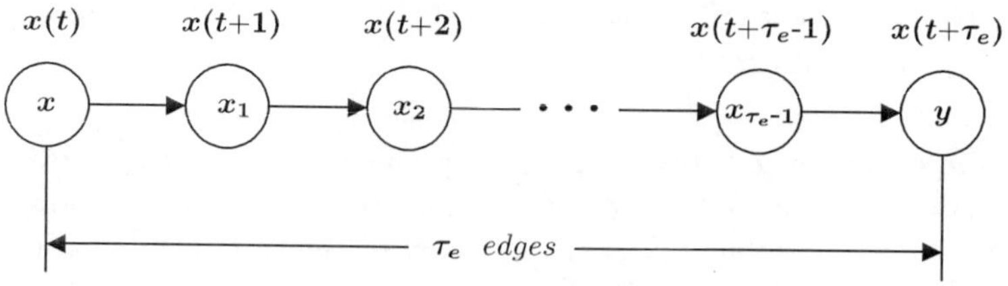

Fig. 2.13 Intermediate states for edge $e = (x, y)$ in G'

Then the set of edges E' is defined as follows:

$$E' = \bigcup_{e\in E} \mathcal{E}^e, \quad \mathcal{E}^e = \{(x, x_1^e),\ (x_1^e, x_2^e),\ \dots,\ (x_{\tau_e-1}^e, y) \mid e = (x, y) \in E\}.$$

We define the cost function $c' : E' \to \mathbb{R}$ in the following way:

$$c'_{x,x^e} = c_{x,y}, \quad \text{if} \quad e = (x, y) \in E;$$

$$c'_{x_1^e,x_2^e} = c_{x_2^e,x_3^e} = \cdots = c_{x_{\tau_e-1},y} = 0.$$

It is evident that between the set of stationary strategies

$$s : x \to y \in X + (x) \quad \text{for} \quad x \in X$$

and the set of stationary strategies

$$s' : x' \to y' \in X'^+(x') \quad \text{for} \quad x' \in X',$$

there exists a bijective mapping such that the corresponding average and discounted costs on G and on C' are the same. So, if s'^* is the optimal stationary strategy of the problem with unit transitions on G', then the optimal stationary strategy s^* on G is determined by fixing $s^*(x) = y$ if $s'^*(x) = x_1^e$, where $e = (x, y)$.

For the stochastic versions of the control problem on $G = (X, E)$, the construction of the auxiliary graph is similar. Here, we should take into account that the set of vertices (states) X are divided into two disjoint subsets X_C and X_N, where X_C corresponds to the set of controllable states and X_N corresponds to the set of uncontrollable states. Moreover, the probability function $p : E_N \to [0, 1]$ on the set $E_N = \{e = (x, y) \in E \,|\, x \in X_N\}$ is defined such that $\sum_{y\in X^+(x)} p_{x,y} = 1$. The graph $G' = (X', E')$, in the case of stochastic control problems, is constructed in the same way as above. Here we only have to be precise when defining the sets X'_C, X'_N and the probability function p' on the set $E'_N = \{e' = (x', y') \in E' \,|\, x' \in X'_N\}$ in G'. To obtain a bijective mapping between the stationary strategies of the problems in the initial graph G and the stationary strategies of the problem in the auxiliary graph, it is necessary to take $X'_C = X_C$, $X'_N = X' \setminus X_C$ and to define the probability function $p' : E' \to [0, 1]$ as follows:

$$p'_{x',y'} = \begin{cases} p_{x,y}, & \text{if } x' = x,\ x' \in X_N \subset X'_N \text{ and } y' = x'_1; \\ 1, & \text{if } x' \in X'_N \setminus X_N. \end{cases}$$

The cost function in G' for the corresponding auxiliary stochastic control problems is defined in the same way as for deterministic problems.

In the following, we extend the approach described above to semi-Markov decision problems, and it is valid for the stochastic control problem in its general form.

Chapter 3
Stochastic Games and Positional Games on Networks

Abstract Stochastic games represent an important class of models in game theory that extend Markov decision processes to competitive situations with more than one decision-maker. Such models may be with finite, countable, or continuum cardinality sets of states Kallenberg ((2011) Markov decision processes. Lecture Notes. University of Leiden, pp 2–5), Puterman ((2014) Markov decision processes: discrete stochastic dynamic programming. Wiley). In this chapter, we consider only stochastic games with finite state and action spaces. We mainly study two classes of games: stochastic games with average payoff optimization criteria and stochastic games with discounted payoff optimization criteria for the players. The main results presented in this chapter are concerned with the existence and determination of stationary Nash equilibria for different classes of stochastic games. By applying the concept of positional games for the Markov decision problems and stochastic control on networks, we formulate a class of stochastic positional games for which Nash equilibria in stationary strategies exist and for which efficient algorithms to determine the optimal stationary strategies of the players can be elaborated.

Keywords Stochastic games · Average stochastic game · Discounted stochastic game · Stochastic positional game · Nash equilibrium · Stationary equilibrium · Non-stationary equilibrium · Positional games on networks

3.1 Foundation and Development of Stochastic Games

Stochastic games, sometimes also called Markov games, were introduced by Shapley [165] in 1953. He considered two-person zero-sum stochastic games for which he proved the existence of the value and the optimal stationary strategies of the players with respect to a discounted payoff criterion. Later, this class of games was extended to general n-person stochastic games with discounted and average payoff criteria (see [53, 55, 63, 143, 174, 189]). The most important results for n-person stochastic games with discounted payoffs were obtained by Fink [55] and Takahashi [174], who proved the existence of stationary Nash equilibria in such

D. Lozovanu, S. W. Pickl, *Markov Decision Processes and Stochastic Positional Games*, International Series in Operations Research & Management Science 349,
https://doi.org/10.1007/978-3-031-40180-0_3

games. It was shown in [53, 160] that the problem of determining stationary Nash equilibria in a general n-person stochastic game with discounted payoffs can be represented as a nonlinear programming problem with linear constraints and the global minimum of the objective function equal to zero.

Mertens and Neyman [134] studied two-person zero-sum stochastic games and proved the existence of uniform ε-optimal strategies for the players, i.e., they showed that for every $\varepsilon > 0$, each of the two players has a strategy that guarantees the discounted value up to ε for every discount factor sufficiently close to 0. These results were extended by Vieilli [183] to zero-sum average stochastic games of two players, and afterward, they were used for studying the problem of the existence of Nash equilibria in non-stationary strategies for arbitrary two-player average stochastic games in [185]. Algorithmic approaches concerned with determining the optimal strategies of the players in some classes of stochastic games can be found in [53, 143, 160, 168, 169].

In general, n-person stochastic games with limiting average payoffs have been studied by many authors [143, 155, 167, 169, 183, 185, 189]; however, the existence of stationary Nash equilibria has only been proved for some classes of such games. Rogers [155] and Sobel [167] showed that stationary Nash equilibria exist for non-zero-sum stochastic games with average payoffs when the transition probability matrices induced by any stationary strategies of the players are unichain. An important class of average stochastic games for which stationary Nash equilibria exist is represented by stochastic positional games studied in [112]. Furthermore, in [112], it was shown that for average stochastic positional games with a unichain property and for two-player zero-sum stochastic positional games, there exist stationary Nash equilibria in pure strategy. The main results concerned with the existence and determination of Nash equilibria in two-player average stochastic games can be found in [134, 143, 169, 183, 185]. In the general case, for an average stochastic game with a given starting state, a stationary Nash equilibrium may not exist. An example of an average game for which equilibria in stationary strategies do not exist is represented by the "Big Match" game introduced by Gillette [63]. Additionally, in [57], an example of a three-player average stochastic game with a fixed starting state, for which a stationary Nash equilibrium does not exist, was constructed. Moreover, in [57], it was shown that for an m-player ($m \geq 3$) average stochastic game, also an ε-equilibrium ($\varepsilon > 0$) may not exist. In general, for an average stochastic game, a non-empty subset of states may exist such that if the game starts in one of them, then a stationary Nash equilibrium exists [54, 175]. However, the problem of determining the initial states in an average stochastic game for which stationary equilibria exist is a difficult problem.

In this chapter, we consider average stochastic games with finite state and action spaces. We show that an arbitrary average stochastic game in stationary strategies can be represented as a game in normal form where each payoff is quasi-monotonic (quasi-concave and quasi-convex) with respect to the strategy of the corresponding player. Furthermore, we show that if the game in normal form has a pure Nash equilibrium, then such an equilibrium corresponds to a stationary Nash equilibrium of the average stochastic game and vice versa.

Based on this result and the results of [36, 37, 64, 154] related to the existence of Nash equilibria in the games with quasi-concave (quasi-convex) payoffs, we formulate conditions for the existence and determination of stationary Nash equilibria in average stochastic games and present the game models in a normal form, allowing us to determine stationary equilibria.

3.2 Nash Equilibria Results for Non-cooperative Games

We present some necessary Nash equilibria existence results for non-cooperative games that we use for stochastic games. These results are mainly concerned with the existence of Nash equilibria in non-cooperative games with quasi-concave and quasi-monotonic payoffs.

A function $f : S \to R^1$ on a convex set $S \subseteq R^n$ is *quasi-concave* [23] if $\forall s', s'' \in S$ and $\forall \lambda \in [0, 1]$ holds $f(\lambda s' + (1-\lambda)s'') \geq \min\{f(s'), f(s'')\}$. If $\forall s', s'' \in S$ and $\forall \lambda \in [0, 1]$ holds $f(\lambda s' + (1-\lambda)s'') \leq \max\{f(s'), f(s'')\}$, then the function $f : S \to R^1$ is called *quasi-convex*. A function $f : S \to R^1$, $S \subseteq R^m$, which is quasi-concave and quasi-convex, is called *quasi-monotonic*. A detailed characterization of quasi-convex, quasi-concave, and quasi-monotonic functions with an application to *linear-fractional programming* problems can be found in [23].

Let $\langle \{S^i\}_{i=\overline{1,m}}, \{f^i(\mathbf{s})\}_{i=\overline{1,m}} \rangle$ be an m-player game in normal form, where $s^i \subseteq R^{n_i}$, $i = 1, 2, \ldots, m$ represent the corresponding sets of strategies of the players $1, 2, \ldots, m$ and $f^i : \prod_{j=1}^m S^i \to R^1$, $i = 1, 2, \ldots, m$ represent the corresponding payoffs of these players. Let $s = (s^1, s^2, \ldots, s^m)$ be a profile of strategies of the players $s \in S = \prod_{j=1}^m S^i$, and define $s^{-i} = (s^1, s^2, \ldots, s^{i-1}, s^{i+1}, \ldots, s^m)$, $S^{-i} = \prod_{j=1 (j \neq i)}^m S^i$, where $s^{-i} \in S^{-i}$. Thus, for an arbitrary $s \in S$, we can write $s = (s^i, s^{-i})$.

Fan [46] extended the well-known equilibrium result of Nash [141] to games with quasi-concave payoffs. He proved the following theorem:

Theorem 3.1 *Let $S^i \subseteq R^{n_i}$, $i = 1, 2, \ldots, m$ be non-empty, convex, and compact sets. If each payoff $f^i : S \to R^1$, $i \in \{1, 2, \ldots, m\}$ is continuous on $S = \prod_{j=1}^m S^i$ and quasi-concave with respect to s^i on S^i, then the game $\langle \{S^i\}_{i=\overline{1,m}}, \{f^i(s)\}_{i=\overline{1,m}} \rangle$ possesses a Nash equilibrium.*

Dasgupta and Maskin [36] considered a class of games with upper semi-continuous, quasi-concave, and *graph-continuous payoffs*.

Definition 3.2 The payoff function $f^i : \prod_{j=1}^m S^i \to R^1$, $i \in \{1, 2, \ldots, m\}$ of the game $\langle \{S^i\}_{i=\overline{1,m}}, \{f^i(s)\}_{i=\overline{1,m}} \rangle$ is *upper semi-continuous* if for any sequence $\{s_k\} \subseteq S = \prod_{j=1}^m S^i$ such that $\{s_k\} \to s$, we have $\limsup_{k \to \infty} f^i(s_k) \leq f^i(s)$. The payoff $f^i : \prod_{j=1}^m S^i \to R^1$ of the game $\langle \{S^i\}_{i=\overline{1,m}}, f^i(s)_{i=\overline{1,m}} \rangle$ is *lower* *semi-continuous* if for any sequence $\{s_k\} \subseteq S = \prod_{j=1}^{m} S^i$ such that $\{s_k\} \to s$, we have $\liminf_{k\to\infty} f^i(s_k) \geq f^i(s)$.

Definition 3.3 The payoff function $f^i : \prod_{j=1}^{m} S^j \to R^1$, $i \in \{1, 2, \ldots, m\}$ of the game $\langle \{S^i\}_{i=\overline{1,m}}, \{f^i(s)\}_{i=\overline{1,m}} \rangle$ is graph-continuous if for all $\overline{s} = (\overline{s}^i, \overline{s}^{-i}) \in S = \prod_{j=1}^{m} S^i$, there exists a function $F^i : S^{-i} \to S^i$ with $F^i(\overline{s}^{-i}) = \overline{s}^i$ such that $f^i(F^i(s^{-i}), s^{-i})$ is continuous in $s^{-i} = \overline{s}^{-i}$.

Dasgupta and Maskin [36] proved the following theorem:

Theorem 3.4 *Let $S^i \subseteq S^{n_i}$, $i = 1, 2, \ldots, m$ be non-empty, convex, and compact sets. If each payoff $f^i : \prod_{j=1}^{m} S^i \to R^1$, $i \in \{1, 2, \ldots, m\}$ is quasi-concave with respect to s^i on S^i, upper semi-continuous with respect to s on $S = \prod_{j=1}^{m} S^i$, and graph-continuous, then the game $\langle \{S^i\}_{i=\overline{1,m}}, \{f^i(s)\}_{i=\overline{1,m}} \rangle$ possesses a Nash equilibrium.*

Proof Let

$$\psi^i(s^{-i}) = \{\hat{s}^i \in S^i \mid f^i(\hat{s}^i, s^{-i}) = \max_{s^i \in S^i} f^i(s^i, s^{-i})\}, \quad i = 1, 2, \ldots, m$$

be the reaction correspondences of the players in the considered game. Then these correspondences exist, and they are compact-valued because $f^i(s), i = 1, 2, \ldots, m$ are upper semi-continuous on S. Moreover, each $\psi^i(s^{-i})$ is convex-valued because each $f^i(s)$ is quasi-concave with respect to $s^i \in S^i$.

Therefore, to prove the theorem, it is sufficient to verify that $\psi^i(s^{-i})$, $i = 1, 2, \ldots, m$ are upper hemi-continuous correspondences.

Consider the sequences $\{s_n^{-i}\} \subseteq S^{-i}$ and $\{s_n^i\} \subseteq S^i$ for $i \in \{1, 2, \ldots, m\}$ such that $s_n^{-i} \to \overline{s}^{-i}$, $s_n^i \to \overline{s}^i$, and $\forall n$, $s_n^i \in \psi^i(s_n^{-i})$. Let us show that $\overline{s}^i \in \psi^i(\overline{s}^{-i})$. Indeed, if $\overline{s}^i \notin \psi^i(\overline{s}^{-i})$ for some $i \in \{1, 2, \ldots, m\}$, then there exists $s^{i*} \in S^i$ such that $f^i(s^{i*}, \overline{s}^{-i}) > f^i(\overline{s}^i, \overline{s}^{-i})$.

Let $\varepsilon = [f^i(s^{i*}, \overline{s}^{-i}) - f^i(\overline{s}^i, \overline{s}^{-i})]/2$. Then there exists the function $F^i : S^{-i} \to S^i$ with $F^i(\overline{s}^{-i}) = s^{i*}$ and $\delta > 0$ such that $\|s^{-i} - \overline{s}^{-i}\| < \delta$ implies $|f^i\big(F^i(s^{-i}), s^{-i}\big) - f^i(s^{i*}, \overline{s}^{-i})| < \varepsilon$.

This means that for n sufficiently large, $|f^i\big(F^i(s_n^{-i}), s_n^{-i}\big) - f^i(s^{i*}, \overline{s}^{-i})| < \varepsilon$ is verified. But $f^i\big(F^i(s_n^{-i}), s_n^{-i}\big) \leq f^i(s_n^i, s_n^{-i})$. Therefore, for n sufficiently large, we have $f^i(s_n^i, s_n^{-i}) \geq f^i\big(F^i(s_n^{-i}), s_n^{-i}\big) \geq f^i(\overline{s}^i, \overline{s}^{-i}) + \varepsilon$. However, f^i is an upper semi-continuous function and $f^i(\overline{s}^i, \overline{s}^{-i}) \geq \limsup_{n\to\infty} f^i(s_n^i, s_n^{-i})$, i.e., we obtain a contradiction. So, $\overline{s}^i \in \psi^i(s^{-i})$, $i = 1, 2, \ldots, m$, and each reaction correspondence ψ^i is upper hemi-continuous. Therefore, we may apply the Kakutani [75] fixed-point theorem for the correspondence $\psi^1 \times \psi^2 \times \cdots \times \psi^m$ to conclude the existence of a Nash equilibrium for the game $\langle \{S^i\}_{i=\overline{1,m}}, \{f^i(s)\}_{i=\overline{1,m}} \rangle$. □

From Theorem 3.4 as a corollary, the following result can be derived:

Theorem 3.5 *Let $S^i \subseteq R^{n_i}$, $i = 1, 2, \ldots, m$ be non-empty, convex, and compact sets. If each payoff $f^i : \prod_{j=1}^{m} S^i \to R^1$, $i \in \{1, 2, \ldots, m\}$ is quasi-concave with respect to s^i on S^i and upper semi-continuous with respect to s on $S = \prod_{j=1}^{m} S^i$ and the functions $\varphi^i(s^-i) = \max_{s^i \in S^i} f^i(s^i, s^-i)$, $i = 1, 2, \ldots, m$ are lower semi-continuous, then the game $\langle \{S^i\}_{i=\overline{1,m}}, \{f^i(s)\}_{i=\overline{1,m}} \rangle$ possesses a Nash equilibrium.*

In the following, we use Theorem 3.4 for the case when each payoff $f^i(s^i, s^{-i})$, $i \in \{1, 2, \ldots, m\}$ is quasi-monotonic with respect to s^i on S^i and graph-continuous. In this case, the reaction correspondences of players

$$\psi^i(s^{-i}) = \{\hat{s}^i \in S^i \,|\, f^i(\hat{s}^i, \mathbf{s}^{-i}) = \max_{s^i \in S^i} f^i(s^i, s^{-i})\}, \quad i = 1, 2, \ldots, m$$

are compact and convex-valued, and therefore, the upper semi-continuous condition for the functions $f^i(s)$, $i = 1, 2, \ldots, m$ in Theorem 3.4 can be released. So, in this case, the theorem can be formulated as follows:

Theorem 3.6 *Let $S^i \subseteq R^{n_i}$, $i = \overline{1, m}$ be non-empty, convex, and compact sets. If each payoff $f^i : \prod_{j=1}^{m} S^i \to R^1, i \in \{1, 2, \ldots, n\}$ is quasi-monotonic with respect to s^i on S^i and graph-continuous, then the game $\langle \{S^i\}_{i=\overline{1,m}}, \{f^i(s)\}_{i=\overline{1,m}} \rangle$ possesses a Nash equilibrium.*

3.3 Formulation of Stochastic Games

We first present the framework of an m-person stochastic game and then specify the formulation of stochastic games with average and discounted payoffs when players use pure and mixed stationary strategies.

3.3.1 The Framework of an m-Person Stochastic Game

A stochastic game with m players consists of the following elements:

- A state space X (which we assume to be finite)
- A finite set $A^i(x)$ of actions with respect to each player $i \in \{1, 2, \ldots, m\}$ for an arbitrary state $x \in X$
- A payoff $r^i_{x,a}$ with respect to each player $i \in \{1, 2, \ldots, m\}$ for each state $x \in X$ and for an arbitrary action vector $a \in \prod_i A^i(x)$
- A transition probability function $p : X \times \prod_{x \in X} \prod_{i=1}^{m} A^i(x) \times X \to [0, 1]$

that gives the probability transitions $p_{x,y}^a$ from an arbitrary $x \in X$ to an arbitrary $y \in X$ for a fixed action vector $a \in \prod_i A^i(x)$, where $\sum_{y\in X} p_{x,y}^a = 1, \ \forall x \in X, \ a \in \prod_i A^i(x)$
- A starting state $x_0 \in X$

The game starts in state x_0 and proceeds in a sequence of stages. At stage t, the players observe state x_t and simultaneously and independently choose actions $a_t^i \in A^i(x_t)$, $i = 1, 2, \ldots, m$. Then nature selects a state $y = x_{t+1}$ according to probability transitions $p_{x_t,y}^{a_t}$ for the given action vector $a_t = (a_t^1, a_t^2, \ldots, a_t^m)$. Such a play of the game produces a sequence of states and actions $x_0, a_0, x_1, a_1, \ldots, x_t, a_t, \ldots,$ defining a stream of stage payoffs $r_t^1 = r_{x_t,a_t}^1, \ r_t^2 = r_{x_t,a_t}^2, \ldots, \ r_t^m = r_{x_t,a_t}^m, \quad t = 0, 1, 2, \ldots$.

The average stochastic game is the game with payoffs of the players

$$\omega_{x_0}^i = \lim_{t\to\infty} \inf \mathbb{E}\left(\frac{1}{t}\sum_{\tau=0}^{t-1} r_\tau^i\right), \quad i = 1, 2, \ldots, m,$$

where $\mathbb{E}$ is the expectation operator with respect to the probability measure in the Markov process induced by actions chosen by players in their position sets and given starting state x_0. Here, $\omega_{x_o}^i$ expresses the *average payoff per transition* of player i in the infinite game. Each player in this game aims to maximize his or her average payoff per transition. In the case $m = 1$, this game becomes the average Markov decision problem with a probability transition function $p : X \times \prod_{x\in X} A(x) \times X \to [0, 1]$ and step rewards $r_x^a = r_{x,a}^1$ in the states $x \in X$ for given actions $a \in A(x) = A^1(x)$.

The discounted stochastic game with given discount factor γ, $0 < \gamma < 1$ is the game with payoffs of the players

$$\sigma_{x_0,\gamma}^i = \mathbb{E}\left(\sum_{\tau=0}^{\infty} \gamma^\tau r_\tau^i\right), \quad i = 1, 2, \ldots, m.$$

In the following, we study stochastic games with average and discounted payoffs when players use pure and mixed stationary strategies of selecting the actions in the states.

3.3.2 Stationary, Non-stationary, and Markov Strategies

We define *stationary, history-dependent, and Markov strategies* in stochastic games similar to Markov decision processes.

A strategy (a policy) of player i, $i \in \{1, 2, \ldots, m\}$ in a stochastic game is a mapping s^i that provides for every state $x_t \in X$ a probability distribution over the set of actions $A^i(x_t)$. These probabilities may depend, in general, not only on x and

t but also on the entire history h in the game up to time t. If these probabilities take only values 0 and 1, then s^i is called *a pure strategy*; otherwise, s^i is called *a mixed strategy*. If these probabilities only depend on the state $x_t = x \in X$ (i.e., s^i does not depend on t), then s^i is called *a stationary strategy*; otherwise, s^i is called a *non-stationary strategy*. If a non-stationary strategy is memoryless, then such a strategy is called *Markov strategy*. Thus, *a pure stationary strategy* of player $i \in \{1, 2, \dots, m\}$ can be regarded as a map

$$s^i : x \to a^i \in A^i(x) \text{ for } x \in X$$

that determines an action $a^i \in A^i(x)$ for each state $x \in X$, i.e., $s^i(x) = a^i$. Obviously, the corresponding sets of pure stationary strategies $\mathbb{S}^1, \mathbb{S}^2, \dots, \mathbb{S}^m$ of the players in the game with finite state and action spaces are finite sets. In the following, we identify a pure stationary strategy $s^i(x)$ of player i with the set of Boolean variables $s^i_{x,a^i} \in \{0, 1\}$, where for a given $x \in X$, $s^i_{x,a^i} = 1$ if and only if player i fixes the action $a^i \in A^i(x)$. So, we can represent the set of pure stationary strategies S^i of player i as the set of solutions to the system

$$\begin{cases} \sum\limits_{a^i \in A^i(x)} s^i_{x,a^i} = 1, & \forall x \in X; \\ s^i_{x,a^i} \in \{0, 1\}, & \forall x \in X, \ \forall a^i \in A^i(x) \end{cases}$$

and in the following take into account that at the same time, s^i is a map for which $s^i(x) = a^i \in A^i(x)$ if $s_{x,a^i} = 1$. If in this system we change the restriction $s^i_{x,a^i} \in \{0, 1\}$ for $x \in X$, $a^i \in A^i(x)$ to the condition $0 \le s^i_{x,a^i} \le 1$, then we obtain the set of stationary strategies in the sense of Shapley [165], where s^i_{x,a^i} is treated as the probability of the choices of the action a^i by player i every time when the state x is reached by any route in the dynamic stochastic game. Thus, we can identify the set of mixed stationary strategies of the players with the set of solutions to the system

$$\begin{cases} \sum\limits_{a^i \in A^i(x)} s^i_{x,a^i} = 1, & \forall x \in X; \\ s^i_{x,a^i} \ge 0, & \forall x \in X, \ \forall a^i \in A^i(x), \end{cases} \tag{3.1}$$

and for a given profile $s = (s^1, s^2, \dots, s^m)$ of mixed strategies $s^1, s^2, \dots, s^m$ of the players, the probability transition $p^s_{x,y}$ from a state x to a state y can be calculated as follows:

$$p^s_{x,y} = \sum_{(a^1, a^2, \dots, a^m) \in A(x)} \prod_{k=1}^{m} s^k_{x,a^k} p^{(a^1,a^2,\dots,a^m)}_{x,y}. \tag{3.2}$$

Here, $a = (a^1, a^2, \ldots, a^m)$ for a given $x \in X$ represents a profile of actions $a^i \in A^i(x)$ and $A(x) = \prod_{i=1}^m A^i(x)$. The set of mixed stationary strategies of player i we denote by S^i.

3.3.3 Stochastic Games in Pure and Mixed Strategies

In the following, we distinguish different versions of stochastic games in pure and mixed stationary strategies:

Games in Pure and Mixed Stationary Strategies
Let $s = (s^1, s^2, \ldots, s^m)$ be a profile of pure stationary strategies of the players, and denote by $a(s) = (a^1(s), a^2(s), \ldots, a^m(s)) \in \prod_{x\in X}\prod_{i=1}^m A^i(x)$ the action vector that corresponds to s and determines the probability distributions $p^s_{x,y} = p^{a(s)}_{x,y}$ in the states $x \in X$. Then the average payoffs per transition $\omega^1_{x_0}(s),\ \omega^2_{x_0}(s), \ldots, \omega^m_{x_0}(s)$ for the players are determined as follows:

$$\omega^i_{x_0}(s) = \sum_{y\in X} q^s_{x_0,y} r^i_{y,a(s)}, \quad i = 1, 2, \ldots, m,$$

where $q^s_{x_o,y}$ represent the limiting probabilities in the states $y \in X$ for the Markov process with a probability transition matrix $P^s = (p^s_{x,y})$ when the transitions start in x_0. So, if for the Markov process with probability matrix P^s the corresponding limiting probability matrix $Q^s = (q^s_{x,y})$ is known, then $\omega^1_x, \omega^2_x, \ldots, \omega^m_x$ can be determined for an arbitrary starting state $x \in X$ of the game. The functions $\omega^1_{x_0}(s),\ \omega^2_{x_0}(s), \ldots, \omega^m_{x_0}(s)$ on $\mathbb{S} = \mathbb{S}^1 \times \mathbb{S}^2 \times \cdots \times \mathbb{S}^m$ define a game in normal form that we denote by $\langle\{\mathbb{S}^i\}_{i=\overline{1,m}},\ \{\omega^i_{x_0}(s)\}_{i=\overline{1,m}}\rangle$. This game corresponds to an *average stochastic game in pure stationary strategies* that in extended form is determined by the tuple $(X,\ \{A^i\}_{i=\overline{1,m}}, \{r^i\}_{i=\overline{1,m}},\ p,\ x_0)$, where $A^i = \cup_{x\in X} A^i(x)$ and r^i is the reward vector of player i.

If an arbitrary profile $s = (s^1, s^2, \ldots, s^m)$ of pure stationary strategies in a stochastic game induces a probability transition matrix P^s that is unichain, then we say that the game possesses a unichain property; shortly, we call it *unichain stochastic game;* otherwise, we call it *multichain stochastic game.*

Let $s = (s^1, s^2, \ldots, s^m)$ be a profile of mixed stationary strategies of the players. Then elements of the probability transition matrix $P^s = (p^s_{x,y})$ in the Markov process induced by s can be calculated according to (3.2). Therefore, if $Q^s = (q^s_{x,y})$ is the limiting probability matrix of P^s, then the average payoffs per transition $\omega^1_{x_0}(s),\ \omega^2_{x_0}(s), \ldots, \omega^m_{x_0}(s)$ for the players are determined as follows:

$$\omega^i_{x_0}(s) = \sum_{y\in X} q^s_{x_0,y} r^i_{y,s}, \quad i = 1, 2, \ldots, m, \tag{3.3}$$

where

$$r^i_{y,s} = \sum_{(a^1,a^2,\dots,a^m)\in A(y)} \prod_{k=1}^{n} s^k_{y,a^k} r^i_{y,(a^1,a^2,\dots,a^m)} \tag{3.4}$$

expresses the average payoff (immediate reward) in the state $y \in X$ of player i when the corresponding stationary strategies $s^1, s^2, \dots, s^m$ have been applied by players $1, 2, \dots, m$ in y.

Let $S^1, S^2, \dots, S^m$ be the corresponding sets of mixed stationary strategies for the players $1, 2, \dots, m$, i.e., each S^i for $i \in \{1, 2, \dots, m\}$ represents the set of solutions to system (3.1). The functions $\omega^1_{x_0}(s), \omega^2_{x_0}(s), \dots, \omega^m_{x_0}(s)$ on $S = S^1 \times S^2 \times \cdots \times S^m$, defined according to (3.3) and (3.4), determine a game in normal form that we denote by $\langle \{S^i\}_{i=\overline{1,m}}, \{\omega^i_{x_0}(s)\}_{i=\overline{1,m}} \rangle$. This game corresponds to an *average stochastic game in mixed stationary strategies* that in extended form is determined by the tuple $(X, \{A^i\}_{i=\overline{1,m}}, \{r^i\}_{i=\overline{1,m}}, p, x_0)$.

In a similar way, stochastic games in pure and mixed stationary strategies can be classified in the case of discounted payoffs for the players.

Discounted Stochastic Games in Stationary Strategies

Let $s = (s^1, s^2, \dots, s^m)$ be a profile of stationary strategies (pure or mixed strategies) of the players. Then the elements of the probability transition matrix $P^s = (p^s_{x,y})$ induced by s can be calculated according to (3.2), and we can find the matrix $W^s = (w^s_{x,y})$, where $W^s = (I - \gamma P^s)^{-1}$. After that, we can find the payoff for the players as follows:

$$\sigma^i_{x_0}(s) = \sum_{y\in X} w_{x_0,y} r^i_{y,s}, \quad i = 1, 2, \dots, m,$$

where $r^i_{y,s}$ is determined according to (3.4). A discounted stochastic game is determined by the tuple $(X, \{A^i\}_{i=\overline{1,m}}, \{r^i\}_{i=\overline{1,m}}, p, \gamma, x_0)$.

Stochastic Games with a Random Starting State

We also consider stochastic games in which the starting state is chosen randomly according to a given distribution $\{\theta_x\}$ on X. So, for a given average stochastic game, we assume that the play starts in the states $x \in X$ with probabilities $\theta_x > 0$, where $\sum_{x\in X} \theta_x = 1$.

If the players use mixed stationary strategies of selecting the actions in the states, then the payoff functions

$$\psi^i_\theta(s^1, s^2, \dots, s^m) = \sum_{x\in X} \theta_x \omega^i_x(s^1, s^2, \dots, s^m), \quad i = 1, 2, \dots, m$$

on $S = S^1 \times S^2 \times \ldots \times S^m$ define a game in normal form $\langle \{S^i\}_{i=\overline{1,m}}, \{\psi^i_\theta(s)\}_{i=\overline{1,m}} \rangle$ that in extended form is determined by the tuple $(X, \{A^i\}_{i=\overline{1,m}},$

$\{r^i\}_{i=\overline{1,m}}, p, \{\theta_x\})$. In the case $\theta_x = 0, \forall x \in X \setminus \{x_0\}$, $\theta_{x_o} = 1$, the considered game becomes a stochastic game with a fixed starting state x_0. In an analogous way, we can specify the game in normal form $\langle\{S^i\}_{i=\overline{1,m}}, \{\psi^i_\theta(s)\}_{i=\overline{1,m}}\rangle$ for the average stochastic game with a random starting state when players use pure stationary strategies to select the actions in the states.

The game in normal form $\langle\{S^i\}_{i=\overline{1,m}}, \{\phi^i_\theta(s)\}_{i=\overline{1,m}}\rangle$ for a discounted stochastic game we obtain by using the payoffs

$$\phi^i_\theta(s^1, s^2, \ldots, s^m) = \sum_{x\in X} \theta_x \sigma^i_x(s^1, s^2, \ldots, s^m), \quad i = 1, 2, \ldots, m,$$

where θ_x for $x \in X$ are the same as for the average stochastic game.

Stationary Nash Equilibria
Let $s = (s^1, s^2, \ldots, s^m) \in S$. Define $s^{-i} = (s^1, s^2, \ldots, s^{i-1}, s^{i+1}, \ldots, s^m)$ as the vector of stationary strategies of all players other than i, and denote $s = (s^i, s^{-i}), i = 1, 2, \ldots, m$. The profile $s^* = (s^{1^*}, s^{2^*}, \ldots, s^{m*})$ is called *stationary Nash equilibrium* for an average stochastic game $\langle\{S^i\}_{i=\overline{1,m}}, \{\omega^i_{x_0}(s)\}_{i=\overline{1,m}}\rangle$ with given starting state x_0 if

$$\omega^i_{x_0}(s^{i*}, s^{-i*}) \geq \omega^i_{x_0}(s^i, s^{-i*}), \quad \forall s^i \in \overline{S}^i, \; i = 1, 2, \ldots, m. \tag{3.5}$$

The profile $s^* = (s^{1^*}, s^{2^*}, \ldots, s^{m*})$ is called *stationary Nash equilibrium* for an average stochastic game $\langle\{S^i\}_{i=\overline{1,m}}, \{\psi^i_\theta(s)\}_{i=\overline{1,m}}\rangle$ when the starting state is chosen randomly according to a given distribution $\{\theta_x\}$ on X if

$$\psi^i_\theta(s^{i*}, s^{-i*}) \geq \psi^i_\theta(s^i, s^{-i*}), \quad \forall s^i \in S^i, \; i = 1, 2, \ldots, m. \tag{3.6}$$

In an analogous way, stationary Nash equilibria for discounted stochastic games with payoffs $\sigma^i_{x_0}(s)$ and $\phi^i_\theta(s)$ are defined.

3.4 Stationary Equilibria for Discounted Stochastic Games

As mentioned in Sect. 3.1, the existence of stationary Nash equilibria for stochastic games with discounted payoffs was proven in [55, 174]. Here, we present a new proof of this result and propose a continuous game model in normal form that allows us to determine all stationary Nash equilibria of a discounted stochastic game with finite state and action spaces.

In Sect. 2.3.4, based on Theorem 2.5, the optimization model (2.33)–(2.35) was formulated that determines all optimal stationary strategies in a discounted Markov decision problem with finite state and action spaces.

Here, based on this theorem, we present the formulation of a normal-form game in stationary strategies for an m-player discounted stochastic game as follows:

Let $S^1, S^2, \dots, S^m$ be the sets of mixed stationary strategies of a discounted Markov decision problem, where each S^i, $i \in \{1, 2, \dots, m\}$ represents the set of solutions to system (3.1). For an arbitrary player $i \in \{1, 2, \dots, m\}$ on $S = S^1 \times S^2 \times \cdots \times S^m$, we define the payoff

$$\begin{cases} \phi_\theta^i(s^1, s^2, \ \dots, \ s^m) = \sum\limits_{x\in X} \ \sum\limits_{(a^1,a^2,\dots,a^m)\in A(x)} \prod\limits_{k=1}^{m} s^k_{x,a^k} \cdot r^i_{x,(a^1,a^2\dots a^m)} \cdot q_x, \\ \qquad\qquad\qquad\qquad\qquad\qquad\qquad\qquad i = 1, 2, \dots, m, \end{cases} \tag{3.7}$$

where q_x, $x \in X$ are uniquely determined by the following system of equations:

$$q_y - \gamma \sum_{x\in X} \ \sum_{(a^1,a^2,\dots,a^m)\in A(x)} \ \prod_{k=1}^{m} s^k_{x,a^k} \cdot p^{(a^1,a^2,\dots,a^m)}_{x,y} \cdot q_x = \theta_y, \qquad \forall y \in X \tag{3.8}$$

for an arbitrary fixed $s = (s^1, s^2, \dots, s^m) \in S = S^1 \times S^2 \times \cdots \times S^m$.

So, we have a game in normal form $\langle \{S^i\}_{i=\overline{1,m}}, \ \{\phi_\theta^i(s)\}_{i=\overline{1,m}} \rangle$, where each payoff function $\phi_\theta^i(s^1, s^2, \ \dots, \ s^m)$ on S is continuous. Additionally, according to Theorem 2.5, each $\phi_\theta^i(s^1, s^2, \ \dots, \ s^m)$ is quasi-monotonic with respect to strategy s^i on S^i, when the remaining players fix their strategies. Therefore, based on this theorem and the results of Dasgupta and Maskin [36] and Debreu [37], we obtain the following theorem:

Theorem 3.7 *The game $\langle \{S^i\}_{i=\overline{1,m}}, \ \{\phi_\theta^i(s)\}_{i=\overline{1,m}} \rangle$ with $\theta_x > 0$, $\forall x \in X$, $\sum_{x\in X} \theta_x = 1$, has a Nash equilibrium $s^* = (s^{1*}, s^{2*}, \dots, s^{m*}) \in S = S^1 \times S^2 \times \cdots \times S^m$ that is a stationary Nash equilibrium of the discounted stochastic game with an arbitrary starting state $x \in X$.*

In [77, 152], a series of iterative algorithms for determining the optimal stationary strategies of the players in a discounted stochastic game with a given discount factor γ, $0 < \gamma < 1$, based on the Kakutani fixed-point theorem, were presented. The efficiency of these algorithms was analyzed in a detailed form in [153].

In general, a stochastic game can be considered with different discounted factors $\gamma^1, \gamma^2, \dots, \gamma^m$ for the players, where γ^i is the discounted factor of player i. In this case, the game in normal form is determined by payoffs on S defined according to (3.7), where q_x, $x \in X$ are uniquely determined by the following system of equations:

$$q_y - \gamma^i \sum_{x \in X} \sum_{(a^1, a^2, \ldots, a^m) \in A(x)} \prod_{k=1}^{m} s^k_{x,a^k} \cdot p^{(a^1, a^2, \ldots, a^m)}_{x,y} \cdot q_x = \theta_y, \qquad \forall y \in X. \tag{3.9}$$

Then Theorem 3.7 holds for the game with different discounted factors for the players because each payoff $\phi^i_\theta(s)$, $i \in \{1, 2, \ldots, m\}$ is continuous in S and quasi-monotonic with respect to $s^i \in S^i$.

Remark 3.8 A particularly important case of a discounted stochastic game is represented by the game in which an arbitrary profile of pure stationary strategies $s = (s^1, s^2, \ldots, s_m)$ for the players induces a transition probability matrix P^s that corresponds to a Markov unichain with an absorbing state $z \in X$, where the rewards of all players in the absorbing state are equal to zero. For such a game, a Nash equilibrium exists for an arbitrary $\gamma > 0$. In the case $\gamma = 1$, the payoffs of the players express the expected total rewards from the starting state to the absorbing state.

3.5 On Nash Equilibria for Average Stochastic Games

In this section, we study the problem of the existence of Nash equilibria in average stochastic games. We show that a unichain average stochastic game in stationary strategies can be represented as a continuous game in normal form where the payoffs are quasi-monotonic with respect to the corresponding strategies of the players. Based on this, we prove the existence of Nash equilibria in stationary strategies for an average stochastic game with unichain property and propose an approach to determine stationary Nash equilibria. For the general case of an average stochastic game (the multichain case), we show that it can also be represented as the game in normal form with convex and compact sets of strategies for the players and quasi-monotonic payoffs with respect to the corresponding strategies. However, the payoffs in such a model may not be continuous, and therefore, Nash equilibria may not exist. Nevertheless, we can see that for some classes of games, such an approach can be used to determine the optimal stationary strategies of the players.

3.5.1 Stationary Equilibria for Unichain Games

In this section, we present the game model in stationary strategies for an average stochastic game with a unichain property.
We show that such a model can be formulated by using the average Markov decision problem from Sect. 2.4.3. So, based on the optimization model of Markov decision problem (2.51), (2.52), we formulate the average stochastic game with a unichain property in stationary strategies as follows:

Let $S = S^1 \times S^2 \times \cdots \times S^m$, where each S^i for $i \in \{1, 2, \ldots, m\}$ represents the set of solutions to system (3.1), i.e., S^i represents the set of mixed stationary strategies for player i. On S, we define the average payoffs for the players as follows:

$$\psi^i(s^1, s^2, \ldots, s^m) = \sum_{x \in X} \sum_{(a^1,a^2,\ldots,a^m) \in A(x)} \prod_{k=1}^{m} s^k_{x,a^k} \cdot r^i_{x,(a^1,a^2,\ldots,a^m)} \cdot q_x,$$
$$i = 1, 2, \ldots, m,$$

where q_x for $x \in X$ are uniquely determined by the following system of linear equations:

$$\begin{cases} \sum\limits_{x \in X} \sum\limits_{(a^1,a^2,\ldots,a^m) \in A(x)} \prod\limits_{k=1}^{m} s^k_{x,a^k} \cdot p^{(a^1,a^2,\ldots,a^m)}_{x,y} \cdot q_x = q_y, \quad \forall y \in X; \\ \sum\limits_{x \in X} q_x = 1, \end{cases}$$

where $s^i \in S^i$, $i = 1, 2, \ldots, m$. The functions $\psi^i(s^1, s^2, \ldots, s^m)$, $i = 1, 2, \ldots, m$ on S define a game in normal form $\langle \{S^i\}_{i=\overline{1,m}}, \{\psi^i(s)\}_{i=\overline{1,m}} \rangle$ that corresponds to a stationary average stochastic game with unichain property, where $\psi^i(s^1, s^2, \ldots, s^n) = \omega^i_x(s^1, s^2, \ldots, s^m)$, $\forall x \in X$, $i = 1, 2, \ldots, n$.

From Lemma 2.7, we obtain the following result:

Lemma 3.9 *Each payoff function $\psi^i(s^i, s^{-i})$, $i \in \{1, 2, \ldots, m\}$ of a unichain stochastic game in normal form $\langle \{S^i\}_{i=\overline{1,m}}, \{\psi^i(s)\}_{i=\overline{1,m}} \rangle$ is quasi-monotonic with respect to $s^i \in \overline{S}^i$ for arbitrary fixed $s^{-i} \in \overline{S}^{-i}$.*

Based on Lemma 3.9 and the results from [37, 45, 46, 64], we obtain the following theorem:

Theorem 3.10 *Let $\langle \{S^i\}_{i=\overline{1,m}}, \{\psi^i(s)\}_{i=\overline{1,m}} \rangle$ be an average stochastic game determined by $(X, \{A^i\}_{i=\overline{1,m}}, \{r^i\}_{i=\overline{1,m}}, p, \overline{x})$. If for an arbitrary $s = (s^1, s^2, \ldots, s^m) \in S$ of the game the transition probability matrix $P^s = (p^s_{x,y})$ corresponds to a Markov unichain, then for the game $\langle \{S^i\}_{i=\overline{1,m}}, \{\psi^i(s)\}_{i=\overline{1,m}} \rangle$,* *there exists a Nash equilibrium* $s^* = (s^{1*}, s^{2*}, \ldots, s^{m*})$, *which is a Nash equilibrium for an arbitrary starting state* $\overline{x} \in X$.

Proof According to Lemma 3.9, each payoff $\psi^i(s^i, s^{-i})$, $i \in \{1, 2, \ldots, m\}$ is quasi-monotonic with respect to $s^i \in S^i$ for fixed $s^{-i} \in S^{-i}$. Additionally, each payoff $\psi^i(s)$, $i \in \{1, 2, \ldots, m\}$ is continuous on S because the stochastic game is unichain. Then according to [37, 45, 46, 64], the game $\langle \{S^i\}_{i=\overline{1,m}}, \{\psi^i(s)\}_{i=\overline{1,m}} \rangle$ possesses a pure Nash equilibrium $s^* \in S$, which is a stationary Nash equilibrium for the unichain average stochastic game with an arbitrary starting state $x \in X$. □

Thus, if we find a pure Nash equilibrium s^* for the game in normal form $\langle \{S^i\}_{i=\overline{1,m}}, \{\psi^i(s)\}_{i=\overline{1,m}} \rangle$, then s^* is a stationary Nash equilibrium for the average stochastic game with unichain property. Note that if it is known that the average stochastic game is unichain, then we can use the continuous game model above to determine the stationary Nash equilibrium. However, as shown in [181], the problem of checking the unichain condition for a stochastic game is NP-hardness.

3.5.2 Some Results for Multichain Stochastic Games

Now we extend the results from Sect. 3.5.1 for the multichain case of an average stochastic game. We show that the normal form of such a game in stationary strategies can be formulated as the game in which the payoffs are quasi-monotonic with respect to the corresponding strategies of the players. Based on this, we formulate conditions for the existence of stationary Nash equilibria in multichain average stochastic games.

The Normal Form of a Game in Stationary Strategies

The multichain average stochastic game in stationary strategies that generalizes the unichain game model from Sect. 3.5.1 is the following:

Let S^i, $i \in \{1, 2, \ldots m\}$ be the set of solutions to the system

$$\begin{cases} \sum\limits_{a_i \in A^i(x)} s^i_{x,a^i} = 1, \quad \forall x \in X; \\ \qquad\quad s^i_{x,a^i} \geq 0, \ \forall x \in X, \ a^i \in A^i(x) \end{cases} \tag{3.10}$$

that determine the set of stationary strategies of player i. Each S^i is a convex compact set and an arbitrary extreme point corresponding to a basic solution s^i of system (3.10), where $s^i_{x,a^i} \in \{0, 1\}$, $\forall x \in X$, $a^i \in A(x)$, i.e., each basic solution to this system corresponds to a pure stationary strategy of player i. On the set $S = S^1 \times S^2 \times \cdots \times S^m$, we define m payoff functions

$$\begin{cases} \psi^i_\theta(s^1, s^2, \ldots, s^m) = \sum\limits_{x \in X} \sum\limits_{(a^1,a^2,\ldots,a^m) \in A(x)} \prod\limits_{k=1}^{m} s^k_{x,a^k} \cdot r^i_{x,(a^1,a^2 \ldots a^m)} \cdot q_x, \\ \qquad\qquad\qquad\qquad\qquad\qquad\qquad\qquad i = 1, 2, \ldots, m, \end{cases} \tag{3.11}$$

where q_x for $x \in X$ are uniquely determined by the following system of linear equations:

$$\begin{cases} q_y - \sum\limits_{x\in X} \sum\limits_{(a^1,a^2,\dots,a^m)\in A(x)} \prod\limits_{k=1}^{m} s^k_{x,a^k} \cdot p_{x,y}^{(a^1,a^2,\dots,a^m)} \cdot q_x = 0, & \forall y \in X; \\ q_y + w_y - \sum\limits_{x\in X} \sum\limits_{(a^1,a^2,\dots,a^m)\in A(x)} \prod\limits_{k=1}^{m} s^k_{x,a^k} \cdot p_{x,y}^{(a^1,a^2,\dots,a^m)} \cdot w_x = \theta_y, & \forall y \in X, \end{cases} \tag{3.12}$$

for a fixed $s = (s^1, s^2, \dots, s^m) \in S$. The functions $\psi^i_\theta(s^1, s^2, \dots, s^m)$, $i = 1, 2, \dots, m$ represent the payoff functions for the average stochastic game in normal form $\langle \{S^i\}_{i=\overline{1,m}}, \{\psi^i_\theta(s)\}_{i=\overline{1,m}} \rangle$. This game is determined by the tuple $(X, \{A^i\}_{i=\overline{1,m}}, \{r^i\}_{i=\overline{1,m}}, p, \{\theta_y\})$, where θ_y for $y \in X$ are given positive values such that $\sum_{y\in X} \theta_y = 1$.

If $\theta_y = 0, \ \forall y \in X \setminus \{x_0\}$ and $\theta_{x_0} = 1$, then we obtain an average stochastic game in normal form $\langle \{S^i\}_{i=\overline{1,m}}, \{\omega^i_{x_0}(s)\}_{i=\overline{1,m}} \rangle$ when the starting state x_0 is fixed, i.e., $\psi^i_\theta(s^1, s^2, \dots, s^n) = \omega^i_{x_0}(s^1, s^2, \dots, s^m)$, $i = 1, 2, \dots, m$. So, in this case, the game is determined by $(X, \{A^i\}_{i=\overline{1,m}}, \{r^i\}_{i=\overline{1,m}}, p, x_0)$.

If $\theta_y > 0, \ \forall y \in X$ and $\sum_{y\in X} \theta_y = 1$, then we obtain an average stochastic game when the play starts in the states $y \in X$ with probabilities θ_y. In this case, for the payoffs of the players in the game in normal form, we have

$$\psi^i_\theta(s^1, s^2, \dots, s^m) = \sum_{y\in X} \theta_y \cdot \omega^i_y(s^1, s^2, \dots, s^m), \quad i = 1, 2, \dots, m. \tag{3.13}$$

This game model represents the game variant of multichain average Markov decision problem (2.80)–(2.82) from Sect. 2.4.7.

The Main Properties of a Game in Normal Form

In [128], the following theorem was proved:

Theorem 3.11 *Let $\langle \{S^i\}_{i=\overline{1,m}}, \{\psi^i_\theta(s)\}_{i=\overline{1,m}} \rangle$ be the game in normal form for the average stochastic game in stationary strategies determined by the tuple $(X, \{A^i\}_{i=\overline{1,m}}, \{r^i\}_{i=\overline{1,m}}, p, \{\theta_y\})$, where $\theta_y > 0, \ \forall y \in X, \ \sum_{y\in X} \theta_y = 1$. If for this game there exists a Nash equilibrium $s^* = (s^{1*}, s^{2*}, \dots, s^{m*})$, then it is a Nash equilibrium for the game in normal form $\langle \{S^i\}_{i=\overline{1,m}}, \{\omega^i_y(s)\}_{i=\overline{1,m}} \rangle$ with an arbitrary $y \in X$, i.e., $s^* = (s^{1*}, s^{2*}, \dots, s^{m*})$ is a stationary Nash equilibrium of the average stochastic game with an arbitrary starting state $y \in X$. Conversely, if for an arbitrary starting state $y \in X$ the corresponding game in normal form $\langle \{S^i\}_{i=\overline{1,m}}, \{\omega^i_y(s)\}_{i=\overline{1,m}} \rangle$ has a Nash equilibrium, then for an arbitrary distribution function $\{\theta_y\}$ on X with $\theta_y > 0, \forall y \in X$ $(\sum_{y\in X} \theta_y = 1)$, the corresponding game in normal form $\langle \{S^i\}_{i=\overline{1,m}}, \{\psi^i_\theta(s)\}_{i=\overline{1,m}} \rangle$ of the average stochastic game* *determined by* $(X, \{A^i\}_{i=\overline{1,m}}, \{r^i\}_{i=\overline{1,m}}, p, \{\theta_y\})$ *has a Nash equilibrium* $s^* = (s^{1*}, s^{2*}, \ldots, s^{m*})$*, which is a Nash equilibrium for each of the game in normal form* $\langle\{S^i\}_{i=\overline{1,m}}, \{\omega^i_y(s)\}_{i=\overline{1,m}}\rangle$ *with the corresponding starting states* $y \in Y$.

Proof ($\Rightarrow$) Let $s^* = (s^{1*}, s^{2*}, \ldots, s^{m*})$ be a Nash equilibrium for the game in normal form $\langle\{S^i\}_{i=\overline{1,m}}, \{\psi^i_\theta(s)\}_{i=\overline{1,m}}\rangle$ determined by the tuple $(X, \{A^i\}_{i=\overline{1,m}}, \{r^i\}_{i=\overline{1,n}}, p, \{\theta_y\})$, where $\theta_y > 0, \forall y \in X, \sum_{y\in X}\theta_y = 1$. Then $(s^{1*}, s^{2*}, \ldots, s^{m*})$ is a Nash equilibrium for the average stochastic game $\langle\{S^i\}_{i=\overline{1,m}}, \{\psi^i_{\theta'}(s)\}_{i=\overline{1,m}}\rangle$ with an arbitrary distribution $\{\theta'_y\}$ on X, where $\theta'_y > 0$, $\forall y \in X$, $\sum_{y\in X}\theta'_y = 1$, i.e.,

$$\psi^i_{\theta'}(s^{i*}, s^{-i*}) \geq \psi^i_{\theta'}(s^i, s^{-i*}), \qquad \forall s^i \in S^i, \quad i = 1, 2, \ldots, m.$$

If here we express $\psi^i_{\theta'}$ via ω^i_y using (3.13), then we obtain

$$\sum_{y\in X}\theta'_y(\omega^i_y(s^{i*}, s^{-i*}) - \omega^i_y(s^i, s^{-i*})) \geq 0, \forall s^i \in S^i,\ i = 1, 2, \ldots, m.$$

This property holds for arbitrary $\theta'_y > 0$, $\forall y \in X$ such that $\sum_{y\in Y}\theta'_y = 1$, and therefore, for an arbitrary $y \in X$, we have

$$\omega^i_y(s^{i*}, s^{-i*}) - \omega^i_y(s^i, s^{-i*}) \geq 0, \qquad \forall s^i \in S^i, \quad i = 1, 2, \ldots, m.$$

So, $(s^{1*}, s^{2*}, \ldots, s^{m*})$ is a Nash equilibrium for each game in normal form $\langle\{S^i\}_{i=\overline{1,m}}, \{\omega^i_y(s)\}_{i=\overline{1,m}}\rangle$ with the corresponding starting state $y \in X$.

($\Leftarrow$) Assume that for each starting state $y \in X$, the average stochastic game $\langle\{S^i\}_{i=\overline{1,m}}, \{\omega^i_y(s)\}_{i=\overline{1,m}}\rangle$ has a stationary Nash equilibrium. Let us show that for the game in normal form $\langle\{S^i\}_{i=\overline{1,m}}, \{\psi^i_\theta(s)\}_{i=\overline{1,m}}\rangle$, determined by $(X, \{A^i\}_{i=\overline{1,m}}, \{r^i\}_{i=\overline{1,m}}, p, \{\theta_y\})$, where $\theta_y > 0, \forall y \in X$, $\sum_{y\in X}\theta_y = 1$, there exists a Nash equilibrium. We prove this using an auxiliary average stochastic game with a new starting state z and the set of states $X \cup \{z\}$, where for an arbitrary state $x \in X$, each player $i \in \{1, 2, \ldots, m\}$ has the same set of actions $A^i(x)$, the same payoffs $r^i_{x,a}$ for $a \in A(x)$, and the same transition probability distributions $p^a_{x,y}$ for $a \in A(x)$ as in the game determined by $(X, \{A^i\}_{i=\overline{1,m}}, \{r^i\}_{i=\overline{1,m}}, p, \{\theta_y\})$; in the state z of the auxiliary game, each player $i \in \{1, 2, \ldots, m\}$ has a single action a^i_z, and $A(z)$ contains a unique profile $a_z = (a^1_z, a^2_z, \ldots, a^m_z)$ for which $p^{a_z}_{z,z} = 0$, $p^{a_z}_{z,y} = \theta_y, \forall y \in X$ and $r^i_{z,a_z} = 0$, $i = 1, 2, \ldots, m$. Obviously, for the auxiliary average stochastic game with starting state z, determined by $(X\cup\{z\}, \{A^i\cup a^i_z\}_{i=\overline{1,m}}, \{r^i, r^i_{z,a_z}\}_{i=\overline{1,m}}, \{p\} \cup \{p^{a_z}_{z,y}\}_{y\in X}, z)$, there exists a stationary Nash equilibrium because a Nash equilibrium exists for an arbitrary average stochastic game $\langle\{S^i\}_{i=\overline{1,m}}, \{\omega^i_y(s)\}_{i=\overline{1,m}}\rangle$ with starting state $y \in Y$. Taking into account that the auxiliary game is equivalent to the average stochastic game determined by

$(X, \{A^i\}_{i=\overline{1,m}}, \{r^i\}_{i=\overline{1,m}}, p, \{\theta_y\})$, where $\theta_y > 0$, $\forall y \in X$, $\sum_{y \in X} \theta_y = 1$, we can see that the considered average stochastic game with a random starting state has a stationary Nash equilibrium $s^* = (s^{1*}, s^{2*}, \ldots, s^{m*})$, which is a stationary Nash equilibrium for the average stochastic game $\langle \{S^i\}_{i=\overline{1,m}}, \{\omega_y^i(s)\}_{i=\overline{1,m}} \rangle$ with an arbitrary starting state $y \in Y$. □

From Theorem 2.19, we can easily obtain the following result:

Lemma 3.12 *For the game in normal form $\langle \{S^i\}_{i=\overline{1,m}}, \{\psi_\theta^i(s)\}_{i=\overline{1,m}} \rangle$ with $\theta_x > 0, \forall x \in X$, $\sum_{y \in X} \theta_y = 1$, each payoff function $\psi_\theta^i(s^1, s^2, \ldots, s^m)$, $i \in \{1, 2, \ldots, m\}$ possesses the property that $\psi_\theta^i(s^i, s^{-i})$ is quasi-monotonic with respect to $s^i \in S^i$ for arbitrary fixed $s^{-i} \in S^{-i}$.*

Proof Indeed, if players $1, 2, \ldots, i-1, i+1, \ldots, m$ fix their stationary strategies $s^k \in S^k$, $k = 1, 2, \ldots, i-1, i+1, \ldots, m$, then we obtain an average decision problem with respect to $s^i \in S^i$ and an average payoff function $\psi_\theta^i(s^i, s^{-i})$. According to Theorem 2.19, $\psi_\theta^i(s^i, s^{-i})$ possesses the property that the value of this function is uniquely determined by $s^i \in S^i$ and it is quasi-monotonic with respect to s^i on S^i. □

Using this lemma, we can prove the following result:

Theorem 3.13 *Let $\langle \{S^i\}_{i=\overline{1,m}}, \{\psi_\theta^i(s)\}_{i=\overline{1,m}} \rangle$ be the normal form of an average stochastic game determined by the tuple $(X, \{A^i\}_{i=\overline{1,m}}, \{r^i\}_{i=\overline{1,m}}, p, \{\theta_x\})$, where $\theta_x > 0$, $\forall x \in X$, $\sum_{y \in X} \theta_y = 1$. If each function ψ_θ^i, $i \in \{1, 2, \ldots, m\}$ is continuous on $S = S^1 \times S^2 \times \cdots \times S^m$, then the game $\langle \{S^i\}_{i=\overline{1,m}}, \{\psi_\theta^i(s)\}_{i=\overline{1,m}} \rangle$ possesses a Nash equilibrium $s^* = (s^{1*}, s^{2*}, \ldots, s^{m*})$, which is a stationary Nash equilibrium for the average stochastic game with an arbitrary starting state $y \in X$.*

Proof Indeed, according to Lemma 3.12, each function $\psi_\theta^i(s^1, s^2, \ldots, s^n)$, $i \in \{1, 2, \ldots, m\}$ satisfies the condition that $\psi_\theta^i(s^i, s^{-i})$ is quasi-monotonic with respect to $s^i \in S^i$ for arbitrary fixed $s^{-i} \in S^{-i}$. In the considered game, each subset S^i is convex and compact, and according to the condition of the theorem, each payoff function $\psi_\theta^i(s^1, s^2, \ldots, s^m)$, $i \in \{1, 2, \ldots, m\}$ is continuous on S. Based on the results from [36, 37, 154, 166], these conditions provide the existence of a Nash equilibrium $s^* = (s^{1*}, s^{2*}, \ldots, s^{m*})$ for the game $\langle \{S^i\}_{i=\overline{1,m}}, \{\psi_\theta^i(s)\}_{i=\overline{1,m}} \rangle$. According to Theorem 3.11, such an equilibrium is a Nash equilibrium for the game $\langle \{S^i\}_{i=\overline{1,m}}, \{\omega_y^i(s)\}_{i=\overline{1,m}} \rangle$ with an arbitrary starting state $y \in X$. □

Remark 3.14 Theorems 3.11 and 3.13 may hold in the case of the game $\langle \{S^i\}_{i=\overline{1,m}}, \{\psi_\theta^i(s)\}_{i=\overline{1,m}} \rangle$, when $\theta_y = 0$ for some $y \in X$; however, in this case, we obtain stationary Nash equilibria only for the games $\langle \{S^i\}_{i=\overline{1,m}}, \{\omega_z^i(s)\}_{i=\overline{1,m}} \rangle$ that start in the states $z \in X^+ = \{z \in X | \theta_z > 0\}$.

Remark 3.15 Theorem 3.13 also holds for the case when the payoffs are not continuous but satisfy the so-called graph-continuous property from [36].

For an average stochastic game, in general, a stationary Nash equilibrium may not exist because the payoffs $\psi_\theta^i(s^1, s^2, \dots, s^m)$ and $\psi_y^i(s^1, s^2, \dots, s^m)$ for $y \in X$ may be discontinuous on S. In the following, we can see that the problem of the existence of stationary Nash equilibria is a difficult problem even for two-player average stochastic games, and in general, for such games, Nash equilibria exist only in the set of history-dependent strategies.

3.5.3 Equilibria for Two-Player Average Stochastic Games

A non-zero-sum average stochastic game of two players is determined by the tuple $(X, \{A^i\}_{i=1,2}, \{r^i\}_{i=1,2}, p, \{\theta_x\})$. So, a normal-form game in stationary strategies in the case of two players is $\langle \{S^i\}_{i=1,2}, \{\psi_\theta^i(s)\}_{i=1,2}\rangle$, where S^1 represents the set of solutions to the following system:

$$\begin{cases} \sum\limits_{a^1 \in A^1(x)} s^1_{x,a^1} = 1, & \forall x \in X; \\ s^i_{x,a^1} \geq 0, & \forall x \in X, \ \forall a^1 \in A^1(x); \end{cases} \tag{3.14}$$

S^2 represents the set of solutions to the following system:

$$\begin{cases} \sum\limits_{a^2 \in A^(x)} s^2_{x,a^1} = 1, & \forall x \in X; \\ s^i_{x,a^2} \geq 0, & \forall x \in X, \ \forall a^2 \in A^(x), \end{cases} \tag{3.15}$$

and

$$\psi_\theta^i(s^1, s^2) = \sum_{x \in X} \sum_{(a^1,a^2) \in A(x)} s^1_{x,a^1} s^2_{x,a^2} r^i_{x,(a^1,a^2)} q_x, \quad i = 1, 2, \tag{3.16}$$

where q_x for $x \in X$ are uniquely determined by the following system of linear equations:

$$\begin{cases} q_y - \sum\limits_{x \in X} \sum\limits_{(a^1,a^2) \in A(x)} s^1_{x,a^1} s^2_{x,a^2} p^{(a^1,a^2)}_{x,y} q_x = 0, & \forall y \in X; \\ q_y + w_y - \sum\limits_{x \in X} \sum\limits_{(a^1,a^2) \in A(x)} s^1_{x,a^1} s^2_{x,a^2} p^{(a^1,a^2)}_{x,y} w_x = \theta_y, & \forall y \in X. \end{cases} \tag{3.17}$$

According to Theorem 3.11, a two-player average stochastic game possesses a stationary Nash equilibrium for an arbitrary starting state $x \in X$ if and only if the game in normal form $\langle \{S^i\}_{i=1,2}, \{\psi_\theta^i(s)\}_{i=1,2}\rangle$ has a Nash equilibrium. If in (3.16) we change $\{\psi_\theta^i(s)\}$ to $\{\psi_y^i(s)\}$, then we obtain the game in normal form in stationary strategies with a given starting state $y \in X$.

A *two-player zero-sum average stochastic game* represents the case of a two-player average stochastic game, where $r(x,a) = r^2_{x,a} = -r^1_{x,a}$ for $x \in X, a = (a^1, a^2) \in A(x) = A^1(x) \times A^2(x)$. Such a game with given starting state x_0 is determined by the tuple $(X, \{A^i\}_{i=1,2}, \{r_{x,a}\}, p, x_0\})$. In the case when the starting state is chosen randomly according to a given distribution $\{\theta_x\}_{x\in X}$, the game is determined by the tuple $(X, \{A^i(x)\}_{i=1,2}, \{r_{x,a}\}, p, \theta\})$. The normal-form game of an average stochastic game in stationary strategies is $\langle\{S^i\}_{i=1,2}, \{\psi_\theta(s)\}_{i=1,2}\rangle$, where S^1 represents the set of solutions to system (3.14), S^2 represents the set of solutions to system (3.15), and the payoff $\psi_\theta(s^1, s^2)$ is defined as follows:

$$\psi_\theta(s^1, s^2) = \sum_{x\in X} \sum_{(a^1,a^2)\in A(x)} s^1_{x,a^1} s^2_{x,a^2} r_{x,(a^1,a^2)} q_x, \quad i = 1, 2, \ldots, m, \tag{3.18}$$

where q_x for $x \in X$ are uniquely determined by the the system of linear equations (3.17).

Obviously, a two-player average stochastic game determined by the tuple $(X, \{A^i(x)\}_{i=1,2}, \{r_{x,a}\}, p, x_0\})$ has a stationary equilibrium if and only if the normal-form game $\langle\{S^i\}_{i=1,2}, \{\psi_{x_0}(s)\}_{i=1,2}\rangle$ has a Nash equilibrium. In general, a two-player average stochastic game may not have a Nash equilibrium in stationary strategies. An example of a two-player zero-sum average stochastic game for which Nash equilibria in stationary strategies do not exist is represented by the *Big Match* game, introduced by Gillette [63] in 1957.

3.5.4 The Big Match and the Paris Match

The Big Match is a zero-sum dynamic game of two players, **1** and **2**, with infinitely many stages, and it is played as follows: In each stage, player **2** chooses number 1 or 2, and player **1** tries to predict the choice of player **2**, winning a point if he or she is correct. This continues as long as player **1** predicts 1. But if he or she ever predicts **2**, all future choices for both players are required to be the same as at given stage choices: if player **1** is correct, then he or she wins a point at each stage thereafter; if he or she is wrong at that stage, he or she wins zero at each stage thereafter. The payoff of player **1** is $\liminf_{t\to\infty}(r_1+r_2+\cdots+r_t)/t$. This game is presented below, where the payoffs of player **2** to player **1** are given in the squares of the table and the asterisks denote transitions to trivial absorbing states with"" the corresponding payoffs (Table 3.1).

Table 3.1 The Big Match

		2	
		1	2
1	1	1	0
	2	0 *	1 *

So, we have a game with the set of states $X = \{x_0, x_1, x_2\}$, where x_1 and x_2 are absorbing states. The game is played in the state x_0 for which the action sets are $A^1(x_0) = \{1, 2\}$ and $A^2(x_0) = \{1, 2\}$. The action $1 \in A^1(x_0)$ for player **1** means that he or she chooses number 1, and the action $2 \in A^1(x_0)$ means that he or she chooses number 2. Similarly, for player **2**, $1 \in A^2(x_0)$ means that player **2** chooses number 1, and $2 \in A^2(x_0)$ means that he or she chooses number 2. For $x_1, x_2 \in X$, we have $A^1(x_1) = A^1(x_2) = \{1\}$ and $A^2(x_1) = A^2(x_2) = \{1\}$. The probabilities $p_{x_k,x_j}^{(a_1,a_2)}$ for $a_1 \in A^1(x_k),\ a_2 \in A(x_k)$ are defined as follows:

$$
\begin{array}{llllll}
p_{x_0,x_0}^{(1,1)} = 1, & p_{x_0,x_0}^{(1,2)} = 1, & p_{x_0,x_0}^{(2,1)} = 0, & p_{x_0,x_0}^{(2,2)} = 0, & p_{x_0,x_1}^{(1,1)} = 0, & p_{x_0,x_1}^{(1,2)} = 0, \\
p_{x_0,x_1}^{(2,1)} = 1, & p_{x_0,x_1}^{(2,2)} = 0, & p_{x_0,x_2}^{(1,1)} = 0, & p_{x_0,x_2}^{(1,2)} = 0, & p_{x_0,x_2}^{(2,1)} = 0, & p_{x_0,x_2}^{(2,2)} = 1, \\
p_{x_1,x_0}^{(1,1)} = 0, & p_{x_1,x_1}^{(1,1)} = 1, & p_{x_1,x_2}^{(1,1)} = 0, & p_{x_2,x_0}^{(1,1)} = 0, & p_{x_2,x_1}^{(1,1)} = 0, & p_{x_2,x_2}^{1,1} = 1.
\end{array}
$$

The rewards $r_{x_k,(a_1,a_2)}$ in a state $x_k \in X$ for $a_1 \in A^1(x_k)$ and $a_2 \in A^2(x_k)$ are defined as follows:

$$
r_{x_0,(1,1)} = 1,\ r_{x_0,(1,2)} = 0,\ r_{x_0,(2,1)} = 0,\ r_{x_0,(2,2)} = 1,\ r_{x_1,(1,1)} = 0,\ r_{x_2,(1,1)} = 1.
$$

In the absorbing states x_1 and x_2, we have $\psi_{x_1}(s^1, s^2) = 0$ and $\psi_{x_2}(s^1, s^2) = 1$ for all strategies s^1 and s^2 of the players. More difficult are the choices of the players in the state x_0 because their choices may lead to the repetition of the same game and each of them may risk making a choice to finish the game in an absorbing state. So, the value of the game, if it exists, is determined by the strategies of the players in the state x_0.

Let S^1 be the set of stationary strategies of player **1**, and let S^2 be the set of stationary strategies of player **2**. Assume that player **1** uses a stationary strategy $s^1 \in S^1$ for which action 1 is chosen with probability p_1 and action 2 is chosen with probability $1 - p_1$ in the state x_0, and assume that player **2** uses a stationary strategy $s^2 \in S^2$ in which action 1 is chosen with probability p_2 and action 2 is chosen with probability $(1 - p_2)$. Thus, $p_1 = s^1_{x_0,1}$, $1 - p_1 = s^1_{x_0,2}$, where $1, 2 \in A^1(x_0)$, and $p_2 = s^2_{x_0,1}$, $1 - p_2 = s^2_{x_0,2}$, where $1, 2 \in A^2(x_0)$.

Denote by $P(p_1, p_2)$ and $\mathbf{r}(p_1, p_2)$ the transition probability matrix and the reward vector, respectively, induced by s^1 and s^2, and let us analyze the following two cases: $p_1 = 1$ and $0 \le p_1 < 1$.

Case 1: $p_1 = 1$

The transition probability matrix $P(p_1, p_2)$, the reward vector $\mathbf{r}(p_1, p_2)$, and the limiting matrix $Q(p_1, p_2)$ are the following:

$$
P(p_1, p_2) = \begin{pmatrix} 1 & 0 & 0 \\ 0 & 1 & 0 \\ 0 & 0 & 1 \end{pmatrix}; \quad \mathbf{r} = (p_2,\ 0,\ 1)^T; \quad Q(p_1, p_2) = \begin{pmatrix} 1 & 0 & 0 \\ 0 & 1 & 0 \\ 0 & 0 & 1 \end{pmatrix}
$$

and therefore, $\psi_{x_0}(s_1, s_2) = p_2$.

Case 2: $0 \le p_1 < 1$

The transition probability matrix $P(p_1, p_2)$, the reward vector $\mathbf{r}(p_1, p_2)$, and the limiting matrix $Q(p_1, p_2)$ are the following:

$$P(p_1, p_2) = \begin{pmatrix} p^1 & (1-p_1)p_2 & (1-p_1)(1-p_2) \\ 0 & 1 & 0 \\ 0 & 0 & 1 \end{pmatrix};$$

$$\mathbf{r} = \Big(p_1(1-p_2) + (1-p_1)(1-p_2),\ 0,\ 1\Big)^T;$$

$$Q(p_1, p_2) = \begin{pmatrix} 0 & p_2 & 1-p_2 \\ 0 & 1 & 0 \\ 0 & 0 & 1 \end{pmatrix},$$

and therefore, $\psi_{x_0}(s^1, s^2) = 1 - p_2$.

Thus, we have

$$\psi_{x_0}(s^1, s^2) = \begin{cases} p_2 & \text{if} \quad p_1 = 1; \\ 1 - p_2 & \text{if} \quad 0 \le p_1 < 1, \end{cases} \tag{3.19}$$

and we obtain

$$\max_{0 \le p_1 \le 1} \psi_{x_0}(s^1, s^2) = \max_{0 \le p_1 \le 1} \{p_2, 1 - p_2\}, \quad i.e., \quad \min_{0 \le p_2 \le 1} \max_{0 \le p_1 \le 1} \psi_{x_0}(s^1, s^2) = \frac{1}{2}$$

and

$$\min_{0 \le p_2 \le 1} \psi_{x_0}(s^1, s^2) = \min_{0 \le p_2 \le 1} \{p_2, 1 - p_2\} = 0,$$

$$i.e., \quad \max_{0 \le p_1 \le 1} \min_{0 \le p_2 \le 1} \psi_{x_0}(s^1, s^2) = 0.$$

So, we have

$$\max_{s^1 \in S^1} \min_{s^2 \in S^2} \psi_{x_0}(s^1, s^2) = 0 < \frac{1}{2} = \min_{s^2 \in S^2} \max_{s^1 \in S^1} \psi_{x_0}(s^1, s^2). \tag{3.20}$$

This means that the *Big Match* does not have the value in the set of stationary strategies. Moreover, considering the Markov strategies $\pi^1 \in \Pi^1$ for player **1** and Markov strategies for player **2**, a similar relation can be derived [176]:

$$\sup_{\pi^1 \in \Pi^1} \inf_{\pi^2 \in \Pi^2} \psi_{x_0}(\pi^1, \pi^2) = 0 < \frac{1}{2} = \inf_{\pi^2 \in \Pi^2} \sup_{\pi^1 \in \Pi^1} \psi_{x_0}(\pi^1, \pi^2),$$

meaning that the Big Match does not have a limiting average value in the set of Markov strategies.

In 1968, Blackwell and Ferguson [17, 18] showed that the value of the Big Match is equal to $\frac{1}{2}$ in the set of non-stationary strategies with perfect monitoring. The results of Blackwell and Ferguson were generalized by Kohlberg [87] for the class of zero-sum repeated games with absorbing states. Kohlberg showed that the limiting average value ν exists for an arbitrary zero-sum repeated game with absorbing states and it is equal to the limit of the average value of the t-stage games, i.e., $\nu = \lim_{t\to\infty} \nu_t$.

Then Mertens and Neyman [134] generalized the Kohlberg result for an arbitrary zero-sum stochastic game with finite state and action spaces. They showed that the limiting average value ν exists in non-stationary strategies for every zero-sum stochastic game. Moreover, they showed that the limiting average value, the limit of the average value of the t-stage game, and the limit of the γ-discounted games are all equal, i.e., $\nu = \lim_{t\to\infty} \nu_n = \lim_{\gamma\downarrow 0} \nu_\gamma$.

Now let us show how to obtain the payoff $\psi_{x_0}(s^1, s^2)$ for the Big Match presented in (3.19) using the general model of a two-player zero-sum game in stationary strategies $\langle \{S^i\}_{i=1,2}, \{\psi_\theta(s)\}_{i=1,2}\rangle$, where the set of strategies S^1, S^2 are defined by (3.14) and (3.15), respectively, and the payoff is defined by (3.17) and (3.18). If for the Big Match example we take $\theta_{x_0} = 1,\ \theta_{x_1} = 0,\ \theta_{x_2} = 0$, then we have

$$\begin{aligned}\psi_{x_0}(s^1, s^2) =& s^1_{x_0,1}s^2_{x_0,1}r_{x_0,(1,1)}q_{x_0} + s^1_{x_0,1}s^2_{x_0,2}r_{x_0,(1,2)}q_{x_0} + s^1_{x_0,2}s^2_{x_0,1}r_{x_0,(2,1)}q_{x_0}\\ &+ s^1_{x_0,2}s^2_{x_0,2}r_{x_0,(2,2)}q_{x_0} + s^1_{x_1,1}s^2_{x_0,1}r_{x_1,(1,1)}q_{x_1} + s^1_{x_2,1}s^2_{x_2,1}r_{x_2,(1,1)}q_{x_2},\end{aligned}$$

where

$$\begin{cases} s^1_{x_0,1} + s^1_{x_0,2} = 1; \\ s^2_{x_0,1} + s^2_{x_0,2} = 1; \\ s^1_{x_0,2},\ s^2_{x_0,1},\ s^2_{x_0,2} \geq 0 \end{cases} \tag{3.21}$$

and $s^1_{x_1,1} = s^2_{x_2,1} = 1$. In this function, q_{x_0}, q_{x_1}, and q_{x_2} represent the limiting probabilities induced by the strategies s^1 and s^2 for the given starting state x_0. The corresponding equations of system (3.16) that determine the limiting probabilities $q_{x_0}, q_{x_1}, q_{x_2}$ are

$$\begin{aligned}q_{x_0} - (s^1_{x_0,1}s^2_{x_0,1}p^{(1,1)}_{x_0,x_0}q_{x_0} + s^1_{x_0,1}s^2_{x_0,2}p^{(1,2)}_{x_0,x_0}q_{x_0} + s^1_{x_0,2}s^2_{x_0,1}p^{(2,1)}_{x_0,x_0}q_{x_0} +\\ s^1_{x_0,2}s^2_{x_0,2}p^{(2,2)}_{x_0,x_0}q_{x_0}) = 0;\end{aligned}$$

$$\begin{aligned}q_{x_1} - (s^1_{x_1,1}s^2_{x_1,1}p^{(1,1)}_{x_1,x_1}q_{x_1} + s^1_{x_0,1}s^2_{x_0,1}p^{(1,2)}_{x_0,x_1}q_{x_0} + s^1_{x_0,1}s^2_{x_0,2}p^{(1,2)}_{x_0,x_1}q_{x_0} +\\ s^1_{x_0,2}s^2_{x_0,1}p^{(2,1)}_{x_0,x_1}q_{x_0} + s^1_{x_0,2}s^2_{x_0,2}p^{(2,2)}_{x_0,x_1}q_{x_0}) = 0;\end{aligned}$$

$$q_{x_2} - \big(s^1_{x_2,1}s^2_{x_2,1}p^{(1,1)}_{x_1,x_1}q_{x_2} + s^1_{x_0,1}s^2_{x_0,1}p^{(1,2)}_{x_0,x_2}q_{x_0} + s^1_{x_0,1}s^2_{x_0,2}p^{(1,2)}_{x_0,x_2}q_{x_0} +$$

$$s^1_{x_0,2}s^2_{x_0,1}p^{(2,1)}_{x_0,x_2}q_{x_0} + s^1_{x_0,2}s^2_{x_0,2}p^{(2,2)}_{x_0,x_2}q_{x_0}\big) = 0;$$

$$q_{x_0} + w_{x_0} - \big(s^1_{x_0,1}s^2_{x_0,1}p^{(1,1)}_{x_0,x_0}w_{x_0} + s^1_{x_0,1}s^2_{x_0,2}p^{(1,2)}_{x_0,x_0}w_{x_0} + s^1_{x_0,2}s^2_{x_0,1}p^{(2,1)}_{x_0,x_0}w_{x_0} +$$

$$s^1_{x_0,2}s^2_{x_0,2}p^{(2,2)}_{x_0,x_0}w_{x_0}\big) = 0;$$

$$q_{x_1} + w_{x_1} - \big(s^1_{x_1,1}s^2_{x_1,1}p^{(1,1)}_{x_1,x_1}w_{x_1} + s^1_{x_0,1}s^2_{x_0,1}p^{(1,2)}_{x_0,x_1}w_{x_0} + s^1_{x_0,1}s^2_{x_0,2}p^{(1,2)}_{x_0,x_1}w_{x_0} +$$

$$s^1_{x_0,2}s^2_{x_0,1}p^{(2,1)}_{x_0,x_1}w_{x_0} + s^1_{x_0,2}s^2_{x_0,2}p^{(2,2)}_{x_0,x_1}w_{x_0}\big) = 0;$$

$$q_{x_2} + w_{x_2} - \big(s^1_{x_2,1}s^2_{x_2,1}p^{(1,1)}_{x_1,x_1}w_{x_2} + s^1_{x_0,1}s^2_{x_0,1}p^{(1,2)}_{x_0,x_2}w_{x_0} + s^1_{x_0,1}s^2_{x_0,2}p^{(1,2)}_{x_0,x_2}w_{x_0} +$$

$$s^1_{x_0,2}s^2_{x_0,1}p^{(2,1)}_{x_0,x_2}w_{x_0} + s^1_{x_0,2}s^2_{x_0,2}p^{(2,2)}_{x_0,x_2}w_{x_0}\big) = 0.$$

By introducing the corresponding values of probability transitions and taking into account that $s^1_{x_1,1} = s^2_{x_1,1} = s^1_{x_2,1} = s^2_{x_2,1} = 1$, we obtain the payoff

$$\psi_{x_0}(s^1, s^2) = \left(s^1_{x_0,1}s^2_{x_2,1} + s^1_{x_0,2}s^2_{x_0,2}\right) q_{x_0} + q_2, \tag{3.22}$$

where for given strategies s^1, s^2, the corresponding limiting probabilities $q_{x_0}, q_{x_1}, q_{x_1}$ can be uniquely determined by the following system:

$$\begin{cases} (1 - s^1_{x_0,1})q_{x_0} = 0; \\ s^1_{x_0,2}s^2_{x_0,1}q_{x_0} = 0; \\ s_{x_0,2}s^2_{x_0,2}q_{x_0} = 0; \\ q_{x_0} + (1 - s^1_{x_0,1})w_{x_0} = 1; \\ q_{x_1} - s^1_{x_0,2}s^2_{x_0,1}w_{x_0} = 0; \\ q_{x_2} - s_{x_0,2}s^2_{x_0,2}w_{x_0} = 0, \end{cases} \tag{3.23}$$

for arbitrary strategies $s^1 \in S^1$ and $s^2 \in S^2$.

From (3.23), we obtain:

If $s^1_{x_0,1} = 1$, then $s_{x_0,2} = 0$, $q_{x_0} = 1$, $q_{x_1} = q_{x_2} = 0$, i.e., in this case, $\psi_{x_0}(s^1, s^2) = s^2_{x_0,1}$.

If $0 \le s^1_{x_0,1} < 1$, then $q_{x_0} = 0$, $w_{x_0} = \frac{1}{1-s^1_{x_0,1}}$ and $q_{x_2} = s^2_{x_0,2} = 1 - s^2_{x_0,1}$, i.e., in this case, $\psi_{x_0}(s^1, s^2) = 1 - s^2_{x_0,1}$.

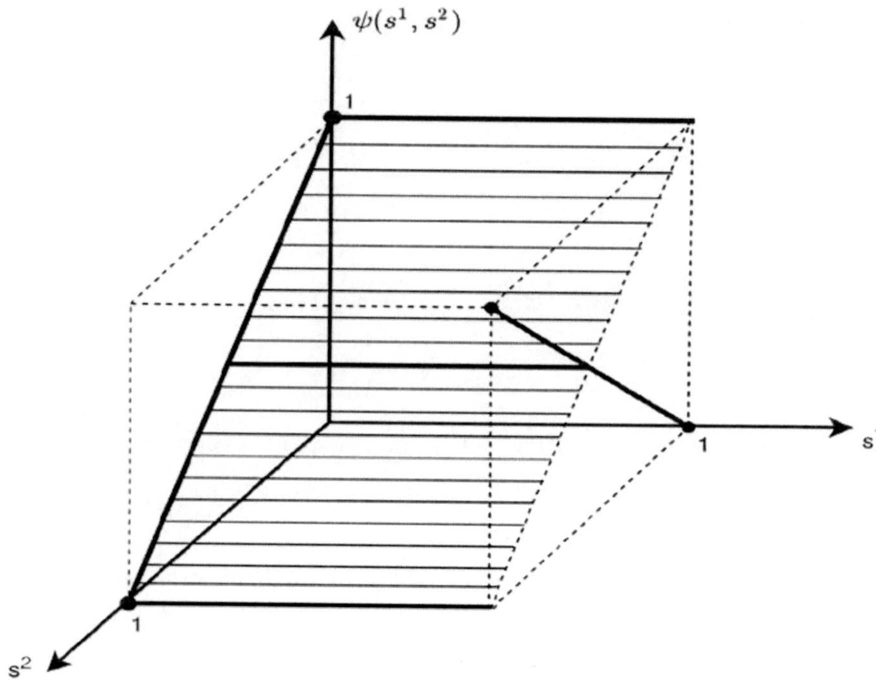

Fig. 3.1 Payoff $\psi(s^1, s^2)$ for the Big Match

So,

$$\psi_{x_0}(s^1, s^2) = \begin{cases} s^2_{x_0,1} & \text{if} \quad s^1_{x_0,1} = 1; \\ 1 - s^2_{x_0,1} & \text{if} \quad 0 \le s^1_{x_0,1} < 1, \end{cases} \tag{3.24}$$

and the relation (3.20) holds.

The graphical representation of the payoff $\psi_{x_0}(s^1, s^2)$ is presented in Fig. 3.1. We can see that the payoff $\psi_{x_0}(s^1, s^2)$ is quasi-monotonic with respect to $s^1 \in S^1$ for a fixed $s^2 \in S^2$ and quasi-monotonic with respect to $s^2 \in S^2$ for a fixed $s^1 \in S^1$. We can also see that $\psi_{x_0}(s^1, s^2)$ is discontinuous at $s^1_{x_0} = 1$.

The non-zero-sum extension of the Big Match was introduced by Sorin [170]. Because of the French origin of the author, it is called the *Paris Match*[177]. The structure of the Paris Match is similar to the Big Match, but players do not pay each other because each of them has their own payoff that they intend to optimize. This game is represented in the table below, where in the square the corresponding stage payoffs of the players are given and the asterisks denote transitions to absorbing states (Table 3.2).

Table 3.2 The Paris Match

		2	
		1	2
1	1	1, 0	0, 1
	2	0, 2 *	1, 0 *

Sorin [170] observed that for the Paris Match, there is a gap between the set of limiting average equilibrium rewards and the set of γ-discounted equilibrium rewards. He showed the existence of ε-equilibrium strategies for the Paris Match. A pair of strategies (s^{1^*}, s^{2^*}) is a limiting average ε-equilibrium ($\varepsilon \geq 0$) if neither player can gain more than ε by unilateral deviation, i.e.,

$$\psi_{x_0}(s^{1^*}, s^{2^*}) \geq \psi_{x_0}(s^1, s^{2^*}) - \varepsilon, \quad \forall s^1 \in S^1,$$

and

$$\psi_{x_0}(s^{1^*}, s^{2^*}) \geq \psi_{x_0}(s^{1^*}, s^2) - \varepsilon, \quad \forall s^2 \in S^2.$$

Vrieze and Thuijsman [190] extended the result of Sorin and proved the existence of an ε-equilibrium for arbitrary non-zero-sum repeated games with absorbing states in the case of finite state and action spaces.

The problem of the existence of ε-equilibria in two-player non-zero-sum average stochastic games in the general case was studied by Vielle [183–185]. Based on the results obtained by Sorin [170], Mertens, and Neyman [134], Vrieze and Thuijsman [190] showed the existence of ε-equilibria in non-stationary strategies for two-player non-zero-sum average stochastic games.

In the following, we discuss an important result concerned with the existence of Nash equilibria in average stochastic games with $m \geq 3$ players obtained by Flesch, Thuijsman, and Vrieze [57].

3.5.5 A Cubic Three-Person Average Game

Flesch, Thuijsman, and Vrieze [57] constructed an example of a three-person game that plays an important role in studying the problem of the existence of equilibria in average stochastic games. The game is represented in Fig. 3.2.

In this cubic three-person game, each player has the choice of two actions. The actions of the players are the following: Player 1, Top, Bottom (T, B); Player 2, Left, Right (L, R); and Player 3, Near, Far (N, F). The two tables in Fig. 3.2 represent two layers of the cube that correspond to the two actions of Player 3 (N, F). Absorbing entries (transitions to absorbing states) are indicated by asterisks (*) and are interpreted as follows:

Once such an entry is selected, the play moves to an absorbing state with these payoffs for the players. For this game, all states are absorbing except one of them,

N

	L	R
T	0, 0, 0	0, 1, 3 *
B	1, 3, 0 *	1, 0, 1 *

F

T	3, 0, 1 *	1, 1, 0 *
B	0, 1, 1 *	0, 0, 0 *

Fig. 3.2 Cubic three-player game

and the play is absorbed by an absorbing state as soon as one of the players chooses his or her second action. Note that the payoffs of the players and the absorbing entries in the considered game are cyclically symmetric ($r^1(i_1, i_2, i_3) = r^2(i_3, i_1, i_2) = r^3(i_2, i_3, i_1)$) for each entry (i_1, i_2, i_3). The author considered such a cyclically symmetric property for the payoffs in order to show the cyclic nature of equilibria for this game that they called *cyclic Markov equilibria*. They also showed that for this game, an ε-equilibrium does not exist but there exists the cyclic Markov equilibrium (s^{1*}, s^{2*}, s^{3*}), where

$$s^{1*} = \left(\frac{1}{2}, 0, 0, \frac{1}{2}, 0, 0, \frac{1}{2}, \dots\right);$$

$$s^{2*} = \left(0, \frac{1}{2}, 0, 0, \frac{1}{2}, 0, 0, \dots\right);$$

$$s^{3*} = \left(0, 0, \frac{1}{2}, 0, 0, \frac{1}{2}, 0, \dots\right).$$

So, for such a strategy, we have

$$\begin{aligned}\psi_{x_0}(s^*) &= \left(\psi^1_{x_0}(s^*, s^{2*}, s^{3*}), \psi^2_{x_0}(s^*, s^{2*}, s^{3*}), \psi^3_{x_0}(s^*, s^{2*}, s^{3*})\right) \\ &= \left[\frac{1}{2}(1, 3, 0) + \left(\frac{1}{2}\right)^2(0, 1, 3) + \left(\frac{1}{2}\right)^3(3, 0, 1)\right] \cdot \sum_{t=0}^{\infty}\left(\frac{1}{2}\right)^{3t} = (1, 2, 1).\end{aligned}$$

The results obtained by Flesch, Thuijsman, and Vrieze [57, 176–179] show that the existing and determining equilibria in average stochastic games, in general, pose a difficult problem, and in the case of $m \geq 3$ players, this remains an open problem. Therefore, this problem is studied for special classes of games. However, in the case of the existence of Nash equilibria for some classes of stochastic games, finding the optimal stationary strategies of the players for most of them poses a difficult problem from a computational complexity point of view [32]. Even such a special case of the problem as determining a stationary equilibrium in a two-player non-zero-sum stochastic game with one state is NP-hard [60, 133]. Therefore, efficient algorithms to determine stationary Nash equilibria in stochastic games (in the case when Nash equilibria exist) can be elaborated only for some special classes of games without large dimensions.

A detailed classification of the algorithms that can be used to determine equilibria in some classes of stochastic games was described in [153, 180]. In general, the computational complexity of determining a pure Nash equilibrium in a non-cooperative game was analyzed in [44, 65].

3.6 Stochastic Positional Games

Stochastic positional games represent an important class of stochastic games for which stationary Nash equilibria exist. This class of games was introduced in [100, 125] as a generalization of the deterministic positional games from [2, 42, 67, 108, 112, 186, 202]. We formulate and study stochastic positional games by applying the concept of positional games to average Markov decision processes with finite state and action spaces. We assume that a Markov decision process is controlled by m players where the set of states is divided into m disjoint subsets such that each subset represents the position set for one of the players and each player controls the Markov process only in his or her position set. In the control process, each player chooses actions in his or her position set in order to maximize his or her average payoff.

We first present the framework of an m-person stochastic positional game and then specify the formulation of stochastic positional games with average and discounted payoffs if the players use pure and mixed stationary strategies of selecting the actions in the states.

3.6.1 The Framework of a Stochastic Positional Game

A stochastic positional game with m players consists of the following elements:

- A state space X (which we assume to be finite)
- A partition $X = X_1 \cup X_2 \cup \cdots \cup X_m$, where X_i represents the position set of player $i \in \{1, 2, \ldots, m\}$
- A finite set $A(x)$ of actions in each state $x \in X$
- A step reward $r^i_{x,a}$ with respect to each player $i \in \{1, 2, \ldots, m\}$ in each state $x \in X$ and for an arbitrary action $a \in A(x)$
- A transition probability function $p : X \times \prod_{x \in X} A(x) \times X \to [0, 1]$ that gives the probability transitions $p^a_{x,y}$ from an arbitrary $x \in X$ to an arbitrary $y \in X$ for a fixed action $a \in A(x)$, where $\sum_{y \in X} p^a_{x,y} = 1, \;\; \forall x \in X, \; a \in A(x)$
- A starting state $x_0 \in X$

The game starts at the moment of time $t = 0$ in the state x_0 where player $i \in \{1, 2, \dots, m\}$, who is the owner of the state position x_0 ($x_0 \in X_i$), chooses an action $a_0 \in A(x_0)$ and determines the rewards $r^1_{x_0,a_0}, r^2_{x_0,a_0}, \dots, r^m_{x_0,a_0}$ for the corresponding players $1, 2, \dots, m$. After that, the game passes to a state $y = x_1 \in X$ according to a probability distribution $\{p^{a_0}_{x_0,y}\}$. At the moment of time $t = 1$, player $k \in \{1, 2, \dots, m\}$, who is the owner of the state position x_1 ($x_1 \in X_k$), chooses an action $a_1 \in A(x_1)$, and players $1, 2, \dots, m$ receive the corresponding rewards $r^1_{x_1,a_1}, r^2_{x_1,a_1}, \dots, r^m_{x_1,a_1}$. Then the game passes indefinitely to a state $y = x_2 \in X$ according to a probability distribution $\{p^{a_1}_{x_1,y}\}$ and so on. Such a play of the game produces a sequence of states and actions $x_0, a_0, x_1, a_1, \dots, x_t, a_t, \dots$, defining a stream of stage rewards $r^1_{x_t,a_t}, r^2_{x_t,a_t}, \dots, r^m_{x_t,a_t}$, $t = 0, 1, 2, \dots$. *The average stochastic positional game* is the game with payoffs of the players

$$\omega^i_{x_0} = \liminf_{t\to\infty} \mathbb{E}\left(\frac{1}{t}\sum_{\tau=0}^{t-1} r^i_{x_\tau,a_\tau}\right), \quad i = 1, 2, \dots, n,$$

where $\mathbb{E}$ is the expectation operator with respect to the probability measure in the Markov process induced by actions chosen by players in their position sets and given starting state x_0. Each player in this game aims to maximize his or her average reward per transition. In the case $m = 1$, this game becomes the average Markov decision problem with given action sets $A(x)$ for $x \in X$, a transition probability function $p : X \times \prod_{x\in X} A(x) \times X \to [0, 1]$, and step rewards $r^a_x = r^1_{x,a}$ for $x \in X$ and $a \in A(x)$. *The discounted stochastic positional game* is the game with payoffs of the players

$$\omega^i_{x_0,\gamma} = \mathbb{E}\left(\sum_{\tau=0}^{\infty} \gamma^\tau r^i_{x_\tau,a_\tau}\right), \quad i = 1, 2, \dots, n.$$

We consider the problem of the existence of Nash equilibria in stochastic positional games with average and discounted payoffs when players use pure and mixed stationary strategies of selecting the actions in the states.

3.6.2 Positional Games in Pure and Mixed Strategies

Strategies

A strategy of player $i \in \{1, 2, \dots, m\}$ in a stochastic positional game we define in a similar way as for the stochastic game from Sect. 3.3.2, taking into account that the strategy of player $i \in \{1, 2, \dots, m\}$ only makes sense in the positions $x \in X_i$. So, a *stationary strategy* of player $i \in \{1, 2, \dots, m\}$ is a mapping s^i that provides a probability distribution over the set of actions $A(x_t)$ for every state $x_t \in X_i$. If these probabilities only take values 0 and 1, then s^i is called a *pure strategy*; otherwise, s^i

is called a *mixed strategy*. If these probabilities only depend on the state $x_t = x \in X_i$ (i.e., s^i does not depend on t), then s^i is called a *stationary strategy*; otherwise, s^i is called a *non-stationary strategy*. Thus, a pure stationary strategy of player i is a mapping

$$s^i : x \to a \in A(x) \text{ for } x \in X_i$$

that determines an action $a \in A(x)$ for each state $x \in X_i$, i.e., $s^i(x) = a$ for $x \in X_i$. This means that we can identify a pure stationary strategy s^i of player i with the set of Boolean variables $s^i_{x,a} \in \{0, 1\}$, where for a given $x \in X_i$, $s^i_{x,a} = 1$ if and only if player i fixes the action $a \in A(x)$.

So, we can represent the set of pure stationary strategies $\mathbb{S}^i$ of player i as the set of solutions to the following system:

$$\begin{cases} \sum\limits_{a \in A(x)} s^i_{x,a} = 1, & \forall x \in X_i; \\ s^i_{x,a} \in \{0, 1\}, & \forall x \in X_i, \ \forall a \in A(x). \end{cases} \tag{3.25}$$

Obviously, the sets of pure strategies $\mathbb{S}^1, \mathbb{S}^2, \ldots, \mathbb{S}^m$ of the players are finite sets. If in system (3.25) we change the restrictions $s^i_{x,a} \in \{0, 1\}$ for $x \in X_i$, $a \in A(x)$ to the conditions $0 \le s^i_{x,a} \le 1$, then we obtain the set of mixed stationary strategies, where $s^i_{x,a}$ is treated as the probability of the choices of the action a by player i every time when the state x is reached by any route in the dynamic stochastic game. Thus, we can identify the set of mixed stationary strategies S^i of player i with the set of solutions to the system

$$\begin{cases} \sum\limits_{a \in A(x)} s^i_{x,a} = 1, & \forall x \in X_i; \\ s^i_{x,a} \ge 0, & \forall x \in X_i, \ \forall a \in A(x), \end{cases} \tag{3.26}$$

and for a given profile $\mathbf{s} = (s^1, s^2, \ldots, s^m) \in S = S^1 \times S^2 \times \cdots \times S^m$ of mixed strategies $s^1, s^2, \ldots, s^m$ of the players, the probability transition matrix $P^s = (p^s_{x,y})$ induced by s can be calculated as follows:

$$p^s_{x,y} = \sum_{a \in A(x)} s^i_{x,a} p^a_{x,y} \quad \text{for } x \in X_i, \ i = 1, 2, \ldots, m. \tag{3.27}$$

In the following, we distinguish stochastic games in pure and mixed stationary strategies.

Average Stochastic Positional Games in Stationary Strategies

Let $s = (s^1, s^2, \ldots, s^m)$ be a profile of stationary strategies (pure or mixed strategies) of the players. Then the elements of the probability transition matrix $P^s = (p^s_{x,y})$ in the Markov process induced by s can be calculated according to (3.27). Therefore, if $Q^s = (q^s_{x,y})$ is the limiting probability matrix of P^s, then the average payoffs per transition $\omega^1_{x_0}(s),\ \omega^2_{x_0}(s), \dots, \omega^m_{x_0}(s)$ for the players are determined as follows:

$$\omega^i_{x_0}(s) = \sum_{k=1}^{m} \sum_{y \in X_k} q^s_{x_0,y} \cdot r^i_{y,s^k}, \quad i = 1, 2, \dots, m, \tag{3.28}$$

where

$$r^i_{y,s^k} = \sum_{a \in A(y)} s^k_{y,a} \cdot r^i_{y,a}, \quad \text{for } y \in X_k,\ k \in \{1, 2, \dots, m\} \tag{3.29}$$

expresses the average reward (step reward) of player i in the state $y \in X_k$ when player k uses strategy s^k.

The functions $\omega^1_{x_0}(s),\ \omega^2_{x_0}(s),\ \dots,\ \omega^m_{x_0}(s)$ on $S = S^1 \times S^2 \times \dots \times S^m$, defined according to (3.28) and (3.29), determine a game in normal form that we denote by $\langle \{S^i\}_{i=\overline{1,m}}, \{\omega^i_{x_0}(s)\}_{i=\overline{1,m}} \rangle$. This game corresponds to the *average stochastic positional game in mixed stationary strategies* that in extended form is determined by the tuple $(\{X_i\}_{i=\overline{1,m}}, \{A(x)\}_{x \in X}, \{r^i\}_{i=\overline{1,m}}, p, x_0)$. The functions $\omega^1_{x_0}(s),\ \omega^2_{x_0}(s), \dots, \omega^m_{x_0}(s)$ on $S = S^1 \times S^2 \times \dots \times S^m$ determine the game $\langle \{S^i\}_{i=\overline{1,m}}, \{\omega^i_{x_0}(s)\}_{i=\overline{1,m}} \rangle$ that corresponds to the *average stochastic positional game in pure strategies*. In an extended form, this game is also determined by the tuple $(\{X_i\}_{i=\overline{1,m}}, \{A(x)\}_{x \in X}, \{r^i\}_{i=\overline{1,m}}, p, x_0)$.

Discounted Stochastic Positional Games in Stationary Strategies

Let $s = (s^1, s^2, \dots, s^m)$ be a profile of stationary strategies (pure or mixed strategies) of the players in a stochastic positional game. Then the elements of probability transition matrix $P^s = (p^s_{x,y})$ induced by s can be calculated according to (3.27), and we can find the matrix $W^s = (w^s_{x,y})$, where $W^s = (I - \gamma P^s)$. After that, we can find the payoff for the players as follows:

$$\sigma^i_{x_0}(s) = \sum_{y \in X} \omega_{x_0,y} \cdot r^i_{y,s}, \quad i = 1, 2, \dots, m,$$

where $r^i_{y,s}$ is determined according to (3.29).

Stochastic Positional Games with a Random Starting State

For stochastic positional games with average payoffs (or discounted payoffs), we consider the games in which the starting state is chosen randomly according to a given distribution $\{\theta_x\}$ on X. So, in the same way as in Sect. 3.6.2 for an average stochastic positional game, we can consider the game with the payoff functions

$$\psi^i_\theta(s) = \sum_{x \in X} \theta_x \cdot \omega^i_x(s), \quad i = 1, 2, \dots, m$$

on S, and we obtain the game in normal form $\langle \{S^i\}_{i=\overline{1,m}}, \{\psi_\theta^i(s)\}_{i=\overline{1,m}} \rangle$ that is determined by $(\{X_i\}_{i=\overline{1,m}}, \{A(x)\}_{x\in X}, \{r^i\}_{i=\overline{1,m}}, p, \{\theta_x\}_{x\in X})$.

The game in normal form $\langle \{S^i\}_{i=\overline{1,m}}, \{\phi_\theta^i(s)\}_{i=\overline{1,m}} \rangle$ for a discounted stochastic positional game we obtain by using the payoffs

$$\phi_\theta^i(s^1, s^2, \dots, s^m) = \sum_{x\in X} \theta_x \cdot \sigma_x^i(s^1, s^2, \dots, s^m), \quad i = 1, 2, \dots, m.$$

3.6.3 Stationary Equilibria for Average Positional Games

The aim of this section is to show that an arbitrary average stochastic positional game possesses a Nash equilibrium in mixed stationary strategies. To prove this, for an average stochastic positional game, we use the normal-form game in stationary strategies $\langle \{S^i\}_{i=\overline{1,m}}, \{\psi_\theta^i(s)\}_{i=\overline{1,m}} \rangle$, where each S^i, $i \in \{1, 2, \dots, m\}$ is the set of solutions to the following system:

$$\begin{cases} \sum\limits_{a\in A(x)} s_{x,a}^i = 1, & \forall x \in X_i; \\ \qquad s_{x,a}^i \geq 0, & \forall x \in X_i,\ a \in A(x), \end{cases} \tag{3.30}$$

representing the set of stationary strategies of player i. Each S^i is a convex compact set, and an arbitrary extreme point corresponds to a basic solution s^i of system (3.30), where $s_{x,a}^i \in \{0, 1\}$, $\forall x \in X_i$, $a \in A(x)$, i.e., each basic solution to this system corresponds to a pure stationary strategy $s^i \in \mathbb{S}^i$ of player i.

On the set $S = S^1 \times S^2 \times \cdots \times S^m$, we define m payoff functions

$$\psi_\theta^i(s^1, s^2, \dots, s^m) = \sum_{k=1}^{m} \sum_{x\in X_k} \sum_{a\in A(x)} s_{x,a}^k \cdot r_{x,a}^i \cdot q_x, \qquad i = 1, 2, \dots, m, \tag{3.31}$$

where q_x for $x \in X$ are uniquely determined by the following system of linear equations:

$$\begin{cases} q_y - \sum\limits_{k=1}^{m} \sum\limits_{x\in X_k} \sum\limits_{a\in A(x)} s_{x,a}^k \cdot p_{x,y}^a \cdot q_x = 0, & \forall y \in X; \\ q_y + w_y - \sum\limits_{k=1}^{m} \sum\limits_{x\in X_k} \sum\limits_{a\in A(x)} s_{x,a}^k \cdot p_{x,y}^a \cdot w_x = \theta_y, & \forall y \in X \end{cases} \tag{3.32}$$

for a fixed profile $s = (s^1, s^2, \dots, s^m) \in S$. The functions $\psi_\theta^i(s^1, s^2, \dots, s^m)$, $i = 1, 2, \dots, m$ represent the payoff functions for the average stochastic game in normal form $\langle \{S^i\}_{i=\overline{1,m}}, \{\psi_\theta^i(s)\}_{i=\overline{1,m}} \rangle$. This game is determined by the tuple $(\{X_i\}_{i=\overline{1,m}}, \{A(x)\}_{x\in X}, \{r^i\}_{i=\overline{1,m}}, p, \{\theta_y\}_{y\in X})$, where θ_y for $y \in X$ are given positive values such that $\sum_{y\in X}\theta_y = 1$.

If $\theta_y = 0$, $\forall y \in X \setminus \{x_0\}$ and $\theta_{x_0} = 1$, then we obtain an average stochastic game in normal form $\langle \{S^i\}_{i=\overline{1,m}}, \{\omega_{x_0}^i(s)\}_{i=\overline{1,m}} \rangle$, when the starting state x_0 is fixed, i.e., $\psi_\theta^i(s^1, s^2, \dots, s^m) = \omega_{x_0}^i(s^1, s^2, \dots, s^m)$, $i = 1, 2, \dots, m$. So, in this case, the game is determined by $(X, \{A^i(x)\}_{i=\overline{1,m}}, \{r^i\}_{i=\overline{1,m}}, p, x_0)$.

If $\theta_y > 0$, $\forall y \in X$ and $\sum_{y\in X}\theta_y = 1$, then we obtain an average stochastic game when the play starts in the states $y \in X$ with probabilities θ_y. In this case, for the payoffs of the players in the game in normal form, we have

$$\psi_\theta^i(s^1, s^2, \dots, s^m) = \sum_{y\in X}\theta_y \cdot \omega_y^i(s^1, s^2, \dots, s^m), \quad i = 1, 2, \dots, m.$$

Let $\langle \{S^i\}_{i=\overline{1,m}}, \{\psi_\theta^i(s)\}_{i=\overline{1,m}} \rangle$ be the non-cooperative game in normal form that corresponds to the average stochastic positional game in stationary strategies determined by $(\{X_i\}_{i=\overline{1,m}}, \{A(x)\}_{x\in X}, \{r^i\}_{i=\overline{1,m}}, p, \{\theta_y\}_{y\in X})$. So, $\mathbf{S}^i, i = 1, 2, \dots, m$ and $\psi_\theta^i(s)$, $i = 1, 2, \dots, m$ are defined according to (3.30)–(3.32).

In [101], the following theorem is proved:

Theorem 3.16 *The game $\langle \{S^i\}_{i=\overline{1,m}}, \{\psi_\theta^i(s)\}_{i=\overline{1,m}} \rangle$ with $\theta_x > 0$, $\forall x \in X$, $\sum \theta_x = 1$ possesses a Nash equilibrium $\mathbf{s}^* = (s^{1*}, s^{2*}, \dots, s^{m*}) \in S$ which is a Nash equilibrium in mixed stationary strategies for the average stochastic positional game determined by $(\{X_i\}_{i=\overline{1,n}}, \{A(x)\}_{x\in X}, \{r^i\}_{i=\overline{1,m}}, p, \{\theta_y\}_{y\in X})$. Moreover, if $\theta_y > 0$, $\forall y \in X$, then $s^* = (s^{1*}, s^{2*}, \dots, s^{m*})$ is a Nash equilibrium in mixed stationary strategies for the average stochastic positional game $\langle \{S^i\}_{i=\overline{1,m}}, \{\omega_y^i(s)\}_{i=\overline{1,m}} \rangle$ with an arbitrary starting state $y \in X$.*

Proof To prove the theorem, we need to verify that $\langle \{S^i\}_{i=\overline{1,m}}, \{\psi_\theta^i(s)\}_{i=\overline{1,m}} \rangle$ satisfies the conditions of Theorem 3.6. So, we have to show that each payoff $\psi_\theta^i(s^i, s^{-i})$ is quasi-monotonic with respect to s^i on a convex and compact set S^i and each payoff function $\psi_\theta^i(s^i, s^{-i})$ is graph-continuous.

Indeed, if players $1, 2, \dots, i-1, i+1, \dots, m$ fix their strategies $\hat{s}^k \in S^k$, $k \neq i$, then we obtain an average Markov decision problem with respect to $s^i \in S^i$ in which it is necessary to maximize the average reward function $\varphi^i(s^i) = \psi_\theta^i(s^i, \hat{s}^{-i})$. According to Theorem 2.19, the function $\varphi^i(s^i) = \psi_\theta^i(s^i, \hat{s}^{-i})$ possesses the property that it is quasi-monotonic with respect to s^i on S^i. Additionally, we can observe that if for the payoff $\psi^i(s^i, s^{-i})$ we consider the function $F^i : S^{-i} \to S^i$ such that

$$F^i(s^{-i}) = \hat{s}^i \in \mathscr{M}^i(s^{-i}) \text{ for } s^{-i} \in S^{-i}, \ i \in \{1, 2, \dots, m\},$$

where

$$\mathcal{M}^i(s^{-i}) = \{\, \hat{s}^i \in S^i \mid \psi_\theta^i(\hat{s}^i, s^{-i})) = \max_{s^i \in S^i} \psi_\theta^i(s^i, s^{-i})\},$$

then the function $\psi_\theta^i(F^i(s^{-i}), s^{-i})$ is continuous at $s^{-i} = \overline{s}^{-i}$ for an arbitrary $(\overline{s}^i, \overline{s}^{-i}) \in S$. So, $\psi_\theta^i(s)$ is graph-continuous, and according to Theorem 3.6, the game $\langle \{S^i\}_{i=\overline{1,m}}, \{\psi_\theta^i(s)\}_{i=\overline{1,m}} \rangle$ possesses a Nash equilibrium $s^* \in \overline{S}$. This Nash equilibrium is a Nash equilibrium in mixed stationary strategies for the average stochastic positional game determined by $(\{X_i\}_{i=\overline{1,m}}, \{A(x)\}_{x\in X}, \{r^i\}_{i=\overline{1,m}}, p, \{\theta_y\}_{y\in X})$. □

Thus, for an arbitrary average stochastic positional game, a Nash equilibrium in mixed stationary strategies exists, and the optimal stationary strategies of the players can be found using the game $\langle \{S^i\}_{i=\overline{1,m}}, \{\psi_\theta^i(s)\}_{i=\overline{1,m}} \rangle$, where S^i and $\psi_\theta^i(s)$, $i = 1, 2, \ldots, m$ are defined according to (3.30)–(3.32).

3.6.4 Average Positional Games on Networks

An average stochastic positional game with given position sets $X_1, X_2, \ldots, X_m$ in the case when the transition probabilities $p_{x,y}^a$ take only 0 and 1 can be represented as an average positional game on a graph $G = (X, E)$ with vertex set $X = X_1 \cup X_2 \cup \cdots \cup X_m$ and edge set E, where $e = (x, y) \in E$ if and only if $p_{x,y}^a = 1$ for $a \in A(x)$. The dynamics of this game are determined by the structure of graph G because in a state position $x \in X$, player i, who is the owner of this position ($x \in X_i$), may select an action $a \in A(x)$ that corresponds to an outgoing directed edge $e = (x, y) \in E(x)$, where $E(x) = \{e = (x, z) \in E | z \in X\}$. Taking into account that in each state $x \in X$, for a given action $a \in A(x)$, the reward $r_{x,a}^i$ is known, we can uniquely define the cost $c_{(x,y)}^i = -r_{x,a}^i$ with respect to each player $i \in \{1, 2, \ldots, m\}$ for each directed edge $e = (x, y)$ if $p_{x,y}^a = 1$. So, an average stochastic positional game (in the case of transition probabilities $p_{x,y}^a$ that take only 0 and 1) can be regarded as the following average positional game on graph G: The game starts at a given starting state x_0 at the moment of time $t = 0$, where player $i_0 \in \{1, 2, \ldots, m\}$, who is the owner of this position, makes a move. A move means that a player selects a neighboring position $x_1 \in X(x_0) = \{y \in X | e = (x_0, y) \in E\}$. Then at the moment of time $t = 1$, player $i_1 \in \{1, 2, \ldots, m\}$, who is the owner of position x_1 ($x_1 \in X_{i_1}$), selects a neighboring position $x_2 \in X(x_1) = \{y \in X | e = (x_1, y) \in E\}$ and so on, indefinitely. In this infinite game, players make the corresponding moves in their position sets in order to minimize their average costs per transition, using pure and mixed stationary strategies of moves. A stationary strategy of moves of player $i \in \{1, 2, \ldots, m\}$ in the game on graph G we use in the same sense as for stochastic positional games, i.e., a stationary strategy of player i is a map s^i that provides a probability distribution on the set of outgoing directed edges $E(x)$ from

x for an arbitrary state $x \in X_i$. If these probabilities take only 0 and 1, then we call such a strategy *pure stationary strategy*; otherwise, we call it *mixed stationary strategy*. So, the set of stationary strategies $\mathbb{S}^i$ of player i we can identify with the set of solutions to the following system:

$$\begin{cases} \sum\limits_{y\in X(x)} s^i_{x,y} = 1, & \forall x \in X_i; \\ s^i_{x,y} \geq 0, & \forall x \in X_i, \ \forall y \in X(x), \end{cases} \tag{3.33}$$

where $X(x) = \{y \in X | (x, y) \in E\}$. An arbitrary basic solution to this system corresponds to a pure stationary strategy of the game on G.

Based on the results of the previous section (Theorem 3.16), we can conclude that the average positional game on G has a mixed stationary equilibrium. However, in general, such a game may not have a stationary equilibrium in pure strategies. In the case of zero-sum games of two players, this game has a value, and players may achieve the value by applying the pure stationary strategies of moves. This fact was shown in [42].

Thus, an average positional game on a network is determined by a tuple $(G = (X, E), \{X_i\}_{i=\overline{1,m}}, \{c^i\}_{i=\overline{1,m}}, x_0)$, where G is a graph with a finite set of states X and a finite set of directed edges E; $X_1, X_2, \dots, X_m$ $(X = X_1 \cup X_2 \cup \dots \cup X_m;\ X_i \cap X_j = \emptyset, i \neq j)$ represent the corresponding position sets of the players; $c^1, c^2, \dots, c^m$ represent m cost functions on E that give to each directed edge $e = (x, y) \in E$ m costs $c^1_e, c^2_e, \dots, c^m_e$ for the corresponding players; and x_0 is the starting position of the game. Graph G is assumed to possess the property that each vertex $x \in X$ contains at least one outgoing directed edge $e = (x, y) \in E$.

We obtain a more general model of an average positional game on graph $G = (X, E)$ if we assume that the set of vertices X is divided into $m+1$ subsets, i.e., $X = X_1 \cup X_2 \cup \dots \cup X_m \cup X^0$, where $X_1, X_2, \dots, X_m$ are the corresponding position sets of the players and X^0 is a set of positions in which the game moves to a next state $y \in X$ randomly, i.e., for each $x \in X^0$ in the set of outgoing directed edges $E(x) = \{(x, y) \in E | y \in X\}$, a transition probability distribution $\{p_{x,y}\}_{y\in X}$ is given, and the game randomly passes from $x \in X^0$ to a next state according to probability distribution $\{p_{x,y}\}_{y\in X}$. For a random move from a state $x \in X_0$ to the next state, the average cost per transition $\mu(x)$ is defined as $\mu(x) = \sum_{y\in X} p_{x,y} c_{x,y}$. This general model of the average positional game on the network is determined by the tuple

$$(G = (X, E), \{X_i\}_{i=\overline{1,m}}, X^0, \{c^i\}_{i=\overline{1,m}}, p, x_0),$$

where p is a given transition probability function defined on the subset of directed edges $E(X^0) = \{e = (x, y) \in E | x \in X^0, y \in X\}$, satisfying the condition $\sum_{e\in E(x)} p_e = 1, \forall x \in X^0$. Such a game represents the game variant of the average control problem on networks from Sect. 2.6. All results mentioned above (for an average positional game on networks) also hold for the general game model.

3.6.5 Pure Stationary Nash Equilibria for Unichain Stochastic Positional Games

The existence of stationary Nash equilibria for an average stochastic positional game with a unichain property can be derived on the basis of the results from [112]. Here, we show that such an arbitrary game possesses a pure stationary Nash equilibrium.

The Normal Form of the Game with a Unichain Property
We consider an m-player average stochastic positional game with a unichain property determined by a tuple $(\{X_i\}_{i=\overline{1,m}}, \{A(x)\}_{x\in X}, \{r^i\}_{i=\overline{1,m}}, p)$. The normal form of this game in stationary strategies can be derived from the game model from previous sections if we consider the unichain property of the game. The values of the payoffs of the players $\omega^i(s^1, s^2, \dots, s^m)$, $i = 1, 2, \dots, m$ do not depend on the starting position but only depend on the strategies of the players.

The normal form of an average stochastic positional game in stationary strategies in the unichain case can be defined as follows:

Let S^i, $i \in \{1, 2, \dots m\}$ be the set of solutions to the system (3.26). On the set $S = S^1 \times S^2 \times \dots \times S^m$, we consider m payoff functions

$$\omega^i(s^1, s^2, \dots, s^m) = \sum_{k=1}^{m} \sum_{x\in X_k} \sum_{a\in A(x)} s^k_{x,a} \cdot r^i_{x,a} \cdot q_x, \qquad i = 1, 2, \dots, m, \tag{3.34}$$

where q_x for $x \in X$ are uniquely determined by the following system of linear equations:

$$\begin{cases} q_y - \sum\limits_{k=1}^{m} \sum\limits_{x\in X_k} \sum\limits_{a\in A(x)} s^k_{x,a} \cdot p^a_{x,y} \cdot q_x = 0, & \forall y \in X; \\ \sum\limits_{y\in X} q_y = 1 \end{cases} \tag{3.35}$$

for an arbitrary fixed profile $s = (s^1, s^2, \dots, s^m) \in S$. The functions $\omega^i(s^1, s^2, \dots, s^m)$, $i = 1, 2, \dots, m$ on the set $S = S^1 \times S^2 \times \dots \times S^m$ determine a game in normal form $\langle \{S^i\}_{i=\overline{1,m}}, \{\omega^i(s)\}_{i=\overline{1,m}} \rangle$ that corresponds to an average stochastic positional game with a unichain property when players use stationary strategies of choosing the action in their position sets. This game represents the positional game variant of the average Markov decision problem, and therefore, based on Theorem 2.19, the payoffs possess the quasi-monotonic property with respect to the corresponding strategies of the players.

Average stochastic positional games with a unichain property represent a special case of unichain average stochastic games from Sect. 3.5.1, for which Rogers [155] proved the existence of stationary Nash equilibria. At the same time, the existence of stationary equilibria for the considered games can also be obtained from Theorem 2.19 based on the results from [36, 46] concerned with the existence of Nash equilibria for the game with quasi-concave (quasi-convex) payoffs. Indeed, the payoffs $\omega^i(s^1, s^2, \ldots, s^m)$, $i = 1, 2, \ldots, m$ are continuous on $S^1 \times S^2 \times \cdots \times S^m$ and quasi-monotonic with respect to the strategy of each player, and therefore, a stationary Nash equilibrium exists. In the following, we can see that for this class of positional games, there exist pure stationary Nash equilibria.

Existence of Pure Stationary Equilibria

We can obtain the existence of Nash equilibria in pure stationary strategies for an average stochastic positional game with a unichain property based on the following result [102]:

Theorem 3.17 *Let an average stochastic positional game be given that is determined by the tuple* $(\{X_i\}_{i=\overline{1,m}}, \{A(x)\}_{x\in X}, \{r^i\}_{i=\overline{1,m}}, p)$. *Assume that for an arbitrary profile* $s = (s^1, s^2, \ldots, s^m)$ *of the game, the transition probability matrix* $P^s = (p^s_{x,y})$ *is unichain. Then there exist the functions*

$$\varepsilon^i : X \to R, \quad i = 1, 2, \ldots, m$$

and the values $\omega^1, \omega^2, \ldots, \omega^m$ *that satisfy the following conditions:*

(1) $r^i_{x,a} + \sum_{y\in X} p^a_{x,y} \cdot \varepsilon^i_y - \varepsilon^i_x - \omega^i \le 0, \ \forall x \in X_i, \quad \forall a \in A(x), \ i = 1, 2, \ldots, m.$

(2) $\max_{a\in A(x)} \{r^i_{x,a} + \sum_{y\in X} p^a_{x,y} \cdot \varepsilon^i_y - \varepsilon^i_x - \omega^i\} = 0, \qquad \forall x \in X_i, \ i = 1, 2, \ldots, m.$

(3) *On each position set* X_i, $i \in \{1, 2, \ldots, m\}$, *there exists a map*

$$s^{i*} : X_i \to \cup_{x\in X_i} A(x)$$

such that

$$s^{i*}(x) = a^* \in \operatorname*{argmax}_{a\in A(x)} \left\{r^i_{x,a} + \sum_{y\in X} p^a_{x,y} \cdot \varepsilon^i_y - \varepsilon^i_x - \omega^i\right\}$$

and

$$r^j_{x,a^*} + \sum_{y\in X} p^{a^*}_{x,y} \cdot \varepsilon^j_y - \varepsilon^j_x - \omega^j = 0, \ \forall x \in X_i, \ j = 1, 2, \ldots, m.$$

The maps $s^{1*}, s^{2*}, \ldots, s^{m*}$ *determine a Nash equilibrium* $s^* = (s^{1*}, s^{2*}, \ldots, s^{m*})$ *for the stochastic positional game and*

$$\omega^i_x(s^{1*}, s^{2*}, \ldots, s^{m*}) = \omega^i, \ \forall x \in X, \ i = 1, 2, \ldots, m.$$

Moreover, $s^* = (s^{1*}, s^{2*}, \ldots, s^{m*})$ *is a pure stationary Nash equilibrium of the average stochastic positional game for an arbitrary starting position* $x \in X$.

Proof According to Theorem 3.11, for the average stochastic positional game determined by $(\{X_i\}_{i=\overline{1,m}}, \{A(x)\}_{x\in X}, \{r^i\}_{i=\overline{1,m}}, p)$, there exists a mixed stationary Nash equilibrium $\overline{s}^* = (\overline{s}^{1^*}, \overline{s}^{2^*}, \ldots, \overline{s}^{m*})$. Taking into account that for this game, the unichain property holds, we have

$$\omega^i = \omega^i_x(\overline{s}^{1^*}, \overline{s}^{2^*}, \ldots, \overline{s}^{m*}), \quad \forall x \in X,\ i = 1, 2, \ldots, m.$$

If $\overline{s}^{i^*}$ is a mixed stationary strategy of player $i \in \{1, 2, \ldots, m\}$, then in a state $x \in X_i$, the strategy $\overline{s}^{i^*}(x)$ is represented as a convex combination of actions determined by a probability distribution $\{\overline{s}^{i^*}_{x,a}\}$ on $A'(x) = \{a \in A(x) |\ \overline{s}^{i^*}_{x,a} > 0\}$. Thus, for a state $x \in X_i$, we have $A(x) = A'(x) \cup A''(x)$, where $A''(x) = \{a \in A(x) |\ \overline{s}^{i^*}_{x,a} = 0\}$.

Let us consider the Markov process induced by the profile of mixed stationary strategies $\overline{s}^* = (\overline{s}^{1^*}, \overline{s}^{2^*}, \ldots, \overline{s}^{m*})$. Then according to (3.2), the elements of transition probability matrix $P^{\overline{s}^*} = (p^{\overline{s}^*}_{x,y})$ of this Markov process are calculated as follows:

$$p^{\overline{s}^*}_{x,y} = \sum_{a\in A(x)} \overline{s}^{i^*}_{x,a} \cdot p^a_{x,y} \quad \text{for } x \in X_i,\ i = 1, 2, \ldots, m. \tag{3.36}$$

This matrix is unichain, and the corresponding step rewards in the states induced by $\overline{\mathbf{s}}$ can be determined according to (3.4) as follows:

$$r^i_{x,\overline{s}^{k*}} = \sum_{a\in A(x)} \overline{s}^{k^*}_{x,a} \cdot r^i_{x,a}, \quad \text{for } x \in X_k,\ k \in \{1, 2, \ldots, m\}. \tag{3.37}$$

Based on Theorem 2.9, for this Markov process, we can write the following reward equations:

$$r^i_{x,\overline{s}^{i*}} + \sum_{y\in X} p^{\overline{s}^{i*}}_{x,y} \cdot \varepsilon^i_y - \varepsilon^i_x - \omega^i = 0, \quad \forall x \in X_i,\ i = 1, 2, \ldots, m, \tag{3.38}$$

and, in general, for an arbitrary position set X_i $i \in \{1, 2, \ldots, m\}$, we have

$$r^j_{x,\overline{s}^{i*}} + \sum_{y\in X} p^{\overline{s}^{k*}}_{x,y} \cdot \varepsilon^j_y - \varepsilon^j_x - \omega^i = 0, \quad \forall x \in X_i,\ j = 1, 2, \ldots, m. \tag{3.39}$$

By introducing (3.36) and (3.37) in (3.38), we obtain

$$\sum_{a\in A(x)} \overline{s}^{i^*}_{x,a} \cdot r^i_{x,a} + \sum_{y\in X} \sum_{a\in A(x)} \overline{s}^{i^*}_{x,a} \cdot p^a_{x,y} \cdot \varepsilon^i_y - \varepsilon^i_x - \omega^i = 0, \quad \forall x \in X_i,\ i = 1, 2, \ldots, m.$$

In these equations, we can set $\varepsilon^i_x = \sum_{a\in A(x)} \overline{s}^{i^*}_{x,a} \cdot \varepsilon_x$ and $\omega^i_x = \sum_{a\in A(x)} \overline{s}^{i^*}_{x,a} \cdot \omega^i_x$.

After these substitutions and some elementary transformations of the equations, we obtain

$$\sum_{a\in A(x)} \overline{s}^{i*}_{x,a}\left(r^i_{x,a}+\sum_{y\in X} p^a_{x,y}\cdot \varepsilon^i_y-\varepsilon^i_x-\omega^i\right)=0,\ \ \forall x\in X_i,\ i=1,2,\dots,m.$$

In a similar way, we can show that for an arbitrary position set $X_i\ i\in\{1,2,\dots,m\}$, the following formula holds:

$$\sum_{a\in A(x)} \overline{s}^{i*}_{x,a}\left(r^j_{x,a}+\sum_{y\in X} p^a_{x,y}\cdot \varepsilon^j_y-\varepsilon^i_x-\omega^j\right)=0,\ \ \forall x\in X_i,\ j=1,2,\dots,m.$$

So, for the Markov process induced by the profile of mixed stationary strategies $\overline{s}^*=(\overline{s}^{1^*},\overline{s}^{2^*},\dots,\overline{s}^*)$, there exist the functions $\varepsilon^i : X\to R,\quad i=1,2,\dots,m$ and the values $\omega^1,\omega^2,\dots,\omega^m$ that satisfy the following conditions:

$$r^i_{x,a}+\sum_{y\in X} p^a_{x,y}\cdot \varepsilon^i_y-\varepsilon^i_x-\omega^i=0,\ \ \forall x\in X_i,\ \forall a\in A'(x),\ \ i=1,2,\dots,m$$

and for an arbitrary position set $X_i\ \ i\in\{1,2,\dots,m\}$, we have

$$r^j_{x,a}+\sum_{y\in X} p^a_{x,y}\cdot \varepsilon^j_y-\varepsilon^j_x-\omega^i=0,\ \ \forall x\in X_i,\ \forall a\in A'(x),\ \ j=1,2,\dots,m.$$

Now let us fix the strategies $\overline{s}^{1^*},\ \overline{s}^{2^*},\ \dots,\ \overline{s}^{i-1^*},\ \overline{s}^{i+1^*},\ \dots,\overline{s}^*$ of the players $1,\ 2,\dots,\ i-1,\ i+1,\dots,\ m$ and consider the problem of determining the maximal average reward per transition with respect to player i $\{1,2,\dots,m\}$. Obviously, if we solve this decision problem, then we obtain the strategy s^{i^*}. We can determine the optimal strategy of this decision problem with an average cost optimization criterion using the bias equations with respect to player i. This means that there exist the functions $\varepsilon^i : X\to R$ and the values ω^i, $i=1,2,\dots,m$ that satisfy the following conditions:

(1) $r^i_{x,a}+\sum\limits_{y\in X} p^a_{x,y}\varepsilon^i_y-\varepsilon^i_x-\omega^i\le 0,\quad \forall x\in X_i,\ \forall a\in A(x).$

(2) $\max\limits_{a\in A(x)}\left\{r^i_{x,a}+\sum\limits_{y\in X} p^a_{x,y}\varepsilon^i_y-\varepsilon^i_x-\omega^i\right\}=0,\quad \forall x\in X_i.$

The values ω^i, $i=1,\ 2,\ \dots,\ m$ are uniquely determined by (3.38) and (3.39). Moreover, these equations determine ε^i_x, $\forall x\in X,\ i=1,2,\dots,m$ up to an additive constant (see [152]). This mean that ω^i, $i=1,2,\dots,m$ and ε^i_x, $\forall x\in X,\ i=1,2,\dots,m$ satisfy conditions (1) and (2).

Thus, for fixed strategies $\overline{s}^{1^*},\ \overline{s}^{2^*},\ \dots,\overline{s}^{i-1^*},\ \overline{s}^{i+1^*},\ \dots,\ \overline{s}^{m*}$ of the players $1,\ 2,\ \dots,\ i-1,i+1,\ \dots,\ m,$ we can determine a pure stationary strategy s^{i^*}

of player i, where

$$s^{i^*}(x) = a^* \in \underset{a \in A(x)}{\operatorname{argmax}} \left\{ r^i_{x,a} + \sum_{y \in X} p^a_{x,y} \varepsilon^i_y - \varepsilon^i_x - \omega^i \right\}, \ for\ x \in X_i,$$

and

$$r^j_{x,a^*} + \sum_{y \in X} p^{a^*}_{x,y} \varepsilon^j_y - \varepsilon^j_x - \omega^j = 0, \ \ \forall x \in X_i, \ \ j = 1, 2, \dots, m.$$

In such a way, we can see that $(\overline{s}^{1^*}, \overline{s}^{2^*}, \dots, \overline{s}^{i-1^*}, s^{i^*}, \overline{s}^{i+1^*}, \dots, \overline{s}^{m*})$ is a Nash equilibrium, where s^{i^*} is a pure stationary strategy. This means that for each player $i \in \{1, 2, \dots, m\}$, there exists a pure stationary strategy s^{i^*} such that $s^* = (s^{1^*}, s^{2^*}, \dots, s^{m*})$ is a Nash equilibrium and such an equilibrium satisfies the conditions (1)–(3) of the theorem. □

Corollary 3.18 *Let an average stochastic positional game be given that is determined by the tuple* $(\{X_i\}_{i=\overline{1,m}}, \{A(x)\}_{x \in X}, \{r^i\}_{i=\overline{1,m}}, p)$. *If there exist the functions*

$$\varepsilon^i : X \to R, \quad i = 1, 2, \dots, m$$

and the values $\omega^1, \omega^2, \dots, \omega^m$ *that satisfy the conditions (1)–(3) of Theorem 3.17, then the game has a Nash equilibrium* $s^* = (s^{1^*}, s^{2^*}, \dots, s^{m*})$ *that is a pure stationary Nash equilibrium of the average stochastic positional game for an arbitrary starting position* $x \in X$.

In the following, we analyze an important class of average positional games in which the transition probabilities $p^a_{x,y}$ take only values 0 and 1, i.e., we consider the deterministic case of average positional games. This class of games in [2, 42, 202] is called mean-payoff games; in [67, 108], this class of games is called cyclic games.

3.6.6 Pure Nash Equilibria Conditions for Cyclic Games

In [2, 67, 108, 112], the following deterministic positional games were considered: Let a dynamical system $\mathbb{L}$ with a finite set of states X be given. The dynamics of the system are described by a directed graph of the state's transitions $G = (X, E)$, where the set of vertices X corresponds to the set of states of the system $\mathbb{L}$ and an arbitrarily directed edge $e = (x, y) \in X$ expresses the possibility of the system's transition from the state $x = x(t)$ to the state $y = x(t+1)$ at every discrete moment of time $t = 0, 1, 2, \dots$. It is assumed that graph G possesses the property that each vertex $x \in X$ contains at least one outgoing directed edge. The starting state of the system is given, i.e., $x_0 = x(0)$, and the dynamics of the system are controlled by m players. For each player $i \in \{1, 2, \dots, m\}$, a cost function $c^i : E \to \mathbb{R}$ is defined, where for $(x, y) = e \in E$, the value $c^i_{x,y}$ expresses the cost of the system's transition from the state $x = x(t)$ to the state $y = x(t+1)$, $t = 0, 1, 2, \dots$. The set of states X is divided into m disjoint subsets $X_1, X_2, \dots, X_m$

$$X = X_1 \cup X_2 \cup \dots \cup X_m \quad (X_i \cap X_j = \emptyset, \ \ \forall i \neq j),$$

where X_i represents the set of positions of the player $i \in \{1, 2, \dots, m\}$. Each player may control the dynamical system only in his or her positions. So, the control process on G is made by the players as follows:

If the starting state $x_0 = x(0)$ belongs to the set of positions X_i, then player i on the set of possible transitions $E(x_0) = \{(x_0, y) \in E \mid y \in X\}$ selects a transition (x_0, x_1) from the state x_0 to the state $x_1 = x(1)$; in general, if at the moment of time t the dynamical system is in the state $x_t = x(t)$ and $x(t) \in X_i$, then the system is controlled by player i, i.e., player i selects the transition from the state x_t to a state $x_{t+1} \in X(x_t)$, where $X(t) = \{y \in X \mid (x_t, y) \in E\}$. In this dynamical process, each player intends to minimize his or her average cost per transition by a trajectory $x_0, x_1, x_2, \dots$, i.e.,

$$F^i = \lim_{t \to \infty} \frac{1}{t} \sum_{\tau=0}^{t-1} c^i_{x_t, x_{t+1}} \quad \to \min, \quad i = 1, 2, \dots, m.$$

In a stricter way, this dynamic game can be formulated in terms of pure stationary strategies. We define the pure stationary strategies of the players as m maps:

$$s^i : x \ \to \ y \in X(x) \ \ for \ \ x \in X_i; \ \ i = 1, 2, \dots, m.$$

Let $s^1, s^2, \dots, s^m$ be an arbitrary set of pure strategies of the players. We denote by $G_s = (X, E_s)$ the subgraph generated by edges $e = (x, s^i(x))$ for $x \in X_i$ and $i = 1, 2, \dots, m$. Obviously, for fixed $s^1, s^2, \dots, s^m$, the graph G_s possesses the property that a unique directed cycle C_s with the set of edges $E(C_s)$ can be reached from x_0. For fixed pure strategies $s^1, s^2, \dots, s^m$ and a fixed starting state, we define the quantities

$$F^i_{x_0}(s^1, s^2, \dots, s^m) = \frac{1}{n(C_s)} \sum_{e \in E(C_s)} c^i_e, \quad i = 1, 2, \dots, m,$$

where $n(C_s)$ is the number of directed edges of the cycle C_s. The graph G is finite, and therefore, the set of pure strategies $\mathbb{S}^i$ of each player $i \in \{1, 2, \dots, m\}$ is a finite set. In such a way, on the set of profiles $\mathbb{S} = \mathbb{S}^1 \times \mathbb{S}^2 \times \dots \times \mathbb{S}^m$, the functions $F^1_{x_0}(s^1, s^2, \dots, s^m)$, $F^2_{x_0}(s^1, s^2, \dots, s^m)$, $\dots$, $F^m_{x_0}(s^1, s^2, \dots, s^m)$ determine a game in normal form which in [67, 108, 112] is called *cyclic game*, and it is denoted by $(G, \{X_i\}_{i=\overline{1,m}}, \{c^i\}_{i=\overline{1,m}}, x_0)$.

It is easy to see that this game represents a particular case of a stochastic positional game from Sect. 3.6.8.

Indeed, we can regard the cyclic game as a stochastic positional game induced by the set of states X with the partition $X = X_1 \cup X_2 \cup \cdots \cup X_m$, the cost functions c^i, $i = 1, 2, \ldots m$, the action sets $A(x)$, and the corresponding probability distributions $\{p_{x,y_k}^{a_k}\}$ for $x \in X$, where for a given state $x \in X$, we have $k(x) = |E(x)|$ actions that correspond to possible transitions through the edges $e_1 = (x, y_1)$, $e_2 = (x, y_2)$, $\ldots$, $e_{r(x)} = (x, y_{k(x)})$, i.e.,

$$p_{x,y_1}^1 = 1,\ p_{x,y_2}^1 = 0,\ p_{x,y_3}^1 = 0,\ \ldots,\ p_{x,y_{k(x)}}^1 = 0;$$

$$p_{x,y_1}^2 = 0,\ p_{x,y_2}^2 = 1,\ p_{x,y_3}^2 = 0,\ \ldots,\ p_{x,y_{k(x)}}^2 = 0;$$

....

$$p_{x,y_1}^{r(x)} = 0,\ p_{x,y_2}^{r(x)} = 0,\ p_{x,y_3}^{r(x)} = 0,\ \ldots,\ p_{x,y_{r(x)}}^{k(x)} = 1.$$

So, cyclic games can be regarded as stochastic positional games with average payoff functions, where $p_{x,y}^a = 0 \vee 1$, $\forall x, y \in X$, $\forall a \in A$, and the rewards are $r_{x,a_k}^i = -c_{e_k}^i$ for $x \in X$, $e_k = (x, y_k)$.

As noted for non-zero-sum cyclic games, a pure Nash equilibrium may not exist. An example of a cyclic game for which a pure Nash equilibrium does not exist was given in [67, 112]. We can obtain a pure Nash equilibrium existence criterion for such games from Theorem 3.10.

Theorem 3.19 *If for an arbitrary profile of the game $s = (s^1, s^2, \ldots, s^m)$, the corresponding graph $G_s = (X, E_s)$ contains a unique directed cycle such that it can be reached from an arbitrary $x \in X$, then in the cyclic game, there exists a Nash equilibrium $s^{1*}, s^{2*}, \ldots, s^{m*}$. Moreover, $s^{1*}, s^{2*}, \ldots, s^{m*}$ is a Nash equilibrium for an arbitrary starting position $x \in X$ of the game.*

We obtain this theorem as a corollary from Theorem 3.10 if we regard cyclic games as a special case of unichain stochastic positional games.

In the following, we formulate a necessary and sufficient condition for the existence of Nash equilibria in so-called ergodic cyclic games with m players. To formulate this result, we need the following definition:

Definition 3.20 Let $s^{1*}, s^{2*}, \ldots, s^{m*}$ be a solution (in the sense of Nash) for a cyclic game determined by a network $(G, \{X_i\}_{i=\overline{1,m}}, \{c^i\}_{i=\overline{1,m}}, x_0)$, where $G = (X, E)$ is a strongly connected directed graph. We call this game an ergodic cyclic game if $s^{1*}, s^{2*}, \ldots, s^{m*}$ represents the solution in the sense of Nash for a cyclic game on the network $(G, \{X_i\}_{i=\overline{1,m}}, \{c^i\}_{i=\overline{1,m}}, x)$ with an arbitrary starting position $x \in X$ and

$$F_x^i(s^{1*}, s^{2*}, \ldots, s^{m*}) = F_y^i(s^{1*}, s^{2*}, \ldots, s^{m*}), \quad \forall x, y \in X,\ i = 1, 2, \ldots, m.$$

Theorem 3.21 *The cyclic game determined by the network* $(G, \{X_i\}_{i=\overline{1,m}}, \{c^i\}_{i=\overline{1,m}}, x_0)$, *where* $G = (X, E)$ *is a strongly connected directed graph, is an ergodic one if and only if there exist m real functions on X*

$$\varepsilon^1 : X \to \mathbb{R}^1,\ \varepsilon^2 : X \to \mathbb{R}^1, \dots,\ \varepsilon^m : X \to \mathbb{R}^1,$$

and m values $\omega^1, \omega^2, \dots, \omega^m$ *such that the following conditions are satisfied:*

(a) $\varepsilon^i_y - \varepsilon^i_x + c^i_{x,y} - \omega^i \ge 0, \quad \forall (x, y) \in E_i$, *where* $E_i = \{e = (x, y) \in E \mid x \in X_i\},\ i = 1, 2, \dots, m.$

(b) $\min\limits_{y \in X_G(x)} \{\varepsilon^i_y - \varepsilon^i_x + c^i_{x,y} - \omega^i\} = 0, \quad \forall x \in X_i,\ i = 1, 2, \dots, m.$

(c) *The subgraph* $G' = (X, E')$ *generated by the edge set* $E' = E^0_1 \cup E^0_2 \cup \dots \cup E^0_m$, $E^0_i = \{e = (x, y) \in E_i \mid \varepsilon^i_y - \varepsilon^i_x + c^i_{x,y} - \omega^i = 0\},\ i = 1, 2, \dots, m$ *has the property that it contains a connected subgraph* $\overline{G}^0 = (X, \overline{E}^0)$ *for which every vertex* $x \in X$ *has only one leaving edge* $e = (x, y) \in \overline{E}^0$ *and besides that*

$$\varepsilon^i_y - \varepsilon^i_x + c^i_{x,y} - \omega^i = 0, \quad \forall (x, y) \in \overline{E}^0,\ i = 1, 2, \dots, m.$$

The optimal solution to the problem can be determined by fixing the maps:

$$s^{1*} : x \to y \in X_{\overline{G}^0}(x) \quad \textit{for} \quad x \in X_1;$$

$$s^{2*} : x \to y \in X_{\overline{G}^0}(x) \quad \textit{for} \quad x \in X_2;$$

$$\vdots$$

$$s^{m*} : x \to y \in X_{\overline{G}^0}(x) \quad \textit{for} \quad x \in X_m,$$

where $X_{\overline{G}^0}(x) = \{y \mid (x, y) \in \overline{E}^0\}$, *and*

$$F^i_{\overline{x}}(s^{1*}, s^{2*}, \dots, s^{m*}) = \omega^i, \quad \forall \overline{x} \in X,\ i = 1, 2, \dots, m.$$

Proof $\Longrightarrow$ Let $(G, \{X_i\}_{i=\overline{1,m}}, \{c^i\}_{i=\overline{1,m}}, x_0)$ be a network that determines an *ergodic cyclic game* with m players, i.e., in this game, there exists a Nash equilibrium $s^{1*}, s^{2*}, \dots, s^{m*}$. Define

$$\omega^i = F^i_{x_0}(s^{1*}, s^{2*}, \dots, s^{m*}), \quad i = 1, 2, \dots, m. \tag{3.40}$$

It is easy to verify that if we change the cost function c^i to $\overline{c}^i = c^i - \omega^i,\ i = 1, 2, \dots, m$, then the obtained network $(G, \{X_i\}_{i=\overline{1,m}}, \{\overline{c}^i\}_{i=\overline{1,m}}, x_0)$ determines a new ergodic cyclic game that is equivalent to the initial one.

For the new game, $s^{1*}, s^{2*}, \ldots, s^{m*}$ is a Nash equilibrium and

$$F^i_{x_0}(s^{1*}, s^{2*}, \ldots, s^{m*}) = 0, \quad i = 1, 2, \ldots, m.$$

This game can be regarded as the dynamic c-game from [19, 109] on the network $(G, \{X_i\}_{i=\overline{1,m}}, \{\overline{c}^i\}_{i=\overline{1,m}}, x_0)$ with a given starting position x_0 and a final position $x_0 \in C_{s^*}$, where C_{s^*} is a directed cycle generated by strategies $s^{1*}, s^{2*}, \ldots, s^{m*}$ such that

$$\sum_{e \in E(C_{s^*})} \overline{c}^i_e = 0, \quad i = 1, 2, \ldots, m.$$

Taking into account that our game is ergodic, we may state, without loss of generality, that x_0 belongs to a directed cycle C_{s^*} generated by strategies $s^{1*}, s^{2*}, \ldots, s^{m*}$. Therefore, the ergodic game with a network $(G, \{X_i\}_{i=\overline{1,m}}, \{\overline{c}^i\}_{i=\overline{1,m}}, x_0)$ can be regarded as the dynamic c-game from [19, 109] on the network $(G, \{X_i\}_{i=\overline{1,m}}, \{\overline{c}^i\}_{i=\overline{1,m}}, x_0)$ with a starting position x_0 and a final position x_0. So, according to Theorem 2 from [19], there exist m real functions

$$\varepsilon^1 : X \to \mathbb{R}^1, \ \varepsilon^2 : X \to \mathbb{R}^1, \ldots, \ \varepsilon^m : X \to \mathbb{R}^1,$$

such that the following conditions are satisfied:

(1) $\varepsilon^i_y - \varepsilon^i_x + \overline{c}^i_{x,y} \geq 0, \quad \forall (x, y) \in E_i, \ i = 1, 2, \ldots, m.$

(2) $\min\limits_{x \in X_G(x)} \{\varepsilon^i_y - \varepsilon^i_x + \overline{c}^i_{x,y}\} = 0, \quad \forall x \in X_i, \ i = 1, 2, \ldots, m.$

(3) The subgraph $G' = (X, E')$, generated by the edge set $E' = E^0_1 \cup E^0_2 \cup \cdots \cup E^0_m$, $E^0_i = \{e = (x, y) \in E_i \mid \varepsilon^i_y - \varepsilon^i_x + \overline{c}^i_{x,y} = 0\}$, $i = 1, 2, \ldots, m$, has the property that it contains a connected subgraph $\overline{G}^0 = (X, \overline{E}^0)$ for which every vertex $x \in X$ has only one leaving edge $e = (x, y) \in \overline{E}^0$ and besides that

$$\varepsilon^i_y - \varepsilon^i_x + \overline{c}^i_{x,y} = 0, \quad \forall (x, y) \in \overline{E}^0, \ i = 1, 2, \ldots, m.$$

If in the conditions (1), (2), and (3) mentioned above we take into account that $\overline{c}^i_{x,y} = \overline{c}^i_{x,y} - \omega^i, \ \forall (x, y) \in E, \ i = 1, 2, \ldots, m$, then we obtain the conditions (a), (b), and (c) from Theorem 3.21.

$\Longleftarrow$ Assume that the conditions (a), (b), and (c) of Theorem 3.21 hold. Then for the network $(G, \{X_i\}_{i=\overline{1,m}}, \{\overline{c}^i\}_{i=\overline{1,m}}, x_0)$, the conditions (1), (2), and (3) are satisfied. It is easy to check that an arbitrary set of strategies $s^{1*}, s^{2*}, \ldots, s^{m*}$, where

$$s^{i*} : x \to y \in X_{\overline{G}^0}(x), \quad i = 1, 2, \ldots, m,$$

is a Nash equilibrium for an ergodic cyclic game on the network $(G, \{X_i\}_{i=\overline{1,m}}, \{\overline{c}^i\}_{i=\overline{1,m}}, x_0)$ and

$$F_x^i(s^{1*}, s^{2*}, \ldots, s^{m*}) = 0, \quad i = 1, 2, \ldots, m.$$

This implies that $s^{1*}, s^{2*}, \ldots, s^{m*}$ represents a Nash equilibrium for the ergodic cyclic game on the network $(G, \{X_i\}_{i=\overline{1,m}}, \{c^i\}_{i=\overline{1,m}}, x_0)$. □

Remark 3.22 The value ω^i, $i = 1, 2, \ldots, m$ coincides with the value of the payoff function $F_x^i(s^{1*}, s^{2*}, \ldots, s^{m*})$, $i = 1, 2, \ldots, m$.

Note that for ergodic zero-sum games, Theorem 3.21 gives necessary and sufficient conditions for the existence of saddle points, i.e., Theorem 3.21 extends results from [67] to cyclic games with m players.

The problem of the existence of Nash equilibria in mixed stationary strategies can also be considered for a cyclic game, where a mixed stationary strategy s^i of player $i \in \{1, 2, \ldots, m\}$ in a state $x \in X_i$ is a map that provides a probability distribution over the set of outgoing directed edges $E(x) = \{e = (x, y) \in E | y \in X\}$ (see [126]). In the case of the cyclic game, Nash equilibria in mixed stationary strategies exist because the game becomes an average positional game for which, according to Theorem 3.11, stationary Nash equilibria exist.

3.6.7 *Pure Stationary Equilibria for Two-Player Zero-Sum Average Positional Games*

In this section, we prove the existence of Nash equilibria in pure stationary strategies for a two-player zero-sum average stochastic positional game and present conditions for determining the optimal pure stationary strategies of the players.

A two-player zero-sum average stochastic game is determined by a tuple $(X = X_1 \cup X_2, \{A(x)\}_{x\in X}, \{r_{x,a}\}_{x\in X,\, a\in A(x)}, p, x_0)$, where X is the set of states of the game, X_1 is the set of positions of the first player, X_2 is the set of positions of the second player, $A(x)$ is the set of actions in a state $x \in X$, $r_{x,a}$ is the step reward in $x \in X$ for a fixed $a \in A(x)$, $p : X \times \cup_{x\in X} A(x) \times X \to [0, 1]$ is a transition probability function that satisfies the condition $\sum_{y\in X} p_{x,y}^a = 1$, $\forall x \in X, a \in A(x)$, and x_0 is the starting state of the game.

The game starts at given initial state x_0, where the player, who is the owner of this position, fixes an action $a_0 \in A(x_0)$. So, if x_0 belongs to the set of positions of the first player, then the action $a_0 \in A(x_0)$ in x_0 is chosen by the first player; otherwise, the action $a_0 \in A(x_0)$ is chosen by the second player. After that, the game passes randomly to a new position according to the probability distribution $\{p_{x_0,y}^{a_0}\}_{y\in X}$. At time moment $t = 1$, the players observe the position $x_1 \in X$. If x_1 belongs to the set of positions of the first player, then the action $a_1 \in A(x_1)$ is chosen by the first player; otherwise, the action is chosen by the second player and so on, indefinitely.

In this process, the first player chooses actions in his or her position set in order to maximize the average reward per transition $\liminf_{t\to\infty} \frac{1}{t}\sum_{\tau=0}^{t} f(x_\tau, a_\tau)$, while the second player chooses the action in his or her position set in order to minimize the average reward per transition $\limsup_{t\to\infty} \frac{1}{t}\sum_{\tau=0}^{t} f(x_\tau, a_\tau)$. Assuming that players choose actions in their state positions independently, we show that for this game, there exists a value ω_{x_0} such that the first player has a strategy of choosing the actions in his or her position set that ensures $\liminf_{t\to\infty} \frac{1}{t}\sum_{\tau=0}^{t} r_{x_\tau,a_\tau} \geq \omega_{x_0}$ and the second player has a strategy of choosing the actions in his or her position set that ensures $\limsup_{t\to\infty} \frac{1}{t}\sum_{\tau=0}^{t} r_{x_\tau,a_\tau} \leq \omega_{x_0}$. Moreover, we show that players can achieve the value ω_{x_0}, applying pure stationary strategies of selecting the actions in their position sets. The pure stationary strategies of the players we define as two maps:

$$s^1 : x \to a \in A(x) \text{ for } x \in X_1; \quad s^2 : x \to a \in A(x) \text{ for } x \in X_2,$$

and the sets of pure stationary strategies of the first player and second player we denote as $S^1 = \{s^1 |\ s^1 : x \to a \in A(x) \text{ for } x \in X_1\}$ and $S^2 = \{s^2 |\ s^2 : x \to a \in A(x) \text{ for } x \in X_1\}$, respectively.

Let s^1, s^2 be arbitrary pure stationary strategies of the players. Then the profile $s = (s^1, s^2)$ determines a Markov process induced by probability distributions $\{p_{x,y}^{s^i(x)}\}_{y\in X}$ in the states $x \in X_i$, $i = 1, 2$ and a given starting state x_0. For this Markov process with step rewards $r_{x,s^i(x)}$, in the states $x \in X_i, i = 1, 2$, we can determine the average reward per transition $\omega_{x_0}(s^1, s^2)$. The function $\omega_{x_0}(s^1, s^2)$ on $S = S^1 \times S^2$ defines an antagonistic game in normal form $\langle S^1, S^2, \omega_{x_0}(s^1, s^2)\rangle$ that in extended form is determined by the tuple $(X = X_1 \cup X_2, \{A(x)\}_{x\in X}, \{r_{x,a}\}_{x\in X,\ a\in A(x)}, p, x_0)$. Taking into account that the strategy sets S^1 and S^2 are finite sets, we can regard $\langle S^1, S^2, \omega_{x_0}(s^1, s^2)\rangle$ as a matrix game, and therefore, for this game, there exist the min-max strategies $\overline{s}^1$, $\overline{s}^2$ of the players and the max-min strategies $\overline{\overline{s}}^1$, $\overline{\overline{s}}^2$ of the players for which

$$\omega_{x_0}(\overline{s}^1, \overline{s}^2) = \min_{s^2\in S^2} \max_{s^1\in S^1} \omega_{x_0}(s^1, s^2); \qquad \omega_{x_0}(\overline{\overline{s}}^1, \overline{\overline{s}}^2) = \max_{s^1\in S^1} \min_{s^2\in S^2} \omega_{x_0}(s^1, s^2).$$

In this section, we show that for the considered two-player zero-sum average stochastic positional game, there exist a pure stationary strategy $s^{1^*} \in S^1$ of the first player and a pure stationary strategy $s^{2^*} \in S^2$ of the second player such that

$$\omega_x(s^{1^*}, s^{2^*}) = \max_{s^1\in S^1} \min_{s^2\in S^2} \omega_x(s^1, s^2) = \min_{s^2\in S^2} \max_{s^1\in S^1} \omega_x(s^1, s^2), \qquad \forall x \in X,$$

i.e., we show that (s^{1^*}, s^{2^*}) is a pure stationary equilibrium of the game for an arbitrary starting position $x \in X$, in spite of the fact that the values of the games with different starting positions may be different.

In the following, we consider the game for which it is necessary to determine the optimal stationary strategies of the players for an arbitrary starting state $x \in X$ and denote such a game as

$$(X = X_1 \cup X_2,\ \{A(x)\}_{x\in X},\ \{r_{x,a}\}_{x\in X,\ a\in A(x)},\ p).$$

Some Auxiliary Results

First, we show that in a two-player zero-sum average stochastic positional game, there exist a strategy $\overline{s}^1 \in S^1$ of the first player and a strategy $\overline{s}^2 \in S^2$ of the second player such that $(\overline{s}^1, \overline{s}^2)$ is a max-min strategy of the game for an arbitrary stating position $x \in X$, i.e.,

$$\omega_x(\overline{s}^1, \overline{s}^2) = \min_{s^2\in S^2} \max_{s^1\in S^1} \omega_x(s^1, s^2), \qquad \forall x \in X.$$

To prove this, we use the version of two-player zero-sum average stochastic positional games in which the starting state is chosen randomly according to a given distribution $\{\theta_x\}$ on X. So, we consider the game in the case when the play starts in a state $x \in X$ with probability $\theta_x > 0$, where $\sum_{x\in X}\theta_x = 1$. We denote this game as $(X = X_1 \cup X_2,\ \{A(x)\}_{x\in X},\ \{r_{x,a}\}_{x\in X,\ a\in A(x)},\ p,\ \{\theta_x\}_{x\in X})$. This game looks more general; however, it can be easily reduced to an auxiliary two-player zero-sum average stochastic positional game with a fixed starting position. Such an auxiliary game is determined by a new tuple obtained from $(X = X_1 \cup X_2,\ \{A(x)\}_{x\in X},\ \{r_{x,a}\}_{x\in X,\ a\in A(x)},\ p)$ by adding to the set of positions of the first player a new state position z that has a unique action $a(z)$, for which the probability transition is $p_{z,x}^{a(z)} = \theta_x,\ \forall x \in X$ and the corresponding step reward $r_{z,a(z)} = 0$. It is evident that for arbitrary strategies of the players in this game, the first player will select the unique action $a(z)$ in position z. If for the obtained game with a given starting position z we consider the normal-form game in pure stationary strategies $\langle \hat{S}^1, \hat{S}^2, \omega_z(s^1, s^2)\rangle$, then for this game, we can determine the min-max strategies of the players $\hat{s}^1, \hat{s}^2$ for which $\hat{\omega}_z(\hat{s}^1, \hat{s}^2) = \sum_{x\in X}\theta_x\omega_x(\overline{s}^1, \overline{s}^2)$. This means that the following lemmas hold:

Lemma 3.23 *For a two-player zero-sum average stochastic positional game determined by a tuple* $(X = X_1 \cup X_2,\ \{A(x)\}_{x\in X},\ \{r_{x,a}\}_{x\in X,\ a\in A(x)},\ p)$, *there exist a strategy* $\overline{s}^2 \in S^2$ *of the second player and a strategy* $\overline{s}^1 \in S^1$ *of the first player such that* $(\overline{s}^1, \overline{s}^2)$ *is a min-max strategy of the game for an arbitrary stating position* $x \in X$, *i.e.,*

$$\omega_x(\overline{s}^1, \overline{s}^2) = \min_{s^2\in S^2} \max_{s^1\in S^1} \omega_x(s^1, s^2), \qquad \forall x \in X.$$

Lemma 3.24 *For a two-player zero-sum average stochastic positional game determined by a tuple* $(X = X_1 \cup X_2,\ \{A(x)\}_{x\in X},\ \{r_{x,a}\}_{x\in X,\ a\in A(x)},\ p)$, *there exist a strategy* $\overline{\overline{s}}^1 \in S^1$ *of the first player and a strategy* $\overline{\overline{s}}^2 \in S^2$ *of the second* *player such that* $(\overline{\overline{s}}^1, \overline{\overline{s}}^2)$ *is a max-min strategy of the game for an arbitrary stating position* $x \in X$, *i.e.,*

$$\omega_x(\overline{\overline{s}}^1, \overline{\overline{s}}^2) = \max_{s^1 \in S^1} \min_{s^2 \in S^2} \omega_x(s^1, s^2), \qquad \forall x \in X.$$

Using these lemmas, we can prove the following theorem:

Theorem 3.25 *Let a two-player zero-sum average stochastic positional game determined by a tuple* $(X = X_1 \cup X_2, \{A(x)\}_{x\in X}, \{r_{x,a}\}_{x\in X}, p)$ *be given. Then the system of equations*

$$\begin{cases} \varepsilon_x + \omega_x = \max\limits_{a\in A(x)} \left\{ r_{x,a} + \sum\limits_{y\in X} p^a_{x,y}\varepsilon_y \right\}, & \forall x \in X_1; \\ \varepsilon_x + \omega_x = \min\limits_{a\in A(x)} \left\{ r_{x,a} + \sum\limits_{y\in X} p^a_{x,y}\varepsilon_y \right\}, & \forall x \in X_2 \end{cases} \tag{3.41}$$

has a solution under the set of solutions to the system of equations

$$\begin{cases} \omega_x = \max\limits_{a\in A(x)} \left\{ \sum\limits_{y\in X} p^a_{x,y}\omega_y \right\}, & \forall x \in X_1; \\ \omega_x = \min\limits_{a\in A(x)} \left\{ \sum\limits_{y\in X} p^a_{x,y}\omega_y \right\}, & \forall x \in X_2, \end{cases} \tag{3.42}$$

i.e., the system of Eqs. (3.42) has such a solution ω^*_x, $x \in X$ *for which there exists a solution* ε^*_x, $x \in X$ *to the system of equations*

$$\begin{cases} \varepsilon_x + \omega^*_x = \max\limits_{a\in A(x)} \left\{ r_{x,a} + \sum\limits_{y\in X} p^a_{x,y}\varepsilon_y \right\}, & \forall x \in X_1; \\ \varepsilon_x + \omega^*_x = \min\limits_{a\in A(x)} \left\{ r_{x,a} + \sum\limits_{y\in X} p^a_{x,y}\varepsilon_y \right\}, & \forall x \in X_2. \end{cases}$$

The optimal pure stationary strategies s^{1^*}, s^{2^*} *of the players can be found by fixing arbitrary maps* $s^{1^*}(x) \in A(x)$ *for* $x \in X_1$ *and* $s^{2^*}(x) \in A(x)$ *for* $x \in X_2$ *such that*

$$s^{1^*}(x) \in \left\{ \operatorname*{argmax}_{a\in A(x)} \left\{ \sum_{y\in X} p^a_{x,y}\omega^*_y \right\} \right\} \bigcap \left\{ \operatorname*{argmax}_{a\in A(x)} \left\{ r_{x,a} + \sum_{y\in X} p^a_{x,y}\varepsilon^*_y \right\} \right\}, \quad x \in X_1,$$

$$s^{2^*}(x) \in \left\{ \operatorname*{argmin}_{a\in A(x)} \left\{ \sum_{y\in X} p^a_{x,y}\omega^*_y \right\} \right\} \bigcap \left\{ \operatorname*{argmin}_{a\in A(x)} \left\{ r_{x,a} + \sum_{y\in X} p^a_{x,y}\varepsilon^*_y \right\} \right\}, \quad x \in X_2,$$

and $\omega_x(s^{1*}, s^{2*}) = \omega^*_x, \ \forall x \in X,$ *i.e.,*

$$\omega_x(s^{1*}, s^{2*}) = \max_{s^1 \in S^1} \min_{s^2 \in S^2} \omega_x(s^1, s^2) = \min_{s^2 \in S^2} \max_{s^1 \in S^1} \omega_x(s^1, s^2), \qquad \forall x \in X.$$

Proof According to Lemma 3.23, for the player in the considered game, there exist the pure stationary strategies $\overline{s}^1 \in S^1, \ \overline{s}^2 \in S^2$ for which

$$\omega_x(\overline{s}^1, \overline{s}^2) = \min_{s^2 \in S^2} \max_{s^1 \in S^1} \omega_x(s^1, s^2), \ \ \forall x \in X.$$

We show that

$$\omega_x(\overline{s}^1, \overline{s}^2) = \max_{s^1 \in S^1} \min_{s^2 \in S^2} \omega_x(s^1, s^2), \ \ \forall x \in X,$$

i.e., we show that $\overline{s}^1 = s^{1*}, \ \overline{s}^2 = s^{2*}$.

Indeed, if we consider the Markov process induced by strategies $\overline{s}^1, \overline{s}^2$, then according to Theorem 2.9, for this process, the system of linear equations

$$\begin{cases} \varepsilon_x + \omega_x = r_{x,a} + \sum\limits_{y \in X} p^a_{x,y} \varepsilon_y, & \forall x \in X_1, \ a = \overline{s}^1(x); \\ \varepsilon_x + \omega_x = r_{x,a} + \sum\limits_{y \in X} p^a_{x,y} \varepsilon_y, & \forall x \in X_2, \ a = \overline{s}^2(x); \\ \omega_x = \sum\limits_{y \in X} p^a_{x,y} \omega_y, & \forall x \in X_1, \ a = \overline{s}^1(x); \\ \omega_x = \sum\limits_{y \in X} p^a_{x,y} \omega_y, & \forall x \in X_2, \ a = \overline{s}^2(x) \end{cases} \tag{3.43}$$

has a basic solution $\varepsilon^*_x, \ \omega^*_x \ (x \in X)$. Now if we assume that in the game, only the second player fixes his or her strategy $\overline{s}^2 \in S^2$, then we obtain a Markov decision problem with respect to the first player, and therefore, according to Theorem 2.9, for this decision problem, the system of linear equations

$$\begin{cases} \varepsilon_x + \omega_x \geq r_{x,a} + \sum\limits_{y \in X} p^a_{x,y} \varepsilon_y, & \forall x \in X_1, \ a \in A(x); \\ \varepsilon_x + \omega_x = r_{x,a} + \sum\limits_{y \in X} p^a_{x,y} \varepsilon_y, & \forall x \in X_2, \ a = \overline{s}^2(x); \\ \omega_x \geq \sum\limits_{y \in X} p^a_{x,y} \omega_y, & \forall x \in X_1, \ a \in A(x); \\ \omega_x = \sum\limits_{y \in X} p^a_{x,y} \omega_y, & \forall x \in X_2, \ a = \overline{s}^2(x) \end{cases}$$

has solutions. We can observe that $\epsilon^*_x, \ \omega^*_x \ (x \in X)$ represents a solution to this system and $\omega_x(\overline{s}^1, \overline{s}^2) = \omega^*_x, \ \forall x \in X$.

Taking into account that $\omega_x(\overline{s}^1, \overline{s}^2) = \min_{s^2 \in S^2} F_x(\overline{s}^1, s^2)$, then for a fixed strategy $\overline{s}^1 \in S^1$, the following system has solutions:

$$\begin{cases} \varepsilon_x + \omega_x = r_{x,a} + \sum\limits_{y \in X} p_{x,y}^a \varepsilon_y, & \forall x \in X_1,\ a = \overline{s}^1(x); \\ \varepsilon_x + \omega_x \le r_{x,a} + \sum\limits_{y \in X} p_{x,y}^a \varepsilon_y, & \forall x \in X_2,\ a \in A(x); \\ \omega_x = \sum\limits_{y \in X} p_{x,y}^a \omega_y, & \forall x \in X_1,\ a = \overline{s}^1(x); \\ \omega_x \le \sum\limits_{y \in X} p_{x,y}^a \omega_y, & \forall x \in X_2,\ a \in A(x), \end{cases}$$

and $\epsilon_x = \epsilon_x^*$, $\omega_x = \omega_x^*$ $(x \in X)$ represent a solution to this system. This means that the following system

$$\begin{cases} \varepsilon_x + \omega_x \ge r_{x,a} + \sum\limits_{y \in X} p_{x,y}^a \varepsilon_y, & \forall x \in X_1,\ a \in A(x); \\ \varepsilon_x + \omega_x \le r_{x,a} + \sum\limits_{y \in X} p_{x,y}^a \varepsilon_y, & \forall x \in X_2,\ a \in A(x); \\ \omega_x \ge \sum\limits_{y \in X} p_{x,y}^a \omega_y, & \forall x \in X_1,\ a \in A(x); \\ \omega_x \le \sum\limits_{y \in X} p_{x,y}^a \omega_y, & \forall x \in X_2,\ a \in A(x) \end{cases}$$

has a solution, which satisfies condition (3.43). Thus, we obtain $\overline{s}^1 = s^{1^*}$, $\overline{s}^2 = s^{2^*}$, and $\omega_x(s^{1^*}, s^{2^*}) = \omega_x^*$, $\forall x \in X$, i.e.,

$$\omega_{\overline{x}}(s^{1^*}, s^{2^*}) = \max_{s^1 \in S^1} \min_{s^2 \in S^2} \omega_{\overline{x}}(s^1, s^2) = \min_{s^2 \in S^2} \max_{s^1 \in S^1} \omega_{\overline{x}}(s^1, s^2), \ \forall \overline{x} \in X.$$

So, the theorem holds. □

Based on Theorem 3.25, we may conclude that the optimal strategies of the players in the considered game can be found if we determine a solution to Eqs. (3.41) and (3.42). A solution to these equations can be determined using iterative algorithms like algorithms for determining the optimal solutions to an average Markov decision problem [152].

Algorithm 3.26 Determining the Optimal Stationary Strategies of the Players in Antagonistic Games with an Average Payoff Function

Preliminary step (Step 0): Fix arbitrary stationary strategies

$$s_0^1 : x \to y \in X(x) \quad \text{for } x \in X_1,$$
$$s_0^2 : x \to y \in X(x) \quad \text{for } x \in X_2,$$

and the profile $s_0 = (s_0^1, s_0^2)$.

General step (Step k, $k \geq 1$):

Determine the matrix $P^{s_{k-1}}$ that corresponds to $s_k = (s^1_{k-1}, s^2_{k-1})$, and find $\omega^{s_{k-1}}$ and $\varepsilon^{s_{k-1}}$, which satisfy the conditions

$$\begin{cases} (P^{s_{k-1}} - I)\omega^{s_{k-1}} = 0; \\ \mu^{s_{k-1}} + (P^{s_{k-1}} - I)\varepsilon^{s_{k-1}} - \omega^{s_{k-1}} = 0. \end{cases}$$

Then find $s_k = (s^1_k, s^2_k)$ such that

$$s^1_k(x) \in \underset{a \in A(x)}{\operatorname{argmax}} \left\{ \sum_{y \in X} p^a_{x,y} \omega_x^{s^1_{k-1}} \right\}, \quad \forall x \in X_1;$$
$$s^2_k(x) \in \underset{a \in A(x)}{\operatorname{argmin}} \left\{ \sum_{y \in X} p^a_{x,y} \omega_x^{s^2_{k-1}} \right\}, \quad \forall x \in X_2,$$

and set $s_k = s_{k-1}$ if

$$s^1_{k-1}(x) \in \underset{a \in A(x)}{\operatorname{argmax}} \left\{ \sum_{y \in X} p^a_{x,y} \omega_x^{s^1_{k-1}} \right\}, \quad \forall x \in X_1;$$
$$s^2_{k-1}(x) \in \underset{a \in A(x)}{\operatorname{argmax}} \left\{ \sum_{y \in X} p^a_{x,y} \omega_x^{s^2_{k-1}} \right\}, \quad \forall x \in X_2.$$

After that, check if $s_k = s_{k-1}$. If $s_k = s_{k-1}$, then go to the next step $k+1$; otherwise, choose a situation $s_k = (s^1_k, s^2_k)$ such that

$$s^1_k(x) \in \underset{a \in A(x)}{\operatorname{argmax}} \left\{ r_{x,a} + \sum_{y \in X} p^a_{x,y} \varepsilon_y^{s^1_{k-1}(x)} \right\} \quad \forall x \in X_1;$$
$$s^2_k(x) \in \underset{a \in A(x)}{\operatorname{argmin}} \left\{ r_{x,a} + \sum_{y \in X} p^a_{x,y} \varepsilon_y^{s^2_{k-1}(x)} \right\} \quad \forall x \in X_2,$$

and set $s_k = s_{k-1}$ if

$$s^1_{k-1}(x) \in \underset{a \in A(x)}{\operatorname{argmax}} \left\{ r_{x,a} + \sum_{y \in X} p^a_{x,y} \varepsilon_y^{s^1_{k-1}(x)} \right\} \quad \forall x \in X_1;$$
$$s^2_{k-1}(x) \in \underset{a \in A(x)}{\operatorname{argmin}} \left\{ r_{x,a} + \sum_{y \in X} p^a_{x,y} \varepsilon_y^{s^2_{k-1}(x)} \right\} \quad \forall x \in X_2.$$

After that, check if $s_k = s_{k-1}$. If $s_k = s_{k-1}$, then stop and set $s^{1*} = s^1_{k-1}$, $s^{2*} = s^2_{k-1}$; otherwise, go to the next step $k+1$.

The results presented above concerned with determining the saddle points for zero-sum average stochastic games generalize the results related to finding the saddle points for *mean-payoff games on graphs* from [42, 67].

We obtain the positional games from [42, 67] from the antagonistic average positional game in the case $p^a_{x,y} \in \{0, 1\}, \forall x, y \in X, \forall a \in A(x)$.

In this case, we have a deterministic positional game that corresponds to a mean-payoff game on graph $G = (X = X_1 \cup X_2, E)$ with $e = (x, y) \in E$ and the costs $c_{x,y} = r^a_{x,y}$ for which $p^a_{x,y} = 1$ for $a \in A(x)$. So, as a corollary from Theorem 3.25, for the mean-payoff on graph $G = (X = X_1 \cup X_2, E)$ with given costs c_e for $e = (x, y) \in E$, we obtain the following result:

Theorem 3.27 *Let a two-player mean-payoff game on graph $G = (X, E)$ with positional sets $X_1, X_2 (X = X_1 \cup X_2)$ and cost function $c : E \to R^1$ be given. Then the system of equations*

$$\begin{cases} \varepsilon_x + \omega_x = \max\limits_{y \in X(x)} \{c_{x,y} + \varepsilon_y\}, & \forall x \in X_1; \\ \varepsilon_x + \omega_x = \min\limits_{y \in X(x)} \{c_{x,y} + \varepsilon_y\}, & \forall x \in X_2 \end{cases} \tag{3.44}$$

has a solution under the set of solutions to the system of equations

$$\begin{cases} \omega_x = \max\limits_{y \in X(x)} \{\omega_y\}, & \forall x \in X_1; \\ \omega_x = \min\limits_{y \in X(x)} \{\omega_y\}, & \forall x \in X_2, \end{cases} \tag{3.45}$$

i.e., the system of Eqs. (3.45) has such a solution ω^*_x, $x \in X$ for which there exists a solution ε^*_x, $x \in X$ of the system of equations

$$\begin{cases} \varepsilon_x + \omega^*_x = \max\limits_{y \in X(x)} \{c_{x,y} + \varepsilon_y\}, & \forall x \in X_1; \\ \varepsilon_x + \omega^*_x = \min\limits_{y \in X(x)} \{c_{x,y} + \varepsilon_y\}, & \forall x \in X_2. \end{cases}$$

The optimal stationary strategies s^{1^*}, s^{2^*} of the players can be found by fixing arbitrary maps $s^{1^*}(x) \in A(x)$ for $x \in X_1$ and $s^{2^*}(x) \in A(x)$ for $x \in X_2$ such that

$$s^{1^*}(x) \in \left\{ \underset{y \in X(x)}{\operatorname{argmax}} \{\omega^*_y\} \right\} \bigcap \left\{ \underset{y \in X(x)}{\operatorname{argmax}} \{c_{x,y} + \varepsilon^*_y\} \right\}, \quad x \in X_1,$$

$$s^{2^*}(x) \in \left\{ \underset{y \in X(x)}{\operatorname{argmin}} \{\omega^*_y\} \right\} \bigcap \left\{ \underset{y \in X(x)}{\operatorname{argmin}} \{c_{x,y} + \varepsilon^*_y\} \right\}, \quad x \in X_2,$$

and $\omega_x(s^{1^*}, s^{2^*}) = \omega^*_x, \forall x \in X$, i.e.,

$$\omega_x(s^{1^*}, s^{2^*}) = \max_{s^1 \in S^1} \min_{s^2 \in S^2} \omega_x(s^1, s^2) = \min_{s^2 \in S^2} \max_{s^1 \in S^1} \omega_x(s^1, s^2), \qquad \forall x \in X.$$

Remark 3.28 Algorithm 3.26 in the case $p_{x,y} \in \{0.1\}, \ \forall x, y \in X$ determines the optimal stationary strategies for the mean-payoff game.

3.6.8 *Pure Stationary Equilibria for Discounted Stochastic Positional Games*

Consider a stochastic positional game with discounted payoffs that is determined by the tuple $(\{X_i\}_{i=\overline{1,n}}, \ \{A(x)\}_{x\in X}, \ \{r^i\}_{i=\overline{1,m}}, p, \gamma)$. We can define the normal-form model in stationary strategies $\langle \{S^i\}_{i=\overline{1,m}}, \ \{\sigma^i_{x_0}(s)\}_{i=\overline{1,m}} \ \rangle$ for such a game in the following way:

Let $S^i, \ i \in \{1, 2, \dots m\}$ be the set of solutions to the system (3.26) that represents the set of stationary strategies of player i. On the set $S = S^1 \times S^2 \times \cdots \times S^m$, we define m payoff functions

$$\sigma^i_\theta(s^1, s^2, \dots, s^m) = \sum_{k=1}^{m} \sum_{x\in X_k} \sum_{a\in A(x)} s^k_{x,a} r^i_{x,a} q_x, \qquad i = 1, 2, \dots, m, \tag{3.46}$$

where q_x for $x \in X$ are uniquely determined by the following system of linear equations:

$$q_y - \gamma \sum_{k=1}^{m} \sum_{x\in X_k} \sum_{a\in A(x)} s^k_{x,a} \, p^a_{x,y} \, q_x = \theta_y, \qquad \forall y \in X, \tag{3.47}$$

for an arbitrary $s = (s^1, s^2, \dots, s^m) \in S = S^1 \times S^2 \times \cdots \times S^m$.

Note that here, the payoff functions $\sigma^i_\theta(s^1, s^2, \dots, s^m)$ defined according to (3.46) and (3.47) for the considered positional game differ from the payoff functions $\sigma^i_\theta(s^1, s^2, \dots, s^m)$ defined by (3.7) and (3.12) in the general case of the game. As a corollary from Theorem 3.7, we can see that for the game $\langle \{S^i\}_{i=\overline{1,m}}, \ \{\sigma^i_\theta(s)\}_{i=\overline{1,m}} \ \rangle$, defined according to (3.26), (3.46), and (3.47), there exists a Nash equilibrium $\mathbf{s} = (s^1, s^2, \dots, s^m) \in S = S^1 \times S^2 \times \cdots \times S^m$ that is a stationary Nash equilibrium of the discounted stochastic positional game with an arbitrary starting state $x \in X$.

Existence of Pure Stationary Equilibria

The existence of Nash equilibria in pure stationary strategies for a discounted stochastic positional game can be derived on the basis of the following theorem that was proved in [127]:

Theorem 3.29 *Let a discounted stochastic positional game be given that is determined by the tuple* $(\{X_i\}_{i=\overline{1,m}}, \{A(x)\}_{x\in X}, \{r^i\}_{i=\overline{1,m}}, p, \gamma)$. *Then there exist the values* σ^i_x *for* $x \in X, \ i = 1, 2, \dots, m$ *that satisfy the following conditions:*

(1) $r^i_{x,a} + \gamma \sum_{y \in X} p^a_{x,y} \sigma^i_y - \sigma^i_x \le 0, \ \ \forall x \in X_i, \ \ \forall a \in A(x), \ i = 1, 2, \dots, m.$

(2) $\max_{a \in A(x)} \{r^i_{x,a} + \gamma \sum_{y \in X} p^a_{x,y} \sigma^i_y - \sigma^i_x\} = 0, \ \ \forall x \in X_i, \ i = 1, 2, \dots, m.$

(3) *On each position set* $X_i, i \in \{1, 2, \dots, m\}$, *there exists a map* $s^{i^*} : X_i \to \cup_{x \in X_i} A(x)$ *such that*

$$s^{i^*}(x) = a^* \in \operatorname*{argmax}_{a \in A(x)} \left\{ r^i_{x,a} + \gamma \sum_{y \in X} p^a_{x,y} \sigma^i_y - \sigma^i_x \right\}$$

and

$$r^j_{x,a^*} + \gamma \sum_{y \in X} p^{a^*}_{x,y} \sigma^j_y - \sigma^j_x = 0, \ \ \forall x \in X_i, \ \ j = 1, 2, \dots, m,$$

where $s^{1^*}, s^{2^*}, \dots, s^{m*}$ *determine a stationary Nash equilibrium* $s^* = (s^{1^*}, s^{2^*}, \dots, s^{m*})$ *for the discounted stochastic positional game determined by* $(\{X_i\}_{i=\overline{1,m}}, \{A(x)\}_{x \in X}, \{r^i\}_{i=\overline{1,m}}, p, \gamma)$ *and* $s^* = (s^{1^*}, s^{2^*}, \dots, s^{m*})$ *is a pure stationary Nash equilibrium for the game with an arbitrary starting position* $x \in X$.

Proof According to Theorem 3.7, for the discounted stochastic positional game determined by $(\{X_i\}_{i=\overline{1,m}}, \{A(x)\}_{x \in X}, \{r^i\}_{i=\overline{1,m}}, p, \gamma)$, there exists a stationary Nash equilibrium $\overline{s}^* = (\overline{s}^{1^*}, \overline{s}^{2^*}, \dots, \overline{s}^*)$. If $\overline{s}^{i^*}$ is a mixed stationary strategy of player $i \in \{1, 2, \dots, m\}$, then for $x \in X_i$, the strategy $\overline{s}^{i^*}(x)$ represents a convex combination of actions determined by the probability distribution $\{\overline{s}^{i^*}_{x,a}\}$ on $A^*(x) = \{a \in A(x) |\ \overline{s}^{i^*}_{x,a} > 0\}$.

Let us consider the Markov process induced by the profile of mixed stationary strategies $\overline{s}^* = (\overline{s}^{1^*}, \overline{s}^{2^*}, \dots, \overline{s}^*)$. Then according to (3.27), the elements of transition probability matrix $P^{\overline{s}^*} = (p^{\overline{s}^*}_{x,y})$ of this Markov process can be calculated as follows:

$$p^{\overline{s}^*}_{x,y} = \sum_{a \in A(x)} \overline{s}^{i^*}_{x,a} p^a_{x,y} \quad \text{for} \ \ x \in X_i, \ \ i = 1, 2, \dots, m, \tag{3.48}$$

and the step rewards in the states induced by $\overline{\mathbf{s}}^*$ can be determined according to (3.29), i.e.,

$$r^i_{x,\overline{s}^{k*}} = \sum_{a \in A(x)} \overline{s}^{k^*}_{x,a} r^i_{x,a}, \ \text{ for } \ x \in X_k, \ k \in \{1, 2, \dots, m\}. \tag{3.49}$$

Based on the optimality Eq. (2.16) (see Sect. 2.3), for this Markov process, we can write the following equations:

$$r^j_{x,\overline{s}^{i*}} + \gamma \sum_{y \in X} p^{\overline{s}^{i^*}}_{x,y} \sigma^j_y - \sigma^j_x = 0, \ \ \forall x \in X_i, \ \forall i, j \in \{1, 2, \dots, m\}. \tag{3.50}$$

From these equations, we uniquely determine $\sigma^j,\ j = 1, 2, \ldots, m$ (see [152] and the results from Sect. 2.3). These values satisfy the condition

$$r^j_{x,a} + \gamma \sum_{y\in X} p^a_{x,y}\sigma^j_y - \sigma^j_x \le 0,\ \ \forall x \in X_i, \forall a \in A(x)\ \forall i, j \in \{1, 2, \ldots, m\}. \tag{3.51}$$

By introducing (3.48) and (3.49) in (3.50), we obtain

$$\sum_{a\in A(x)} \overline{s}^{i*}_{x,a} r^j_{x,a} + \gamma \sum_{y\in X}\sum_{a\in A(x)} \overline{s}^{i*}_{x,a} p^a_{x,y}\sigma^j_y - \sigma^j = 0,\ \forall x \in X_i, \forall i, j \in \{1, 2, \ldots, m\}.$$

In these equations, we can set $\sigma^j_x = \sum_{a\in A(x)} \overline{s}^{i*}_{x,a}\sigma^j_x$. After these substitutions and some elementary transformations of the equations, we obtain

$$\sum_{a\in A(x)} \overline{s}^{i*}_{x,a}\left(r^j_{x,a} + \gamma \sum_{y\in X} p^a_{x,y}\sigma^j_y - \sigma^j_x\right) = 0,\ \ \forall x \in X_i,\ \forall i, j \in \{1, 2, \ldots, m\}.$$

So, for the Markov process induced by the profile of mixed stationary strategies $\overline{s}^* = (\overline{s}^{1*}, \overline{s}^{2*}, \ldots, \overline{s}^*)$, there exist the values $\sigma^i_x,\ x \in X,\ i = 1, 2, \ldots, m$ that satisfy the following condition:

$$r^j_{x,a} + \gamma \sum_{y\in X} p^a_{x,y}\sigma^j_y - \sigma^j_x = 0,\ \ \forall x \in X_i,\ \forall a \in A^*(x),\ j = 1, 2, \ldots, m. \tag{3.52}$$

Now let us fix the strategies $\overline{s}^{1*}, \overline{s}^{2*}, \ldots, \overline{s}^{i-1*}, \overline{s}^{i+1*}, \ldots, \overline{s}^*$ of the players $1, 2, \ldots, i-1, i+1, \ldots, m$ and consider the problem of determining the maximal expected total discounted reward with respect to player $i \in \{1, 2, \ldots, m\}$. Obviously, if we solve this decision problem, then we obtain strategy $\overline{s}^{i*}$. However, for this decision problem, there exists also a pure optimal strategy s^{i*}. If we write the optimality equations for the discounted Markov decision problem with respect to player i, then we can see that there exist the values ω^i_x for $x \in X$ such that

(1) $r^i_{x,a} + \gamma \sum\limits_{y\in X} p^a_{x,y}\sigma^i_y - \sigma^i_x \le 0,\quad \forall x \in X_i,\ \forall a \in A(x)$

(2) $\max\limits_{a\in A(x)} \left\{r^i_{x,a} + \gamma \sum\limits_{y\in X} p^a_{x,y}\sigma^i_y - \omega^i_x\right\} = 0,\quad \forall x \in X_i$

We can observe that $\sigma^i_x,\ x \in X$, determined by (3.50), satisfy the conditions (1) and (2) above and (3.52) holds. So, if for an arbitrary $i \in \{1, 2, \ldots, m\}$ we fix a map $s^{i*} : X_i \to \cup_{x\in X_i} A(x)$ such that

$$s^{i*}(x) = a^* \in \underset{a\in A(x)}{\operatorname{argmax}} \left\{r^i_{x,a} + \gamma \sum_{y\in X} p^a_{x,y}\sigma^i_y - \sigma^i_x\right\},\ \ \forall x \in X$$

and

$$r^j_{x,a^*} + \gamma \sum_{y \in X} p^{a^*}_{x,y} \sigma^j_y - \sigma^j_x = 0, \;\; \forall x \in X_i, \;\; j = 1, 2, \ldots, m,$$

then we obtain a Nash equilibrium in pure stationary strategies. □

Remark 3.30 Theorem 3.29 also holds for a stochastic positional game with different discount factors $\gamma_1, \gamma_2, \ldots, \gamma_m$. In the case of different discount factors for the players, in the conditions (1)–(3) of the theorem, γ should be replaced by γ_i.

Based on Theorem 3.29, we can formulate saddle point conditions for *discounted stochastic antagonistic positional games*. We obtain such games from discounted stochastic positional games in the case $m = 2$ for $r_{x,a} = r^1_{x,a} = -r^2_{x,a}$. As a corollary from Theorem 3.29, we obtain the following saddle point condition for a discounted stochastic antagonistic game:

Corollary 3.31 *Let a discounted stochastic antagonistic positional game be given that is determined by the tuple* $(X_1, X_2, \{A(x)\}_{x \in X}, \{r_{x,a}\}, p, \gamma)$. *Then there exist the values* σ_x *for* $x \in X$ *that satisfy the following conditions:*

(1) $\max\limits_{a \in A(x)} \left\{ r_{x,a} + \gamma \sum\limits_{y \in X} p^a_{x,y} \sigma_y - \sigma_x \right\} = 0, \qquad \forall x \in X_1.$

(2) $\min\limits_{a \in A(x)} \left\{ r_{x,a} + \gamma \sum\limits_{y \in X} p^a_{x,y} \sigma_y - \sigma_x \right\} = 0, \qquad \forall x \in X_2.$

The optimal stationary strategies s^{1*}, s^{2*} *of the players in the game can be found by fixing the maps*

$$s^{1*}(x) = a^* \in \operatorname*{argmax}_{a \in A(x)} \left\{ r_{x,a} + \gamma \sum_{y \in X} p^a_{x,y} \sigma_y - \sigma_x \right\}, \quad \forall x \in X_1;$$

$$s^{2*}(x) = a^* \in \operatorname*{argmin}_{a \in A(x)} \left\{ r_{x,a} + \gamma \sum_{y \in X} p^a_{x,y} \sigma_y - \sigma_x \right\}, \quad \forall x \in X_2,$$

where $\sigma_x(s^{1*}, s^{2*}) = \sigma_x, \forall x \in X$ *and*

$$\sigma_x = \max_{s^1 \in S^1} \min_{s^2 \in S^2} \sigma_x(s^1, s^2) = \min_{s^2 \in S^2} \max_{s^1 \in S^1} \sigma_x(s^1, s^2), \; \forall x \in X.$$

3.6.9 Pure Nash Equilibria for Discounted Games on Networks

In a similar way as for an average positional game on a network determined by the tuple $(G = (X, E), \{X_i\}_{i=\overline{1,m}}, X^0, \{c^i\}_{i=\overline{1,m}}, x_0)$ (see Sect. 3.6.4), we can consider a positional game on a network with discounted payoffs determined by the same tuple and a given discount factor γ. This allows us to formulate the game variant of the discounted control problem from Sect. 2.7 that is determined by the tuple $(G = (X, E), \{X_i\}_{i=\overline{1,m}}, X^0, \{c^i\}_{i=\overline{1,m}}, p, \gamma, \overline{x})$. By applying the pure and mixed stationary strategies for the players as defined in Sect. 2.7, we obtain the corresponding discounted positional games in pure and mixed stationary strategies on the network.

From Theorem 3.29, we can derive pure Nash equilibria conditions in the term of potential transformation for a discounted positional game on networks.

Theorem 3.32 *Let a discounted stochastic positional game on the network* $(G = (X, E), \{X_i\}_{i=\overline{1,m}}, X^0, \{c^i\}_{i=\overline{1,m}}, p, \gamma, \overline{x})$ *with a discount factor* γ, $0 < \gamma < 1$ *be given, where* $X = \bigcup_{i=1}^{m} X_i \cup X^0$. *Then there exist the values* σ_x^i, $i = 1, 2, \dots m$ *for* $x \in X$ *that satisfy the following conditions:*

(1) $c_{x,y}^i + \gamma\sigma_y^i - \sigma_x^i \geq 0, \ \forall x \in X_i, \ \forall y \in X(x), \ \ i = 1, 2, \dots, m.$

(2) $\min\limits_{y \in X(x)} \{c_{x,y}^i + \gamma\sigma_y^i - \sigma_x^i\} = 0, \quad \forall x \in X_i, \quad i = 1, 2, \dots, m.$

(3) $\mu_x^i + \gamma \sum\limits_{y \in X(x)} p_{x,y}\sigma_y^i - \sigma_x^i = 0, \ \ \forall x \in X^0, \quad i = 1, 2, \dots, m.$

(4) *On each position set* X_i, $i \in \{1, 2, \dots, m\}$, *there exists a map* $s^{i^*} : X_i \to X$ *such that*

$$s^{i^*}(x) = y^* \in \underset{y \in X(x)}{\operatorname{argmin}} \left\{\mu_x^i + \gamma\sigma_y^i - \sigma_x^i\right\}, \ \forall x \in X_i$$

and

$$c_{x,y^*}^j + \gamma\sigma_{y^*}^j - \sigma_x^j = 0, \ \ \forall x \in X_i, \ \ j = 1, 2, \dots, m,$$

where $s^{1^*}, s^{2^*}, \dots, s^{m*}$ *determines a pure stationary Nash equilibrium for the discounted stochastic positional game on the network* $(G = (X, E), X^0, \{X_i\}_{i=\overline{1,m}}, \{c^i\}_{i=\overline{1,m}}, \gamma, p, \overline{x})$ *and*

$$\widehat{F}_{\overline{x}}^i(s^{1^*}, s^{2^*}, \dots, s^{m*}) = \sigma_{\overline{x}}^i, \ \ \forall \overline{x} \in X, \ \ i = 1, 2, \dots, m.$$

Moreover, the strategy profile $s^* = (s^{1^*}, s^{2^*}, \dots, s^{m*})$ *is a Nash equilibrium for an arbitrary starting position* $x \in X$.

In the case $m = 2$ and $c = c^2 = -c^1$ from Theorem 3.32, we obtain the following result:

Corollary 3.33 *Let a discounted antagonistic stochastic positional game on the network* $(G, X_1, X_2, X^0, c, p, \gamma, \overline{x})$ *with a discount factor* $0 < \gamma < 1$*, where* $G = (X, E)$ *is the graph of the states' transitions of the dynamical system and* $X = X_1 \cup X_2 \cup X^0$*, be given. Then there exist the values* σ_x *for* $x \in X$ *that satisfy the following conditions:*

(1) $\max\limits_{y\in X(x)} \{c_{x,y} + \gamma\sigma_y - \sigma_x\} = 0, \qquad \forall x \in X_1.$

(2) $\min\limits_{y\in X(x)} \{c_{x,y} + \gamma\sigma_y - \sigma_x\} = 0, \qquad \forall x \in X_2.$

(3) $\mu_x + \gamma \sum\limits_{y\in X(x)} p_{x,y}\sigma_y - \sigma_x = 0, \qquad \forall x \in X^0.$

The optimal stationary strategies s^{1^*}, s^{2^*} *in the antagonistic game that satisfy the condition*

$$\widehat{F}_{\overline{x}}(s^{1^*}, s^{2^*}) = \max_{s^1\in\mathbb{S}^1} \min_{s^2\in\mathbb{S}^2} \widehat{F}_{\overline{x}}(s^1, s^2) = \min_{s^2\in\mathbb{S}^2} \max_{s^1\in\mathbb{S}^1} \widehat{F}_{\overline{x}}(s^1, s^2), \quad \forall \overline{x} \in X$$

can be found by fixing

$$s^{1^*}(x) = y^* \in \operatorname*{argmax}_{y\in X(x)}\{c_{x,y} + \gamma\sigma_y - \sigma_x\}, \quad \forall x \in X_1$$

and

$$s^{2^*}(x) = y^* \in \operatorname*{argmin}_{y\in X(x)}\{c_{x,y} + \gamma\sigma_y - \sigma_x\}, \quad \forall x \in X_2.$$

3.7 Single-Controller Stochastic Games

In a single-controller stochastic game, the transition probabilities depend only on the actions of one player; however, the rewards in the states of the players depend on the actions of all players. We assume that a "single controller" in such a game is player 1. This means that the transition probability distributions $\{p^a_{x,y}\}_{y\in X}$ in the states $x \in X$ depend only on the actions $a^1 \in A^1(x)$ of player 1 and $\{p^a_{x,y}\}_{y\in X}$ are independent of the actions of the remaining players. The problem of the existence of stationary Nash equilibria in single-controller stochastic games was studied in [49, 50, 52, 77, 188]. In particular, the existence of stationary Nash equilibria for two-player zero-sum single-controller stochastic games with average payoff was proved, and algorithms for determining the value and the stationary strategies of the players were proposed. The existence of stationary Nash equilibria in single-controller stochastic games with discounted payoffs follows from the existence of stationary equilibria of discounted stochastic games in the general case.

3.7.1 Single-Controller Discounted Stochastic Games

For a single-controller discounted stochastic games, we consider the normal-form game in stationary strategies $\langle \{S^i\}_{i=\overline{1,m}}, \ \{\phi^i_\theta(s)\}_{i=\overline{1,m}} \ \rangle$. We can obtain such a normal-form game from the general game model from Sect. 3.4. Using (3.7) and (3.8), we obtain the following payoffs for the normal-form single-control model in stationary strategies:

$$\phi^i_\theta(s) = \sum_{x\in X} \ \sum_{(a^1,a^2,\dots,a^m)\in A(x)} \ \prod_{k=1}^{m} s^k_{x,a^k} \cdot r^i_{x,(a^1,a^2\dots a^m)} \cdot q_x, \quad i = 1,2,\dots,m, \tag{3.53}$$

where $q_x, \ x \in X$ are uniquely determined by the following system of equations:

$$q_y - \gamma \sum_{x\in X} \ \sum_{a^1\in A^1(x)} p^{a^1}_{x,y} \cdot q_x \cdot s_{x,a^1} = \theta_y, \qquad \forall y \in X \tag{3.54}$$

for an arbitrary fixed $s = (s^1, s^2, \dots, s^m) \in S = S^1 \times S^2 \times \dots \times S^m$.

Recall that here, $\theta_y > 0, \ \forall y \in X$ and $\sum_{y\in X} \theta_y = 1$, i.e., we have a model of a single-controller game when the starting state is chosen randomly according to distribution $\{\theta_y\}_{y\in V}$. If we take $\theta_{x_0} = 1$ and $\theta_y = 0, \ \forall y \in X\setminus\{x_0\}$, then we obtain a single-controller game with given starting state x_0. If we find a Nash equilibrium for the normal-form game $\langle \{S^i\}_{i=\overline{1,m}}, \ \{\phi^i_\theta(s)\}_{i=\overline{1,m}} \ \rangle$, then we determine a stationary Nash equilibrium for the single-controller discounted stochastic game with an arbitrary starting state.

3.7.2 Single-Controller Average Stochastic Games

We can formulate a normal-form game in stationary strategies $\langle \{S^i\}_{i=\overline{1,m}}, \{\psi^i_\theta(s)\}_{i=\overline{1,m}} \ \rangle$ for a single-controller average stochastic game by using the normal-form games from Sect. 3.5. The payoffs $\psi^i_\theta(s), \ i = 1,2,\dots,m$ in this case are defined as follows:

$$\psi^i_\theta(s) = \sum_{x\in X} \ \sum_{(a^1,a^2,\dots,a^m)\in A(x)} \ \prod_{k=1}^{m} s^k_{x,a^k} \cdot r^i_{x,(a^1,a^2\dots a^m)} \cdot q_x, \quad i = 1,2,\dots,m, \tag{3.55}$$

where q_x for $x \in X$ are uniquely determined by the following system of linear equations:

$$\begin{cases} q_y - \sum\limits_{x\in X} \sum\limits_{a^1\in A^1(x)} s^1_{x,a^1} \cdot p^{a^1}_{x,y} \cdot q_x = 0, & \forall y \in X; \\ q_y + w_y - \sum\limits_{x\in X} \sum\limits_{a^1\in A^1(x)} s^1_{x,a^1} \cdot p^{a^1}_{x,y} \cdot w_x = \theta_y, & \forall y \in X, \end{cases} \tag{3.56}$$

for a fixed $s = (s^1, s^2, \dots, s^m) \in S$.

In this model, $\theta_y > 0$ for $y \in X$ makes as much sense as the model with discounted payoffs; in the case $\theta_{x_0} = 1$, $\theta_y = 0, \forall y \in X \setminus \{x_0\}$, we obtain a single-controller game with given starting state x_0. For the unichain case of a single-controller average game, the payoffs are also represented by (3.55); however, q_x for $x \in X$ are uniquely determined by the following system of linear equations:

$$\begin{cases} q_y - \sum\limits_{x\in X} \sum\limits_{a^1\in A^1(x)} s^1_{x,a^1} \cdot p^{a^1}_{x,y} \cdot q_x = 0, \quad \forall y \in X; \\ \sum\limits_{x\in X} q_x = 1, \end{cases} \tag{3.57}$$

for $s^i \in S^i, i = 1, 2, \dots, m$. Obviously, in the unichain case, stationary Nash equilibria exist.

An important class of single-controller average stochastic games is represented by two-player zero-sum games for which there exists the value that players can achieve by applying stationary strategies (see [49, 50, 52]).

The payoff for the normal form of zero-sum single-controller games in stationary strategies $\langle S^1, S^2, \psi_\theta(s^1, s^2)\rangle$ is

$$\psi_\theta(s^1, s^2) = \sum_{x\in X} \sum_{a^1\in A^1(x)} \sum_{a^2\in A^2(x)} s^1_{x,a^1} s^2_{x,a^2} \cdot r_{x,(a^1,a^2)} \cdot q_x,$$

where q_x for $x \in X$ are uniquely determined by (3.56) for given $(s^1, s^2) \in S^1 \times S^2$ (in the unichain case, q_x for $x \in X$ are uniquely determined by (3.57)). In the case $\theta_{x_0} = 1$ and $\theta_y = 0, \forall y \in X \setminus x_0$, we have the game in normal form with given starting state x_0.

The main approaches and algorithms for determining the value and the strategies of the players for two-player zero-sum single-controller stochastic games were analyzed in [153].

3.8 Switching Controller Stochastic Games

In a switching controller stochastic game, the transition probabilities in a state $x \in X$ are controlled only by one player. This means that the set of states can be divided into several disjoint subsets such that each player controls the transition probability in one of these subsets; however, the rewards of the players in the states depend on the actions of all players.

In the case of switching controller stochastic games with discounted payoffs, stationary Nash equilibria exist because in general, there exist stationary equilibria for a discounted stochastic game.

The problem of the existence and determination of Nash equilibria in a switching stochastic game with average payoffs is a more difficult problem; however, in the case of two-person stochastic games with a switching control, Thuijsman and Raghavan [180] showed the existence of a limiting average ε-equilibrium. Here, we show the existence of stationary Nash equilibria for an m-player switching controller average stochastic game with additive rewards in the states.

3.8.1 *Formulation of Switching Controller Stochastic Games*

A switching stochastic game with m players consists of the following elements:

- A state space X (which we assume to be finite)
- A partition $X = X_1 \cup X_2 \cup \cdots \cup X_m$, where X_i represents the position set of player $i \in \{1, 2, \ldots, m\}$
- A finite set $A^i(x)$ of actions with respect to each player $i \in \{1, 2, \ldots, m\}$ for an arbitrary state $x \in X$
- A payoff $r^i_{x,a}$ with respect to each player $i \in \{1, 2, \ldots, m\}$ for each state $x \in X$ and for an arbitrary action vector $a \in \prod_i A^i(x)$
- A transition probability function $p : X_i \times \prod_{x \in X} A^i(x) \times X \rightarrow [0, 1]$ with respect to each player $i \in \{1, 2, \ldots, m\}$ that gives the probability transitions $p^{a^i}_{x,y}$ from an arbitrary $x \in X_i$ to an arbitrary $y \in X$ for a given action $a^i \in A^i(x)$, where $\sum_{y \in X} p^{a^i}_{x,y} = 1, \ \ \forall x \in X_i, \ a^i \in A^i(x)$
- A starting state $x_0 \in X$

The game starts in the state x_0 and then proceeds in the same way as in the general case of a stochastic game. At stage t, the players observe state x_t and simultaneously and independently choose actions $a^i_t \in A^i(x_t)$, $i = 1, 2, \ldots, m$, and determine the rewards $r^1_{x_t,a_t}, \ r^2_{x_t,a_t}, \ldots, \ r^m_{x_t,a_t}$ for given action vector $a_t = (a^1_t, a^2_t, \ldots, a^m_t)$. Then nature selects a state $y = x_{t+1}$ according to transition probability distribution $\{p^{a^i_t}_{x_t,y}\}_{y \in X}$ if x_t belongs to the set of states X_i controlled by player i, i.e., $a^i_t \in A^i(x_t)$. Such a play of the game produces a sequence of states and actions $x_0, a_0, x_1, a_1, \ldots, x_t, a_t, \ldots,$ defining a stream of stage payoffs $r_t^1 = r^1_{x_t,a_t},\ r_t^2 = r^2_{x_t,a_t}, \ldots,\ r_t^m = r^m_{x_t,a_t},\ t = 0, 1, 2, \ldots.$

The average switching controller stochastic game is the game with payoffs of the players

$$\omega^i_{x_0} = \lim_{t\to\infty} \inf \mathbb{E}\left(\frac{1}{t}\sum_{\tau=0}^{t-1} r^i_\tau\right), \quad i = 1, 2, \ldots, m,$$

where $\mathbb{E}$ is the expectation operator with respect to the probability measure in the Markov process induced by actions chosen by players in their position sets and given starting state x_0. *The discounted switching controller stochastic game* with given discount factor $\gamma,\ 0 < \gamma < 1$ is the game with payoffs of the players

$$\sigma^i_{x_0,\gamma} = \mathbb{E}\left(\sum_{\tau=0}^{\infty} \gamma^\tau r^i_\tau\right), \quad i = 1, 2, \ldots, m.$$

We study the considered switching controller stochastic games in the cases when players use stationary and history-dependent strategies as defined in Sect. 3.3.2.

3.8.2 Discounted Switching Controller Stochastic Games

For a discounted switching controller stochastic game, we formulate a normal-form game in stationary strategies $\langle \{S^i\}_{i=\overline{1,m}},\ \{\phi^i_\theta(s)\}_{i=\overline{1,m}} \rangle$ by using the model from Sect. 3.4 (see model (3.7), (3.8)). If we specify the payoffs for the switching controller game model, then we obtain

$$\begin{cases} \phi^i_\theta(s^1, s^2, \ldots, s^m) = \sum\limits_{x\in X}\ \sum\limits_{(a^1,a^2,\ldots,a^m)\in A(x)}\ \prod\limits_{k=1}^{m} s^k_{x,a^k} \cdot r^i_{x,(a^1,a^2\ldots a^m)} \cdot q_x, \\ \qquad\qquad i = 1, 2, \ldots, m, \end{cases} \tag{3.58}$$

where $q_x,\ x \in X$ are uniquely determined by the following system of equations:

$$q_y - \gamma \sum_{k=1}^{m} \sum_{x\in X_k} \sum_{a^k\in A^k(x)} s^k_{x,a^k} \cdot p^{a^k}_{x,y} \cdot q_x = \theta_y, \qquad \forall y \in X \tag{3.59}$$

for an arbitrary fixed $s = (s^1, s^2, \ldots, s^m) \in S = S^1 \times S^2 \times \cdots \times S^m$. Taking into account the results from Sect. 3.4, we may conclude that for a discounted switching controller stochastic game, stationary Nash equilibria exist, and such equilibria can be found by using the normal-form game model above with payoffs (3.58), (3.59).

3.8.3 Average Switching Controller Stochastic Games

The problem of the existence and determination of stationary equilibria for average switching controller stochastic games is more complicated than for the discounted switching controller games. The normal-form game in stationary strategies $\langle \{S^i\}_{i=\overline{1,m}},\ \{\psi_\theta^i(s)\}_{i=\overline{1,m}}\rangle$ for an average switching controller stochastic game can be formulated by specifying the models from Sect. 3.5. The payoffs for such a normal-form game are

$$\psi_\theta^i(s)=\sum_{x\in X}\ \sum_{(a^1,a^2,\ldots,a^m)\in A(x)}\ \prod_{k=1}^{m} s^k_{x,a^k}\cdot r^i_{x,(a^1,a^2\ldots a^m)}\cdot q_x,\ i=1,2,\ldots,m, \tag{3.60}$$

where q_x for $x\in X$ are uniquely determined by the following system of linear equations:

$$\begin{cases} q_y-\sum\limits_{k=1}^{m}\sum\limits_{x\in X_k}\sum\limits_{a^k\in A^k(x)} s^k_{x,a^k}\cdot p^{a^k}_{x,y}\cdot q_x=0, & \forall y\in X;\\ q_y+w_y-\sum\limits_{k=1}^{m}\sum\limits_{x\in X_k}\sum\limits_{a^k\in A^k(x)} s^k_{x,a^k}\cdot p^{a^k}_{x,y}\cdot w_x=\theta_y, & \forall y\in X \end{cases} \tag{3.61}$$

for a fixed $s=(s^1,\ s^2,\ldots,\ s^m)\in S$. In the unichain case, (3.60) and (3.61) are represented as

$$\psi^i(s)=\sum_{x\in X}\ \sum_{(a^1,a^2,\ldots,a^m)\in A(x)}\ \prod_{k=1}^{m} s^k_{x,a^k}\cdot r^i_{x,(a^1,a^2\ldots a^m)}\cdot q_x,\ i=1,2,\ldots,m, \tag{3.62}$$

and

$$\begin{cases} q_y-\sum\limits_{k=1}^{m}\sum\limits_{x\in X_k}\sum\limits_{a^k\in A^k(x)} s^k_{x,a^k}\cdot p^{a^k}_{x,y}\cdot q_x=0, & \forall y\in X\\ \sum\limits_{y\in X} q_y=1 \end{cases} \tag{3.63}$$

respectively, and we obtain the normal-form game $\langle \{S^i\}_{i=\overline{1,m}},\ \{\psi^i(s)\}_{i=\overline{1,m}}\rangle$ for a switching controller average stochastic game with a unichain property. A Nash equilibrium for this game is a stationary equilibrium for the switching average stochastic game with a unichain property.

The problem of the existence of limiting average equilibria (in "almost stationary" behavior strategies) for perfect information stochastic games and for stochastic games with Additive Rewards and Additive Transitions (ARAT) were studied by Thuijsman and Tughavan [180].

The additive property of the rewards and transition probabilities means that

$$r^i_{x,(a^1,a^2,\ldots,a^m)} = \sum_{i=1}^{m} r^i_{x,a^i} \text{ and } p^{(a^1,a^2,\ldots,a^m)}_{x,y} = \sum_{i=1}^{m} p^{a^i}_{x,y}.$$

For two-person stochastic games with switching control, they showed the existence of limiting average ε-equilibria.

3.9 Stochastic Games with a Stopping State

In this section, we consider a class of stochastic games in which the play stops as soon as a given state $z \in X$ is reached and players select actions in their feasible action sets in order to reach the stopping state with a maximal expected total discounted reward. A simple game with a stopping state for which stationary equilibria exist is represented by the game in which the transition probability matrix P^s of an arbitrary strategy profile $s = (s^1, s^2, \ldots, s^m)$ corresponds to a Markov chain with an absorbing state z, where $p^s_{z,z} = 1, \forall s \in S$ and $r^i_{z,s} = 0, i = 1, 2, \ldots, m$. It is easy to show that for such a game, there exists a Nash equilibrium not only for $0 < \gamma < 1$ but also for $\gamma = 1$. If the rewards are not positive for some special cases of the games with a stopping state, Nash equilibria may exist also when γ is greater than 1. We present Nash equilibria existence results for some classes of discounted stochastic positional games that we use for positional games on networks with total transition costs.

Based on the results from Sects. 3.4, 3.6.8, and 3.6.9, we show how to derive pure stationary equilibria conditions for a discounted stochastic positional game that is determined by a tuple $(X, A, \{X_i\}_{i=\overline{1,m}}, \{c^i\}_{i=\overline{1,m}}, p, \gamma, z)$, where $z \in X$ is the stopping state with $p^a_{z,z} = 1, \forall a \in A$; $r^i_{z,a} = 0,\ i = 1, 2, \ldots, m$, and for a discounted stochastic game on a network that is determined by the tuple $(G = (X, E), \{X_i\}_{i=\overline{1,m}}, X^0, \{c^i\}_{i=\overline{1,m}}, p, \gamma, z)$, where $z \in X$ is a stopping vertex containing a single outgoing directed edge (a loop) $e = (z, z)$ with $c^i_{z,z} = 0,\ i = 1, 2, \ldots, m$.

3.9.1 Stochastic Positional Games with a Stopping State

The main result that characterizes stationary Nash equilibria in a discounted stochastic game with a stopping state is the following:

Theorem 3.34 *Let a discounted stochastic positional game with a stopping state be given that is determined by the tuple* $(X, A, \{X_i\}_{i=\overline{1,m}}, \{r^i\}_{i=\overline{1,m}}, p, \gamma, z)$, *where* $\gamma = 1$ *and* $p^a_{z,z} = 1, \forall a \in A(z),\ r^i_{z,a} = 1, 2, \ldots, m.$

Assume that the reward functions r^i, $i = 1, 2, \ldots, m$ satisfy the condition

$$r^i_{x,a} < 0,\ \forall x \in X \setminus \{z\},\ a \in A(x),\ i = 1, 2, \ldots, m;\ r^i_{z,z} = 0,\ i = 1, 2, \ldots, m,$$

and for the considered game, there exists a strategy profile $s = (s^1, s^2, \ldots, s^m)$ for which the transition probability matrix P^s corresponds to a Markov chain with the absorbing state $z \in X$. Then there exist the values σ^i_x, $i = 1, 2, \ldots, m$ for $x \in X$ that satisfy the following conditions:

(1) $\sigma^i_z = 0,\ i = 1, 2, \ldots, m.$

(2) $r^i_{x,a} + \sum\limits_{y \in X} p^a_{x,y}\sigma^i_y - \sigma^i_x \geq 0,\ \ \forall x \in X_i \setminus \{z\},\ \forall a \in A(x),\ i = 1, 2, \ldots, m.$

(3) $\min\limits_{a \in A(x)} \left\{ r^i_{x,a} + \sum\limits_{y \in X} p^a_{x,y}\sigma^i_y - \sigma^i_x \right\} = 0, \qquad \forall x \in X_i \setminus \{z\},\ i = 1, 2, \ldots, m.$

(4) *On each position set X_i, $i \in \{1, 2, \ldots, m\}$, there exists a map $s^{i*} : X_i \to A$ such that*

$$s^{i*}(x) = a^* \in \operatorname*{argmin}_{a \in A(x)} \left\{ \mu^i_{x,a} + \sum_{y \in X} p^a_{x,y}\sigma^i_y - \sigma^i_x \right\},\ \forall x \in X_i$$

and

$$r^j_{x,a^*} + \sum_{y \in X} p^{a^*}_{x,y}\sigma^j_y - \sigma^j_x = 0,\ \ \forall x \in X_i,\ \ j = 1, 2, \ldots, m,$$

where $s = (s^{1*}, s^{2*}, \ldots, s^{m*})$ is a pure Nash equilibrium for the stochastic positional game determined by $(X, A, \{X_i\}_{i=\overline{1,m}}, \{r^i\}_{i=\overline{1,m}}, p, z)$ and $\sigma^i_x = \sigma_x{}^i\,(s^{1*}, s^{2*}, \ldots, s^{m*})$ $\forall x \in X$, $i = 1, 2, \ldots, m$ represent the corresponding values of the payoffs. The strategy profile $s^* = (s^{1*}, s^{2*}, \ldots, s^{m*})$ is a pure Nash equilibrium of the game for an arbitrary starting position $x \in X$.

Proof According to Theorem 3.29, for the discounted stochastic positional game determined by $(X, A, \{X_i\}_{i=\overline{1,m}}, \{c^i\}_{i=\overline{1,m}}, p, \gamma, \overline{x}, z)$, there exists a pure stationary Nash equilibrium $s^*(\gamma) = (s^{1*}(\gamma), s^{2*}(\gamma), \ldots, s^{m*}(\gamma))$ for an arbitrary $\gamma \in (0, 1)$. If for a strategy profile $s = (s^1, s^2, \ldots, s^m)$, the transition probability matrix P^s does not generate a Markov chain with the absorbing state z, then taking into account that

$$r^i_{x,a} < 0, \forall x \in X \setminus \{z\},\ y \in X,\ i = 1, 2, \ldots, m;\ r^i_{z,a} = 0,\ i = 1, 2, \ldots, m,$$

we have

$$\sigma^i_x(s^1, s^2, \ldots, s^m) = \lim_{\gamma \to 1} \sigma^{\gamma}{}^i_x(s^1, s^2, \ldots, s^m) = -\infty,\ i = 1, 2,\ \ldots,\ m.$$

However, for a strategy profile $s = (s^1, s^2, \ldots, s^m)$ that generates a Markov chain with absorbing state z, there exist the finite values

$$\sigma_x^i(s^1, s^2, \ldots, s^m) = \sigma^{\gamma}{}_x^i(s^1, s^2, \ldots, s^m)\; i = 1, 2, \ \ldots, \ m$$

for an arbitrary $\gamma \in (0, 1]$ because $r_{z,z}^i = 0$.

This means that for an arbitrary γ close to 1, for the game, there exists the corresponding stationary Nash equilibrium $s^{1^*}(\gamma), s^{2^*}(\gamma), \ldots, \ s^{m*}(\gamma)$ that induces a Markov chain s with the absorbing state z.

So, we can see that for the stochastic positional game with stopping state z in the case $\gamma = 1$, there exists a strategy $s^* = (s^{1^*}, s^{2^*}, \ldots, s^{m*})$ for which

$$\sigma_{\overline{x}}^{\underline{i}}(s^{1^*}, s^{2^*}, \ldots, s^{m*}) = \lim_{\gamma \to 1} \sigma_x^i(s^{1^*}, s^{2^*}, \ldots, s^{m*}), \quad i = 1, 2, \ldots, m,$$

and $s^* = (s^{1^*}, s^{2^*}, \ldots, s^{m*})$ is a pure stationary Nash equilibrium.

The conditions (1)–(4) of the theorem can be obtained from the conditions of Theorem 3.29 for $\gamma = 1$ in the case if an arbitrary Nash equilibrium solution s^* of the game generates a Markov chain with the absorbing state z.

For the absorbing state, we obtain $\sigma_z^i = 0, \ i = 1, 2, \ldots, m$ because $r_{z,a}^i = 0$, $i = 1, 2, \ldots, m$. □

Remark 3.35 Theorem 3.34 can be extended for discounted stochastic positional games in the case when $r_{x,a}^i < 0, \forall x \in X \setminus \{z\}, a \in A(x)$ and $\gamma > 1$. For this, it is necessary to assume that there exists at least a strategy profile $s = (s^1, s^2, \ldots, s^m)$, where the matrix P^s corresponds to a Markov chain with the absorbing state $z \in X$, $(p_{z,z}^a = 1, a \in A(z)$ and $r_{x,a}^i = 0, \ i = 1, 2, \ldots, m)$.

3.9.2 Positional Games on Networks with a Stopping State

The conditions for the existence of pure stationary Nash equilibria in a discounted stochastic positional game on a network with a stopping state can be derived by using the conditions from Theorem 3.34.

Theorem 3.36 *Let a discounted stochastic positional game on a network be given that is determined by the tuple* $(G = (X, E), \ \{X_i\}_{i=\overline{1,m}}, \ X^0, \ \{c^i\}_{i=\overline{1,m}}, \ p, \ \gamma, \ z)$ *with* $\gamma = 1$ *and stopping state* $z \in X$*, where the cost functions* c^i, $i = 1, 2, \ldots, m$ *satisfy the condition*

$$c_{x,y}^i > 0, \ \forall x \in X \setminus \{z\}, \ y \in X(x), \ i = 1, 2, \ldots, m; \ c_{z,z}^i = 0, \ i = 1, 2, \ldots, m.$$

Additionally, assume that for the considered game, there exists a strategy profile $s = (s^1, s^2, \ldots, s^m)$ *that generates a Markov chain with an absorbing state* z.

Then there exist the values ε_x^i, $i = 1, 2, \ldots, m$ *for* $x \in X$ *that satisfy the following conditions:*

(1) $\varepsilon_z^i = 0, \quad i = 1, 2, \ldots, m.$
(2) $c_{x,y}^i + \varepsilon_y^i - \varepsilon_x^i \geq 0, \quad \forall x \in X_i \setminus \{z\}, \quad \forall y \in X(x), \quad i = 1, 2, \ldots, m.$
(3) $\min\limits_{y \in X(x)} \{c_{x,y}^i + \varepsilon_y^i - \varepsilon_x^i\} = 0, \qquad \forall x \in X_i \setminus \{z\}, \quad i = 1, 2, \ldots, m.$
(4) *For each* $x \in X^0$, *the following equation holds:*

$$\mu_x^i + \sum_{y \in X} p_{x,y} \varepsilon_y^i - \varepsilon_x^i = 0, \; i = 1, 2, \ldots, m.$$

(5) *On each position set* X_i, $i \in \{1, 2, \ldots, m\}$, *there exists a map* $s^{i^*} : X_i \to X$ *such that*

$$s^{i^*}(x) = y^* \in \underset{y \in X(x)}{\operatorname{argmin}} \left\{ c_{x,y}^i + \varepsilon_y^i - \varepsilon_x^i \right\}, \quad \forall x \in X_i$$

and

$$c_{x,y^*}^j + \varepsilon_{y^*}^j - \varepsilon_x^j = 0, \forall x \in X_i, \; i, j = 1, 2, \ldots, m,$$

where $s^* = (s^{1^*}, s^{2^*}, \ldots, s^{m*})$ *is a pure Nash equilibrium for the stochastic positional game* $(X, A, \{X_i\}_{i=\overline{1,m}}, \{c^i\}_{i=\overline{1,m}}, p, z)$ *with a stopping state* z, *where*

$$\sigma_x^i(s^{1^*}, s^{2^*}, \ldots, s^{m*}) = \varepsilon_x^i, \quad \forall x \in X, \quad i = 1, 2, \ldots, m,$$

and $s^* = (s^{1^*}, s^{2^*}, \ldots, s^{m*})$ *is a pure Nash equilibrium for an arbitrary starting state* $\overline{x} \in X$ *of the game.*

The proof of this theorem follows from the proof of Theorem 3.34.

3.10 Nash Equilibria for Dynamic *c*-Games on Networks

In [19, 22, 110, 116], the following dynamic game on networks was studied: Let $G = (X, E)$ be a directed graph of the states' transitions of the dynamical system $\mathbb{L}$ with a given partition $X_1, X_2, \ldots, X_m$ of the vertex set X, where the corresponding subsets X_i, $i = 1, 2, \ldots, m$ are regarded as position sets of m players $1, 2, \ldots, m$. Graph G possesses the property that each vertex contains at least one leaving directed edge, and on the edge set E, m cost functions are defined:

$$c^i : E \to \mathbb{R}, \quad i = 1, 2, \ldots, m.$$

These functions represent the corresponding costs $c^1_{x,y}, c^2_{x,y}, \ldots, c^m_{x,y}$ for each directed edge $e = (x, y) \in E$ for the players if the dynamical system makes a transition from the state x to the state y through the directed edge $e = (x, y)$. Additionally, in G, two vertices x_0 and x_f are distinguished, where x_0 represents the starting position of the game and x_f is a position in which the game stops.

Thus, the game starts in the position x_0 at the moment of time $t = 0$. If x_0 belongs to the set of positions X_{k_0} of the player $k_0 \in \{1, 2, \ldots, m\}$, then the move is done by player k_0. "Move" means that player k_0 fixes an outgoing directed edge $e_1 = (x_0, x_1)$ and the dynamical system makes a transition from state x_0 to state $x_1 \in X(x_0)$. After the first move, the dynamical system is in the state x_1 at the moment of time $t = 1$. If x_1 belongs to the set of positions X_{k_1} of the player k_1, then the move is made by player $k_1 \in \{1, 2, \ldots, m\}$, i.e., player k_1 selects a new position x_2 for the state's transition of the system and so on. The game stops at the moment of time t if the state x_f is reached, i.e., if $x_t = x_f$. This game may be finite or infinite. If the final state x_f is reached at a finite moment of time t, then the game is finite; otherwise, the game is infinite. Each player in the game aims to minimize his or her own integral cost $\sum_{\tau=0}^{t-1} c^i_{e_t}$. In this game, we are looking for a Nash equilibrium.

In [110, 116], the considered game is called *dynamic c-game*, and it is denoted by $(G, \{X_i\}_{i=\overline{1,m}}, \{c^i\}_{i=\overline{1,m}}, x_0, x_f)$. Below, we formulate conditions for determining Nash equilibria in a dynamic c-game. We show that if in G the vertex x_f is attainable from the vertex x_0 and the cost functions are positive, then a Nash equilibrium exists. Moreover, we show that the optimal strategies of the players in the game can be found in the set of stationary strategies $\mathbb{S}_1, \mathbb{S}_2, \ldots, \mathbb{S}_m$. We define the stationary strategies of the players $1, 2, \ldots, m$ as m maps:

$$s^i : x \to y \in X(x) \quad \text{for} \quad x \in X_i, \quad i = 1, 2, \ldots, m.$$

In the terms of stationary strategies, the dynamic c-game in normal form with payoff functions

$$H^1_{x_0x_f}(s^1, s^2, \ldots, s^m),\ H^2_{x_0x_f}(s^1, s^2, \ldots, s^m),\ \ldots,\ H^m_{x_0x_f}(s^1, s^2, \ldots, s^m)$$

of the players $1, 2, \ldots, m$ can be defined in the following way: Let $s^1, s^2, \ldots, s^m$ be a fixed set of strategies of the players $1, 2, \ldots, m$. Then in G, the subset of directed edges $E^i_s = \{(x, s^i(x)) \in E \mid x \in X_i\}$ corresponds to the set of transitions of the dynamical system in the states $x \in X_i$ controlled by player $i \in \{1, 2, \ldots, m\}$. The subset $E_s = \bigcup_{i=1}^m E^i_s$ in G generates a subgraph $G_s = (X, E_s)$ in which either a unique directed path $P_s(x_0, x_f)$ from x_0 to x_f exists or such a directed path in G does not exist.

If $s^1, s^2, \ldots, s^m$ generate in G a subgraph G_s, which contains a unique directed path $P_s(x_0, x_f)$ from x_0 to x_f, then we put

$$H^i_{x_0 x_f}(s^1, s^2, \ldots, s^m) = \sum_{e \in E(P_s(x_0, x_f))} c^i_e, \quad i = 1, 2, \ldots, m, \tag{3.64}$$

where $E(P_s(x_0, x_f))$ represents the set of directed edges of the directed path $P_s(x_0, x_f)$.

If in G_s there is no directed path from x_0 to x_f, then a unique directed cycle C_s with a set of edges $E(C_s)$ can be obtained if we pass through directed edges from x_0. Therefore, there exist a unique directed cycle C_s, which we can obtain from x_0, and a unique directed path $P'_s(x_0, x')$, which connects x_0 and C_s (the vertex x' is a unique common vertex of $P'_s(x_0, x')$ and C_s). In this case, $H^i_{x_0 x_f}(s^1, s^2, \ldots, s^m)$, $i = 1, 2, \ldots, m$ are defined as follows:

$$H^i_{x_0 x_f}(s^1, s^2, \ldots, s^m) = \begin{cases} +\infty, & \text{if } \sum\limits_{e \in E(C_s)} c^i_e > 0; \\ \sum\limits_{e \in E(P'_s(x_0, x'))} c^i_e, & \text{if } \sum\limits_{e \in E(C_s)} c^i_e = 0; \\ -\infty, & \text{if } \sum\limits_{e \in E(C_s)} c^i_e < 0. \end{cases} \tag{3.65}$$

It is easy to observe that the dynamic c-game represents a particular case of the network model of a stochastic positional game with a stopping state $z = x_f$. We obtain the dynamic c-game from a stochastic positional game in the case $X^0 = \emptyset$. Therefore, from Theorem 3.34, we obtain the following result:

Theorem 3.37 *Let $(G, \{X_i\}_{i=\overline{1,m}}, \{c^i\}_{i=\overline{1,m}}, x_0, x_f)$ be a dynamic network for which the vertex x_f in G is attainable from every $x \in X$. Assume that the vectors $c^i = (c^i_{e_1}, c^i_{e_2}, \ldots, c^i_{e_{|E|}})$, $i \in \{1, 2, \ldots, m\}$ have positive and constant components. Then in a dynamic c-game on the network $(G, \{X_i\}_{i=\overline{1,m}}, \{c^i\}_{i=\overline{1,m}}, x_0, x_f)$ for the players $1, 2, \ldots, m$, there exists an optimal solution $s^{1*}, s^{2*}, \ldots, s^{m*}$ in the sense of Nash that satisfies the following properties:*

- *The graph $G_{s^*} = (X, E_{s^*})$ generated by $s^{1*}, s^{2*}, \ldots, s^{m*}$ has the structure of a directed tree with a sink vertex x_f.*
- *$s^{1*}, s^{2*}, \ldots, s^{m*}$ represent the solution to a dynamic c-game on the network $(G, \{X_i\}_{i=\overline{1,m}}, \{c^i\}_{i=\overline{1,m}}, x, x_f)$ with an arbitrary starting position $x \in X$ and a given final position x_f.*

This theorem was formulated and proved in [19]. Theorem 3.36 represents a generalization of this theorem for stochastic positional games. In [19], the following result was also proved:

Theorem 3.38 *Let $(G, \{X_i\}_{i=\overline{1,m}}, \{c^i\}_{i=\overline{1,m}}, x_0, x_f)$ be a network for which the vertex x_f in G is attainable from every $x \in X$. Assume that the vectors $c^i = (c^i_{e_1}, c^i_{e_2}, \ldots, c^i_{e_{|E|}})$, $i \in \{1, 2, \ldots, m\}$ have positive and constant components. Then on the vertex set X of the network game, there exist m real functions*

$$\varepsilon^1 : X \to \mathbb{R}^1, \ \ \varepsilon^2 : X \to \mathbb{R}^1, \ \ \ldots, \ \ \varepsilon^m : X \to \mathbb{R}^1,$$

which satisfy the following conditions:

(1) $\varepsilon^i_y - \varepsilon^i_x + c^i_{x,y} \geq 0, \quad \forall (x, y) \in E_i, \ i = 1, 2, \ldots, m,$ *where* $E_i = \{e = (x, y) \in E \mid x \in X_i, \ y \in X\}$.
(2) $\min\limits_{y \in X(x)} \{\varepsilon^i_y - \varepsilon^i_x + c^i_{x,y}\} = 0, \quad \forall x \in X_i, \quad i = 1, 2, \ \ldots, \ m.$
(3) *The subgraph $G^0 = (X, E^0)$ generated by the edge set $E^0 = E^0_1 \cup E^0_2 \cup \cdots \cup E^0_m$, $E^0_i = \{e = (x, y) \in E_i \mid \varepsilon^i_y - \varepsilon^i_x + c^i_{x,y} = 0\}$, $i = 1, 2, \ldots, m$ has the property that the vertex x_f is attainable from any vertex $x \in X$, and G^0 contains a subgraph $\overline{G}^0 = (X, \overline{E}^0)$, $\overline{E}^0 \subset E$, which possesses the same property, and besides that*

$$\varepsilon^i_y - \varepsilon^i_x + c^i_{x,y} = 0, \ \forall (x, y) \in \overline{E}^0, \quad i = 1, 2, \ldots, m.$$

If $\varepsilon^1, \varepsilon^2, \ldots, \varepsilon^m$ are arbitrary real functions, which satisfy the conditions (1)–(3), *then the optimal solution characterized by Nash strategies in the dynamic c-game on the network $(G, \{X_i\}_{i=\overline{1,m}}, \{c^i\}_{i=\overline{1,m}}, x_0, x_f)$ can be found as follows: Choose in $\overline{G}^0$ an arbitrarily directed tree $GT = (X, E^*)$ with sink vertex x_f and fix the following maps in GT:*

$$s^{1^*} : x \to y \in X_{GT}(x) \ \ \textit{for} \ \ x \in X_1;$$
$$s^{2^*} : x \to y \in X_{GT}(x) \ \ \textit{for} \ \ x \in X_2;$$
$$\vdots$$
$$s^{m*} : x \to y \in X_{GT}(x) \ \ \textit{for} \ \ x \in X_m,$$

where $X_{GT}(x) = \{y \in X \mid (x, y) \in E^*\}$.

Proof According to Theorem 3.37, for the considered dynamic c-game $(G, \{X_i\}_{i=\overline{1,m}}, \{c^i\}_{i=\overline{1,m}}, x_0, x_f)$, there exists an optimal solution characterized by Nash strategies $s^{1^*}, s^{2^*}, \ldots, s^{m*}$ of the players $1, 2, \ldots, m$, and these strategies generate in G a directed tree $GT_{s^*} = (X, E_{s^*})$ with sink vertex x_f. In this tree, we find the functions

$$\varepsilon^1 : X \to \mathbb{R}^1, \ \ \varepsilon^2 : X \to \mathbb{R}^1, \ \ \ldots, \ \ \varepsilon^m : X \to \mathbb{R}^1,$$

where $\varepsilon_x^i = H^i_{xx_f}(s^{1^*}, s^{2^*}, \ldots, s^{m*})$, $\forall x \in X, i = 1, 2, \ldots, m$. It is easy to verify that $\varepsilon^1, \varepsilon^2, \ldots, \varepsilon^m$ satisfy the conditions (1) and (2).

Additionally, we can see that in G^0, there exists graph $\overline{G}^0 = (X, \overline{E}^0)$, which satisfies condition (3), because $GT \subseteq \overline{G}^0$. Moreover, if in $\overline{G}^0$, a directed tree $GT_{s'} = (X, E_{s'})$, which is different from GT_{s^*} with a sink vertex, is chosen, then $GT_{s'}$ generates another optimal solution characterized by a Nash solution $s'^1, s'^2, \ldots, s'^m$. Now let us show that if

$$\varepsilon^1 : X \to \mathbb{R}^1, \ \ \varepsilon^2 : X \to \mathbb{R}^1, \ \ldots, \ \varepsilon^m : X \to \mathbb{R}^1$$

are arbitrary functions, which verify the conditions (1)–(3), then an arbitrary directed tree $GT = (X, E_{s^*})$ of $\overline{G}^0$ generates the maps:

$$s^{1^*} : x \to y \in X_{GT}(x) \quad \text{for} \quad x \in X_1;$$
$$s^{2^*} : x \to y \in X_{GT}(x) \quad \text{for} \quad x \in X_2;$$
$$\vdots$$
$$s^{m*} : x \to y \in X_{GT}(x) \quad \text{for} \quad x \in X_m,$$

which correspond to an optimal solution characterized by a Nash solution.

We use the induction on the number m of the players in the dynamic c-game. In the case that $m = 1$, the statement is true, because $X_1 = X$ and the conditions (1)–(3) for positive c_e^1 provide the existence of a tree $GT = (X, E_{s^*})$ of optimal paths, corresponding to the solution s^{1^*} for the problem of finding the shortest paths from $x \in X$ to x_f in G.

Assume that the statement holds for $m \le k$, $k \ge 1$, and let us prove it for $m = k + 1$. We consider that the first player fixes his or her strategy $s^1 = s^{1^*}$ and consider the problem of finding an optimum through Nash strategies in the network game with respect to other players. The obtained game in the positional form can be interpreted as a c-game with $m - 1$ players, since the positions of the first player can be considered to be the positions of any other player. Furthermore, we consider them as the positions of the second player.

Thus, if we fix $s^1 = s^{1^*}$, then we obtain a new game with $m - 1$ players, where we may consider the position of the first player as the position of any other player; we consider these positions as the positions of the second player. This means that we obtain a new c-game $(G^1, X_2^1, X_3, \ldots, X_m, \ c_1^2, c_1^3, \ldots, c_1^m, x_0, x_f)$, where X_2^1, G^1, and the functions c_1^i, $i = 2, \ldots, m$ are defined as follows:

$$X_2^1 = X_1 \cup X_2, \ \ G = \left(X, (E \setminus E_1) \cup E^1_{s^{1*}}\right),$$

and the cost function c_1^2 on $(E \setminus E_1) \cup E^1_{s^{1*}}$ is induced by the cost function c^2.

In the normal form, this game is determined by the payoff functions

$$H^2_{x_0x_f}(s^{1^*}, s^2, \ldots, s^m),\ H^3_{x_0x_f}(s^{1^*}, s^2, \ldots, s^m),\ \ldots,\ H^m_{x_0x_f}(s^{1^*}, s^2, \ldots, s^m),$$

where $s^2 \in \mathbb{S}_2$, $s^3 \in \mathbb{S}_3$, ..., $s^m \in \mathbb{S}_m$; $\mathbb{S}_2, \mathbb{S}_3, \ldots, \mathbb{S}_m$ are the respective sets of feasible stationary strategies of the players $2, 3, \ldots, m$.

In this new dynamic c-game $(G^1, X_2^1, X_3, \ldots, X_m, c_1^2, c_1^3, \ldots, c_1^m, x_0, x_f)$, we consider $m-1$ functions

$$\varepsilon^2 : X \to \mathbb{R}^1, \quad \varepsilon^3 : X \to \mathbb{R}^1, \quad \ldots, \quad \varepsilon^m : X \to \mathbb{R}^1$$

that satisfy the following conditions:

(1) $\varepsilon_y^i - \varepsilon_x^i + c^i_{1(x,y)} \geq 0, \quad \forall (x,y) \in E_i^1, \quad i = 2, \ldots, m,$ where
$E_2^1 = \{e = (x,y) \in E^1 \mid x \in X_2^1,\ y \in X\}$,
$E_i^1 = \{e = (x,y) \in E^1 \mid x \in X_i,\ y \in X\}, \quad i = 3, \ldots, m.$

(2) $\min\limits_{y \in X_{G^1}(x)} \{\varepsilon_y^2 - \varepsilon_x^2 + c^2_{1(x,y)}\} = 0, \quad \forall x \in X_2^1,$

$\min\limits_{y \in X_{G^1}(x)} \{\varepsilon_y^2 - \varepsilon_x^2 + c^i_{1(x,y)}\} = 0, \quad \forall x \in X_i, \quad i = 3, \ldots, m.$

(3) The subgraph $G^{1^0} = (X, E^{1^0})$, generated by the edge set $E^{1^0} = E_2^{1^0} \cup E_3^0 \cup \cup \cdots \cup E_m^0$, $E_2^{1^0} = \{e = (x,y) \in E_2^1 \mid \varepsilon_y^2 - \varepsilon_x^2 + c^2_{1(x,y)} = 0\}$, $E_i^0 = \{e = (x,y) \in E_i \mid \varepsilon_y^i - \varepsilon_x^i + c^i_{1(x,y)} = 0\}$, $i = 2, \ldots, m$, has the property that the vertex x_f is attainable from any vertex $x \in X$, and G^{1^0} contains a subgraph $\overline{G}^{1^0} = (X, \overline{E}^{1^0})$ which possesses the same property, and besides that

$$\varepsilon_y^i - \varepsilon_x^i + c^i_{1(x,y)} = 0, \quad \forall (x,y) \in \overline{E}^{1^0}, \quad i = 2, \ldots, m.$$

According to the induction assumption, in the dynamic c-game $(G^1, X_2^1, X_3, \ldots, X_m, c_1^2, c_1^3, \ldots, c_1^m, x_0, x_f)$, the stationary strategies $\overline{s}^{2^*}, s^{3^*}, \ldots, s^{m*}$ induced by the directed tree $GT = (X, E_{s^*})$,

$$\overline{s}^{2^*} : x \to y \in X_{GT}(x) \quad \text{for} \quad x \in X_2^1;$$
$$s^{3^*} : x \to y \in X_{GT}(x) \quad \text{for} \quad x \in X_3;$$
$$\vdots$$
$$s^{m*} : x \to y \in X_{GT}(x) \quad \text{for} \quad x \in X_m,$$

where $\overline{s}^{2^*}(x) = s^{1^*}(x)$ for $x \in X_1$ and $\overline{s}^{2^*}(x) = s^{2^*}(x)$ for $x \in X_2$, determine a Nash equilibrium.

Thus,

$$H^i_{xx_f}(s^{1*}, s^{2*}, s^{3*}, \dots, s^{i-1*}, s^{i*}, s^{i+1*}, \dots, s^{m*})$$
$$\le H^i_{xx_f}(s^{1*}, s^{2*}, s^{3*}, \dots, s^{i-1*}, s^{i}, s^{i+1*}, \dots, s^{m*}), \quad \forall s^i \in \mathbb{S}_i,\ 2 \le i \le m.$$

Also, it is easy to verify that

$$H^1_{xx_f}(s^{1*}, s^{2*}, \dots, s^{m*}) \le H^1_{xx_f}(s^{1}, s^{2*}, \dots, s^{m*}), \quad \forall s^1 \in S_1$$

because for fixed $s^{2*}, s^{3*}, \dots, s^{m*}$ in G, the problem of finding

$$\min_{s^1 \in \mathbb{S}_1} H^1_{xx_f}(s^1, s^{2*}, \dots, s^{m*}) \text{ for } x \in X$$

becomes the problem of finding the shortest paths from x to x_f in the graph $G' = (X, E')$, generated by a set E_1 and the edges $(x, s_i^*(x))$, $x \in X_i$, $i = 2, 3, \dots, m$ with costs c^1_e on the edges $e \in E'$. On this graph, the following condition is satisfied:

$$\varepsilon^1_y - \varepsilon^1_x + c^1_{x,y} \ge 0; \quad \forall (x, y) \in E',$$

which implies

$$H^1_{xx_f}(s^{1*}, s^{2*}, \dots, s^{m*}) \le H^1_{xx_f}(s^{1}, s^{2*}, \dots, s^{m*}), \quad \forall s^1 \in \mathbb{S}_1$$

because $H^1_{xx_f}(s^{1*}, s^{2*}, \dots, s^{m*}) = \varepsilon^1_x$, $\forall x \in X$.

Hence, $s^{1*}, s^{2*}, \dots, s^{m*}$ is a Nash equilibrium for the dynamic c-game. □

Remark 3.39 Let

$$\varepsilon^1 : X \to \mathbb{R}^1, \ \ \varepsilon^2 : X \to \mathbb{R}^1, \ \ \dots, \ \ \varepsilon^m : X \to \mathbb{R}^1$$

be arbitrary real functions on X in G, and $\overline{c}^1, \overline{c}^2, \dots, \overline{c}^m$ are m new cost functions on the edges $e \in E$ obtained from $c^1, c^2, \dots, c^m$ as follows:

$$\overline{c}^i_{x,y} = \varepsilon^i_y - \varepsilon^i_x + c^i_{x,y}, \quad \forall (x, y) \in E,\ i = \overline{1, m}. \tag{3.66}$$

Then the dynamic c-games determined on the networks $(G, X_1, X_2, \dots, X_m, c^1, c^2, \dots, c^m, x_0, x_f)$ and $(G, X_1, X_2, \dots, X_m, \overline{c}^1, \overline{c}^2, \dots, \overline{c}^m, x_0, x_f)$ are equivalent because the payoff functions $H^i_{xx_f}(s^1, s^2, \dots, s^m)$ and $\overline{H}^i_{xx_f}(s^1, s^2, \dots, s^m)$, $i = 1, 2, \dots, m$ in such games differ only by a constant, i.e.,

$$H^i_{xx_f}(s^1, s^2, \dots, s^m) = \overline{H}^i_{xx_f}(s^1, s^2, \dots, s^m) + \varepsilon^i(x_f) - \varepsilon^i(x).$$

In [112, 116], the transformation (3.66) is called the *potential transformation* of the edges' costs of the players in the game.

Remark 3.40 The conditions of Theorem 3.38 ensure the existence of the optimal stationary strategies $s^{1^*}, s^{2^*}, \ldots, s^{m*}$ of the players $1, 2, \ldots, m$ for every starting position $x \in X$ in a dynamic c-game on the network $(G, X_1, X_2, \ldots, X_m, c^1, c^2, \ldots, c^m, x, x_f)$ with positive and constant cost functions $c^1, c^2, \ldots, c^m$. If $c^1, c^2, \ldots, c^m$ are arbitrary constant functions, then the conditions of Theorem 3.38 represent necessary and sufficient conditions for the existence of optimal stationary strategies $s^{1^*}, s^{2^*}, \ldots, s^{m*}$ in the dynamic c-game on the network $(G, X_1, X_2, \ldots, X_m, c^1, c^2, \ldots, c^m, x, x_f)$ for every starting position $x \in X$.

On the basis of the obtained results, we can propose the following algorithm to determine Nash equilibria in the considered dynamic game with constant costs on the edges of the network:

Algorithm 3.41 Determining Nash Equilibria for the Dynamic c-Game on an Acyclic Network

Let us consider a dynamic c-game for which graph $G = (X, E)$ has the structure of a directed acyclic graph with sink vertex x_f. Then a Nash equilibrium for the dynamic c-game on the network can be found as follows:

Preliminary step (Step 0): Fix $X^0 = \{x_f\}$ and put $\varepsilon^i(x_f) = 0, i = 1, 2, \ldots, m$.

General step (Step k, $k \geq 1$): If $X \setminus X^{k-1} = \emptyset$, then stop; otherwise, find a vertex $x^k \in X \setminus X^{k-1}$ for which $X_G(x^k) \subseteq X^{k-1}$, where $X_G(x^k) = \{y \in X \mid (x^k, y) \in E\}$. If $x^k \in X_{i_k}, i_k \in \{1, 2, \ldots, m\}$, then find an edge (x^k, y^k) for which

$$\varepsilon^{i_k}_{y^k} + c^{i_k}_{x^k, y^k} = \min_{y \in X_G(x^k)} \left\{ \varepsilon^{i_k}_y + c^{i_k}_{x^k, y} \right\}.$$

After that, put

$$\varepsilon^i_{x^k} = \varepsilon^i_{y^k} + c^i_{x^k, y^k}, \quad i = 1, 2, \ldots, m$$

and

$$X^k = X^{k-1} \cup \{x^k\}.$$

Then go to the next step.

If the functions $\varepsilon^i, i = 1, 2, \ldots, m$ are known, then the optimal strategies of the players $s^{1^*}, s^{2^*}, \ldots, s^{m*}$ can be found as follows:

Find a tree $GT_{s^*} = (X, E_{s^*})$ in the graph $\overline{G}^0 = (X, \overline{E}^0)$, and fix the strategies

$$s^i(x) : x \to y \in X_i, \quad (x, y) \in E_{s^*}, \quad i = 1, 2, \ldots, m.$$

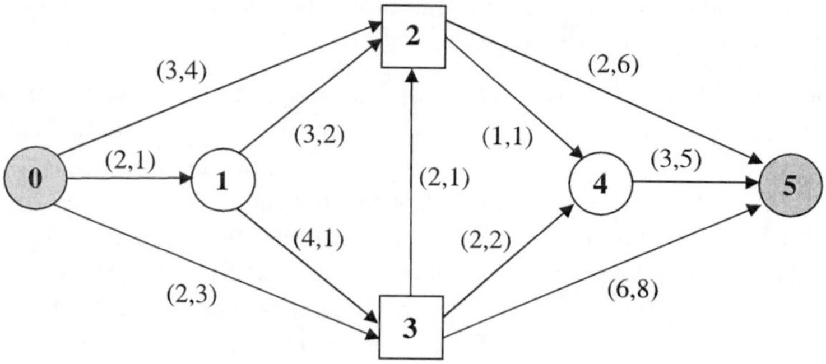

Fig. 3.3 Network for the acyclic c-game

Example Let a dynamic c-game on an acyclic network with two players represented in Fig. 3.3 be given, i.e., the network consists of a directed graph $G = (X, E)$ with partition $X = X_1 \cup X_2$, $X_1 = \{0, 1, 4, 5\}$, $X_2 = \{2, 3\}$, a starting position $x_0 = 0$, a final position $x_f = 5$, and the costs of the players 1 and 2 as given in parenthesis in Fig. 3.3, where the positions of the first player are represented by circles and the positions of the second player are represented by squares.

We consider the problem of determining optimal stationary strategies of the players in this dynamic c-game with an arbitrary starting position $x \in X$ and a fixed stopping position $x_f = 5$.

If we apply Algorithm 3.41, we obtain the following steps:

Step 0.
$X^0 = \{5\}$, $\varepsilon_5^1 = 0$, $\varepsilon_5^2 = 0$.

Step 1.
$X \setminus X^0 \neq \emptyset$; therefore, find a vertex $x^1 \in X \setminus X^0$ such that $X_G(x^1) \subseteq X^0$, i.e., $x^1 = 4$. Vertex 4 belongs to the set of positions of the first player, and we calculate

$$\varepsilon^1(4) = \varepsilon_5^1 + 3 = 3; \quad \varepsilon_4^2 = \varepsilon_5^2 + 5 = 5;$$
$$X^1 = X^0 \cup \{4\} = \{5, 4\}.$$

Step 2.
$X \setminus X^1 \neq \emptyset$, and find vertex $x^2 \in X \setminus X^1$ such that $X_G(x^2) \subseteq X^1$, i.e., $x^2 = 2$. Vertex 2 belongs to the set of positions of the second player, and we calculate

$$\min\left\{\varepsilon_5^2 + c_{2,5}^2, \varepsilon_4^2 + c_{2,4}^2\right\} = \min\{6, 6\} = 6.$$

So, we obtain this minimum for the edges $(2, 4)$ and $(2, 5)$. Here, we can fix an arbitrary edge from $\{(2, 4), (2, 5)\}$. For example, we fix edge $(2, 5)$.

Then at step 2, we obtain

$$\varepsilon_2^2 = 6; \quad \varepsilon_2^1 = \varepsilon_5^1 + c_{2,5}^1 = 2;$$
$$X^2 = X^1 \cup \{2\} = \{2, 4, 5\}.$$

Step 3.
$X \setminus X^2 \neq \emptyset$; $x^3 = 3$. Vertex 3 belongs to the set of positions of the second player; therefore, we find

$$\min\left\{\varepsilon_2^2 + c_{3,2}^2,\ \varepsilon_4^2 + c_{3,4}^2,\ \varepsilon^2(5) + c_{(3,5)}^2\right\} = 7.$$

So, we obtain this minimum for $e = (3, 2)$. We calculate

$$\varepsilon_3^2 = \varepsilon_2^2 + c_{3,2}^2 = 7; \quad \varepsilon_3^1 = \varepsilon_2^1 + c_{3,2}^1 = 4;$$
$$X^3 = X^2 \cup \{3\} = \{2, 3, 4, 5\}.$$

Step 4.
$X \setminus X^3 \neq \emptyset$; $x^4 = 1$. Vertex 1 belongs to the set of positions of the first player; therefore, we find

$$\min\left\{\varepsilon_2^1 + c_{1,2}^1,\ \varepsilon_2^1 + c_{1,3}^1\right\} = 5.$$

So, we obtain this minimum for $e = (1, 2)$. We calculate

$$\varepsilon_1^1 = \varepsilon_2^1 + c_{1,2}^1 = 5; \quad \varepsilon_1^2 = \varepsilon_2^2 + c_{1,2}^2 = 8;$$
$$X^4 = X^3 \cup \{1\} = \{1, 2, 3, 4, 5\}.$$

Step 5.
$X \setminus X^4 \neq \emptyset$; $x^5 = 0$. Vertex 1 belongs to the set of positions of the first player; therefore, we find

$$\min\left\{\varepsilon_2^1 + c_{0,2}^1,\ \varepsilon_1^1 + c_{0,1}^1,\ \varepsilon_3^1 + c_{0,3}^1\right\} = 5.$$

We determine

$$\varepsilon_0^1 = \varepsilon_2^1 + c_{0,2}^1 = 5; \quad \varepsilon_0^2 = \varepsilon_2^2 + c_{0,2}^2 = 10;$$
$$X^5 = X^4 \cup \{0\} = \{0, 1, 2, 3, 4, 5\}.$$

Step 6.
$X \setminus X^5 = \emptyset$; stop.

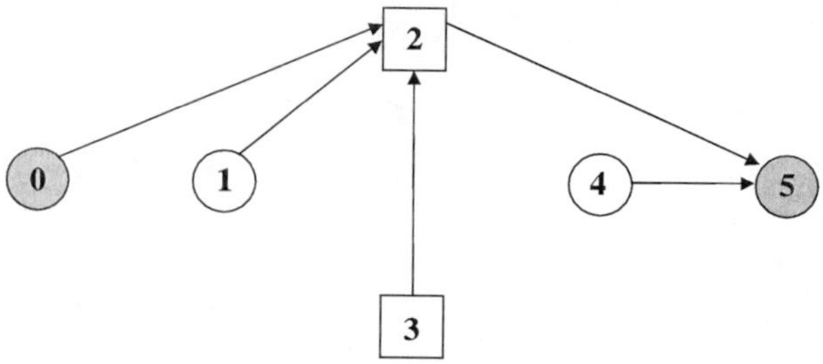

Fig. 3.4 Graph of optimal strategies of players in case 1

Thus, according to Theorem 3.38, we can determine a tree of the optimal paths $GT_{s^*} = (X, E_{s^*})$ that corresponds to the optimal stationary strategies of the players:

$$s^{1^*} : 0 \to 2; \quad 1 \to 2; \quad 4 \to 5;$$
$$s^{2^*} : 2 \to 5; \quad 3 \to 2.$$

The graph $GT_{s^*} = (X, E_{s^*})$, generated by the corresponding edges that determine ε_x^i, is represented in Fig. 3.4. Note that at step 2, the minimal value ε_2^2 is not uniquely determined because $\varepsilon_5^2 + c_{2,5}^2 = \varepsilon_4^2 + c_{2,4}^2 = 6$. Therefore, if at this step we select the directed edge $(2, 4)$, then in the following, the calculation procedure leads to another set of optimal stationary strategies of the players. Below is the result of the algorithm with the mentioned alternative at step 2.

Step 0: $\varepsilon_5^1 = 0, \quad \varepsilon_5^2 = 0.$
Step 1: $\varepsilon_4^1 = 3, \quad \varepsilon_4^2 = 5.$
Step 2: $\varepsilon_2^1 = 4, \quad \varepsilon_2^2 = 6.$
Step 3: $\varepsilon_3^1 = 6, \quad \varepsilon_3^2 = 7.$
Step 4: $\varepsilon_1^1 = 7, \quad \varepsilon_1^2 = 8.$
Step 5: $\varepsilon_0^1 = 7, \quad \varepsilon_0^2 = 10.$

In this case, the corresponding directed tree $GT_{s^*} = (X, E_{s^*})$ is represented in Fig. 3.5. This directed tree corresponds to the optimal stationary strategies:

$$s^{1^*} : 0 \to 2; \quad 1 \to 2; \quad 4 \to 5;$$
$$s^{2^*} : 2 \to 4; \quad 3 \to 2.$$

Algorithm 3.42 Determining All Stationary Strategies of the Players on Arbitrary Networks

Let us consider a dynamic c-game with m players, and let the directed graph G have an arbitrary structure, i.e., G may contain directed cycles. Moreover, we consider that for x_f, there are no leaving edges $(x_f, x) \in E$. We show that in this case, the problem can be reduced to the problem of finding optimal strategies in an auxiliary game with a network without directed cycles.

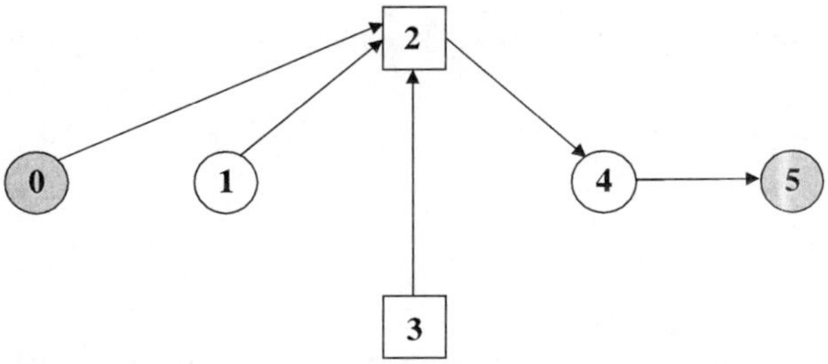

Fig. 3.5 Graph of optimal strategies of players in case 2

We construct an auxiliary directed graph $\overline{G} = (Z, \overline{E})$ without directed cycles, where Z and $\overline{E}$ are defined as follows:

$$Z = Z^0 \cup Z^1 \cup Z^2 \cup \cdots \cup Z^{|X|-1},$$

where

$$Z^j = \left\{z_0^j, z_1^j, z_2^j, \ldots, z_{|X|-1}^j\right\}, \quad j = 0, 1, 2, \ldots, |X| - 1;$$

so, $Z^0, Z^1, \ldots, Z^{|X|-1}$ represent the copies of the set X;

$$\overline{E} = E^0 \cup E^1 \cup E^2 \cup \cdots \cup E^{|X|-2} \cup E^f,$$

where

$$E^j = \left\{\left(z_k^j, z_l^{j+1}\right) \mid (x_k, x_l) \in E\right\}, \quad j = 0, 1, 2, \ldots, |X| - 2;$$

$$E^f = \left\{(z_k^j, z_f^{|X|-1}) \mid (x_k, x_f) \in E, \ j = 0, 1, 2, \ldots, |X| - 3\right\}.$$

It is easy to observe that in this graph, the vertex $z_f^{|X|-1}$ is attainable from any $z_k^0 \in Z^0$. If we delete in $\overline{G}$ all vertices z_k^i, for which there is no directed path from z_k^i to z^i_f, then we obtain a directed acyclic graph $\overline{G}' = (Z', \overline{E}')$ with sink vertex $z_f^{|X|-1}$. In the following, we divide the vertex set Z' into m subsets $Z_1', Z_2', \ldots, Z_m'$ corresponding to the position sets of the players $1, 2, \ldots, m$, respectively:

$$Z_1' = \left\{z_k^j \in Z' \mid x_k \in X_1, \ j = 0, 1, 2, \ldots, |X| - 1\right\};$$

$$Z_2' = \left\{z_k^j \in Z' \mid x_k \in X_2, \ j = 0, 1, 2, \ldots, |X| - 1\right\};$$

$$\vdots$$

$$Z_m' = \left\{z_k^j \in Z' \mid x_k \in X_m, \ j = 0, 1, 2, \ldots, |X| - 1\right\}.$$

On the edge set $\overline{E}'$, we define the cost functions as follows:

$$\overline{c}^i_{z_k^j, z_l^{j+1}} = c^i_{x_k, x_l}, \ \forall \left(z_k^j, z_l^{j+1}\right) \in E^j, \ j = 0, 1, 2, \dots, |X| - 2, \ i = 1, 2, \dots, m;$$

$$\overline{c}^i_{z_k^j, z_f^{|X|-1}} = c^i_{x_k, x_f}, \ \forall \left(z_k^j, z_f^{|X|-1}\right) \in E^f, \ j = 0, 1, 2, \dots, |X| - 3.$$

After that, we consider a dynamic c-game on the network $(\overline{G},' Z_1', Z_2', \dots, Z_m', \overline{c}^1, \overline{c}^2, \dots, \overline{c}^m, z_0^0, z_f^{|X|-1})$, where $\overline{G}'$ is a directed acyclic graph with sink vertex $z_f^{|X|-1}$. If we use Algorithm 3.42, then we determine the values $\varepsilon^i_{z_k^j}$, $\forall z_k^j \in Z'$, $i = 1, 2, \dots, m$. It is easy to observe that if we put $\varepsilon^i_{x_f} = 0$, $i = 1, 2, \dots, m$, and $\varepsilon^i_{x_k} = \varepsilon^i_{z_k^{|X|-1}}$, $\forall x_k \in X \setminus \{x_f\}$, $i = 1, 2, \dots, m$, then we obtain functions $\varepsilon^i : X \to \mathbb{R}$ that satisfy the conditions (1)–(3) from Theorem 3.38. Thus, we find the tree $GT = (X, E_s)$ which corresponds to the optimal strategies $s_1^*, s_2^*, \dots, s_m^*$ of the players in our dynamic c-game.

Algorithm 3.42 is inconvenient because of the great number of vertices in the auxiliary network.

Therefore, we present a simpler algorithm for finding optimal strategies of the players:

Algorithm 3.43 Determining Nash Equilibria for the Dynamic c-Game with an Arbitrary Network

Preliminary step: Assign a set of labels $\varepsilon_x^1, \varepsilon_x^2, \dots,$ to each vertex $x \in X$ ε_x^m as follows:

$$\begin{aligned} \varepsilon^i_{x_f} &= 0, && \forall i = 1, 2, \dots, m, \\ \varepsilon^i_x &= \infty, && \forall x \in X \setminus \{x_f\}, \ i = 1, 2, \dots, m. \end{aligned}$$

General step (step k $(k \geq 1)$): For each vertex $x \in X \setminus \{x_f\}$, change the labels $\varepsilon^i(x)$, $i = 1, 2, \dots, m$ as follows: If $x \in X_k$, then find vertex $\overline{x}$ for which

$$\varepsilon^k_{\overline{x}} + c^k_{x,\overline{x}} = \min_{y \in X(x)} \left\{\varepsilon^k_y + c^k_{x,y}\right\}.$$

If $\varepsilon^k_x > \varepsilon^k_{\overline{x}} + c^k_{x,\overline{x}}$, then replace $\varepsilon^i(x)$ with $\varepsilon^i_{\overline{x}} + c^i_{x,\overline{x}}$, $i = 1, 2, \dots, p$.

If $\varepsilon^k_x \leq \varepsilon^k_{\overline{x}} + c^k_{x,\overline{x}}$, then do not change the labels. Repeat the general step $n - 1$ times. Then the labels ε^i_x, $i = 1, 2, \dots, m$ for $x \in X$ become constant.

Remark 3.44 In the algorithm, the labels $\varepsilon^i(x)$, $i = 1, 2, \dots, m$ may become constant after less than $n - 1$ steps. So, the algorithm stops if the labels become constant.

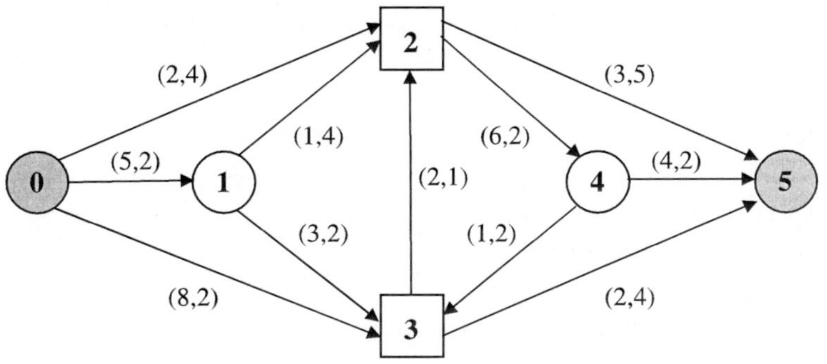

Fig. 3.6 Network for a c-game containing cycles

Let us state that these labels satisfy the conditions of Theorem 3.38. Hence, using the labels ε_x^i, $i = 1, \dots, m$, $x \in X$ and Theorem 3.38, we construct an optimal solution characterized by Nash strategies of the players $1, 2, \dots, m$. Algorithm 3.43 has the computational complexity $O(m|X|^2|E|)$.

Example Let a dynamic c-game of two players on the network represented by Fig. 3.6 be given. This network consists of a directed graph $G = (X, E)$ with sink vertex $x_f = 5$, given partition $X = X_1 \cup X_2$, $X_1 = \{0, 1, 4, 5\}$, $X_2 = \{2, 3\}$, and the costs for the players 1 and 2 written in parenthesis in Fig. 3.6. We are looking for optimal stationary strategies of the players in the dynamic c-game with an arbitrary starting position $x \in X$ and fixed stopping state $x_f = 5$.

Step 0 (Preliminary step).

Fix $\varepsilon_5^1 = 0$, $\varepsilon_5^2 = 0$;

$$\varepsilon_0^1 = \varepsilon_1^1 = \varepsilon_2^1 = \varepsilon_3^1 = \varepsilon_4^1 = \infty;$$

$$\varepsilon_0^2 = \varepsilon_1^2 = \varepsilon_2^2 = \varepsilon_3^2 = \varepsilon_4^2 = \infty.$$

We repeat the general step five times. At each step, we examine each vertex $x \in X$ and update its labels ε_x^1, ε_x^2 by using the condition of the algorithm; we also examine the vertices according to their numerical order.

Step 1.

Vertex $0 \in X_1$; therefore, calculate $\varepsilon_0^1 = \infty$; this implies $\varepsilon_0^2 = \infty$;
vertex $1 \in X_1$; therefore, calculate $\varepsilon_0^1 = \infty$; this implies $\varepsilon_1^2 = \infty$;
vertex $2 \in X_2$; therefore, calculate
$\varepsilon_5^2 + c_{2,5}^2 = \min\{\varepsilon_5^2 + c_{2,5}^2, \varepsilon_4^2 + c_{2,4}^2\} = \min\{5, \infty\} = 5$;
so, $\varepsilon_2^2 = 5$; this implies $\varepsilon_2^1 = \varepsilon_5^1 + c_{2,5}^1 = 3$;
vertex $3 \in X_2$; therefore, calculate
$\varepsilon_5^2 + c_{3,5}^2 = \min\{\varepsilon_5^2 + c_{3,5}^2, \varepsilon_2^2 + c_{3,2}^2\} = \min\{0 + 4, 5 + 1\} = 4$;
so, $\varepsilon_3^2 = 4$; this implies $\varepsilon_3^1 = \varepsilon_5^1 + c_{3,5}^1 = 0 + 2 = 2$;
vertex $4 \in X_1$; therefore, calculate
$\varepsilon_3^1 + c_{4,3}^1 = \min\{\varepsilon_5^1 + c_{4,5}^1, \varepsilon_3^1 + c_{4,3}^1\} = \min\{0 + 4, 2 + 1\} = 3$;

so, $\varepsilon_4^1 = 3$; this implies $\varepsilon_4^2 = \varepsilon_3^2 + c_{4,3}^2 = 4 + 2 = 6$;
vertex $5 \in X_1$; $\varepsilon_5^1 = 0$, $\varepsilon_5^2 = 0$.

Step 2.

Vertex $0 \in X_1$; therefore, calculate $\varepsilon_0^1 = \infty$; this implies $\varepsilon_0^2 = \infty$;
$\varepsilon_2^1 + c_{0,2}^1 = \min\{\varepsilon_2^1 + c_{0,2}^1, \varepsilon_1^1 + c_{0,1}^1, \varepsilon_3^1 + c_{0,3}^1\} = \min\{3+2, \infty+5, 2+8\} = 5$;
so, $\varepsilon_0^1 = 5$; this implies $\varepsilon_2^2 = \varepsilon_2^2 + c_{0,2}^2 = 4 + 5 = 9$;
vertex $1 \in X_1$; therefore, calculate
$\varepsilon_2^1 + c_{1,2}^1 = \min\{\varepsilon_2^1 + c_{1,2}^1, \varepsilon_3^1 + c_{1,3}^1\} = \min\{3 + 1, 2 + 3\} = 4$;
so, $\varepsilon_1^1 = 4$; this implies $\varepsilon_1^2 = \varepsilon_2^2 + c_{1,2}^2 = 5 + 4 = 9$;
vertex $2 \in X_2$; therefore, calculate
$\varepsilon_5^2 + c_{2,5}^2 = \min\{\varepsilon_5^2 + c_{2,5}^2, \varepsilon_4^2 + c_{2,4}^2\} = \min\{5, 8\} = 5$;
so, $\varepsilon_2^2 = 5$; this implies $\varepsilon_2^1 = \varepsilon_5^1 + c_{2,5}^2 = 3$;
vertex $3 \in X_2$; therefore, calculate
$\varepsilon_5^2 + c_{3,5}^2 = \min\{\varepsilon_5^2 + c_{3,5}^2, \varepsilon_2^2 + c_{3,2}^2\} = \min\{4, 6\} = 4$;
so, $\varepsilon_3^2 = 4$; this implies $\varepsilon_3^1 = \varepsilon_5^1 + c_{3,5}^1 = 2$;
vertex $4 \in X_1$; therefore, calculate
$\varepsilon_3^1 + c_{1,3}^1 = \min\{\varepsilon_5^1 + c_{4,5}^1, \varepsilon_3^1 + c_{4,3}^1\} = \min\{4, 3\} = 3$;
so, $\varepsilon_4^1 = 3$ and $\varepsilon_4^2 = 6$;
vertex $5 \in X_1$; $\varepsilon_5^1 = 0$, $\varepsilon_5^2 = 0$.

Step 3.

Vertex $0 \in X_1$; therefore, calculate
$\varepsilon_1^1 + c_{0,1}^1 = \min\{\varepsilon_2^1 + c_{0,2}^1, \varepsilon_1^1 + c_{0,1}^1, \varepsilon_3^1 + c_{0,3}^1\} = \min\{5, 9, 10\} = 5$;
so, $\varepsilon_0^1 = 5$ and $\varepsilon_0^2 = \varepsilon_2^2 + c_{0,2}^2 = 9$;
vertex $1 \in X_1$; therefore, calculate
$\varepsilon_2^1 + c_{1,2}^1 = \min\{\varepsilon_2^1 + c_{1,2}^1, \varepsilon_3^1 + c_{1,3}^1\} = \min\{3 + 1, 2 + 3\} = 4$;
so, $\varepsilon_1^1 = 4$ and $\varepsilon_1^2 = \varepsilon_2^2 + c_{1,2}^2 = 5 + 4 = 9$;
vertex $2 \in X_2$; therefore, calculate
$\varepsilon_5^2 + c_{2,5}^2 = \min\{\varepsilon_5^2 + c_{2,5}^2, \varepsilon_4^2 + c_{2,4}^2\} = \min\{5, 8\} = 5$;
so, $\varepsilon_2^2 = 5$ and $\varepsilon_2^1 = 3$;
vertex $3 \in X_2$; therefore, calculate
$\varepsilon_5^2 + c_{3,5}^2 = \min\{\varepsilon_5^2 + c_{3,5}^2, \varepsilon_2^2 + c_{3,2}^2\} = \min\{4, 6\} = 4$;
so, $\varepsilon_3^2 = 4$ and $\varepsilon^1(3) = 2$;
vertex $4 \in X_1$; therefore, calculate
$\varepsilon_3^1 + c_{4,3}^1 = \min\{\varepsilon_5^1 + c_{4,5}^1, \varepsilon_3^1 + c_{4,3}^1\} = \min\{4, 3\} = 3$;
so, $\varepsilon_4^1 = 4$ and $\varepsilon_4^2 = 2$.

After step 3, we observe that the labels coincide with the labels after step 2. So, the labels become constant and we finish the algorithm.

Finally, we have obtained

$$\varepsilon_0^1 = 5,\ \ \varepsilon_1^1 = 4,\ \ \varepsilon_2^1 = 3,\ \ \varepsilon_3^1 = 2,\ \ \varepsilon_4^1 = 3,\ \ \varepsilon_5^1 = 0;$$
$$\varepsilon_0^2 = 9,\ \ \varepsilon_1^2 = 9,\ \ \varepsilon_2^2 = 5,\ \ \varepsilon_3^2 = 4,\ \ \varepsilon_4^2 = 6,\ \ \varepsilon_5^2 = 0.$$

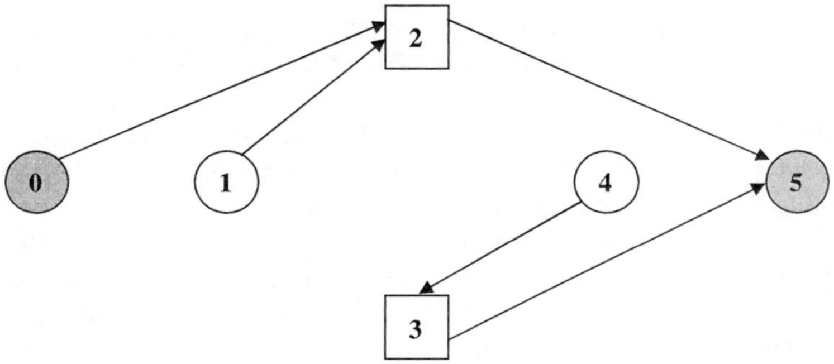

Fig. 3.7 Tree of optimal strategies

Thus, if we make the potential transformation of the edges' costs, then we can select the tree $GT_{s^*} = (X, E_{s^*})$ with zero cost of the edges that satisfy the conditions of Theorem 3.38. This tree is represented in Fig. 3.7. So, the optimal stationary strategies of the players are the following:

$$s^{1^*} : 0 \to 2; \quad 1 \to 2; \quad 4 \to 3;$$

$$s^{2^*} : 2 \to 5; \quad 3 \to 5.$$

In [109, 110, 112], the non-stationary version of a dynamic c-game was considered. A similar result for the non-stationary game was obtained, and algorithms for determining optimal strategies of the players were derived. Moreover, in [112], this game model was formulated and studied with the condition that the stopping position x_f should be reached at the moment of time $t(x_f)$, such that $t_1 \leq t(x_f) \leq t_2$, where t_1 and t_2 are given. However, efficient polynomial-time algorithms for determining Nash equilibria in such games can be derived only for the non-stationary case.

3.11 Two-Player Zero-Sum Positional Games on Networks

In the previous section, we studied dynamic c-games with positive cost functions on the edges. Therefore, we cannot use those results for zero-sum games. In the following, we study zero-sum games of two players with arbitrary cost functions on the edges and propose polynomial-time algorithms for their solving. The main results related to this problem were obtained in [42, 67, 96–98, 116, 202]. First, we study a max-min path problem on networks, which generalizes classical combinatorial problems of the shortest and the longest paths in weighted directed graphs.

This max-min path problem arises as an auxiliary problem when seeking optimal stationary strategies of the players in cyclic games. In addition, we use the considered dynamic c-game for studying and solving zero-sum control problems with an alternate player's control [116]. The main results are concerned with the existence of polynomial-time algorithms for determining max-min paths in networks as well as with an elaboration of such algorithms.

Let $G = (X, E)$ be a directed graph with a vertex set X and an edge set E. Assume that G contains a vertex $x_f \in X$ such that it is attainable from each vertex $x \in X$, i.e., x_f is a sink vertex in G. On the edge set E, a function $c : E \to \mathbb{R}$, which assigns a cost c_e to each edge $e \in E$, is given. In addition, the vertex set is divided into two disjoint subsets X_1 and X_2 ($X = X_1 \cup X_2$, $X_1 \cap X_2 = \emptyset$), which we regard as position sets of two players.

On G, we consider a game of two players. The game starts at the position $x_0 \in X$. If $x_0 \in X_1$, then the move is done by the first player; otherwise, it is done by the second one. The move indicates the passage from the position x_0 to the neighboring position x_1 through the edge $e_1 = (x_0, x_1) \in E$. After that, if $x_1 \in X_1$, then the move is done by the first player; otherwise, it is done by the second one and so on. As soon as the final position is reached, the game is over. The game can be finite or infinite. If the final position x_f is reached in finite time, the game is finite.

In case the final position x_f is not reached, the game is infinite. The first player in this game has the aim to maximize $\sum_i c_{e_i}$, while the second one has the aim to minimize $\sum_i c_{e_i}$.

The considered game in normal form in terms of stationary strategies can be defined as follows: We identify the stationary strategies s^1 and s^2 of the players with the maps

$$s^1 : x \to y \in X(x) \quad \text{for} \quad x \in X_1;$$
$$s^2 : x \to y \in X(x) \quad \text{for} \quad x \in X_2,$$

where $X(x)$ represents the set of extremities of edges $e = (x, y) \in E$, i.e., $X(x) = \{y \in X \mid e = (x, y) \in E\}$. Since G is a finite graph, the sets of strategies of the players

$$\mathbb{S}^1 = \{s^1 : x \to y \in X(x) \quad \text{for} \quad x \in X_1\};$$
$$\mathbb{S}^2 = \{s^2 : x \to y \in X(x) \quad \text{for} \quad x \in X_2\}$$

are finite sets. The payoff function $H_{x_0}(s^1, s^2)$ on $\mathbb{S}^1 \times \mathbb{S}^2$ is defined as follows:

Let in G be a subgraph $G_s = (X, E_s)$ generated in the edges of the form $(x, s^1(x))$ for $x \in X_1$ and $(x, s^2(x))$ for $x \in X_2$. Then either a unique directed path $P_s(x_0, x_f)$ from x_0 to x_f exists in G_s or such a path does not exist in G_s.

In the second case, in G_s, there exists a unique directed cycle C_s, which can be reached from x_0. For given s^1 and s^2, we set

$$H_{x_0}(s^1, s^2) = \sum_{e \in E(P_s(x_0, x_f))} c_e,$$

if in G_s there exists a directed path $P_s(x_0, x_f)$ from x_0 to x_f, where $E(P_s(x_0, x_f))$ is a set of edges of the directed path $P_s(x_0, x_f)$. If in G there is no directed path from x_0 to x_f, then we define $H_{x_0}(s^1, s^2)$ as follows:

Let $P'_s(x_0, y')$ be a directed path, which connects the vertex x_0 with the cycle C_s, and $P'_s(x_0, y')$ has no other common vertices with C_s except y'. Then we put

$$H_{x_0}(s^1, s^2) = \begin{cases} +\infty & \text{if } \sum\limits_{e \in E(C_s)} c_e > 0; \\ \sum\limits_{e \in E(P'_s(x_0, y'))} c_e & \text{if } \sum\limits_{e \in E(C_s)} c_e = 0; \\ -\infty & \text{if } \sum\limits_{e \in E(C_s)} c_e < 0. \end{cases}$$

This game is related to zero-sum positional games of two players, and it is determined by the graph G with the sink vertex x_f, the partition $X = X_1 \cup X_2$, the cost function $c : E \to \mathbb{R}$, and the starting position x_0. We denote the network, which determines this game, by $(G, X_1, X_2, c, x_0, x_f)$. In case the dynamic c-game is considered for an arbitrary starting position $x \in X$, we use the notation (G, X_1, X_2, c, x_f).

In [116, 118], it was shown that if G does not contain directed cycles, then for every $x \in X$, the following equality holds:

$$v(x) = \max_{s^1 \in \mathbb{S}^1} \min_{s^2 \in \mathbb{S}^2} H_x(s^1, s^2) = \min_{s^2 \in \mathbb{S}^2} \max_{s^1 \in \mathbb{S}^1} H_x(s^1, s^2), \tag{3.67}$$

which means the existence of optimal strategies of the players in the considered game. Moreover, in [116, 118], it was shown that in G, there exists a tree $GT^* = (X, E^*)$ with the sink vertex x_f, which gives the optimal strategies of the players in the game for an arbitrary starting position $x_0 \in X$. The strategies of the players are obtained by fixing

$$s^{1^*}(x) = y \ \text{ if } \ (x, y) \in E^* \text{ and } x \in X_1 \setminus \{x_f\};$$

$$s^{2^*}(x) = y \ \text{ if } \ (x, y) \in E^* \text{ and } x \in X_2 \setminus \{x_f\}.$$

In the general case, for an arbitrary graph G, equality (3.67) may fail to hold. Therefore, we formulate necessary and sufficient conditions for the existence of optimal strategies of the players in this game and propose a polynomial-time algorithm for determining the tree of max-min paths from every $x \in X$ to x_f. Furthermore, we show that our max-min path problem on the network can be regarded as a zero-value ergodic cyclic game. Thus, the proposed algorithm can be used for solving such games.

In [67, 97, 116], the formulated game on network $(G, X_1, X_2, c, x_0, x_f)$ is called the *dynamic c-game*. Some preliminary results related to this problem were obtained in [116, 118]. More general models of positional games on networks with p players were studied in [19, 107–109, 114–116].

The considered max-min path problem can be used for the zero-sum control problem with an alternate players' control (see Sect. 3.14).

3.11.1 An Algorithm for Games on Acyclic Networks

The formulated problem for acyclic networks was studied in [109, 116, 118].

Let $G = (X, E)$ be a finite directed graph without directed cycles and a given sink vertex x_f. The partition $X = X_1 \cup X_2$ $(X_1 \cap X_2 = \emptyset)$ of the vertex set of G is given, and the cost function $c : E \to \mathbb{R}$ on the edges is defined. We consider the dynamic c-game on G with a given starting position $x \in X$.

It is easy to observe that for the fixed strategies of players $s^1 \in \mathbb{S}^1$ and $s^2 \in \mathbb{S}^2$, the subgraph $G_s = (X, E_s)$ has a structure of a directed tree with sink vertex $x_f \in X$. This means that the value $H_x(s^1, s^2)$ is uniquely determined by the sum of edge costs of the unique directed path $P_s(x, x_f)$ from x to x_f. In [116, 118], it was proved that for an acyclic c-game on network (G, X_1, X_2, c, x, x_f), there exist the strategies of players s^{1^*} and s^{2^*} such that

$$\begin{aligned} v(x) = H_x(s^{1^*}, s^{2^*}) &= \max_{s^1 \in \mathbb{S}^1} \min_{s^2 \in S^2} H_x(s^1, s^2) \\ &= \min_{s^2 \in \mathbb{S}^2} \max_{s^1 \in \mathbb{S}^1} H_x(s^1, s^2) \qquad (3.68) \end{aligned}$$

and s^{1^*} and s^{2^*} do not depend on a starting position $x \in X$, i.e., (3.68) holds for every $x \in X$.

The equality (3.68) is evident in case if $\text{ext}(c, x) = 0, \forall x \in X \setminus \{x_f\}$, where

$$\text{ext}(c, x) = \begin{cases} \max\limits_{y \in X(x)} c_{(x,y)}, & x \in X_1; \\ \min\limits_{y \in X(x)} c_{(x,y)}, & x \in X_2. \end{cases}$$

In this case, $v(x) = 0, \forall x \in X$, and the optimal strategies of the players can be obtained by fixing the maps $s^{1^*} : X_1 \setminus \{x_f\} \to X$ and $s^{2^*} : X_2 \setminus \{x_f\} \to X$ such that $s^{1^*} \in \text{VEXT}(c, x)$ for $x \in X_1 \setminus \{x_f\}$ and $s^{2^*} \in \text{VEXT}(c, x)$ for

$x \in X_2 \setminus \{x_f\}$, where

$$\mathrm{VEXT}(c, x) = \{y \in X(x) \mid c_{(x,y)} = \mathrm{ext}(c, x)\}.$$

If the network $(G, X_1, X_2, c, x_0, x_f)$ has the property that $\mathrm{ext}(c, x) = 0, \forall x \in X \setminus \{x_f\}$, then it is called the *network in canonical form*. So, for the acyclic c-game on the network in canonical form, equality (3.68) holds, and $v(x) = 0, \forall x \in X$.

In the general case, equality (3.68) can be proved by using properties of the potential transformation $c'_{(x,y)} = c_{(x,y)} + \varepsilon_y - \varepsilon_x$ on the edges $e = (x, y)$ of the network, where $\varepsilon : X \to \mathbb{R}$ is an arbitrary real function on X (the potential transformation for antagonistic positional games was introduced in [67, 109]). The fact is that such a transformation of the costs on the edges of the acyclic network in a c-game does not change the optimal strategies of the players, even though the values $v(x)$ of positions $x \in X$ are changed to $v(x) + \varepsilon_{x_f} - \varepsilon_x$. It means that for an arbitrary function $\varepsilon : X \to \mathbb{R}$, the optimal strategies of the players in acyclic c-games on the networks $(G, X_1, X_2, c, x_0, x_f)$ and $(G, X_1, X_2, c', x_0, x_f)$ are the same.

We assume that the vertices $x \in X$ of the acyclic graph G are numbered $1, 2, \ldots, |X|$ such that if $x > y$, then in G, there is no directed path from y to x. For acyclic graphs, such a numbering of the vertices is possible and can be made starting from a sink vertex. Therefore, we can use the following recursive formula:

$$\varepsilon(x) = \begin{cases} \max\limits_{y \in X(x)} \{c_{(x,y)} + \varepsilon_y\} & \text{for } x \in X_1 \setminus \{x_f\}; \\ \min\limits_{y \in X(x)} \{c_{(x,y)} + \varepsilon_y\} & \text{for } x \in X_2 \setminus \{x_f\} \end{cases} \tag{3.69}$$

to tabulate the values $\varepsilon_x, \forall x \in X$ starting with $\varepsilon(x_f) = 0$. It is evident that the transformation $c'_{(x,y)} = c_{(x,y)} + \varepsilon_y - \varepsilon_x$ satisfies the condition $\mathrm{ext}(c', x) = 0$, $\forall x \in X$. This means that the following theorem holds:

Theorem 3.45 *For an arbitrary acyclic network* $(G, X_1, X_2, c, x_0, x_f)$ *with a sink vertex* x_f, *there exists a function* $\varepsilon : X \to \mathbb{R}$ *which determines the potential transformation* $c'_{(x,y)} = c_{(x,y)} + \varepsilon_y - \varepsilon_x$ *on the edges* $e = (x, y)$ *such that the network* $(G, X_1, X_2, c, x_0, x_f)$ *has a canonical form. The values* $\varepsilon_x, x \in X$, *which determine function* $\varepsilon : X \to \mathbb{R}$, *can be found by using recursive formula* (3.69).

On the basis of this theorem, the following algorithm for determining optimal strategies of the players in the c-game was proposed in [109].

Algorithm 3.46 Determining Optimal Strategies for the Players on an Acyclic Network

1. Determine the values $\varepsilon(x), x \in X$ according to the recursive formula (3.69) and the corresponding potential transformation $c'_{(x,y)} = c_{(x,y)} + \varepsilon_y - \varepsilon_x$ on the edges $(x, y) \in E$.

2. Fix arbitrary maps $s^{1^*}(x) \in \mathrm{VEXT}(c', x)$ for $x \in X_1 \setminus \{x_f\}$ and $s^{2^*}(x) \in \mathrm{VEXT}(c', x)$ for $x \in X_2 \setminus \{x_f\}$.

Remark 3.47 The values $\varepsilon_x,\ x \in X$ represent the values of the acyclic c-game on $(G, X_1, X_2, c, x_0, x_f)$ with starting position x, i.e., $\varepsilon(x) = v(x), \forall x \in X$. Algorithm 3.46 needs $O(|X|^2)$ elementary operations because the tabulation of the values $\varepsilon(x),\ x \in X$, using formula (3.69) for acyclic networks, requires this number of operations.

3.11.2 The Main Results for the Games on Arbitrary Networks

First, we give an example that shows that equality (3.67) may fail to hold. In Fig. 3.8, the network with the starting position $x_0 = 0$ and the final position $x_f = 3$ is given where the positions of the first player are represented by circles and the positions of the second player are represented by squares; the values of the cost functions on the edges are given alongside.

It is easy to observe that

$$\max_{s^1 \in \mathbb{S}^1} \min_{s^2 \in \mathbb{S}^2} H_0(s^1, s^2) = 2,$$

$$\min_{s^2 \in \mathbb{S}^2} \max_{s^1 \in \mathbb{S}^1} H_0(s^1, s^2) = 3,$$

i.e., the game does not have a value.

The following theorem gives conditions for the existence of a saddle point with finite $v(x)$ for each $x \in X$ in the c-game.

Theorem 3.48 *Let $(G, X_1, X_2, c, x_0, x_f)$ be an arbitrary network with the sink vertex $x_f \in X$. In addition, assume that $\sum_{e \in E(C_s)} c_e \neq 0$ for every directed cycle C_s from G. Then for the c-game on $(G, X_1, X_2, c, x_0, x_f)$, condition (3.67) with*

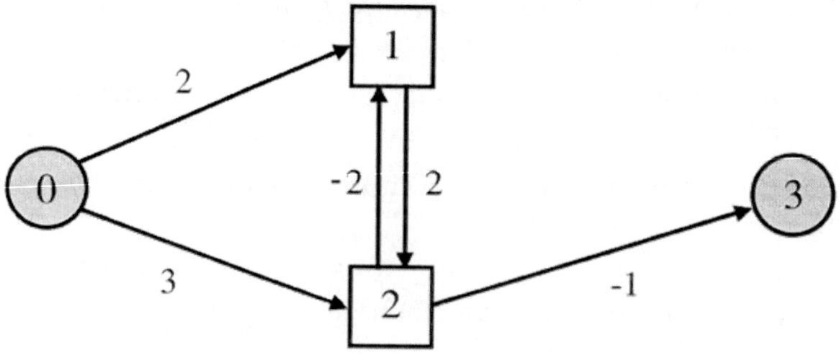

Fig. 3.8 Game on the network without a value

finite $v(x)$ *holds for every* $x \in X$ *if and only if there exists a function* $\varepsilon : X \to \mathbb{R}$ *that determines the potential transformation* $c'_{(x,y)} = c_{(x,y)} + \varepsilon_y - \varepsilon_x$ *on edges* $(x, y) \in E$ *such that* $\text{ext}(c', x) = 0, \forall x \in X$. *Moreover, if in G there exists the potential transformation* $c'_{(x,y)} = c_{(x,y)} + \varepsilon_y - \varepsilon_x$ *on edges* $(x, y) \in E$ *such that* $\text{ext}(c', x) = 0, \forall x \in X \setminus \{x_f\}$, *then* $v(x) = \varepsilon_x - \varepsilon_{x_f}, \forall x \in X$.

Proof $\Longrightarrow$ Let us consider that $\sum_{e\in E(C_s)} c_e \neq 0$ for every directed cycle C_s in G and condition (3.67) holds for every $x \in X$. Moreover, we consider that $v(x)$ is a finite value for every $x \in X$.

Taking into account that the potential transformation does not change the cost of the cycles, we can see that this transformation also does not change the optimal strategies of the players, although values $v(x)$ of the positions $x \in X$ are changed to $v(x) - \varepsilon_x + \varepsilon_{x_f}$. It is easy to observe that if we put $\varepsilon_x = v(x)$ for $x \in X$, then the function $\varepsilon : X \to \mathbb{R}$ determines the potential transformation $c'_{(x,y)} = c_{(x,y)} + \varepsilon_y - \varepsilon_x$ on the edges $(x, y) \in E$ such that $\text{ext}(c', x) = 0, \forall x \in X$.

$\Longleftarrow$ Let us consider that there exists the potential transformation $c'_{(x,y)} = c_{(x,y)} + \varepsilon_y - \varepsilon_x$ on the edges $(x, y) \in E$ such that $\text{ext}(c', x) = 0, \forall x \in X$. The value $v(x)$ of the game after the potential transformation is zero for every $x \in X$, and the optimal strategies of the players can be found by fixing s^{1*} and s^{2*} such that $s^{1*}(x) \in \text{VEXT}(c', x)$ for $x \in X_1 \setminus \{x_f\}$ and $s^{2*}(x) \in \text{VEXT}(c', x)$ for $x \in X_2 \setminus \{x_f\}$.
Since the potential transformation does not change the optimal strategies of the players, we put $v(x) = \varepsilon_x - \varepsilon_{x_f}$ and obtain (3.67). □

Corollary 3.49 *If for every directed cycle* C_s *in* G *the condition* $\sum_{e\in E(G_s)} c_e \neq 0$ *and equality* (3.67) *hold, then there exists the potential transformation* $\varepsilon : X \to \mathbb{R}$ *such that* $\text{ext}(c', x) = 0$, $\varepsilon(x_f) = 0$, *and* $v(x) = \varepsilon_x, \forall x \in X$.

Corollary 3.50 *If for every directed cycle* C_s *in* G *the condition* $\sum_{e\in E(G_s)} c_e \neq 0$ *holds, then the existence of the potential transformation* $c'_{(x,y)} = c_{(x,y)} + \varepsilon_y - \varepsilon_x$ *on edges* $(x, y) \in E$ *such that*

$$\text{ext}(c', x) = 0, \forall x \in X \tag{3.70}$$

represents necessary and sufficient conditions for the validity of the equality (3.67) *for every* $x \in X$. *In case there exists a cycle* C_s *with* $\sum_{e\in E(C_s)} c_e = 0$ *in* G, *condition* (3.70) *becomes only necessary for the validity of the equality* (3.67) *for every* $x \in X$.

Corollary 3.51 *If in the c-game there exist the strategies* s^{1*} *and* s^{2*}, *for which* (3.67) *holds for every* $x \in X$, *and these strategies generate a tree* $T_{s^*} = (X, E_{s^*})$ *with the sink vertex* x_f *in* G, *then there exists the potential transformation* $c'_{(x,y)} = c_{(x,y)} + \varepsilon_y - \varepsilon_x$ *on the edges* $(x, y) \in E$ *such that graph* $G^0 = (X, E^0)$, *generated by the set of the edges* $E^0 = \{(x, y) \in E \mid c'_{(x,y)} = 0\}$, *contains the tree* T_{s^*} *as a subgraph.*

Taking into account the results mentioned above, we propose an algorithm for determining the optimal strategies of the players in a c-game based on the construction of the tree of max-min paths. This algorithm works if such a tree in G exists.

3.11.3 Determining the Optimal Strategies of the Players

We consider the dynamic c-game determined by the network (G, X_1, X_2, c, x_f), where graph G has a sink vertex x_f. First, we assume that for an arbitrary vertex, there exists the value $v(x)$, which satisfies condition (3.70) and $v(x) \neq \pm\infty$. So, we assume that in G, there exists a tree of max-min paths from $x \in X$ to x_f. We show that there exists a polynomial-time algorithm for determining the optimal strategies of the players in the considered game. In this section, we propose such an algorithm based on the reduction to an auxiliary dynamic c-game with an acyclic network $(\overline{G}, W_1, W_2, \overline{c}, w_f^0)$, where graph $\overline{G} = (W, \overline{E})$ with $W = W_1 \cup W_1$ is obtained from $G = (X, E)$ as follows:

The set of vertices W consists of $n - 1$ copies of the vertex set X and the sink vertex w_f^0, i.e.,

$$W = \left\{w_f^0\right\} \cup W^1 \cup W^2 \cup \cdots \cup W^{n-1},$$

where $W^i = \{w_0^i, w_1^i, \ldots, w_{n-1}^i\}$, $i = 1, 2, \ldots, n - 1$. Here, $W^i \cap W^j = \emptyset$ for $i \neq j$ and vertices $w_k^i \in W^i$, $i = 1, 2, \ldots, n - 1$ correspond to vertex x_k from $X = \{x_0, x_1, x_2, \ldots, x_{n-1}\}$.

The set of edges $\overline{E}$ is defined as follows:

$$\overline{E} = E^0 \cup E^1 \cup E^2 \cup \cdots \cup E^{n-1};$$

$$E^i = \left\{(w_k^{i+1}, w_l^i) \mid (x_k, x_l) \in E\right\}, \ i = 1, 2, \ldots, n - 2;$$

$$E^0 = \left\{(w_k^i, w_f^0) \mid (x_k, x_f) \in E, \ i = 1, 2, \ldots, n - 1\right\}.$$

In $\overline{G}$, the edge subset $E^i \subseteq \overline{E}$ connects vertices of the set W^{i+1} to vertices of the set W^i by the edges (w_k^{i+1}, w_l^i) if there exists a directed edge (x_k, x_l) in G. Moreover, in $\overline{G}$, each vertex w_k^i, $i = 1, 2, \ldots, n - 1$ is connected to a sink vertex w_f^0 by the edge (w_k^i, w_f^0) if there exists a directed edge (x_k, x_f) in G.

The subsets W_1, W_2 and the cost function $\overline{c}\colon \overline{E} \to \mathbb{R}$ are defined as follows:

$$W_1 = \left\{ w_k^i \in W \,|\, x_k \in X_1 \right\}, \quad W_2 = \{ w_k^i \in W \,|\, x_k \in X_2 \};$$

$$\overline{c}_{(w_k^{i+1}, w_l^i)} = c_{(x_k, x_l)}, \ (x_k, x_l) \in E \ \text{ and } \ (w_k^{i+1}, w_l^i) \in E^i; \ \ i = 1, 2, \dots, n-2;$$

$$\overline{c}_{(w_k^i, w_f^0)} = c_{(x_k, x_f)}, \ (x_k, x_f) \in E \ \text{ and } \ (w_k^i, w_f^0) \in E^0; \ \ i = 1, 2, \dots, n-1.$$

From $\overline{G}$, we delete all vertices w_k^i for which there are no directed paths from w_k^i to w_f^0. For the obtained directed graph, we preserve the same notation, and we keep in mind that $\overline{G}$ does not contain such vertices.

Let us consider the dynamic c-game determined by the acyclic network $(\overline{G}, W_1, W_2, \overline{c}, w_f^0)$ with sink vertex w_f^0. So, we consider the problem of determining the values $v'(w_k^i)$ of the game for every $w_k^i \in W$.

We show that if $v'(w_k^1), v'(w_k^2), \dots, v'(w_k^{n-1})$ are the corresponding values of the vertices $w_k^1, w_k^2, \dots, w_k^{n-1}$ in the auxiliary game, then there exists $i \in \{1, n-1\}$ such that $v(x_k) = v'(w_k^i)$. We seek the vertex w_k^i among $w_k^{n-1}, w_k^{n-2}, \dots, w_k^2, w_k^1$ that starts with the highest-level set W^{n-1}.

We consider in $\overline{G}$ the max-min path

$$P_{\overline{G}}\left(w_k^{n-1}, w_f^0\right) = \left\{ w_k^{n-1}, w_{k_1}^{n-2}, w_{k_2}^{n-3}, \dots, w_{k_r}^{n-r-1}, w_f^0 \right\}$$

from w_k^{n-1} to w_f^0 generated by directed edges $e = (w_{k_i}^{n-i-1}, w_{k_{i+1}}^{n-i})$ for which

$$\varepsilon'_{(w_{k_{i+1}}^{n-i})} - \varepsilon'_{(w_{k_i}^{n-i-1})} + \overline{c}_{(w_{k_i}^{n-i-1}, w_{k_{i+1}}^{n-i})} = 0,$$

where

$$\varepsilon'_{(w_k^j)} = v'\left(w_k^j\right), \quad \forall w_k^j \in \left\{ w_k^{n-1}, w_{k_1}^{n-2}, w_{k_2}^{n-3}, \dots, w_{k_r}^{n-r-1}, w_f^0 \right\}.$$

In G, the directed path $P_{\overline{G}}(w_k^{n-1}, w_f^0)$ corresponds to a directed path

$$P_G(x_k, x_f) = \{ x_k, x_{k_1}, x_{k_2}, \dots, x_{k_r}, x_f \}$$

from x_k to x_f. Furthermore, in G, we also consider the subgraph $G_k^{n-1} = (X_k^{n-1}, E_k^{n-1})$ induced by the set of vertices $X_k^{n-1} = \{x_k, x_{k_1}, x_{k_2}, \dots, x_{k_r}, x_f\}$. For vertices x_{k_i} and x_k, we put $v(x_{k_i}) = v'(w_{k_i}^{n-i-1})$, $v(x_k) = v'(w_k^{n-1})$ and determine

$\varepsilon_{x_{k_i}} = v(x_{k_i})$, $\varepsilon_{x_k} = v(x_k)$, respectively. Then we verify if in G_k^{n-1}, the following condition holds:

$$\text{ext}(c', z) = 0, \ \forall z \in X_k^{n-1}, \tag{3.71}$$

where $c'_{(z,x)} = \varepsilon_x - \varepsilon_z + c_{(z,x)}$ for $e = (z, x) \in E_k^{n-1}$.

If condition (3.71) holds and G_k^{n-1} does not contain directed cycles, then we may conclude that for the dynamic c-game on G with starting position x_k, the equation $v(x_k) = v'(w'_k)$ holds. Note that for every vertex x_{k_i} of the directed path $P_0(x_{k_i}, x_f)$, we obtain $v(x_{k_i}) = v'(w_{k_i}^{n-i-1})$. If the condition mentioned above does not take place, then $v(x_k) \neq v'(w_k^{n-1})$, and we delete w_k^{n-1} from $\overline{G}$. After that, we consider the vertex w_k^{n-2}, construct the graph $G_k^{n-2} = (X_k^{n-2}, E_k^{n-2})$, and, in the same way, verify if $v(x_k) = v'(w_k^{n-1})$. Finally, we can see that at least for a vertex w_k^i, the directed path $P_{\overline{G}}(w_k^i, w_f)$ does not contain a directed cycle and condition (3.71) holds, i.e., $v(x_k) = v(w_k^i)$. In such a way, we obtain $v(x_k)$ for every $x_k \in X$. If $v(x)$ is known for every $x \in X$, then we fix $\varepsilon_x = v(x)$ and define the potential transformation $c'_{(z,x)} = c_{(z,x)} + \varepsilon_x - \varepsilon_z$ on edges $(z, x) \in E$. After that, find graph $G^0 = (V, E^0)$ generated by the set of edges $E^0 = \{(z, x) \in E \mid c'_{(z,x)} = 0\}$.

In G^0, we fix an arbitrary tree $T^* = (V, E^*)$, which determines the optimal strategies of the players as follows:

$$s^{1*}(z) = x \ \text{ if } \ (z, x) \in E^* \ \text{ and } \ z \in X_1 \setminus \{x_f\};$$

$$s^{2*}(z) = x \ \text{ if } \ (z, x) \in E^* \ \text{ and } \ z \in V_B \setminus \{x_f\}. \qquad \square$$

The correctness of the algorithm is based on the following theorem:

Theorem 3.52 *Let $v(x_k)$ be the value of the vertex x_k in the dynamic c-game on G, and let*

$$P_G(x_k, x_f) = \{x_k, x_{k_1}, x_{k_2}, \dots, x_{k_r}, x_f\}$$

be the max-min path from x_k to x_f in G. Then $v'(w_k^{r+1}) = v(x_k)$.

Proof The construction described above allows us to conclude that between the set of max-min directed paths from x_k to x_f with no more than $r+1$ edges in G and the set of max-min directed paths from w_k^{r+1} to w_f^0 with no more than $r+1$ edges in $\overline{G}$, there exists a bijective mapping, which preserves the sum of costs of the edges. Therefore, $v'(w_k^{r+1}) = v(x_k)$. $\square$

Remark 3.53 If $P_G(x_k, x_f) = \{x_k, x_{k_1}, x_{k_2}, \dots, x_{k_r}, x_f\}$ is the max-min path from x_k to x_f in G, then in $\overline{G}$, several vertices $w_k^{r+i} \in W$ for which $v'(w_k^{r+i}) = v(x_k)$, where $i \geq 1$, may exist. If $v'(w_k^{r+i}) = v(x_k)$, then in $\overline{G}$, the max-min path $P_{\overline{G}}(w_k^{r+1}, w_f^0) = \{w_k^{r+i}, w_{k_1}^{r+i-1}, w_{k_2}^{r+i-2}, \ldots, w_{k_r}^{i}, w_f^0\}$ corresponds to the max-min path $P_G(x_k, x_f)$ in G.

It is easy to observe that the running time of the algorithm is $O(|X|^4)$. Indeed, the values of the positions of the game on the auxiliary acyclic network can be calculated in time $O(N^2)$, where N is the number of vertices of the auxiliary network. Taking into account that $N \approx X^2$ for our auxiliary network, we can observe that the running time of the algorithm is $O(|X|^4)$.

Note that the proposed algorithm can also be applied for the c-game if the tree of max-min paths in G may not exist but there exists a max-min path from a given vertex $x_0 \in X$ to x_f. Then in G, the algorithm finds the max-min path with a given starting position x_0 and a final position x_f.

An important problem for the dynamic c-game is how to determine vertices $x \in X$ for which $v(x) = +\infty$ and vertices $x \in X$ for which $v(x) = -\infty$. Taking into account that the final position x_f in such games cannot be reached, we may delete vertices x of the graph G for which there exist max-min paths from x to x_f. In order to specify the algorithm in this case, we need to study the infinite dynamic c-game, where graph G has no sink vertex x_f. This means that the outcome of the game is a cycle that may have positive, negative, or zero-sum costs of the edges. In order to determine the outcome of the game in this case, we can use the same approach based on the reduction of the acyclic c-game.

The algorithm for finding the optimal strategies of the players in an infinite dynamic c-games is similar to the algorithm for finding the optimal strategies of the players in cyclic games. We describe such an algorithm in Sect. 3.13.5, and we can state that for an arbitrary position $x \in X$, the value of the cyclic game is positive if and only if the value $v(x)$ of the infinite dynamic c-game is positive. In addition, we can state that efficient polynomial-time algorithms for solving cyclic games can be elaborated if a polynomial-time algorithm for solving the infinite dynamic c-game exists.

In the following, we give an example that illustrates the details of the algorithm proposed above.

Example Consider the dynamic c-game determined by the network (G, X_1, X_2, c, x_f) given in Fig. 3.9. The position set X_1 of the first player is represented by circles, and the position set X_2 of the second player is represented by squares; $x_f = 0$. The costs of the edges are given alongside.

The auxiliary acyclic network for our dynamic c-game is represented in Fig. 3.10.

Each vertex in Fig. 3.10 is represented by double numbers, where the first one represents the number of the copy in G and the second one corresponds to the number of the vertex in G. Alongside the edges, their costs are given, and alongside the vertices, the values of the dynamic c-game on an auxiliary network are given.

Let us fix vertex $w_k^{n-1} = 33$ as the starting position of the dynamic c-game on the auxiliary network. Then we obtain $v'(33) = -5$. In order to verify if $v(3) = -5$, we find the max-min path $P_{\overline{G}}(33, 00) = \{33, 22, 11, 00\}$ from 33 to 00 and the values $v'(33) = -5$, $v'(22) = 1$, $v'(11) = 0$, and $v'(00) = 0$. The path $P_{\overline{G}}(33, 00)$ in G

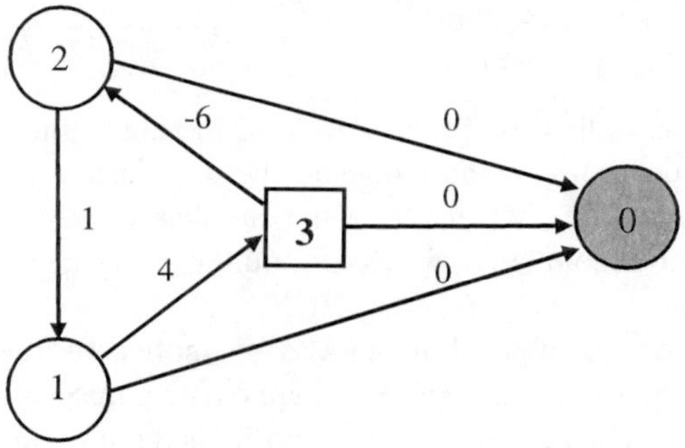

Fig. 3.9 Network for the dynamic c-game

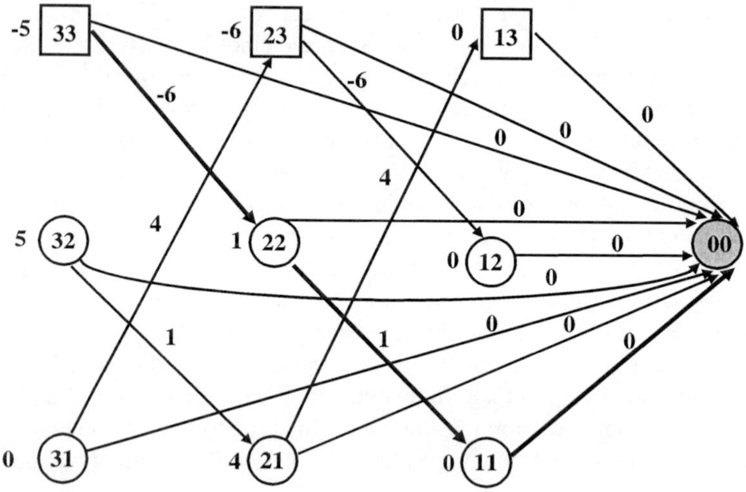

Fig. 3.10 Auxiliary acyclic network for the dynamic c-game

corresponds to the path $P_G(3,0) = \{3, 2, 1, 0\}$. For vertices 3, 2, 1, 0 in G, we fix $\varepsilon_3 = v'(33) = -5$, $\varepsilon_2 = v'(22) = 1$, $\varepsilon_1 = v'(11) = 0$, and $\varepsilon_0 = v'(00) = 0$. After that, find the graph $G_3^3 = (X_3^3, E_3^3)$ generated by the set of vertices $X^3 = \{3, 2, 1, 0\}$. In this case, graph G_3^3 coincides with graph G. Then we make a potential transformation $c'_{(x,y)} = \varepsilon_y - \varepsilon_x + c_{(x,y)} = 0$ with given $\varepsilon_3 = -5$, $\varepsilon_2 = 1$, $\varepsilon_1 = 0$, $\varepsilon_0 = 0$,

$$c'_{(1,0)} = \varepsilon_0 - \varepsilon_1 + c_{(1,0)} = 0 - 0 + 0 = 0,$$

$$c'_{(2,0)} = \varepsilon_0 - \varepsilon_2 + c_{(2,0)} = 0 - 1 + 0 = -1,$$

$$c'_{(3,0)} = \varepsilon_0 - \varepsilon_3 + c_{(3,0)} = 0 - (-5) + 0 = 5,$$

$$c'_{(1,3)} = \varepsilon_3 - \varepsilon_1 + c_{(1,3)} = -5 - 0 + 4 = -1,$$

$$c'_{(2,1)} = \varepsilon_1 - \varepsilon_2 + c_{(2,1)} = 0 - 1 + 1 = 0,$$

$$c'_{(3,2)} = \varepsilon_2 - \varepsilon_3 + c_{(3,2)} = 1 - (-5) - 6 = 0.$$

So, after the potential transformation $c'_{(x,y)} = \varepsilon_y - \varepsilon_x + c_{(x,y)}$, $\forall(x, y) \in E$, we obtain the network given in Fig. 3.11 with new costs on the edges. If we select the tree with zero cost edges, we obtain the tree of max-min paths represented in Fig. 3.12. If we start with vertex $w_k^{n-1} = 32$, then we obtain the subgraph $G_2^3 = (X_2^3, E_2^3)$, which coincides with graph $G = (X, E)$, for which we determine $\varepsilon_2 = v'(32) = 5$, $\varepsilon_1 = v'(21) = 4$, $\varepsilon_3 = v'(13) = 0$, and $\varepsilon_0 = v'(00) = 0$. It is easy to see that in this case, the condition

$$\text{extr}(c', x) = 0, \ \forall x \in X$$

is not satisfied.

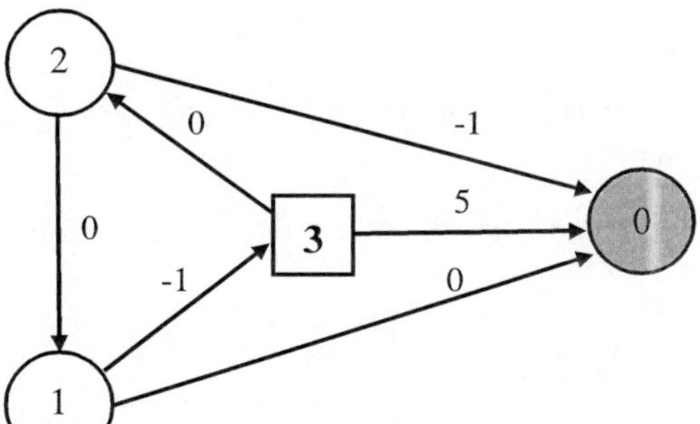

Fig. 3.11 Network of the game after potential transformation of costs

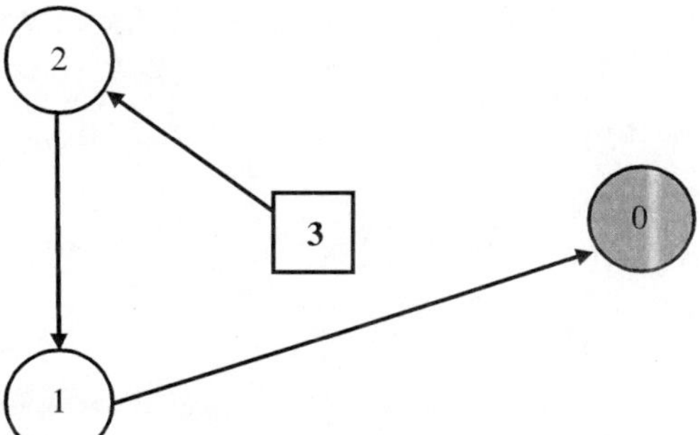

Fig. 3.12 Graph of optimal strategies

3.11.4 An Algorithm for Zero-Sum Dynamic c-Games

In this section, we describe an algorithm for solving zero-sum dynamic c-games based on a special recursive procedure for the calculation of the values $v(x)$. From a practical point of view, the proposed algorithm may be more useful than the algorithm from the previous section, although its computational complexity is $O(|X|^3 \sum_{e\in E} |c_e|)$ ($c : E \to \mathbb{R}$ is an integer function).

We assume that in G, there exists the tree of max-min paths.

Preliminary step (Step 0): Set $X^* = \{x_f\}$, $\varepsilon_{x_f} = 0$.

General step (Step k): Find the set of vertices

$$X' = \{z \in X \setminus X^* \mid (z, x) \in E, x \in X^*\}.$$

For each $z \in X'$, we calculate

$$\varepsilon_z = \begin{cases} \max\limits_{x\in O_{X^*}(z)} \{\varepsilon_x + c_{(z,x)}\}, & z \in X_1 \cap X'; \\ \min\limits_{x\in O_{X^*}(z)} \{\varepsilon_x + c_{(z,x)}\}, & z \in X_2 \cap X', \end{cases} \tag{3.72}$$

where $O_{X^*}(z) = \{x \in X^* \mid (z, x) \in E\}$, and then complete the following points (a) and (b):

(a) Fix $\beta(z) = \varepsilon_x$ for $z \in X' \cup X^*$, and then for every $x \in X' \cup X^*$, calculate

$$\beta(z) = \begin{cases} \max\limits_{x\in O_{X^*\cup X'}(z)} \{\varepsilon_x + c_{(z,x)}\}, & z \in X_1 \cap (X' \cup X^*); \\ \min\limits_{x\in O_{X^*\cup X'}(z)} \{\varepsilon_x + c_{(z,x)}\}, & z \in X_2 \cap (X' \cup X^*). \end{cases} \tag{3.73}$$

(b) Check if $\beta(z) = \varepsilon_z$ for every $z \in X' \cup X^*$. If this condition is not satisfied, then fix $\varepsilon_z = \beta(z)$ for every $z \in X' \cup X^*$, and go to point (a).

If $\beta(z) = \varepsilon_z$ for every $z \in X' \cup X^*$, then in $X' \cup X^*$, we find the subset

$$Y^k = \left\{z \in X^* \cup X' \,\middle|\, \mathrm{extr}_{x\in O_{X^*\cup X'}(z)}\{\varepsilon_x - \varepsilon_z + c_{(z,x)}\} = 0\right\},$$

where

$$\mathrm{ext}_{x\in O_{X^*\cup X'}(z)}\{\varepsilon_x - \varepsilon_z + c_{(z,x)}\} =$$

$$= \begin{cases} \max\limits_{x\in O_{X^*\cup X'}(z)} \{\varepsilon_x - \varepsilon_z + c_{(z,x)}\}, & z \in (X' \cup X^*) \cap X_1; \\ \min\limits_{x\in O_{X^*\cup X'}(z)} \{\varepsilon_x - \varepsilon_z + c_{(z,x)}\}, & z \in (X' \cup X^*) \cap X_2. \end{cases}$$

After that, we replace X^* with Y^k and check if $X^* = X$. If $X^* \neq X$, then go to the next step. If $X^* = X$, then define the potential transformation $c'_{(z,x)} = c_{(z,x)}+\varepsilon_x-\varepsilon_z$ on edges $(z, x) \in E$, and find the graph $G^0 = (X, E^0)$, generated by the set of edges $E^0 = \{(z, x) \in E \mid c'(z, x) = 0\}$. In G^0, fix an arbitrary tree $T^* = (X, E^*)$, which determines the optimal strategies of the players as follows:

$$s^{1*}(z) = x \text{ if } (z, x) \in E^* \text{ and } z \in X_1 \setminus \{x_f\};$$

$$s^{2*}(z) = x \text{ if } (z, x) \in E^* \text{ and } z \in X_2 \setminus \{x_f\}.$$

Let us show that this algorithm finds the tree of max-min paths $T^* = (X, E^*)$ if such a tree exists in G.

Theorem 3.54 *If in G there exists the tree of max-min paths $T^* = (X, E^*)$ with sink vertex x_f, then the algorithm finds it using $O\left(|X|^3 \sum_{e\in E} |c_e|\right)$ elementary operations.*

Proof Consider the set Y^{k-1} obtained after $k-1$ steps of the algorithm, and assume that at step k after points (a) and (b), the condition

$$\beta(z) = \varepsilon_z \quad \text{for every } \; z \in X' \cup X^*$$

holds. This condition is equivalent to the condition

$$\text{ext}_{x\in O_{X^*\cup X'}(z)}\{\varepsilon_x - \varepsilon_z + c_{(z,x)}\} = 0, \;\; \forall z \in X' \cup X^*$$

that involves $Y^{k-1} \subset Y^k$.

Therefore, in the following, we can see that if for every step k of the algorithm, the corresponding calculation procedure (3.73) is convergent, then $Y^0 \subset Y^1 \subset Y^2 \subset \cdots \subset Y^r = X$, where $r < n$. This means that after $r < n$ steps, the algorithm finds the values $\varepsilon(x)$ for $x \in X$ and the potential transformation $c'_{(y,x)} = \varepsilon_x - \varepsilon_y + c_{(y,x)}$ for edges $e = (y, x) \in E$ such that $\text{ext}(c', y) = 0$, $\forall x \in X$, i.e., the algorithm constructs the tree $T^* = (X, E^*)$. So, for a complete proof of the theorem, we have to show the convergence of the calculation procedure based on formula (3.73) for an arbitrary step k of the algorithm.

Assume that at step k of the algorithm, the condition

$$\text{ext}_{x\in O_{X^*\cup X'}(z)}\{\varepsilon_x - \varepsilon_z + c_{(z,x)}\} \neq 0 \;\text{ for every } \; z \in X'$$

holds. Consider the set of edges $E' = \{e = (z, x') \in E \mid \beta(z) = \varepsilon_{x'} + c_{(z,x')}$, $z \in X',\; x' \in x \in O_{X^*\cup X'}(z)\}$, where x' corresponds to vertex z such that

$$\varepsilon(x') + c(z, x') = \begin{cases} \max\limits_{x\in O_{X^*\cup X'}(z)} \{\varepsilon_x + c_{(z,x)}\}, \; z \in X_1 \cap (X' \cup X^*); \\ \min\limits_{x\in O_{X^*\cup X'}(z)} \{\varepsilon_x + c_{(z,x)}\}, \; z \in X_2 \cap (X' \cup X^*). \end{cases}$$

The calculation based on (a) and (b) can be treated as follows: Players improve the values ε_z of vertices $z \in X'$ using passages from z to corresponding vertices $x' \in O_{X^* \cup X'}(z)$. At each iteration of this calculation procedure, the players can improve their income by $\beta(z) - \varepsilon(z)$ units for every position $z \in X$.

Denote by $\widetilde{X}$ the subset of vertices $z' \in X'$ for which in $G' = (X', E')$, there exist directed paths from $z' \in \widetilde{X}$ to vertices from X^{k-1}. Then the improvements of the players mentioned above are possible for an arbitrary vertex $z \in \widetilde{X}$. This means that if procedure (a)–(b) at step k is applied, then after using one iteration of this procedure, we obtain $\beta(z) = \varepsilon_z, \forall z \in \widetilde{X}$. In the following, we can see that in order to achieve $\varepsilon_z = \beta(z)$ for the rest of the vertices $z \in X' \setminus \widetilde{X}$, it is necessary to apply more than one iteration.

Let us consider the subset $\widetilde{X}' = X' \setminus \widetilde{X}$ in G'. Then in G', there are no directed edges $e = (z, x')$ such that $z \in \widetilde{X}'$ and $x' \in X^{k-1}$. Without loss of generality, we may consider that in G, the subset $\widetilde{X}'$ generates a directed cycle C'. Denote by $n(C')$ the number of vertices of this cycle, and assume that the sum of its edges is equal to θ (θ may be positive or negative). We can see that if we apply formula (3.73), after $n(G)$ iterations of the calculation procedure, the values ε_z of the vertices $z \in C'$ will decrease at least by $|\theta|$ units if $\theta < 0$; if $\theta > 0$, then these values will increase by θ. Therefore, the first player will preserve passages from vertices $z \in C'$ to vertices x' of the cycle C' if $\beta(z) - \varepsilon(z) > 0$; otherwise, the first player will change the passage from one vertex $z^0 \in C'$ to a vertex $x'' \in O_{X^* \cup X'}(z)$, which may belong to X^{k-1}.

In an analogous way, the second player will preserve passages from vertices $z \in C'$ to vertices x' of cycle C' if $\beta(z) - \varepsilon(z) < 0$; otherwise, the second player will change the passage from one vertex $z^0 \in X$ to a vertex x'', which may belong to X^{k-1}. So, if in G there exists the tree of max-min paths, then after a finite number of iterations of the procedure (a)–(b), we obtain $\beta(z) = \varepsilon(z)$ for $z \in X'$. Taking into account that the values $\beta(z)$ decrease (or increase) after $n(G)$ iterations by integer units $|\theta|$, we may conclude that the number of iterations of the procedure is comparable with $|X|^2 \cdot \max_{z \in X'} |\beta(z) - \varepsilon_z|$. In the worst case, these quantities are limited by $|X|^2 \sum_{e \in E} |c_e|$. This involves that the computational complexity of the algorithm is $O\left(|X|^3 \sum_{e \in E} |c(e)|\right)$. □

Remark 3.55 The algorithm for an acyclic network can be applied without points (a) and (b) because the condition $\beta(z) = \varepsilon_z, \forall z \in X'$ holds at every step k. In general, the version of the algorithm without the points (a) and (b) can be used if $Y^{k-1} \neq Y^k$ at every step k. In this case, the running time of the algorithm is $O(|X|^3)$. The algorithm described above can be modified for the dynamic c-game in general form if the network contains vertices x for which $v(x) = \pm\infty$. In order to detect such vertices in point (a), it is necessary to introduce a new condition which allows us to select vertices $z \in X'$ with great values $\beta(z)$ (positive and negative). But in this case, the algorithm becomes more difficult than the algorithm for finite games.

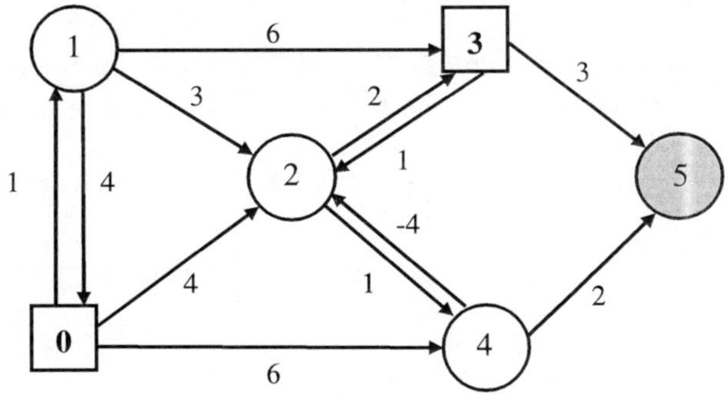

Fig. 3.13 Network with sink vertex x_f=5

Below, we present two examples that illustrate the details of the algorithm. The first example illustrates the work of the algorithm if it is not necessary to use points (a) and (b). The second example illustrates the details of the recursive calculation procedure in points (a) and (b).

Example 1 Consider the problem of determining the optimal stationary strategies on the network that may contain cycles. The corresponding network with sink vertex $x_f = 5$ is given in Fig. 3.13. On this network, the positions of the first player are represented by circles, and the positions of the second player are represented by squares, i.e., $X_1 = \{1, 2, 4, 5\}$, and $X_2 = \{0, 3\}$.

The values of the cost functions of the edges are given in parenthesis alongside them. We can see that for the given network, there exists a tree of max-min paths, which can be found by using the algorithm.

Step 0.
$X^* = \{5\}$; $\varepsilon_5 = 0$.
Step 1.
Find the set of vertices $X' = \{3, 4\}$ for which there exist directed edges (3, 5) and (4, 5) from vertices 3 and 4 to vertex 5. Then we calculate values $\varepsilon_3 = 3$ and $\varepsilon_4 = 2$ according to (3.72). It is easy to check that for vertices 3 and 4, the following condition holds:

$$\text{ext}_{y \in X_{X^* \cup X'}(x)}\{\varepsilon_y - \varepsilon_x + c_{(x,y)}\} = 0.$$

So, $Y^1 = \{3, 4, 5\}$. Therefore, if we replace X^* with Y^1, after step 1, we obtain $X^* = \{3, 4, 5\}$.

Step 2.
Find the set of vertices $X' = \{0, 1, 2\}$ for which there exist directed edges from vertices $x \in X'$ to vertices $y \in X^*$. Then according to (3.72), we calculate

$$\varepsilon_2 = \max_{y \in X^*(2)} \{\varepsilon_3 + c_{(2,3)},\ \varepsilon_4 + c_{(2,4)}\} = \max\{5, 3\} = 5;$$

$$\varepsilon_1 = \varepsilon_3 + c_{(1,3)} = 9;$$

$$\varepsilon_0 = \varepsilon_4 + 6 = 8.$$

So, $\varepsilon_0 = 8,\ \varepsilon_1 = 9, \varepsilon_2 = 5,\ \varepsilon_3 = 3,\ \varepsilon_4 = 2$, and $\varepsilon_5 = 0$.
It is easy to check that $Y^2 = \{0, 2, 3, 4, 5\}$. Indeed,

$$\text{ext}_{y \in X_{X^* \cup X'}(3)} \{\varepsilon_y - \varepsilon_3 + c_{(3,y)}\} =$$
$$= \min\{\varepsilon_5 - \varepsilon_3 + c_{(3,5)},\ \varepsilon_2 - \varepsilon_3 + c_{(3,2)}\} = \min\{0 - 3 + 3,\ 5 - 3 + 1\} = 0;$$

$$\text{ext}_{y \in X_{X^* \cup X'}(2)} \{\varepsilon_y - \varepsilon_3 + c_{(2,y)}\} =$$
$$= \max\{\varepsilon_3 - \varepsilon_2 + c_{(2,3)},\ \varepsilon_4 - \varepsilon_2 + c_{(2,4)}\} =$$
$$= \max\{3 - 5 + 2,\ 2 - 5 + 1\} = 0;$$

$$\text{ext}_{y \in X_{X^* \cup X'}(1)} \{\varepsilon_y - \varepsilon_1 + c_{(1,y)}\} =$$
$$= \max\{\varepsilon_3 - \varepsilon_1 + c_{(1,3)},\ \varepsilon_2 - \varepsilon_1 + c_{(1,2)},\ \varepsilon_0 - \varepsilon_1 + c_{(0,1)}\} =$$
$$= \max\{3 - 9 + 6,\ 3 - 9 + 5,\ 8 - 9 + 4\} = 3;$$

$$\text{ext}_{y \in X_{X^* \cup X'}(0)} \{\varepsilon_y - \varepsilon_0 + c_{(0,y)}\} =$$
$$= \min\{\varepsilon_4 - \varepsilon_0 + c_{(0,4)},\ \varepsilon_2 - \varepsilon_0 + c_{(0,2)},\ \varepsilon_1 - \varepsilon_0 + c_{(0,1)}\} =$$
$$= \min\{2 - 8 + 6,\ 5 - 8 + 4,\ 9 - 8 + 1\} = 0;$$

$$\text{ext}_{y \in X_{X^* \cup X'}(4)} \{\varepsilon_y - \varepsilon_4 + c_{(4,y)}\} =$$
$$= \max\{\varepsilon_5 - \varepsilon_4 + c_{(4,5)},\ \varepsilon_2 - \varepsilon_4 + c_{(4,2)}\} =$$
$$= \max\{0 - 2 + 2,\ 5 - 2 - 4\} = 0.$$

So, the set of vertices for which $\text{ext}_{y \in X_{X^* \cup X'}(x)} \{\varepsilon_y - \varepsilon_x + c_{(x,y)}\} = 0$ consists of vertices 0, 2, 3, 4, 5.
Step 3.
Find the set of vertices $X' = \{1\}$, and calculate

$$\varepsilon(1) = \max_{y \in X_{X^*}(1)} \{\varepsilon_y + c_{(1,y)}\} = \max\{\varepsilon_3 + c_{(1,3)},\ \varepsilon_2 + c_{(1,2)},\ \varepsilon_0 + c_{(1,0)}\} =$$
$$= \max\{3 + 6,\ 5 + 3,\ 8 + 4\} = 12.$$

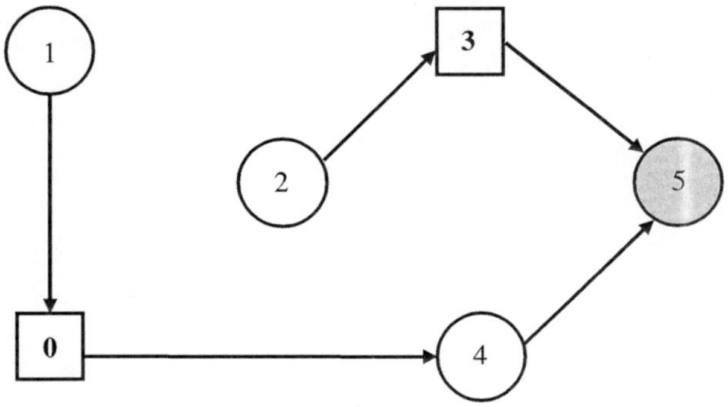

Fig. 3.14 Tree of optimal strategies

Now we can see that the obtained values $\varepsilon_0 = 8$, $\varepsilon_1 = 12$, $\varepsilon_2 = 5$, $\varepsilon_3 = 3$, $\varepsilon_4 = 2$, and $\varepsilon_5 = 0$ satisfy the conditions

$$\varepsilon_y - \varepsilon_x + c_{(x,y)} \leq 0 \text{ for every } (x, y) \in E,\ x \in X_1;$$

$$\varepsilon_y - \varepsilon_x + c_{(x,y)} \geq 0 \text{ for every } (x, y) \in E,\ x \in X_2.$$

The directed tree $GT = (X, E^*)$ generated by edges $(x, y) \in E$ for which $\varepsilon_y - \varepsilon_x + c_{(x,y)} = 0$ is represented in Fig. 3.14.

The optimal strategies of the players are:

$$s^1 : 1 \to 0;\ \ 2 \to 3;\ \ 4 \to 5;$$

$$s^2 : 0 \ \to 4;\ \ 3 \to 5.$$

Example 2 Consider the problem of determining the tree of max-min paths $T^* = (X, E^*)$ for the network given in Fig. 3.10 with the same costs of the edges as in the previous section.

If we apply the algorithm described above, then we use only one step ($k = 1$). But this step consists of two items (a) and (b), which make calculations based on formula (3.73). In Table 3.3, the values $\beta(0)$, $\beta(1)$, $\beta(2)$, and $\beta(3)$ at each iteration of the calculation procedure based on formula (3.73) are given.

We can see that the convergence of the calculation procedure is obtained at iteration 14. Therefore, we conclude that $\varepsilon_0 = 0$, $\varepsilon_1 = 0$, $\varepsilon_2 = 1$, and $\varepsilon_3 = -5$. If we make a potential transformation, we obtain the network in Fig. 3.11. In Fig. 3.12, the tree of max-min paths $T^* = (X, E^*)$ is presented.

Table 3.3 The results of the iteration procedure for $\beta(0), \beta(1), \beta(2), \beta(3)$

i	$\beta(0)$	$\beta(1)$	$\beta(2)$	$\beta(3)$
0	0	0	0	0
1	0	4	1	−6
2	0	0	5	−5
3	0	0	1	−1
4	0	3	1	−5
5	0	0	4	−5
6	0	0	1	−2
7	0	2	1	−5
8	0	0	3	−5
9	0	0	1	−3
10	0	1	1	−5
11	0	0	2	−5
12	0	0	1	−4
13	0	0	1	−5
14	0	0	1	−5

3.12 Acyclic l-Games on Networks

An acyclic l-game on networks was introduced in [96, 109] as an auxiliary problem for studying and solving special cyclic games, which we consider in the next section.

3.12.1 Problem Formulation

Let $(G, X_1, X_2, c, x_0, x_f)$ be a network where $G = (X, E)$ represents a directed acyclic graph with the sink vertex $x_f \in X$. On E, a function $c\colon E \to \mathbb{R}$ is defined, and on X, a partition $X = X_1 \cup X_2$ $(X_1 \cap X_2 = \emptyset)$ is given, where X_1 and X_2 correspond to the sets of positions of players 1 and 2, respectively.

We consider the following acyclic game from [109].

Again, we define the strategies of the players as maps

$$s^1\colon x \to y \in X(x) \quad \text{for} \quad x \in X_1 \setminus \{x_f\};$$

$$s^2\colon x \to y \in X(x) \quad \text{for} \quad x \in X_2 \setminus \{x_f\}.$$

We define the payoff function $\overline{H}_{x_0}\colon S^1 \times S^2 \to \mathbb{R}$ in this game as follows:

Let $s^1 \in S^1$ and $s^2 \in S^2$ be fixed strategies of the players. Then graph $G_s = (X, E_s)$, generated by the edges $(x, s^1(x))$, $x \in X \setminus \{x_f\}$ and $(x, s^2(x))$, $x \in X \setminus \{x_f\}$, has the structure of a directed tree with the sink vertex x_f.

Therefore, it contains a unique directed path $P_s(x_0, x_f)$ with $n(P_s(x_0, x_f))$ edges. We put

$$\overline{H}_{x_0}(s^1, s^2) = \frac{1}{n(P_s(x_0, x_f))} \sum_{e \in E(P_s(x_0, x_f))} c_e.$$

The payoff function $\overline{H}_{x_0}(s^1, s^2)$ on $S^1 \times S^2$ defines a game in normal form, which is determined by the network $(G, X_1, X_2, c, x_0, x_f)$.

We consider the problem of finding the strategies s^{1^*} and s^{2^*} for which

$$\overline{v}(x_0) = \overline{H}_{x_0}(s^{1^*}, s^{2^*}) = \max_{s^1 \in \mathbb{S}^1} \min_{s^2 \in \mathbb{S}^2} \overline{H}_{x_0}(s^1, s^2).$$

3.12.2 The Main Properties of Acyclic l-Games

First of all, let us show that for the considered max-min problem, there exists a saddle point.

Denote

$$\overline{\overline{v}}(x_0) = \overline{H}_{x_0}(s^{1^0}, s^{2^0}) = \min_{s^2 \in \mathbb{S}^2} \max_{s^1 \in \mathbb{S}^1} \overline{H}_{x_0}(s^1, s^2)$$

and let us show that $\overline{v}(x_0) = \overline{\overline{v}}(x_0)$.

Theorem 3.56 *For an arbitrary acyclic l-game, the following equality holds:*

$$\overline{v}(x_0) = \overline{H}_{x_0}(s^{1^*}, s^{2^*}) = \max_{s^1 \in \mathbb{S}^1} \min_{s^2 \in \mathbb{S}^2} \overline{H}_{x_0}(s^1, s^2) = \min_{s^2 \in \mathbb{S}^2} \max_{s^1 \in \mathbb{S}^1} \overline{H}_{x_0}(s^1, s^2).$$

Proof Let us note the following property of an acyclic l-game, determined by $(G, X_1, X_2, c, x_0, x_f)$: If the cost function c is changed to $c' = c + h$ (h is an arbitrary real number), then we obtain an equivalent acyclic l-game determined by $(G, X_1, X_2, c', x_0, x_f)$ for which $\overline{v}'(x_0) = \overline{v}(x_0) + h$ and $\overline{\overline{v}}''(x_0) = \overline{\overline{v}}(x_0) + h$. If we denote by $\overline{H}'_{x_0}(s^1, s^2)$ the payoff function of the l-game after the transformation mentioned above, then we have

$$\overline{H}_{x_0}(s^1, s^2) = \overline{H}'_{x_0}(s^1, s^2) + h, \ \forall s^1 \in \mathbb{S}^1, \ \forall s^2 \in \mathbb{S}^2.$$

It is easy to observe that if $h = -\overline{v}(x_0)$, then for the acyclic l-game with network $(G, X_1, X_2, c', x_0, x_f)$, we obtain $\overline{v}'(x_0) = 0$. This means that an acyclic l-game becomes an acyclic c-game with max-min value of the game $\overline{H}'_{x_0}(s^{1^0}, s^{2^0})$. Therefore, if we regard it as an acyclic c-game after the transformation of the game, the following property holds:

$$0 = \overline{v}'(x_0) = \max_{s^1 \in \mathbb{S}^1} \min_{s^2 \in \mathbb{S}^2} \overline{H}'_{x_0}(s^1, s^2) = \min_{s^2 \in \mathbb{S}^2} \max_{s^1 \in \mathbb{S}^1} \overline{H}'_{x_0}(s^1, s^2) = 0.$$

Taking into account that

$$\overline{H}'_{x_0}(s^1, s^2) = \overline{H}_{x_0}(s^1, s^2) - \overline{v}(x_0),$$

we obtain

$$\min_{s^2 \in \mathbb{S}^2} \max_{s^1 \in \mathbb{S}^1} \left(\overline{H}_{x_0}(s^1, s^2) - \overline{v}(x_0)\right) = \max_{s^1 \in \mathbb{S}^1} \min_{s^2 \in \mathbb{S}^2} \left(\overline{H}_{x_0}(s^1, s^2) - \overline{v}(x_0)\right) =$$

$$= \overline{\overline{v}}(x_0) - \overline{v}(x_0),$$

i.e., $\overline{\overline{v}}(x_0) - \overline{v}(x_0) = 0$. So, $\overline{\overline{v}}(x_0) = \overline{v}(x_0)$. □

Theorem 3.57 *Let an acyclic l-game determined by the network* $(G, X_1, X_2, c, x_0, x_f)$ *with the starting position* x_0 *be given. Then there exists the value* $\overline{v}(x_0)$, *and the function* $\varepsilon \colon X \to \mathbb{R}$ *determines the potential transformation* $c'_{(x,y)} = c_{(x,y)} + \varepsilon_x - \varepsilon_y$ *of the costs on the edges* $e = (x, y) \in E$ *such that the following conditions hold:*

(a) $\overline{v}(x_0) = \text{ext}(c', x), \ \forall x \in X \setminus \{x_f\}$.
(b) $\varepsilon_{x_0} = \varepsilon_{x_f}$.

The optimal strategies of the players in an acyclic l-game can be found as follows: Fix the arbitrary maps $s^{1*} \colon X_1 \setminus \{x_f\} \to X$ *and* $s^{2*} \colon X_2 \setminus \{x_f\} \to X$ *such that* $s^{1*}(x) \in \text{VEXT}(c', x)$ *for* $x \in X_1 \setminus \{x_f\}$ *and* $s^{2*}(x) \in \text{VEXT}(c', x)$ *for* $x \in X_2 \setminus \{x_f\}$.

Proof The proof of the theorem follows from Theorem 3.45 if we regard the acyclic l-game as an acyclic c-game on the network $(G, X_1, X_2, c', x_0, x_f)$ with the cost function $c' = c - \overline{v}(x_0)$. □

Corollary 3.58 *The difference* $\varepsilon_x - \varepsilon_{x_0}$, $x \in X$ *represents the costs of the max-min path from* x *to* x_f *in the acyclic c-game on the network* $(G, X_1, X_2, c', x_0, x_f)$ *with* $c'_{(x,y)} = c_{(x,y)} - \overline{v}(x_0), \ \forall (x, y) \in E$.

3.12.3 An Algorithm for Solving Acyclic l-Games

The algorithm, which is described below, is based on the results from Sect. 3.12.2. In this algorithm, we use the following properties:

1. The value $\overline{v}(x_0)$ of an acyclic l-game on the network $(G, X_1, X_2, c, x_0, x_f)$ is non-negative if and only if the value $v(x_0)$ of an acyclic l-game on the network

$(G, X_1, X_2, c, x_0, x_f)$ is non-negative; moreover, $\overline{v}(x_0) = 0$ if and only if $v(x_0) = 0$.

2. If $M^1 = \min_{e \in E} c_e$ and $M^2 = \max_{e \in E} c_e$, then $M^1 \leq \overline{v}(x_0) \leq M^2$.
3. If on the network $(G, X_1, X_2, c, x_0, x_f)$ the cost function $c : E \to \mathbb{R}$ is changed to the function $c^h : E \to \mathbb{R}$, where

$$c_e^h = c_e - h, \ \forall e \in E \tag{3.74}$$

(h is an arbitrary constant), then the acyclic l-games on $(G, X_1, X_2, c, x_0, x_f)$ and $(G, X_1, X_2, c^h, x_0, x_f)$ have the same optimal strategies s^{1^*} and s^{2^*}, respectively. In addition, the values $\overline{v}(x_0)$ and $\overline{v}_h(x_0)$ of these games differ by a constant h: $\overline{v}_h(x_0) = \overline{v}(x_0) - h$. So, the acyclic l-games on $(G, X_1, X_2, c, x_0, x_f)$ and $(G, X_1, X_2, c^h, x_0, x_f)$ are equivalent.

According to the properties mentioned above, if $\overline{v}(x_0)$ is known, then the acyclic l-game can be reduced to the acyclic c-game by using transformation (3.69) with $h = \overline{v}(x_0)$. After that, we can find the optimal strategies in the game with network $(G, X_1, X_2, c^h, x_0, x_f)$ by using Algorithm 3.46. The most important phase in the proposed algorithm is represented by the problem of finding the value h for which $\overline{v}_h(x_0) = 0$. Taking into account properties 1 and 2, we look for this value by using the dichotomy method on segment $[M^1, M^2]$ such that at each step of this method, we solve a dynamic c-game with network $(G, X_1, X_2, c^k, x_0, x_f)$, where $c^k = c - h_k$. The main idea of the general step of the algorithm is the following: We make a transformation (3.74) with $h = h_k$, where h_k is a midpoint of the segment $[M_k^1, M_k^2]$ at step k. After that, we apply Algorithm 3.46 for the dynamic c-game on network $(G, X_1, X_2, c^{h_k}, x_0, x_f)$ and find $v_{h_k}(x_0)$. If $v_{h_k}(x_0) > 0$, then we fix segment $[M_{k+1}^1, M_{k+1}^2]$, where $M_{k+1}^1 = M_k^1$ and $M_{k+1}^2 = \left(M_k^1 + M_k^2\right)/2$; otherwise, we put $M_{k+1}^1 = \left(M_k^1 + M_k^2\right)/2$ and $M_{k+1}^2 = M_k^2$. If $v_{h_k}(x_0) = 0$, then stop. In the following, there is a detailed description of the algorithm.

Algorithm 3.59 Determining the Value and Optimal Strategies of the Players for the Acyclic l-Game

Let us assume that the cost function $c : E \to \mathbb{R}$ is integer and $\max_{e \in E} |c_e| \neq 0$.

Preliminary step (step 0): Find the value $v(x_0)$ and optimal strategies s^{1^*} and s^{2^*} of the dynamic c-game on $(G, X_1, X_2, c, x_0, x_f)$ by using Algorithm 3.46. If $v(x_0) = 0$, then fix s^{1^*} and s^{2^*} as the solution to the l-game, put $\overline{v}(x_0) = 0$, and stop; otherwise, fix $M_1^1 = \min_{e \in E} c_e$, $M_1^2 = \max_{e \in E} c_e$, $L = \max_{e \in E} |c_e| + 1$.

General step (step k, $k \geq 1$): Find $h_k = \left(M_k^1 + M_k^2\right)/2$, and make the transformation of the edges' costs

$$c_e^k = c_e - h_k \quad \text{for} \quad e \in E.$$

Solve the dynamic c-game on the network $(G, X_1, X_2, c^k, x_0, x_f)$, and find the value $v_k(x_0)$ and the optimal strategies s^{1^*} and s^{2^*}.

If $v_k(x_0) = 0$, then fix the optimal strategies s^{1^*} and s^{2^*}, and put $\overline{v}(x_0) = h_k$. If $|v_k(x_0)| \leq 1/\left(4|X|^2 L\right)$, then fix s^{1^*} and s^{2^*}, find $\overline{v}(x_0) = \overline{H}_{x_0}(s^{1^*}, s^{2^*})/n(P_{s^*}(x_0, x_f))$, and stop. If $v_k(x_0) > 1/\left(4|X|^2 L\right)$, then fix $M^1_{k+1} = M^1_k$, $M^2_{k+1} = h_k$, and go to step $k+1$. If $v_k(x_0) < -1/\left(4|X|^2 L\right)$, then fix $M^1_{k+1} = h_k$, $M^2_{k+1} = M^2_k$, and go to step $k+1$.

Theorem 3.60 *Let $(G, X_1, X_2, c, x_0, x_f)$ be a network with an integer cost function $c : E \to \mathbb{R}$ and $L = \max_{e \in E} |c_e|$. Then Algorithm 3.59 correctly finds the value $\overline{v}(x_0)$ and optimal strategies s^{1^*} and s^{2^*} in the acyclic l-game. The running time of the algorithm is $O(|X|^2 \log L + 2|X|^2 \log |X|)$.*

Proof Let $(G, X_1, X_2, c^k, x_0, x_f)$ be a network after the final step k of Algorithm 3.59. If we solve the corresponding acyclic c-game, then

$$|v_k(x_0)| \leq \frac{1}{4|X|^2 L}$$

and the values ε^k_x, $x \in X$, determined according to Algorithm 3.46, represent the approximate solution to the system

$$\begin{cases} \varepsilon_y - \varepsilon_x + c^k_{(x,y)} \leq 0 & \text{for} \quad x \in X_1, \ (x, y) \in E; \\ \varepsilon_y - \varepsilon_x + c^k_{(x,y)} \geq 0 & \text{for} \quad x \in X_2, \ (x, y) \in E; \\ \varepsilon_{x_0} = \varepsilon_{x_f}. \end{cases}$$

This means that ε^k_x, $x \in X$, and h_k represents the approximate solution to the system

$$\begin{cases} \varepsilon_y - \varepsilon_x + c_{(x,y)} \leq h_k & \text{for} \quad x \in X_1, \ (x, y) \in E; \\ \varepsilon_y - \varepsilon_x + c_{(x,y)} \geq h_k & \text{for} \quad x \in X_2, \ (x, y) \in E; \\ \varepsilon_{x_0} = \varepsilon_{x_f}. \end{cases}$$

According to [84, 85], the exact solution $h = \overline{v}(x)$, ε_x, $x \in X$ of this system can be obtained from h_k, ε^k_x, $x \in X$ by using the special round-off method in time $O(\log(L + 1))$. Therefore, the strategies s^{1^*} and s^{2^*} after the final step k of the algorithm correspond to the optimal solution to the acyclic l-game.

Taking into account that the tabulation of the values $\varepsilon(x)$ and $x \in X$ in G needs $O(|X|^2)$ operations, and the number of iterations of the algorithm is $O(\log L + 2 \log |X|)$, we can observe that the running time of the algorithm is $O(|X|^2 \log L + 2|X|^2 \log |X|)$. □

3.13 Determining the Optimal Strategies for Cyclic Games

Cyclic games were studied in [42, 67, 80, 96, 98, 138, 202]. Later in [99, 112, 116], this game was used for studying the game variant of the infinite horizon discrete control problem with an average cost criterion by a trajectory. Here, we show that the problem of finding optimal strategies of the players in a cyclic game is tightly connected to the problem of finding optimal strategies of the players in the dynamic c-game and the acyclic l-game. Based on these results, we propose algorithms to determine the value and the optimal strategies in a cyclic game.

3.13.1 Problem Formulation and the Main Properties

Let $G = (X, E)$ be a finite directed graph in which every vertex $x \in X$ has at least one leaving edge $e = (x, y) \in E$. On the edge set E, a function $c\colon E \to \mathbb{R}$ is given, which assigns a cost c_e to each edge $e \in E$. In addition, the vertex set X is divided into two disjoint subsets X_1 and X_2 ($X = X_1 \cup X_2$, $X_1 \cap X_2 = \emptyset$), which we regard as position sets of the two players. On G, we consider the following two-person game from [42, 67, 112, 186, 202]: The game starts at position $x_0 \in X$. If $x_0 \in X_1$, then the move is done by the first player; otherwise, it is done by the second one. This move means that it passes from position x_0 to the neighboring position x_1 through the edge $e_1 = (x_0, x_1) \in E$. After that, if $x_1 \in X_1$, then the move is done by the first player; otherwise, it is done by the second one and so on, indefinitely. The first player has the aim to maximize $\lim_{t\to\infty} \inf \frac{1}{t} \sum_{i=1}^{t} c_{e_i}$, while the second player has the aim to minimize $\lim_{t\to\infty} \sup \frac{1}{t} \sum_{i=1}^{t} c_{e_i}$. In [42], it was proved that for this game, there exists a value $\overline{v}(x_0)$ such that the first player has a strategy of moves that ensures $\lim_{t\to\infty} \inf \frac{1}{t} \sum_{i=1}^{t} c_{e_i} \geq \overline{v}(x_0)$ and the second player has a strategy of moves that ensures $\lim_{t\to\infty} \sup \frac{1}{t} \sum_{i=1}^{t} c_{e_i} \leq \overline{v}(x_0)$. Furthermore, in [42], it was shown that the players can achieve the value $\overline{v}(x_0)$ by applying the strategies of moves that do not depend on t. This means that the considered game can be formulated in terms of stationary strategies. Such a statement of the game in [67] is called a *cyclic game*. In [186, 202], this game is called *parity game*.

The strategies of the players in the cyclic game are defined as maps

$$s^1\colon x \to y \in X(x) \quad \text{for} \quad x \in X_1;$$
$$s^2\colon x \to y \in X(x) \quad \text{for} \quad x \in X_2,$$

where $X(x) = \{y \in X \mid e = (x, y) \in E\}$.

Since G is a finite graph, the sets of strategies of the players

$$\mathbb{S}^1 = \{s^1 : x \to y \in X(x) \quad \text{for} \quad x \in X_1\};$$
$$\mathbb{S}^2 = \{s^2 : x \to y \in X(x) \quad \text{for} \quad x \in X_2\}$$

are finite sets.

The payoff function $\overline{H}_{x_0} : \mathbb{S}^1 \times \mathbb{S}^2 \to \mathbb{R}$ in the cyclic game is defined as follows:

Let $s^1 \in \mathbb{S}^1$ and $s^2 \in \mathbb{S}^2$ be fixed strategies of the players. Denote by $G_s = (X, E_s)$ the subgraph of G generated by the edges of the form $(x, s^1(x))$ for $x \in X_1$ and $(x, s^2(x))$ for $x \in X_2$. Then G_s contains a unique directed cycle C_s, which can be reached from x_0 through the edges $e \in E_s$. We consider that the value $\overline{H}_{x_0}(s^1, s^2)$ is equal to the mean edges cost of cycle C_s, i.e.,

$$\overline{H}_{x_0}(s^1, s^2) = \frac{1}{n(C_s)} \sum_{e \in E(C_s)} c_e,$$

where $E(C_s)$ represents the set of edges of cycle C_s and $n(C_s)$ is a number of the edges of C_s. So, the cyclic game is uniquely determined by the network (G, X_1, X_2, c, x_0), where x_0 is the given starting position of the game. If we consider the problem of finding the optimal strategies of the players for an arbitrary starting position $x \in X$, then we will use the notation (G, X_1, X_2, c). In [42, 67], it was proved that there exist the strategies $s^{1^*} \in S^1$ and $s^{2^*} \in S^2$ such that

$$\begin{aligned} \overline{v}(x) = \overline{H}_x(s^{1^*}, s^{2^*}) &= \max_{s^1 \in \mathbb{S}^1} \min_{s^2 \in \mathbb{S}^2} \overline{H}_x(s^1, s^2) \\ &= \min_{s^2 \in \mathbb{S}^2} \max_{s^1 \in \mathbb{S}^1} \overline{H}_x(s^1, s^2), \quad \forall x \in X. \end{aligned}$$

So, the optimal strategies s^{1^*}, s^{2^*} of the players in cyclic games do not depend on a starting position x_0, although for different positions $x, y \in X$, the values $\overline{v}(x)$ and $\overline{v}(y)$ may be different. It means that the positions set X can be divided into several classes $X = X^1 \cup X^2 \cup \cdots \cup X^k$ according to the values of positions $\overline{v}^1, \overline{v}^2, \ldots, \overline{v}^k$, i.e., $x, y \in X^i$ if and only if $\overline{v}^i = \overline{v}(x) = \overline{v}(y)$. In the case $k = 1$, the network (G, X_1, X_2, c) is called the *ergodic network* [67]. In [112, 116], it was shown that every cyclic game with an arbitrary network (G, X_1, X_2, c, x_0) with given starting position x_0 can be reduced to a cyclic game on an auxiliary ergodic network (G', X'_A, X'_B, c'). It is well known [80, 186, 202] that the decision problem associated with the cyclic game is in $NP \cap$ co-NP. Some exponential and pseudo-polynomial algorithms for finding the value and the optimal strategies of the players in cyclic games were proposed in [202]. The computational complexity of the problem of determining the optimal stationary strategies for stochastic games was studied in [32]. Our aim is to propose polynomial-time algorithms for determining optimal strategies of the players in cyclic games. We argue such algorithms based on the results that were announced in [116, 118].

3.13.2 Some Preliminary Results

First, we need to remember some preliminary results from [67, 99, 109, 112, 116, 118]. Let (G, X_1, X_2, c) be a network with the properties described in Sect. 3.13.1. In an analogous way, as it is in the case for dynamic c-games, here, we denote

$$\text{ext}(c, x) = \begin{cases} \max\limits_{y \in X(x)} c_{(x,y)} & \text{for } x \in X_1, \\ \min\limits_{y \in X(x)} c_{(x,y)} & \text{for } x \in X_2, \end{cases}$$

$$\text{VEXT}(c, x) = \{y \in X(x) \mid c_{(x,y)} = \text{ext}(c, x)\}.$$

We use the potential transformation $c'_{(x,y)} = c_{(x,y)} + \varepsilon(y) - \varepsilon(x)$ for costs on the edges $e = (x, y) \in E$, where $\varepsilon\colon X \to \mathbb{R}$ is an arbitrary function on the vertex set X. In [67], it was shown that the potential transformation does change neither the value nor the optimal strategies of the players in cyclic games.

Theorem 3.61 *Let (G, X_1, X_2, c) be an arbitrary network with the properties described in Sect.* 3.13.1. *Then there exists the value $\overline{v}(x)$, $x \in X$, and the function $\varepsilon\colon X \to \mathbb{R}$ determines the potential transformation $c'_{(x,y)} = c_{(x,y)} + \varepsilon(y) - \varepsilon(x)$ for the costs on the edges $e = (x, y) \in E$ such that the following properties hold:*

(a) $\overline{v}(x) = \text{ext}(c', x)$ *for* $x \in X$.
(b) $\overline{v}(x) = \overline{v}(y)$ *for* $x \in X_1 \cup X_2$ *and* $y \in \text{VEXT}(c', x)$.
(c) $\overline{v}(x) \geq \overline{v}(y)$ *for* $x \in X_1$ *and* $y \in X_G(x)$.
(d) $\overline{v}(x) \leq \overline{v}(y)$ *for* $x \in X_2$ *and* $y \in X_G(x)$.
(e) $\max\limits_{e \in E} |c'_e| \leq 2|X| \max\limits_{e \in E} |c_e|$.

The values $\overline{v}(x)$, $x \in X$ on the network (G, X_1, X_2, c) are uniquely determined, and the optimal strategies of the players can be found in the following way: Fix the arbitrary strategies $s^{1*}\colon X_1 \to X$ and $s^{2*}\colon X_2 \to X$ such that $s^{1*}(x) \in \text{VEXT}(c', x)$ for $x \in X_1$ and $s^{2*}(x) \in \text{VEXT}(c', x)$ for $x \in X_2$.

This theorem follows from Theorem 3.27. In [67], a constructive proof of Theorem 3.61 was given. In [112], it was shown that the conditions of this theorem can be obtained from the continuous optimal mean cost cycle problem in a weighted directed graph. Furthermore, we use Theorem 3.61 in the case of the ergodic network (G, X_1, X_2, c), i.e., we use the following corollary:

Corollary 3.62 *Let (G, X_1, X_2, c) be an ergodic network. Then there exists the value $\overline{v}$, and the function $\varepsilon\colon X \to \mathbb{R}$ determines the potential transformation $c'_{(x,y)} = c_{(x,y)} + \varepsilon_y - \varepsilon_x$ for the costs of the edges $e = (x, y) \in E$ such that $\overline{v} = \text{ext}(c', x)$ for $x \in X$. The optimal strategies of the players can be found as follows: Fix arbitrary strategies $s^{1*}\colon X_1 \to X$ and $s^{2*}\colon X_2 \to X$ such that $s^{1*}(x) \in \text{VEXT}(c', x)$ for $x \in X_1$ and $s^{2*}(x) \in \text{VEXT}(c', x)$ for $x \in X_2$.*

3.13.3 The Reduction of Cyclic Games to Ergodic Ones

Let us consider an arbitrary network (G, X_1, X_2, c, x_0) with a given starting position $x_0 \in X$ that determines a cyclic game. In [112, 116, 118], it was shown that this game can be reduced to a cyclic game on an auxiliary ergodic network $(G', W_1, W_2, \overline{c})$, $G' = (W, E')$ with the same value $\overline{v}(x_0)$ of the game as the initial one, where $x_0 \in W = X \cup U \cup Z$.

The graph $G' = (W, E')$ is obtained from G if each edge $e = (x, y)$ is changed by a triple of edges $e^1 = (x, u)$, $e^2 = (u, z)$, $e^3 = (z, y)$ with the costs $\overline{c}_{e^1} = \overline{c}_{e^2} = \overline{c}_{e^3} = c_e$. Here, $u \in U$, $z \in Z$, and $x, y \in X$; $W = X \cup U \cup Z$. In addition, in G', each vertex u is connected with x_0 by the edge (u, x_0) with the cost $\overline{c}_{(u,x_0)} = M$ (M is of great value), and each vertex z is connected with x_0 by the edge (z, x_0) with the cost $\overline{c}_{(z,x_0)} = -M$. In $(G', W_1, W_2, \overline{c})$, the sets W_1 and W_2 are defined as follows: $W_1 = X_1 \cup Z$; $W_2 = X_2 \cup U$.

It is easy to observe that this reduction can be done in linear time.

3.13.4 An Algorithm for Ergodic Cyclic Games

Let us consider an ergodic zero-value cyclic game determined by the network (G, X_1, X_2, c, x_0), where $G = (X, E)$. Then according to Theorem 3.61, there exists the function $\varepsilon : X \to \mathbb{R}$ that determines the potential transformation $c'_{(x,y)} = c_{(x,y)} + \varepsilon_y - \varepsilon_x$ on the edges $(x, y) \in E$ such that

$$\text{ext}(c, x) = 0, \quad \forall x \in X. \tag{3.75}$$

This means that if x_f is a vertex of the cycle C_{s^*} determined by optimal strategies s^{1*} and s^{2*}, then the problem of finding the function $\varepsilon : X \to \mathbb{R}$, which determines the canonic potential transformation, is equivalent to the problem of finding the values ε_x, $x \in X$ in a max-min path problem on G with the sink vertex x_f, where $\varepsilon_{x_f} = 0$.

So, in order to solve the zero-value cyclic game, we fix a vertex $x \in X$ as a sink vertex ($x_f = x$) at each time step and solve a max-min path problem on G with the sink vertex x_f. If for given $x_f = x$ the function $\varepsilon : X \to \mathbb{R}$ (obtained on the basis of the algorithm from Sects. 3.11.3 and 3.11.4) determines the potential transformation that satisfies (3.75), then we fix s^{1*} and s^{2*} such that $s^{1*}(x) \in \text{VEXT}(c', x)$ for $x \in X_1$ and $s^{2*}(x) \in \text{VEXT}(c', x)$ for $x \in X_2$. If for given x the function $\varepsilon : X \to \mathbb{R}$ does not satisfy (3.75), then we select another vertex $x \in X$ as a sink vertex and so on. This means that the optimal strategies of the players in zero-value ergodic cyclic games can be found in time $O(|X|^4)$.

Example Consider the ergodic zero-sum cyclic game determined by the network given in Fig. 3.15 with starting position $x_0 = 0$. Positions of the first player are represented by circles, and positions of the second player are represented by squares, i.e., $X_1 = \{1, 2, 4, 5\}$ and $X_2 = \{0, 3\}$.

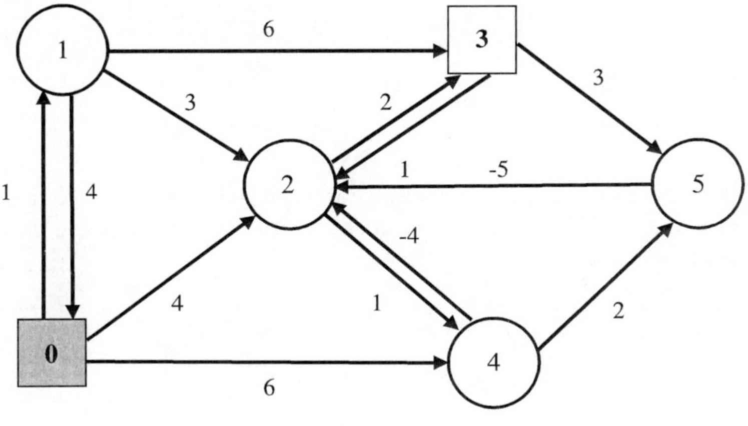

Fig. 3.15 Network for the ergodic zero-sum cyclic game

The network in Fig. 3.15 is obtained from the network in Fig. 3.15 by adding the edge $(5, 2)$ with the cost $c_{(5,2)} = -5$. It is easy to check that the value of a cyclic game on this network for an arbitrary fixed starting position is equal to zero. The max-min mean cycle, which determines a way at zero cost, is $2 \to 3 \to 5 \to 2$. Therefore, if we fix a vertex of this cycle as a sink vertex (e.g., $x = 5$), then we can find the potential function $\varepsilon : X \to \mathbb{R}$, which determines the potential transformation $c'_{(x,y)} = \varepsilon_y - \varepsilon_x + c_{(x,y)}$ such that $ext(c', x) = 0, \forall x \in X$. This function $\varepsilon : X \to \mathbb{R}$ can be found by applying the algorithm from the examples in Sects. 3.11.3 and 3.11.4, i.e., we find the costs of min-max paths from every $x \in X$ to vertex 5. So, $\varepsilon_0 = 8$, $\varepsilon_1 = 12$, $\varepsilon_2 = 5$, $\varepsilon_3 = 3$, $\varepsilon_4 = 2$, and $\varepsilon_5 = 0$. After the potential transformation, we obtain the network with the following costs of edges:

$$c'_{(3,5)} = \varepsilon_5 - \varepsilon_3 + c_{(3,5)} = 0 - 3 + 3 = 0.$$

$$c'_{(4,5)} = \varepsilon_5 - \varepsilon_4 + c_{(4,5)} = 0 - 2 + 2 = 0.$$

$$c'_{(5,2)} = \varepsilon_2 - \varepsilon_5 + c_{(5,2)} = 5 - 0 - 5 = 0.$$

$$c'_{(2,3)} = \varepsilon_3 - \varepsilon_2 + c_{(2,3)} = 3 - 5 + 2 = 0.$$

$$c'_{(3,2)} = \varepsilon_2 - \varepsilon_3 + c_{(3,2)} = 5 - 3 + 1 = 3.$$

$$c'_{(4,2)} = \varepsilon_2 - \varepsilon_4 + c_{(4,2)} = 5 - 2 - 4 = -1.$$

$$c'_{(2,4)} = \varepsilon_4 - \varepsilon_2 + c_{(2,4)} = 2 - 5 + 1 = -2.$$

$$c'_{(0,4)} = \varepsilon_4 - \varepsilon_0 + c_{(0,4)} = 2 - 8 + 6 = 0.$$

$$c'_{(0,2)} = \varepsilon_2 - \varepsilon_0 + c_{(0,2)} = 5 - 8 + 4 = 1.$$

$$c'_{(1,3)} = \varepsilon_3 - \varepsilon_1 + c_{(1,3)} = 3 - 12 + 6 = -3.$$

$$c'_{(1,2)} = \varepsilon_2 - \varepsilon_1 + c_{(1,2)} = 5 - 12 + 3 = -4.$$
$$c'_{(1,0)} = \varepsilon_0 - \varepsilon_1 + c_{(1,0)} = 8 - 12 + 4 = 0.$$
$$c'_{(0,1)} = \varepsilon_1 - \varepsilon_0 + c_{(0,1)} = 12 - 8 + 1 = 5.$$

The network after the potential transformation is given in Fig. 3.16. We can see that $\text{ext}(c', x) = 0, \forall x \in X$. Therefore, the edges with zero cost determine the optimal strategies of the players:

$$s^{1^*} : 1 \to 0;\ \ 2 \to 3;\ \ 4 \to 5;\ \ 5 \to 2;$$
$$s^{2^*} : 0 \to 4;\ \ 3 \to 5.$$

The graph $G_{s^*} = (X, E_{s^*})$ generated by these strategies is represented in Fig. 3.17.

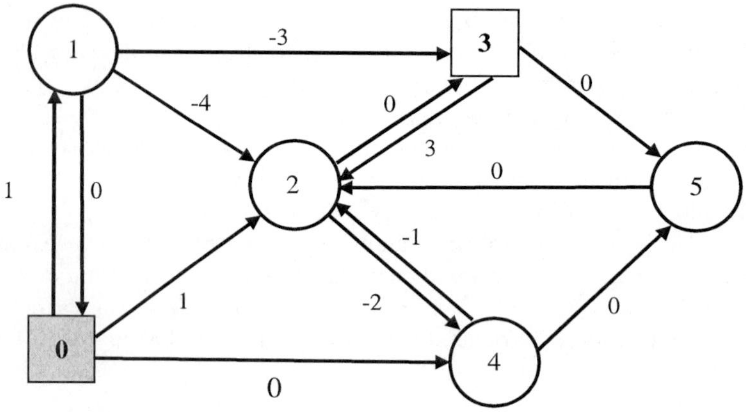

Fig. 3.16 Network after potential transformation

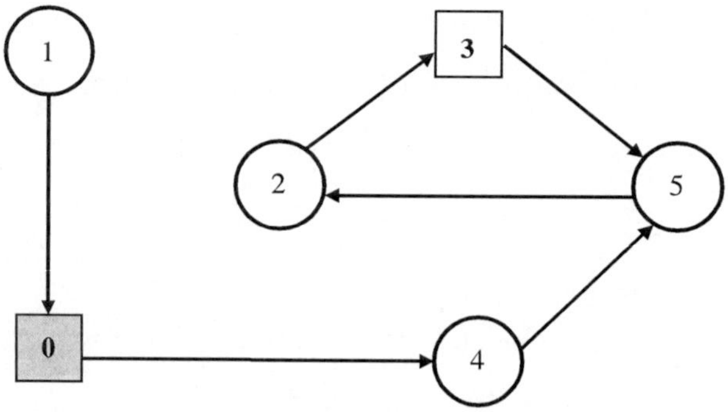

Fig. 3.17 Network induced by optimal strategies of the players

3.13.5 An Algorithm Based on the Reduction of Acyclic l-Games

Based on the results from previous sections, we can propose the following algorithm for determining the optimal strategies of the players in an ergodic cyclic game: We consider an acyclic game on the ergodic network (G, X_1, X_2, c, x_0) with the given starting position x_0. The graph $G = (X, E)$ is considered to be strongly connected and $X = \{x_0, x_1, x_2, \dots, x_{n-1}\}$. Assume that x_0 belongs to the cycle C_{s^*} determined by the optimal strategies of the players s^{1^*} and s^{2^*}. If there are several of these cycles in G, we consider one of them with the minimum number of edges.

We construct an auxiliary acyclic graph $GT_r = (\overline{W}_r, \overline{E}_r)$, where

$$\overline{W}_r = \left\{w_0^0\right\} \cup W^1 \cup W^2 \cup \dots \cup W^r, \quad W^i \cap W^j = \emptyset, \ i \neq j$$

$$W^i = \left\{w_0^i, w_1^i, \dots, w_{n-1}^i\right\}, \quad i = 1, 2, \dots, r$$

$$\overline{E}_r = E^0 \cup E^1 \cup E^2 \cup \dots \cup E^{r-1}$$

$$E^i = \left\{\left(w_k^{i+1}, w_l^i\right) \mid (x_k, x_l) \in E\right\}, \quad i = 1, 2, \dots, r-1$$

$$E^0 = \left\{\left(w_k^i, w_0^0\right) \mid (x_k, x_0) \in E, \quad i = 1, 2, \dots, r\right\}$$

The vertex set $\overline{W}_r$ of GT_r is obtained from X if it is doubled r times and then a sink vertex w_0^0 is added. The edge subset $E^i \subseteq \overline{E}$ in GT_r connects the vertices of set W^{i+1} and the vertices of set W^i in the following way:

If there exists an edge $(x_k, x_l) \in E$ in G, then in GT_r we add the edge (w_k^{i+1}, w_l^i). The edge subset $E^0 \subseteq \overline{E}$ in GT_r connects the vertices $w_k^i \in W^1 \cup W^2 \cup \dots \cup W^r$ with the sink vertex w_0^0, i.e., if there exists an edge $(x_k, x_0) \in E$, then we add the edges $(w_k^i, w_0^0) \in E^0$, $i = 1, 2, \dots, r$ in GT_r.

After that, we define the acyclic network $(GT_r', W_1, W_2, c', w_0^0)$, $GT_r' = (W_r, E_r)$, where GT_r' is obtained from GT_r by deleting the vertices $w_k^i \in \overline{W}_r$ from which the vertex w_0^0 cannot be attainable. The sets W_1, W_2 and the cost function $c' \colon E_r \to \mathbb{R}$ are defined as follows:

$$W_1 = \left\{w_k^i \in W_r \mid x_k \in X_1\right\}, \quad W_2 = \left\{w_k^i \in W_r \mid x_k \in X_2\right\}.$$

$$c'_{(w_k^{i+1}, w_l^i)} = c_{(x_k, x_l)} \text{ if } (x_k, x_l) \in E \text{ and } \left(w_k^{i+1}, w_l^i\right) \in E^i; \ i = 1, 2, \dots, r-1.$$

$$c'_{(w_k^i, w_0^0)} = c_{(x_k, x_0)} \text{ if } (x_k, x_0) \in E \text{ and } \left(w_k^i, w_0^0\right) \in E^0; \ i = 1, 2, \dots, r.$$

Now we consider the acyclic c-game on the acyclic network $(GT_r', W_1, W_2, c', w_0^r, w_0^0)$ with the sink vertex w_0^0 and the starting position w_0^r.

Lemma 3.63 *Let $\overline{v} = \overline{v}(x_0)$ be a value of the ergodic cyclic game on G, and the number of edges of the max-min cycle C_{s^*} in G is equal to r. In addition, let $\overline{v}_r(w_0^r)$ be the value of the l-game on (GT_r', W_1, W_2, c') with the starting position w_0^r. Then $\overline{v}(x_0) = \overline{v}_r(w_0^r)$.*

Proof It is evident that there exists a bijective mapping between the set of cycles with no more than r edges (which contains the vertex x_0) in G and the set of directed paths with no more than r edges from w_0^r to w_0^0 in GT_r'. Therefore, $\overline{v}(x_0) = \overline{v}_r(w_0^r)$. □

Based on this lemma, we can propose the following algorithm for finding the optimal strategies of the players in cyclic games:

Algorithm 3.64 Determining the Optimal Stationary Strategies of the Players in Cyclic Games with a Known Vertex of a Max-Min Cycle of the Network

We construct the acyclic networks (GT_r', W_1, W_2, c'), $r = 2, 3, \ldots, n$ and for each of them solve an l-game. In such a way, we find the values $\overline{v}_2(w_0^2)$, $\overline{v}_3(w_0^3)$, $\ldots$, $\overline{v}_n(w_0^n)$ for these l-games.

Then we consecutively fix $\overline{v} = \overline{v}_2(w_0^2), \overline{v}_3(w_0^3), \ldots, \overline{v}_n(w_0^n)$ and each time solve the c-game on the network (G, X_1, X_2, c''), where $c'' = c - \overline{v}$. By fixing the values $\varepsilon'(x_k) = v(x_k)$ for $x_k \in X$ each time, we check if the condition

$$\text{ext}(c^r, x_k) = 0, \quad \forall x_k \in X$$

is satisfied, where $c^r_{(x_k, x_l)} = c''_{(x_k, x_l)} + \varepsilon(x_l) - \varepsilon(x_k)$. We find r for which this condition holds and fix the respective maps s^{1^*} and s^{2^*} such that $s^{1^*}(x_k) \in \text{VEXT}(c'', x_k)$ for $x_k \in X_1$ and $s^{2^*}(x_k) \in \text{VEXT}(c'', x_k)$ for $x_k \in X_2$. So, s^{1^*} and s^{2^*} represent the optimal strategies of the players in cyclic games on G.

Remark 3.65 Algorithm 3.64 finds the value $\overline{v}(x_0)$ and optimal strategies of the players in time $O(|X|^5 \log L + 4|X|^3 \log |X|)$ because Algorithm 3.59 needs $O(|X|^4 \log L + 4|X|^2 \log |X|)$ elementary operations for solving an acyclic l-game on the network (GT_r', W_1, W_2, c'), where $L = \max_{e \in E} |c_e| + 1$.

In the general case, if the belonging of x_0 to the max-min cycle is unknown, then we use the following algorithm:

Algorithm 3.66 Determining the Optimal Strategies of the Players in Ergodic Cyclic Games (General Case)

Preliminary step (step 0): Fix $Y_1 = X$.

General step (step k): Select a vertex $y \in Y_1$, fix $x_0 = y$, and apply Algorithm 3.64. If there exists $r \in \{2, 3, \ldots, n\}$ such that $\text{ext}(c^r, x) = 0$, $\forall x \in X$, then fix $s^{1^*} \in \text{VEXT}(c^k, x)$ for $x \in X_1$ and $s^{2^*} \in \text{VEXT}(c^k, x)$ for $x \in X_2$, and stop; otherwise, put $Y_{k+1} = Y_k \setminus \{y\}$, and go to the next step $k + 1$.

Remark 3.67 Algorithm 3.66 finds the value $\overline{v}$ and optimal strategies of the players in time $O(|X|^6 \log L + 4|X|^4 \log |X|)$ because in the worst case, Algorithm 3.64 is repeated $|X|$ times.

The algorithm for solving cyclic games allows us to determine the sign value $v(x_0)$ in an infinite dynamic c-game on G with starting position x_0. In order to determine $\text{sign}(v(x_0))$, we solve a cyclic game with starting position x_0 and determine $\overline{v}(x_0)$. Then $\text{sign}(v(x_0)) = \text{sign}(\overline{v}(x_0))$.

3.13.6 A Dichotomy Method for Cyclic Games

In this section, we describe an approach for solving cyclic games considering that there exist efficient algorithms for solving a dynamic c-game (including infinite dynamic c-games).

Consider the ergodic cyclic game determined by the ergodic network (G, X_1, X_2, c, x_0), where the value of the game may be different from zero. Graph G is assumed to be strongly connected.

First, we show how to determine the value of the game and optimal strategies of the players in case the vertex x_0 belongs to a max-min cycle in G induced by optimal strategies of the players.

With our ergodic cyclic game, we associate a dynamic c-game determined by an auxiliary network $(\overline{G}, X_1, X_2 \cup \{x_0'\}, \overline{c}, x_0, x_0')$, where graph $\overline{G} = (X \cup \cup \{x_0'\}, \overline{E})$ is obtained from G by adding a copy x_0' of vertex x_0 together with copies $e' = (x, x_0')$ of the edges $e = (x, x_0) \in E$ with costs $\overline{c}_{e'} = c_e$. So, for x_0', there are no leaving edges (x_0', x). It is evident that if the value $\overline{v} = \overline{v}(x_0)$ of the ergodic cyclic game on (G, X_1, X_2, c, x_0) is known, then the problem of finding the optimal strategies of the players is equivalent to the problem of finding optimal strategies of the players in a dynamic c-game on the network $(\overline{G}, X_1, X_2 \cup \{x_0'\}, \overline{c}', x_0, x_0')$ with the cost functions

$$\overline{c}_e' = \overline{c}_e - \overline{v}(x_0) \quad \text{for} \quad e \in \overline{E}.$$

Moreover, if s^{1*} and s^{2*} are optimal strategies of the players in the dynamic c-game on $(\overline{G}, X_1, X_2 \cup \{x_0'\}, \overline{c}', x_0, x_0')$, then the optimal strategies $\overline{s}_A^*$ and $\overline{s}_B^*$ of the players in the ergodic cyclic game can be found as follows:

$$\overline{s}_1^*(x) = s^1(x) \quad \text{for} \quad x \in X_1 \quad \text{if} \quad s^1(x) \neq x_0';$$

$$\overline{s}_2^*(x) = s^2(x) \quad \text{for} \quad x \in X_2 \quad \text{if} \quad s^2(x) \neq x_0'$$

and

$$\overline{s}_1^*(x) = x_0 \quad \text{if} \quad s^1(x) = x_0';$$
$$\overline{s}_2^*(x) = x_0' \quad \text{if} \quad s^2(x) = x_0'.$$

It is easy to observe that for the considered problems, the following properties hold:

1. The value $\overline{v}(x_0)$ of the ergodic cyclic game on the network (G, X_1, X_2, c, x_0) is non-negative if and only if the value $v(x_0)$ of the dynamic c-game on the network $(\overline{G}, X_1, X_2 \cup \{x_0'\}, \overline{c}, x_0, x_0')$ is non-negative; moreover, $\overline{v}(x_0) = 0$ if and only if $v(x_0) = 0$.
2. If $M^1 = \min_{e \in E} c_e$ and $M^2 = \max_{e \in E} c_e$, then $M^1 \leq \overline{v}(x_0) \leq M^2$.
3. If in the network (G, X_1, X_2, c, x_0) the cost function $c : E \to \mathbb{R}$ is changed to $c' = c + h$, then the optimal strategies of the players in the ergodic cyclic game on the network (G, X_1, X_2, c', x_0) do not change even though the value $\overline{v}(x_0)$ is changed to $\overline{v}'(x_0) = \overline{v}(x_0) + h$.

Based on these properties, we look for the unknown value $\overline{v}(x_0) = v(x)$, which we denote by h, using the dichotomy method on the segment $[M^1, M^2]$ such that at each step of this method, we solve a dynamic c-game with network $(\overline{G}, X_1, X_2 \cup \{x_0'\}, \overline{c}^h, x_0, x_0')$, where $\overline{c}^h = \overline{c} - h$. So, the main idea of the general step of the algorithm is the following: We make the transformation

$$\overline{c}^k = \overline{c} - h_k \quad \text{for} \quad e \in E,$$

where h_k is a midpoint of the segment $[M_k^1, M_k^2]$ at step k.

After that, we apply the algorithm from Sect. 3.11.4 for the dynamic c-game on the network $(\overline{G}, X_1, X_2 \cup \{x_0'\}, \overline{c}^k, x_0, x_0')$ and find $v_{h_k}(x_0)$. If $v_{h_k}(x_0) > 0$, then we fix the segment $[M_{k+1}^1, M_{k+1}^2]$, where $M_{k+1}^1 = M_k^1$ and $M_{k+1}^2 = \left(M_k^1 + M_k^2\right)/2$; otherwise, we put $M_{k+1}^1 = \left(M_k^1 + M_k^2\right)/2$ and $M_{k+1}^2 = M_k^2$. If $v_{h_k}(x_0) = 0$, then stop.

So, using the dichotomy method in an analogous way to the acyclic l-game, we determine the value of the acyclic game. If this value of the dynamic c-game is known, then we determine the strategies of the players by using algorithms from Sect. 3.11.3 or Sect. 3.11.4.

In the case where x_0 does not belong to a max-min cycle determined by optimal strategies of the players in the cyclic game, we solve $|X|$ problems by fixing the starting position $x_0 = x$ for $x \in X$ each time. Then at least for a position $x_0 = x \in X$, we obtain the value of the cyclic game and the optimal strategies of the players.

3.14 Multi-Objective Control Based on the Concept of Non-cooperative Games: Nash Equilibria

In this section, we formulate a multi-objective discrete control model that is based on the dynamic non-cooperative concept from [7].

Consider a dynamical system $\mathbb{L}$ with a finite set of states X, where at every time step t, the state of $\mathbb{L}$ is $x(t) \in X$. For the system $\mathbb{L}$, two states $x_0,\ x_f \in X$ are given, where x_0 represents the starting state of $\mathbb{L}$, i.e., $x_0 = x(0)$, and x_f represents the state into which the dynamical system must be brought, i.e., x_f is the final state of $\mathbb{L}$. We assume that the dynamical system should reach the final state x_f at the moment of time $t(x_f)$ such that $t_1 \leq t(x) \leq t_2$, where t_1 and t_2 are given. The dynamics of the system $\mathbb{L}$ are controlled by m players, and they are described as follows:

$$x(t+1) = g_t(x(t),\ u^1(t),\ u^2(t),\ \ldots,\ u^m(t)), \quad t = 0, 1, 2, \ldots, \tag{3.76}$$

where

$$x(0) = x_0$$

is a starting point of the system $\mathbb{L}$ and $u^i(t) \in \mathbb{R}^{n_i}$ represents the vector of the control parameters of the player $i, i \in \{1, 2, \ldots, m\}$.

The state $x(t+1)$ of the system $\mathbb{L}$ at time step $t+1$ is uniquely obtained if the state $x(t)$ at time step t is known, and the players $1, 2, \ldots, m$ fix their vectors of the control parameters $u^1(t), u^2(t), \ldots, u^m(t)$ independently. For each player $i,\ i \in \{1, 2, \ldots, m\}$, the admissible sets $U_t^i(x(t))$ for the vectors of the control parameters $u^i(t)$ are given, i.e.,

$$u^i(t) \in U_t^i(x(t)), \quad t = 0, 1, 2, \ldots;\ i = 1, 2, \ldots, m. \tag{3.77}$$

We assume that $U_t^i(x(t)),\ t = 0, 1, 2, \ldots;\ \ i = 1, 2, \ldots, m$ are non-empty finite sets and

$$U_t^i(x(t)) \cap U_t^j(x(t)) = \emptyset, \quad i \neq j,\ \ t = 0, 1, 2, \ldots.$$

Let us consider that the players $1, 2, \ldots, m$ fix their vectors of the control parameters

$$u^1(t), u^2(t), \ldots, u^m(t); \quad t = 0, 1, 2, \ldots$$

and the starting state $x(0) = x_0$ and the final state x_f are known. Then for the fixed vectors of the control parameters $u^1(t), u^2(t), \ldots, u^m(t)$, either a unique trajectory

$$x_0 = x(0), x(1), x(2), \ldots, x(t(x_f)) = x_f$$

from x_0 to x_f exists, and $t(x_f)$ represents the time moment when the state x_f is reached, or such a trajectory from x_0 to x_f does not exist.

We denote by

$$F^i_{x_0x_f}(u^1(t), u^2(t), \dots, u^m(t)) = \sum_{t=0}^{t(x_f)-1} c^i_t(x(t), g_t(x(t), u^1(t), u^2(t), \dots, u^m(t)))$$

the integral time cost of the system's transition from x_0 to x_f for player i, $i \in \{1, 2, \dots, m\}$ if the vectors $u^1(t), u^2(t), \dots, u^m(t)$ satisfy condition (3.77) and generate a trajectory

$$x_0 = x(0), x(1), x(2), \dots, x(t(x_f)) = x_f$$

from x_0 to x_f such that

$$t_1 \le t(x_f) \le t_2;$$

otherwise, we put

$$F^i_{x_0x_f}(u^1(t), u^2(t), \dots, u^m(t)) = \infty.$$

Note that $c^i_t(x(t), g_t(x(t), u^1(t), u^2(t), \dots, u^m(t))) = c^i_t(x(t), x(t+1))$ represents the cost of the system's passage from state $x(t)$ to state $x(t+1)$ at the stage $[t, t+1]$ for player i.

Problem 3.68 Find vectors of control parameters

$$u^{1^*}(t), u^{2^*}(t), \dots, u^{i-1^*}(t), u^{i^*}(t), u^{i+1^*}(t), \dots, u^{m^*}(t)$$

that satisfy the condition

$$F^i_{x_0x_f}(u^{1^*}(t), u^{2^*}(t), \dots, u^{i-1^*}(t), u^{i^*}(t), u^{i+1^*}(t), \dots, u^{m^*}(t)) \le$$

$$\le F^i_{x_0x_f}(u^{1^*}(t), u^{2^*}(t), \dots, u^{i-1^*}(t), u^{i}(t), u^{i+1^*}(t), \dots, u^{m^*}(t))$$

$$\forall u^i(t) \in \mathbb{R}^{m_i}, \quad i = 0, 1, 2, \dots, m.$$

So, we consider the problem of finding the solution in the sense of Nash [7, 112, 139, 141].

The problems formulated above can be regarded as mathematical models for dynamical systems controlled by several players who do not inform each other which vectors of control parameters they use in the control process.

An important particular case of Problem 3.68 is represented by the zero-sum control problem of two players with the given costs

$$c_t(x(t),\ x(t+1)) = c_t^2(x(t),\ x(t+1)) = -c_t^1(x(t),\ x(t+1))$$

of the system's passage from state $x(t)$ to state $x(t+1)$, which determine the payoff function

$$F_{x_0x_f}(u^1(t),\ u^2(t)) = F^2_{x_0x_f}(u^1(t),\ u^2(t)) = -F^1_{x_0x_f}(u^1(t),\ u^2(t)).$$

In this case, we look for a saddle point $(u^{1*}(t),\ u^{2*}(t))$ of the function $F_{x_0x_f}(u^1(t),\ u^2(t))$ [142], i.e., we consider the following max-min control problem:

Problem 3.69 Find vectors of control parameters $u^{1^*}(t)$, $u^{2^*}(t)$ such that

$$\begin{aligned} F_{x_0x_f}(u^{1^*}(t),\ u^{2^*}(t)) &= \max_{u^1(t)} \min_{u^2(t)} F_{x_0x_f}(u^1(t),\ u^2(t)) \\ &= \min_{u^2(t)} \max_{u^1(t)} F_{x_0x_f}(u^1(t),\ u^2(t)). \end{aligned}$$

So, for this max-min control problem, we are looking for a saddle point [142].

The results obtained in previous sections allow us to formulate conditions for the existence of Nash equilibria in such dynamic games. Moreover, we describe a class of game-theoretic control problems for which the dynamic programming technique can be used to determine Nash equilibria.

3.14.1 Stationary and Non-stationary Control Models

The multi-objective control model formulated above is related to a non-stationary case. In this model, the functions g_t^i, $t = 0, 1, 2, \ldots$ may be different for different moments of time, and the players in the control process can change their vectors of control parameters for an arbitrary state $x = x(t)$ at different moments of time t. Additionally, for a given state $x = x(t)$, the admissible sets $U_t^i(x(t))$, $i = 1, 2, \ldots, m$ can be different for different moments of time t. Moreover, the costs of the system's transition $c_t(x(t), x(t+1))$ from state $x = x(t)$ to state $y = x(t+1)$ are varying in time for given x and y.

Stationary versions of the considered control problems correspond to the case when the functions g_t^i, $t = 0, 1, 2, \ldots$ do not change in time, i.e., $g_t^i \equiv g^i$, $t = 0, 1, 2, \ldots$, and the players preserve the same vectors of control parameters in time for given states $x \in X$. Additionally, we consider that the admissible sets $U_t^i(x(t))$, $t = 0, 1, 2, \ldots$ for vectors of control parameters do not change in time, i.e., $U_t^i(x(t)) = U^i(x)$, $t = 0, 1, 2, \ldots,\ \ i = 1, 2, \ldots, m$.

In general, for non-stationary control problems, the players can use non-stationary strategies, although the functions g_t^i, $t = 0, 1, 2, \dots$ and the admissible sets of the control parameters $U_t^i(x(t))$, $t = 0, 1, 2, \dots$ may not change in time, i.e., $g_t^i \equiv g^i$, $t = 0, 1, 2, \dots$ and $U_t^i(x(t)) = U^i(x)$, $t = 0, 1, 2, \dots, i = 1, 2, \dots, m$.

3.14.2 Infinite Horizon Multi-Objective Control Problems

For the control problems with an infinite time horizon, the final state is not given, and the control process is made indefinitely for discrete moments of time $t = 0, 1, 2, \dots$. Mainly two classes of multi-objective problems in this topic are considered.

In the first class of problems, each player $i \in \{1, 2, \dots, m\}$ has the aim to minimize his or her own objective function

$$
\begin{aligned}
&F_{x0}^i(u^1(t), u^2(t), \dots, u^m(t)) \\
&\quad = \lim_{\tau \to \infty} \frac{1}{\tau} \sum_{t=0}^{\tau} c_t^i(x(t), g_t(x(t), u^1(t), u^2(t), \dots, u^m(t)))
\end{aligned}
$$

that expresses the average cost per transition by a trajectory determined by all players together.

For the second class, each player $i \in \{1, 2, \dots, m\}$ has to minimize the discounted objective function

$$
\begin{aligned}
&\widehat{F}_{x0}^i(u^1(t), u^2(t), \dots, u^m(t)) \\
&\quad = \sum_{t=0}^{\tau} \gamma^t c_t^i(x(t), g_t(x(t), u^1(t), u^2(t), \dots, u^m(t)))
\end{aligned}
$$

with a given discount factor γ, where $0 < \gamma < 1$.

As noted for the single-objective control problems with infinite time horizon and constant transition costs, there exists the optimal stationary control. Based on the results obtained for positional games, we may derive conditions and algorithms to determine Nash equilibria for the stationary game control problem with average and discounted objective functions.

If $m = 2$ and

$$
c_t^2(x(t), g_t(x(t), u^1(t), u^2(t))) = -c_t^1(x(t), g_t(x(t), u^1(t), u^2(t))),
$$

then we obtain a zero-sum control problem with infinite time horizon. For such a game, we are looking for a saddle point.

3.15 Hierarchical Control and Stackelberg's Optimization Principle

Now we use the concept of hierarchical control introduced in [187] and assume that in (3.76), for an arbitrary state $x(t)$ at every moment of time, the players fix their vectors of control parameters successively one after another according to their numerical order. Moreover, we assume that each player fixing his or her vectors of control parameters informs posterior players which vector of control parameters has been chosen at the given moment of time for a given state. So, we consider the following hierarchical control process:

Let $\mathbb{L}$ be a dynamical system with a finite set of states X and a fixed starting point $x(0) = x_0 \in X$. The dynamics of system $\mathbb{L}$ are defined by the system of difference equations (3.76), and they are controlled by p players using the corresponding vectors of the control parameters $u^1(t), u^2(t), \dots, u^m(t)$.

For each vector of control parameters $u^i(t)$, the feasible set (3.77) is defined for an arbitrary state $x(t)$ at every discrete moment of time t. Additionally, we assume that for an arbitrary state $x(t) \in X$ at every moment of time t, the players fix their vectors of control parameters successively one after another according to a given order. For simplicity, we consider that the players fix their vectors of control parameters in the order corresponding to their numbers. Each player, after fixing his or her vectors of control parameters, informs posterior players which vector of control parameters has been chosen at the given moment of time for a given state. Finally, if the vectors of control parameters $u^1(t), u^2(t), \dots, u^m(t)$ and the starting state $x(0) = x_0$ are known, then the cost $F^i_{x_0x_f}(u^1(t), u^2(t), \dots, u^m(t))$ of the system's passage from the starting state x_0 to the final state x_f for player $i \in \{1, 2, \dots, m\}$ is defined in the same way as in Sect. 3.14.

In this hierarchical control process, we are looking for Stackelberg strategies [111, 112, 187], i.e., we consider the following hierarchical control problem:

Problem 3.70 Find vectors of control parameters $u^{1^*}(t), u^{2^*}(t), \dots, u^{m^*}(t)$, for which

$$u^{1^*}(t) = \operatorname*{argmin}_{\substack{u^1(t)\in U^1 \\ (u^i(t)\in R_i(u^1,\dots,u^{i-1}))_{2\le i\le m}}} F^1_{x_0x_f}(u^1(t), u^2(t), \dots, u^m(t));$$

$$u^{2^*}(t) = \operatorname*{argmin}_{\substack{u^2(t)\in R_2(u^{1^*}) \\ \left(u^i(t)\in R_i(u^{1^*},u^2,\dots,u^{i-1})\right)_{3\le i\le m}}} F^2_{x_0x_f}(u^{1*}(t), u^2(t), \dots, u^m(t));$$

$$u^{3^*}(t) = \operatorname*{argmin}_{\substack{u^3(t)\in R_3(u^{1^*},u^{2^*}) \\ \left(u^i(t)\in R_i(u^{1^*},u^{2^*},u^3,\dots,u^{i-1})\right)_{4\le i\le m}}} F^3_{x_0x_f}(u^{1^*}(t), u^{2^*}(t), \dots, u^m(t));$$

$$\vdots$$

$$u^{m^*}(t) = \operatorname*{argmin}_{u^m(t)\in R_m(u^{1^*},u^{2^*},\ldots,u^{m-1^*})} F^m_{x_0x_f}(u^{1^*}(t), u^{2^*}(t), \ldots, u^{m-1^*}(t), u^m(t)),$$

where $R_k(u^1, u^2, \ldots, u^{k-1})$ represents the best responses of player k if the players $1, 2, \ldots, k-1$ have already fixed their vectors $u^1(t),\ u^2(t),\ \ldots,\ u^{k-1}(t)$, i.e.,

$$R_2(u^1) = \operatorname*{argmin}_{\substack{u^2(t)\in U^2 \\ (u^i(t)\in R_i(u^1,\ldots,u^{i-1}))_{3\le i\le m}}} F^2_{x_0x_f}(u^1(t), u^2(t), \ldots, u^m(t));$$

$$R_3(u^1, u^2) = \operatorname*{argmin}_{\substack{u^3(t)\in U^3 \\ (u^i(t)\in R_i(u^1,\ldots,u^{i-1}))_{4\le i\le m}}} F^3_{x_0x_f}(u^1(t), u^2(t), \ldots, u^m(t));$$

$$\vdots$$

$$R_m(u^1, u^2, \ldots, u^{m-1}) = \operatorname*{argmin}_{u^m(t)\in U^m} F^m_{x_0x_f}(u^1(t),\ u^2(t),\ \ldots,\ u^m(t));$$

$$U^i = \prod_{x(t)}\prod_t U^i_t(x(t)), \quad t = 0, 1, 2, \ldots;\ \ i = 0, 1, 2, \ldots, m.$$

It is easy to observe that if the solution $u^{1^*}(t),\ u^{2^*}(t),\ \ldots,\ u^{m^*}(t)$ of Problem 3.68 does not depend on the order of fixing the vectors of control para-meters of the players $1, 2, \ldots, m$, then $u^{1^*}(t),\ u^{2^*}(t),\ \ldots,\ u^m(t)$ is a solution in the sense of Nash.

If $c^2_t(x(t),\ x(t+1)) = -c^1_t(x(t),\ x(t+1)) = c_t(x(t),\ x(t+1))$, then we obtain the max-min control problem of two players with the payoff functions

$$F_{x_0x_f}(u^1(t),\ u^2(t)) = F^2_{x_0x_f}(u^1(t),\ u^2(t)) = -F^1_{x_0x_f}(u^1(t),\ u^2(t)).$$

In this case, we are looking for the vector of the control parameters $u^{1^*}(t)$ and $u^{2^*}(t)$ such that

$$F_{x_0x_f}(u^{1^*}(t),\ u^{2^*}(t)) = \max_{u^1(t)}\min_{u^2(t)} F_{x_0x_f}(u^1(t),\ u^2(t)).$$

For the considered class of problems, we also develop an algorithm based on dynamic programming.

3.16 Multi-Objective Control Based on the Concept of Cooperative Games: Pareto Optima

We consider a dynamical system $\mathbb{L}$, which is controlled by m players $1, 2, \ldots, m$, and formulate the control model, which is based on the concept of cooperative games. Assume that the players coordinate their actions in the control processes by using common vectors of control parameters $u(t) = (u^1(t), u^2(t), \ldots, u^m(t))$ (see [21, 27, 89, 139, 140, 144, 149]). So, the dynamics of the system are described by the following system of difference equations:

$$x(t+1) = g_t(x(t), u(t)), \quad t = 0, 1, 2, \ldots,$$

where

$$x(0) = x_0 \quad \text{and} \quad u(t) \in U_t(x(t)), \quad t = 0, 1, 2, \ldots.$$

Additionally, we assume that system $\mathbb{L}$ should reach the final state at the time moment $t(x_f)$ such that $t_1 \le t(x_f) \le T_2$.

Let

$$u(0), u(1), u(2), \ldots, u(t-1), \ldots$$

be a player's control which generates a trajectory

$$x(0), x(1), x(2), \ldots, x(t), \ldots.$$

Then either this trajectory passes through state x_f at the finite moment $T(x_f)$ or it does not pass through x_f. We denote by

$$F^i_{x_0 x_f}(u(t)) = \sum_{t=0}^{t(x_f)-1} c^i_t(x(t), g_t(x(t), u(t))), \quad i = 1, 2, \ldots, m$$

the integral time cost of the system's passage from x_0 to x_f if

$$t_1 \le t(x_f) \le t_2;$$

otherwise, we put

$$F^i_{x_0 x_f}(u(t)) = \infty.$$

Here, $c^i_t(x(t), g_t(x(t), u(t))) = c^i_t(x(t), x(t+1))$ represents the cost of the system's passage from state $x(t)$ to state $x(t+1)$ at the stage $[t, t+1]$ for player i, $i \in \{1, 2, \ldots, m\}$.

Problem 3.71 Find vectors of control parameters $u^*(t)$ such that there is no other control vector $u(t) \neq u^*(t)$ for which

$$\left(F^1_{x_0x_f}(u(t)),\ F^2_{x_0x_f}(u(t)),\ \dots,\ F^m_{x_0x_f}(u(t))\right) \leq$$
$$\leq (F^1_{x_0x_f}(u^*(t)),\ F^2_{x_0x_f}(u^*(t)),\ \dots,\ F^m_{x_0x_f}(u^*(t)))$$

and for any $i_0 \in \{1, 2, \dots, m\}$

$$F^{i_0}_{x_0x_f}(u(t)) < F^{i_0}_{x_0x_f}(u^*(t)).$$

So, we consider the problem of finding a Pareto solution [139, 140, 144].

Unlike Nash equilibria, Pareto optima for multi-objective discrete control always exist if there is an admissible solution $u(t)$, $t = 0, 1, 2, \dots, t(x_f)$ which generates a trajectory $x_0 = x(0),\ x(1),\ x(2),\ \dots,\ x(t(x_f)) = x_f$ from x_0 to x_f. Therefore, to obtain a Pareto solution for a multi-objective control problem, the standard linear convolution method of criteria [43, 140, 149] can be used. However, as it was shown in [24–26], for discrete problems, the linear convolution method may not find all Pareto optima, and in this case, it is necessary to apply the nonlinear convolution procedure of criteria for the multi-objective discrete control problems.

In a similar way, the multi-objective problems for stochastic Markov decision processes were formulated by applying the Pareto solution concept (see [40, 62, 192, 195, 196, 198]). To determine a solution for such problems, the convolution method of criteria can also be used.

3.17 Alternate Players' Control Conditions and Nash Equilibria for Dynamic Games in Positional Form

In order to formulate the theorem of the existence of Nash equilibria for the considered multi-objective control problem from Sect. 3.14, we apply the following condition:

We assume that an arbitrary state $x(t) \in X$ of the dynamical system $\mathbb{L}$ at the time moment t represents a position $(x, t) \in X \times \{0, 1, 2, \dots\}$ of one of the players $i \in \{1, 2, \dots, m\}$. This means that in the control process, the next state $x(t+1) \in X$ is determined (chosen) by player i if the dynamical system $\mathbb{L}$ at the time moment t has the state $x(t)$ that corresponds to the position (x, t) of player i. This situation corresponds to the case if the expression

$$g_t(x(t),\ u^1(t),\ u^2(t),\ \dots,\ u^{i-1}(t),\ u^i(t),\ u^{i+1}(t),\ \dots,\ u^m(t))$$

in (3.76) for a given position (x, t) of player i only depends on the control vector $u^i(t)$, i.e.,

$$g_t(x(t),\ u^1(t),\ u^2(t),\ \ldots,\ u^{i-1}(t),\ u^i(t),\ u^{i+1}(t),\ \ldots,\ u^m(t)) = g_t^i(x(t),\ u^i(t)).$$

So, the notations (x, t) and $x(t)$ have the same meaning.

Definition 3.72 We say that the alternate players' control condition is satisfied for multi-objective control problems if for any fixed $(x, t) \in X \times \{0, 1, 2, \ldots\}$, the Eqs. (3.76) only depend on one of the vectors of control parameters. The multi-objective control problems with such an additional condition are called game-theoretic control models in positional form.

The following lemma presents a necessary and sufficient condition that holds for the alternate players' control.

Lemma 3.73 *The alternate players' control condition for the multi-objective control problem holds if and only if at every time step* $t = 0, 1, 2, \ldots$ *for the set of states* X*, there exists a partition*

$$X = X_1(t) \cup X_2(t) \cup \cdots \cup X_m(t); \quad (X_i(t) \cap X_j(t) = \emptyset,\ i \neq j) \tag{3.78}$$

such that the Eqs. (3.76) *can be represented as follows:*

$$\begin{aligned} x(t+1) = g_t^i(x(t),\ u^i(t)) \ \ &\textit{if} \ \ x(t) \in X_i(t); \\ t = 0, 1, 2, \ldots; \ \ &i = 1, 2, \ldots, m, \end{aligned} \tag{3.79}$$

i.e.,

$$\begin{aligned} &g_t(x(t),\ u^1(t),\ u^2(t),\ \ldots,\ u^i(t),\ u^{i+1}(t),\ \ldots,\ u^m(t)) \\ &= g_t^i(x(t),\ u^i(t)) \ \textit{if} \ x(t) \in X_i(t); \quad t = 0, 1, 2, \ldots; \ i = 1, 2, \ldots, m. \end{aligned}$$

Here, $X_i(t)$ *corresponds to the set of the positions of player* i *at the time step* t *(note that some of* $X_i(t)$ *in* (3.78) *can be empty sets).*

Proof $\Rightarrow$ Let us assume that the alternate players' control condition holds for a multi-objective control problem. Then for a fixed time step t, the Eqs. (3.76) depend on only one of the vectors of control parameters $u^i(t)$, $i \in \{1, 2, \ldots, m\}$. Therefore, if we denote by $X_i(t)$ the set of states of the dynamical system that corresponds to the positions of player i at time step t, Eq. (3.76) can be regarded as a solution that satisfies (3.79).

$\Leftarrow$ Let us assume that the partition (3.78) is given for any $t = 0, 1, 2, \ldots$, and the expression in (3.76) is represented in the form (3.79). This means that at every time step t, this equation depends on only one of the vectors of the control parameters.

□

Based on these results, we can prove the important fact that the set of the positions can be characterized in the following way:

Corollary 3.74 *If the alternate players' control condition for the multi-objective control problem holds, then the set of the positions* $Z_i \subseteq X \times \{0, 1, 2, \dots\}$ *of player* i *can be represented as follows:*

$$Z_i = \bigcup_t (X_i(t), t), \quad i = 1, 2, \dots, m.$$

Let us assume that the alternate players' control for the problem from Sect. 3.14 holds. Then the set of the possible system's transitions of the dynamical system $\mathbb{L}$ can be described by a directed graph $\overline{G} = (Z, \overline{E})$ with the set of vertices $Z = \bigcup_{i=1}^{m} Z_i$, where $Z_i, i = 1, 2, \dots, m$ represents the set of the positions of player i. An arbitrary vertex $z \in Z$ in $\overline{G}$ corresponds to a position (x, t) of one of the players $i \in \{1, 2, \dots, m\}$, and a directed edge $e = (z', z'')$ reflects the possibility of the system's transition from state $z' = (x, t)$ to state $z'' = (y, t + 1)$ determined by $x(t)$ and the control vector $u^i(t) \in U_t^i(x(t))$ such that

$$y = x(t + 1) = g_t^i(x(t), u^i(t)) \quad \text{if} \quad x(t) \in Z_i.$$

We associate the costs $c^i((x, t), (y, t + 1)) = c_t^i(x(t), g_t^i(x(t), u^i(t))), i = 1, 2, \dots, m$ with the edges $((x, t), (y, t + 1))$ of graph G. Graph $\overline{G}$ is represented in Fig. 3.18.

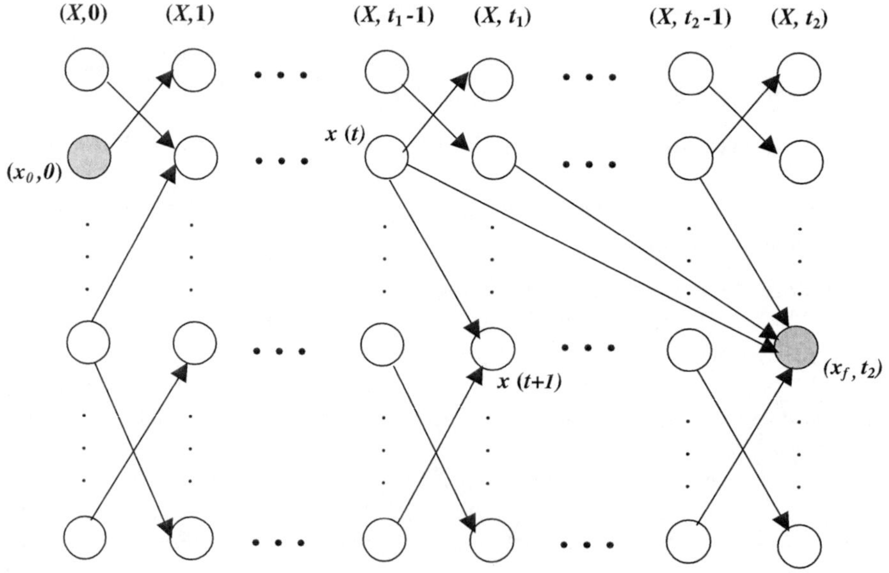

Fig. 3.18 Network of the multi-objective control problem

This graph contains $t_2 + 1$ copies of the set of states $X(t) = (X, t)$, where $X(t) = X_1(t) \cup X_2(t) \cup \cdots \cup X_m(t)$, $t = 0, 1, 2, \ldots, t_2$. In $\overline{G}$, there are also the edges $((x, t), (x_f, t_2))$ if $t_1 - 1 \le t \le t_2 - 1$, and for a given position $(x, t) = x(t) \in X_i(t)$ of the player i, there exists a control $u^i(t) \in U_t^i(x(t))$ such that

$$x_f = x(t+1) = g_t^i(x(t), u^i(t)).$$

With these edges $((x, t), (x_f, t_2))$, we associate the costs $c^i((x, t), (x_f, t_2)) = c_t^i(x(t), g_t^i(x(t), u^i(t)))$, $i = 1, 2, \ldots, m$. It is easy to observe that $\overline{G}$ is a directed acyclic graph in which an arbitrarily directed path from $(x_0, 0)$ to (x_f, t_2) contains $t(x_f)$ edges such that $t_1 \le t(x_f) \le t_2$. So, a directed path from $(x_0, 0)$ to (x_f, t_2) corresponds to a feasible trajectory of the dynamical system from x_0 to x_f. This means that our multi-objective problem with an alternate players' control condition can be regarded as a dynamic non-cooperative game on a network. Based on such a representation of the dynamics of system $\mathbb{L}$ in [107, 109, 110, 112], the following result is justified:

Theorem 3.75 *Let us assume that for the multi-objective control problem, there exists a trajectory*

$$x_0 = x(0), \ x(1), \ x(2), \ \ldots, \ x(t(x_f)) = x_f$$

from a starting state x_0 to a final state x_f generated by the vectors of control parameters

$$u^1(t), \ u^2(t), \ \ldots, \ u^m(t), \quad t = 0, 1, 2, \ldots, t(x_f) - 1,$$

where $u^i(t) \in U_t^i(x(t))$, $i = 1, 2, \ldots, m$, $t = 0, 1, 2, \ldots, t(x_f) - 1$, and $t_1 \le t(x_f) \le t_2$. Moreover, we assume that the alternate players' control condition is satisfied. Then for this problem, there exists an optimal solution $u^{1^*}(t), u^{2^*}(t), \ldots, u^{m^*}(t)$ in the sense of Nash.

The correctness of this theorem follows from Theorem 3.37 because the problem of determining Nash equilibria for the game-theoretic control problem with an alternate players' control condition is equivalent to the problem of determining Nash equilibria in the dynamic c-game on an auxiliary constructed time-expanded network.

As an important result from Theorem 3.75, we obtain the following corollary:

Corollary 3.76 *Assume that for any $u^1(t) \in U_t^1(x(t))$, $t = 0, 1, 2, \ldots$ in the max-min control problem, there exists a control $u^2(t) \in U_t^2(x(t))$, $t = 0, 1, 2, \ldots, t(x_f) - 1$ such that $u^1(t)$ and $u^2(t)$ generate a trajectory*

$$x_0 = x(0), \ x(1), \ x(2), \ \ldots, \ x(t(x_f)) = x_f$$

from a starting state x_0 to a final state x_f, where $t_1 \le t(x_f) \le t_2$. Moreover, we assume that the alternate players' control condition is satisfied.

Then for the payoff function $F_{x_0x_f}(u^1(t), u^2(t))$ in the max-min control problem, there exists a saddle point $(u^{1^*}(t), u^{2^*}(t))$, i.e.,

$$F_{x_0x_f}(u^{1^*}(t), u^{2^*}(t)) = \max_{u^1(t)} \min_{u^2(t)} F_{x_0x_f}(u^1(t), u^2(t))$$
$$= \min_{u^2(t)} \max_{u^1(t)} F_{x_0x_f}(u^1(t), u^2(t)).$$

All results related to the existence of theorems and algorithms for solving the problems on networks can be applied to the problems from Sects. 3.9–3.11.

3.18 Stackelberg Solutions for Hierarchical Control Problems

We consider the hierarchical control problem from Sect. 3.15. In order to develop a dynamic programming technique for determining a Stackelberg solution, we study the static case of the hierarchical problem considered in [111] and analyze the computational complexity of this problem. Additionally, we formulate the hierarchical control problem on the network and propose a dynamic programming algorithm for its solving. Based on these results, we extend the dynamic programming technique for the hierarchical control problem from Sect. 3.15.

3.18.1 Stackelberg Solutions for Static Games

Let a static game of m players $\Gamma = (\{\mathbb{S}i\}_{i=\overline{1,m}}, \{F_i\}_{i=\overline{1,m}})$ be given, where $\mathbb{S}_i$, $i = 1, 2, \ldots, m$ represent non-empty finite sets of the strategies of the players and

$$F_i : \mathbb{S}_1 \times \mathbb{S}_2 \times \mathbb{S}_m \to \mathbb{R}^1, \quad i = 1, 2, \ldots, m$$

are the corresponding payoff functions in Γ. We consider the problem of determining a Stackelberg solution in this game, i.e., we are seeking strategies $s^{1*}, s^{2*}, \ldots, s^{m*}$ such that we have

$$s^{1*} = \operatorname*{argmin}_{\substack{s^1 \in \mathbb{S}^1, \\ (s^i \in R_2(s^1,\ldots,s^{i-1}))_{2\le i\le m}}} F_1(s^1, s^2, \ldots, s^m);$$

$$s^{2*} = \operatorname*{argmin}_{\substack{s^2 \in R_2(s^{1*}), \\ \left(s^i \in R_i(s^{1*},s^2,\ldots,s^{i-1})\right)_{3\le i\le m}}} F_2(s^{1*}, s^2, \ldots, s^m);$$

$$s^{3*} = \operatorname*{argmin}_{\substack{s^3 \in R_3(s^{1*}, s^{2*}), \\ \left(s^i \in R_i(s^{1*}, s^{2*}, s^3, \ldots, s^{i-1})\right)_{4 \le i \le m}}} F_3(s^{1*}, s^{2*}, \ldots, s^m);$$

$$\vdots$$

$$s^{m*} = \operatorname*{argmin}_{s^m \in R_m(s^{1*}, s^{2*}, \ldots, s^{m-1*})} F_m(s^{1*}, s^{2*}, \ldots, s^{m-1*}, s^m),$$

where $R_k(s^1, s^2, \ldots, s^{k-1})$ is the set of the best responses of player k if players $1, 2, \ldots, k-1$ have already fixed their strategies $s^1, s^2, \ldots, s^{k-1}$, i.e.,

$$R_2(s^1) = \operatorname*{argmin}_{\substack{s^2 \in \mathbb{S}_2, \\ (s^i \in R_i(s^1, \ldots, s^{i-1}))_{3 \le i \le m}}} F_2(s^1, s^2, \ldots, s^m);$$

$$R_3(s^1, s^2) = \operatorname*{argmin}_{\substack{s^3 \in \mathbb{S}_3, \\ (s^i \in R_i(s^1, s^2, \ldots s^{i-1}))_{4 \le i \le m}}} F_3(s^1, s^2, \ldots, s^m);$$

$$\vdots$$

$$R_m(s^1, s^2, \ldots, s^{m-1}) = \operatorname*{argmin}_{s^m \in \mathbb{S}_m} F_m(s^1, s^2, \ldots, s^m).$$

In this game, the players fix their strategies successively according to their numerical order. Therefore, if the order of fixing the strategies of the players is changed, then the best responses of the players will correspond to a Stackelberg solution with respect to a new order of the players.

Lemma 3.77 *Let $s^{1*}, s^{2*}, \ldots, s^{m*}$ be a Stackelberg solution for the game Γ. If this solution remains the same for an arbitrary order of fixing strategies of the players, then $s^{1*}, s^{2*}, \ldots, s^{m*}$ is a Nash equilibrium.*

Proof Assume that $s^{1*}, s^{2*}, \ldots, s^{i-1*}, s^{i*}, s^{i+1*}, \ldots, s^{m*}$ is a Stackelberg solution for an arbitrary order of fixing strategies of the players. Then we may consider that an arbitrary player i fixes his or her strategy in the last order, and therefore,

$$F_i(s^{1*}, s^{2*}, \ldots, s^{i-1*}, s^{i*}, s^{i+1*}, \ldots, s^{m*}) \le$$
$$\le F_i(s^{1*}, s^{2*}, \ldots, s^{i-1*}, s^i, s^{i+1*}, \ldots, s^{m*}), \quad \forall s^i \in \mathbb{S}_i, \ i = 1, 2, \ldots, m.$$

So, $s^{1*}, s^{2*}, \ldots, s^{i-1*}, s^{i*}, s^{i+1*}, \ldots, s^{m*}$ is a Nash equilibrium. □

The computational complexity of determining pure Nash equilibria in discrete games was studied in [44, 65]. Based on the results from [44], we can conclude that finding a Stackelberg solution in the considered games is NP-hard if the number of players m acts as the input data parameter of the problem.

In case m is fixed (i.e., m does not act as an input data parameter of the problem), then a Stackelberg solution can be found in polynomial time. In the case of a small number of players, especially in the case of two or three players, exhaustive search allows us to determine Stackelberg strategies for finite games with large dimensions. Indeed, if we calculate $s^{1*}, s^{2*}, \ldots, s^{m*}$ according to the condition from the definition of a Stackelberg solution, we use $|\mathbb{S}_1| \times |\mathbb{S}_2| \times \cdots \times |\mathbb{S}_m|$ steps. So, in the case of two players, we can determine a Stackelberg solution using $O(|\mathbb{S}_1||\mathbb{S}_2|)$ elementary operations (here, we do not take into account the number of operations for calculating the values $F_i(s^1, s^2)$ for given (s^1, s^2)). We can use this fact for solving hierarchical control problems with two or three players.

3.18.2 Hierarchical Control on Networks

Let $G = (X, E)$ be the graph of the states' transitions for a discrete-time system $\mathbb{L}$. So, X corresponds to the set of states of $\mathbb{L}$, and an arbitrarily directed edge $e = (x, y) \in E$ means the possibility of system $\mathbb{L}$ to pass from state $x = x(t)$ to state $y = x(t+1)$ at every time moment $t = 0, 1, 2, \ldots$. Assume that the system $\mathbb{L}$ is controlled by m players, and on the edge set, the following m functions are defined:

$$c^i : E \to \mathbb{R}^1, \quad i = 1, 2, \ldots, m,$$

which assign m costs $c_e^1, c_e^2, \ldots, c_e^m$ to each edge $e \in E$. For player i, the quantity c_e^i expresses the cost of system $\mathbb{L}$ to pass through edge $e = (x, y)$ from state $x = x(t)$ to state $y = x(t+1)$ at every time moment $t = 0, 1, 2, \ldots$. On G, the players use only stationary strategies and intend to minimize their integral-time costs by a trajectory $x_0 = x(0), x(1), x(2), \ldots, x(t(x_f)) = x_f$ from a starting state x_0 to a final state x_f, where $t_1 \leq t(x_f) \leq t_2$. We define the stationary strategies of the players as m multivalued functions

$$\begin{aligned}
s^1 &: x \to X_1^{j_1}(x) \in A^1(x) \quad \text{for} \quad x \in X \setminus \{x_f\}; \\
s^2 &: x \to X_2^{j_2}(x) \in A^2(x) \quad \text{for} \quad x \in X \setminus \{x_f\}; \\
&\vdots \\
s^m &: x \to X_m^{j_m}(x) \in A^m(x) \quad \text{for} \quad x \in X \setminus \{x_f\},
\end{aligned}$$

which satisfy the condition

$$|s^1(x) \cap s^2(x) \cap \cdots \cap s^m(x)| = 1, \quad \forall x \in X, \tag{3.80}$$

where $A^i(x)$, $i = 1, 2, \dots, m$ are given sets of subsets from $X(x) = \{y \in X \mid e = (x, y) \in E\}$, i.e., $A^i(x) = \{X_i^1(x), X_i^2(x), \dots, X_i^{K_i(x)}(x)\}$, $X_i^j(x) \subseteq X_i(x)$, $j = 1, 2, \dots, K_i(x)$.

The strategies $s^1(x), s^2(x), \dots, s^m(x)$ for a given $x \in X \setminus \{x_f\}$ correspond to vectors of control parameters $u^1(t), u^2(t), \dots, u^m(t)$ at a given state $x = x(t)$ in the control Problem 3.69 and reflect the fact that the set of control parameters at the given state $x = x(t)$ uniquely determines the next state $y = x(t+1) \in X$ at every time moment $t = 0, 1, 2, \dots$. Therefore, here, we use condition (3.80) and consider that $y = \bigcap_{i=1}^m s^i(x)$.

If $|\cap s^j(x)| \neq 1$, then the set of strategies $s^1, s^2, \dots, s^m$ is not feasible. In the following, we consider that the players use only feasible strategies.

An arbitrary set $X_i^j(x) \in A^i(x)$ in our control problem represents a possible set of the next states $y = x(t+1) \in X$ in which player i prefers to transfer system $\mathbb{L}$ if at the moment of time t, the state of the dynamical system is $x = x(t)$. This set for control Problem 3.69 can be treated as a set of possible next states $y = x(t+1) \in X$ if player i fixes a feasible vector of control parameters $u^i(t) \in U_t^i(x(t))$.

Therefore, if we treat $X_i^j(x)$ as preferences of the next possible sets of the states from $A^i(x)$ for player $i \in \{1, 2, \dots, m\}$, then the unique next state y represents the intersection of preferences $s^1(x), s^2(x), \dots, s^m(x)$ of the players $1, 2, \dots, m$, i.e., $y = \bigcap_{i=1}^m s^i(x)$, where $s^i : x \to X_i^{j_i} \in A^i(x)$ for $x \in X \setminus \{x_f\}$, $i = 1, 2, \dots, m$. Let $s^1, s^2, \dots, s^m$ be a fixed set of feasible strategies of the players $1, 2, \dots, m$. Denote by $G_s = (X, E_s)$ the subgraph generated by the edges $e = (x, y) \in E$ such that $y = \bigcap_{i=1}^m s^i(x)$ for $x \in X \setminus \{x_f\}$. Then in G_s, either a unique directed path $P_s(x_0, x_f)$ in G_s exists or such a path does not exist. Therefore, for $s^1, s^2, \dots, s^m$ and fixed starting and final states $x_0, x_f \in X$, we can define the quantities

$$\widehat{H}^1_{x_0 x_f}(s^1, s^2, \dots, s^m),\ \widehat{H}^2_{x_0 x_f}(s^1, s^2, \dots, s^m),\ \dots,\ \widehat{H}^m_{x_0 x_f}(s^1, s^2, \dots, s^m)$$

as follows: We put

$$\widehat{H}^i_{x_0 x_f}(s^1, s^2, \dots, s^m) = \sum_{e \in E(P_s(x_0, x_f))} c^i_e, \quad i = 1, 2, \dots, m$$

if

$$t_1 \le |E(P_s(x_0, x_f))| \le t_2;$$

otherwise, we put

$$\widehat{H}^i_{x_0 x_f}(s^1, s^2, \dots, s^m) = +\infty.$$

Note that in this control process, the players fix their strategies successively one after another according to their numerical order at each time moment $t = 0, 1, 2, \dots$

for every state $x \in X \setminus \{x_f\}$. Additionally, we assume that each player fixing his or her strategies informs posterior players which strategy has been chosen. In the considered hierarchical control problem, we are looking for Stackelberg stationary strategies, i.e., we are looking for strategies $s^{1^*}, s^{2^*}, \ldots, s^{m*}$, for which

$$s^{1^*} = \operatorname*{argmin}_{\substack{s^1 \in \mathbb{S}_1, \\ (s^i \in R_i(s^1,\ldots,s^{i-1}))_{2\le i\le m}}} \widehat{H}^1_{x_0x_f}(s^1, s^2, \ldots, s^m);$$

$$s^{2^*} = \operatorname*{argmin}_{\substack{s^2 \in R_2(s^{1^*}), \\ \left(s^i \in R_3(s^{1^*}, s^2,\ldots,s^{i-1})\right)_{3\le i\le m}}} \widehat{H}^2_{x_0x_f}(s^{1^*}, s^2, \ldots, s^m);$$

$$s^{3^*} = \operatorname*{argmin}_{\substack{s^3 \in R_3(s^{1^*}, s^{2^*}), \\ \left(s^i \in R_i(s^{1^*}, s^{2^*}, s^3,\ldots,s^{i-1})\right)_{4\le i\le m}}} \widehat{H}^3_{x_0x_f}(s^{1^*}, s^{2^*}, \ldots, s^m);$$

$$\vdots$$

$$s^{m*} = \operatorname*{argmin}_{s^m \in R_m(s^{1^*}, s^{2^*},\ldots,s^{m-1^*})} \widehat{H}^m_{x_0x_f}(s^{1^*}, s^{2^*}, \ldots, s^{m-1^*}, s^m),$$

where $R_k(s^1, s^2, \ldots, s^{k-1})$ is the set of best responses of player k if the players $1, 2, \ldots, k-1$ have already fixed their strategies $s_1, s_2, \ldots, s_{k-1}$, i.e.,

$$R_2(s^1) = \operatorname*{argmin}_{\substack{s^2 \in \mathbb{S}_2, \\ (s^i \in R_i(s^1,\ldots,s^{i-1}))_{3\le i\le m}}} \widehat{H}^2_{x_0x_f}(s^1, s^2, \ldots, s^m);$$

$$R_3(s^1, s^2) = \operatorname*{argmin}_{\substack{s^3 \in \mathbb{S}_3, \\ (s^i \in R_i(s^1,\ldots,s^{i-1}))_{4\le i\le m}}} \widehat{H}^3_{x_0x_f}(s^1, s^2, \ldots, s^m);$$

$$\vdots$$

$$R_m(s^1, s^2, \ldots, s^{m-1}) = \operatorname*{argmin}_{s^m \in \mathbb{S}_m} \widehat{H}^m_{x_0x_f}(s^1, s^2, \ldots, s^m),$$

where $\mathbb{S}_1, \mathbb{S}_2, \ldots, \mathbb{S}_m$ represent the corresponding admissible sets of stationary strategies of the players $1, 2, \ldots, m$.

Remark 3.78 In general, the stationary strategies $s^1, s^2, \ldots, s^m$ of the players in the hierarchical control problem on G can be defined as arbitrary multi-value functions

$$s^i : x \to X_i^{j_i}(x) \in A^i(x) \text{ for } x \in X \setminus \{x_f\}, \quad i = 1, 2, \ldots, m.$$

If the conditions (3.80) for $x \in X \setminus \{x_f\}$ do not take place, i.e., if at least for a state $x \in X \setminus \{x_f\}$, the condition

$$|s^1(x) \cap s^2(x) \cap \cdots \cap s^m(x)| \neq 1$$

holds, then we put $\widehat{H}^i_{x_0 x_f}(s^1, s^2, \ldots, s^m) = +\infty$.

So, the hierarchical control problem is determined by the dynamic network $(G, A^1, A^2, \ldots, A^m, c^1, c^2, \ldots, c^m, x_0, x_f, T_1, T_2)$, where $A^i = \bigcup_{x \in X} A^i(x)$ and $c^i = (c^i_{e_1}, c^i_{e_2}, \ldots, c^i_{e_{|E|}})$, $i = 1, 2, \ldots, m$. In the case $T_2 = \infty$, $T_1 = 0$, we denote the corresponding network by $(G, A^1, A^2, \ldots, A^m, c^1, c^2, \ldots, c^m, x_0, x_f)$. The following theorem allows us to describe a class of multi-objective hierarchical control problems for which an arbitrary Stackelberg solution is also a Nash equilibrium:

Theorem 3.79 *Let the hierarchical control problem on the network $(G, A^1, A^2, \ldots, A^m, c^1, c^2, \ldots, c^m, x_0, x_f)$ be given, where G has the property that for an arbitrary vertex $x \in X$, there exists a directed path from x to x_f. Additionally, assume that the sets $A^1(x), A^2(x), \ldots, A^m(x)$ satisfy the following condition: For an arbitrary vertex $x \in X \setminus \{x_f\}$, there exists $i_x \in \{1, 2, \ldots, m\}$ such that $A^{i_x}(x) = \{\{y\} \mid y \in X_G(x)\}$ and $A^{i_x}(x) = \{X_G(x)\}$ if $i \in \{1, 2, \ldots, m\} \setminus \{i_x\}$. Then for the hierarchical control problem on G, there exists a Nash equilibrium.*

Proof First of all, it is easy to observe that if the conditions of the theorem hold, then in the multi-objective control problem on G, there exist stationary strategies $s^1, s^2, \ldots, s^m$ that generate a trajectory $x_0, x_1, x_2, \ldots, x_f$ from a starting state x_0 to a final state x_f for an arbitrary given starting vertex $x_0 = x \in X$. This means that a Stackelberg solution for the hierarchical control problem on G exists. Additionally, we can see that the dynamic c-game from Sect. 3.10 (the multi-objective control problem in positional form) represents a particular case of the problem formulated above if the sets $A^1(x), A^2(x), \ldots, A^m(x)$ satisfy the condition that for an arbitrary $x \in X \setminus \{x_f\}$, there exists $i_x \in \{1, 2, \ldots, m\}$ such that $A^{i_x}(x) = \{\{y\} \mid y \in X_G(x)\}$ and $A^i(x) = \{X_G(x)\}$ if $i \in \{1, 2, \ldots, m\} \setminus \{i_x\}$. In this case, a Stackelberg solution to the hierarchical control problem does not depend on the order of fixing strategies by the players, and therefore, based on Lemma 3.77, an arbitrary Stackelberg solution to the multi-objective control problem on G is a Nash equilibrium. □

3.18.3 Optimal Stackelberg Strategies on Acyclic Networks

We consider the hierarchical control problem on an acyclic network $(G, A^1, A^2, \ldots, A^m, c^1, c^2, \ldots, c^m, x_0, x_f)$, i.e., $G = (X, E)$ is a directed acyclic graph and $t_1 = 0$, $t_2 = \infty$. We also assume that in G, vertex x_f is attainable from every vertex $x \in X$.

Algorithm 3.80 Determining Stackelberg Strategies on Acyclic Networks

Preliminary step (Step 0): Fix $X^0 = \{x_f\}$, $E^0 = \emptyset$, and put $\varepsilon^i(x_f) = 0$, $i = 1, 2, \dots, m$.

General step (Step k, $k \geq 1$): If $X \setminus X^{k-1} = \emptyset$, then stop; otherwise, find a vertex $x^k \in X \setminus X^{k-1}$ for which $X_G(x^k) \subseteq X^{k-1}$, where $X_G(x^k) = \{y \in X \mid (x^k, y) \in E\}$. With respect to vertex x^k, we consider the static problem of finding Stackelberg strategies $s^{1^*}(x^k), s^{2^*}(x^k), \dots, s^{m*}(x^k)$ in the game

$$\Gamma(x^k) = (S_1(x^k),\ S_2(x^k),\ \dots,\ S_m(x^k),\ F_1,\ F_2,\ \dots,\ F_m),$$

where the sets of strategies of the players $S_1(x^k), S_2(x^k), \dots, S_m(x^k)$ and the payoff functions $F_1, F_2, \dots, F_m$ are defined as follows:

$$S_i(x^k) = \left\{ s^i(x^k) \mid s^i : x^k \to X_i^{j_j}(x^k) \in A^i(x^k) \right\}, \quad i = 1, 2, \dots m;$$

$$F_i(s^1, s^2, \dots, s^m) = \begin{cases} \varepsilon^i(y) + c^i_{(x^k, y)}, & \text{if} \quad y = \bigcap_{i=1}^{m} s^i(x^k); \\ +\infty, & \text{if} \quad \left| \bigcap_{i=1}^{m} s^i(x^k) \right| \neq 1. \end{cases} \tag{3.81}$$

After that, find a Stackelberg solution $s^{1^*}(x^k),\ s^{2^*}(x^k),\ \dots,\ s^{m*}(x^k)$ for the static game $\Gamma(x^k)$, and determine the vertex $y^* = \bigcap_{i=1}^{m} s^{i^*}(x^k)$.

Then calculate

$$\varepsilon^i(x^k) = \varepsilon^i(y^*) + c^i_{(x^k, y^*)}, \quad i = 1, 2, \dots, m$$

and fix $\widehat{H}_{x^k x_f}(s^{1^*}, s^{2^*}, \dots, s^{m*}) = \varepsilon^i(x^k)$, $i = 1, 2, \dots, m$.

Put $X^k = X^{k-1} \cup \{x^k\}$, $E^k = E^{k-1} \cup \{(x^k, y)\}$, $GT^k = (X^k, E^k)$, and go to the next step.

This algorithm determines Stackelberg stationary strategies $s^{1^*}, s^{2^*}, \dots, s^{m*}$ for an arbitrary starting position $x_0 = x$ and the fixed final position x_f. The corresponding optimal values of integral costs of the system's passage from a starting state $x_0 = x$ to a final state x_f are $\widehat{H}_{x_0 x_f}(s^{1^*}, s^{2^*}, \dots, s^{m*})$. The algorithm determines the tree $GT^{|X|-1} = (X, E^{|X|-1})$ of optimal pure strategies with a sink vertex x_f that gives Stackelberg strategies for an arbitrary starting position $x_0 = x \in X$. It is easy to observe that if for a given starting position x_0 of the considered dynamic game a Nash equilibrium exists, then the algorithm determines this equilibrium. Note that the proposed algorithm can also be adapted for the problem when for different moments of time and for different states, the order of fixing stationary strategies of the players can be different. We should take this order of fixing strategies of the players into account by calculating the values $h_i^*(x^k, y)$ for a given state $x^k(t)$.

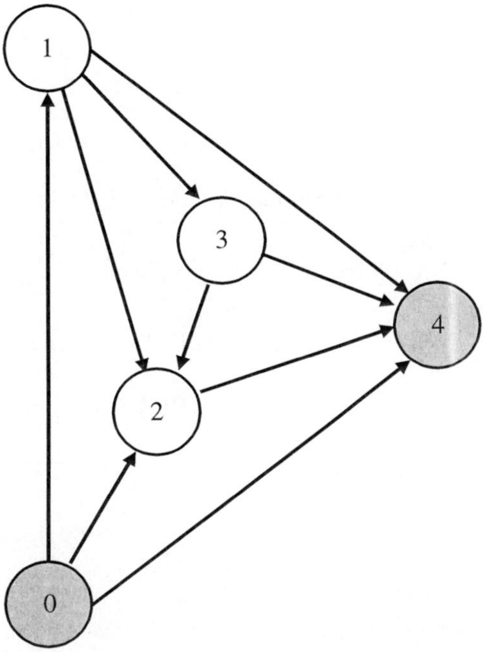

Fig. 3.19 Network for the hierarchical control problem

Example Consider a hierarchical control problem on a network with two players where the corresponding graph $G = (X, E)$ is presented in Fig. 3.19. This network has the structure of a directed acyclic graph with a given starting vertex $x_0 = x(0)$ and final vertex $x_f = 4$. At each vertex, a set of subsets $A^i(x) = \{X_i^j\}$ is given, where

$$A^1(0) = \{\{1, 2\}, \{1, 4\}\}; \quad A^2(0) = \{\{2, 4\}, \{1, 4\}\};$$
$$A^1(1) = \{\{2, 3\}, \{3, 4\}\}; \quad A^2(1) = \{\{2, 4\}, \{2, 3\}\};$$
$$A^1(2) = \{\{4\}\}; \quad A^2(2) = \{\{4\}\};$$
$$A^1(3) = \{\{2\}, \{2, 4\}\}; \quad U^2(2) = \{\{4\}, \{2, 4\}\}.$$

On the edges $e \in E$, there are defined cost functions $c^1 : E \to \mathbb{R}^1$ and $c^2 : E \to \mathbb{R}^1$, where

$$c^1_{(0,1)} = 2; \quad c^1_{(0,2)} = 1; \quad c^1_{(0,4)} = 5;$$
$$c^2_{(0,1)} = 3; \quad c^2_{(0,2)} = 2; \quad c^2_{(0,4)} = 6;$$
$$c^1_{(1,2)} = 3; \quad c^1_{(1,3)} = 4; \quad c^1_{(1,4)} = 3;$$
$$c^2_{(1,2)} = 3; \quad c^2_{(1,3)} = 1; \quad c^2_{(1,4)} = 5;$$

$$c^1_{(3,2)} = 1; \quad c^1_{(3,4)} = 2;$$

$$c^2_{(3,2)} = 1; \quad c^2_{(3,4)} = 4;$$

$$c^1_{(2,4)} = 1; \quad c^2_{(2,4)} = 2.$$

If we use Algorithm 3.80, then we obtain:

Step 0.
Fix $X^0 = \{4\}$, $\varepsilon^1(4) = 0$, $\varepsilon^2(4) = 0$, $E^0 = \emptyset$.

Step 1.
$X \setminus X^0 \neq \emptyset$ and $X_G(2) \subseteq X^0$; therefore, fix $x^1 = 2$, and solve the static game

$$\Gamma(2) = (S_1(2),\ S_2(2),\ F_1,\ F_2),$$

where

$$S_1(2) = \{s^1 : 2 \to \{4\}\}, \quad S_2(2) = \{s^2 : 2 \to \{4\}\}$$

and $F_1(s^1, s^2) = 1$; $F_2(s^1, s^2) = 2$. For this game, we have a trivial solution $s^{1*}(2) = \{4\}$; $s^{2*}(2) = \{4\}$. We calculate$\varepsilon^1(2) = 0 + c^1_{(2,4)} = 1$; $\varepsilon^2(2) = 0 + c^2_{(2,4)} = 2$ and put $\widehat{H}^1_{2,4}(s^{1*}, s^{2*}) = 1, \widehat{H}^2_{2,4}(s^{1*}, s^{2*}) = 2$. Fix $X^1 = X^0 \cup \{2\} = \{2, 4\}$; $E^1 = E^0 \cup \{(2, 4)\} = \{(2, 4)\}$; $GT^1 = (\{2, 4\},\ \{(2, 4)\})$.

Step 2.
$X \setminus X^1 \neq \emptyset$ and $X_G(3) \subseteq X^1$; therefore, fix $x^2 = 3$, and solve the static game

$$\Gamma(3) = (S_1(3),\ S_2(3),\ F_1,\ F_2),$$

where

$$S_1(3) = \left\{s^1_1 : 3 \to \{2\}; \quad s^1_2 : 3 \to \{2, 4\}\right\},$$

$$S_2(3) = \left\{s^2_1 : 3 \to \{2, 4\}; \quad s^2_2 : 3 \to \{4\}\right\}$$

and $F_i(s^1_j, s^2_j)$ are defined according to (3.81), i.e.,

$$F_1\left(s^1_1, s^2_1\right) = 2; \quad F_2\left(s^1_1, s^2_1\right) = 3 \quad \left(s^1_1(3) \cap s^2_1(3) = 2\right);$$

$$F_1\left(s^1_1, s^2_2\right) = F_2\left(s^1_1, s^2_2\right) = \infty \qquad \left(s^1_1(3) \cap s^2_2(3) = \emptyset\right);$$

$$F_1\left(s^2_1, s^1_2\right) = F_2\left(s^2_1, s^1_2\right) = \infty \quad \left(s^2_1(3) \cap s^1_2(3) = \{2, 4\},\right.$$

i.e., $|\{2, 4\}| \neq 1)$;

$$F_1\left(s_1^2, s_2^2\right) = 2; \quad F_2\left(s_1^2, s_2^2\right) = 4 \quad \left(s_1^2(3) \cap s_2^2(3) = 4\right).$$

If we solve this game, we find a Stackelberg solution $s^{1*}(3) = \{2\}$; $s^{2*}(3) = \{2, 4\}$. We calculate $\varepsilon^1(3) = \varepsilon^1(2) + c^1_{(3,2)} = 2$; $\varepsilon^2(3) = \varepsilon^2(2) + c^2_{(3,2)} = 3$ and put $\widehat{H}^1_{2,4}(s^{1*}, s^{2*}) = 1$, $\widehat{H}^2_{2,4}(s^{1*}, s^{2*}) = 2$.
Fix $X^2 = X^1 \cup \{3\} = \{2, 3, 4\}$; $E^2 = E^1 \cup \{(3, 2)\} = \{(2, 4), (3, 2)\}$; $GT^2 = (\{2, 3, 4\}, \{(3, 2), (2, 4)\})$.

Step 3.
$X \setminus X^2 \neq \emptyset$ and $X_G(1) \subseteq X^2$; therefore, fix $x^3 = 1$, and solve the static game

$$\Gamma(1) = (S_1(1), S_2(1), F_1, F_2),$$

where

$$S_1(1) = \left\{s_1^1 : 1 \to \{2, 3\}; \quad s_2^1 : 1 \to \{3, 4\}\right\},$$

$$S_2(3) = \left\{s_1^2 : 1 \to \{2, 4\}; \quad s_2^2 : 1 \to \{3, 4\}\right\}$$

and $F_i(s_j^1, s_j^2)$ are defined according to (3.81), i.e.,

$$F_1\left(s_1^1, s_1^2\right) = 4; \quad F_2\left(s_1^1, s_1^2\right) = 5 \quad \left(s_1^1(1) \cap s_1^2(1) = 2\right);$$

$$F_1\left(s_1^1, s_2^2\right) = F_2\left(s_1^1, s_2^2\right) = \infty \quad \left(s_1^1(1) \cap s_2^2(1) = \emptyset\right);$$

$$F_1\left(s_2^1, s_1^2\right) = 3; \quad F_2\left(s_2^1, s_1^2\right) = 5 \quad \left(s_2^1(1) \cap s_1^2(1) = 4\right);$$

$$F_1\left(s_2^1, s_2^2\right) = 5; \quad F_2\left(s_2^1, s_2^2\right) = 3 \quad \left(s_2^1(1) \cap s_2^2(1) = 3\right).$$

If we solve this game, we find a Stackelberg solution $s^{1*}(1) = s_1^1(1)$; $s^{2*}(1) = s_1^2(1)$, i.e., $s^{1*}(1) = \{2, 3\}$; $s^{2*}(1) = \{2, 4\}$. We calculate
$\varepsilon^1(1) = \varepsilon^1(2) + c^1_{(3,2)} = 4$; $\varepsilon^2(1) = \varepsilon^2(1) + c^2_{(3,2)} = 5$ and put $\widehat{H}^1_{2,4}(s^{1*}, s^{2*}) = 4$, $\widehat{H}^2_{2,4}(s^{1*}, s^{2*}) = 5$.
Fix $X^3 = X^2 \cup \{1\} = \{1, 2, 3, 4\}$; $E^3 = E^2 \cup \{(1, 2)\} = \{(1, 2), (2, 4), (3, 2)\}$; $GT^3 = (\{1, 2, 3, 4\}, \{(1, 2), (3, 2), (2, 4)\})$.

Step 4.
$X \setminus X^3 \neq \emptyset$ and $X_G(0) \subseteq X^3$; therefore, fix $x^4 = 0$, and solve the static game

$$\Gamma(0) = (S_1(0), S_2(0), F_1, F_2),$$

where

$$S_1(0) = \left\{s_1^1 : 0 \to \{1, 2\};\ s_2^1 : 0 \to \{1, 4\}\right\},$$

$$S_2(0) = \left\{s_1^2 : 0 \to \{2, 4\};\ s_2^2 : 0 \to \{1, 4\}\right\},$$

and $F_i(s_j^1, s_j^2)$ are defined according to (3.81), i.e.,

$$F_1\left(s_1^1, s_1^2\right) = 2;\quad F_2\left(s_1^1, s_1^2\right) = 4 \quad \left(s_1^1(0) \cap s_1^2(0) = 2\right);$$

$$F_1\left(s_1^1, s_2^2\right) = F_2\left(s_1^1, s_2^2\right) = \infty \quad \left(s_1^1(0) \cap s_2^2(0) = \emptyset\right);$$

$$F_1\left(s_2^1, s_1^2\right) = 5;\quad F_2\left(s_2^1, s_1^2\right) = 6 \quad \left(s_2^1(0) \cap s_1^2(0) = 4\right);$$

$$F_1\left(s_2^1, s_2^2\right) = F_2\left(s_2^1, s_2^2\right) = \infty \quad \left(s_2^1(0) \cap s_2^2(0) = \{1, 4\},\right.$$

$$\left.|\{1, 4\}| \neq 1\right).$$

If we solve this game, we find a Stackelberg solution $s^{1^*} = s_1^1;\ s^{2^*} = s_2^2$, i.e., $s^{1^*} = \{1, 2\};\ s^{2^*} = \{2, 4\}$. We calculate $\varepsilon^1(0) = \varepsilon^1(2) + c^1_{(0,2)} = 2$; $\varepsilon^2(0) = \varepsilon^2(0) + c^2_{(0,2)} = 4$ and put $\widehat{H}^1_{2,4}(s^{1^*}, s^{2^*}) = 2$, $\widehat{H}^2_{2,4}(s^{1^*}, s^{2^*}) = 4$. Fix $X^4 = X^3 \cup \{0\} = \{0, 1, 2, 3, 4\}$; $E^4 = E^3 \cup \{(0, 2)\} = \{(0, 2), (1, 2), (2, 4), (3, 2)\}$; $GT^4 = (\{0, 1, 2, 3, 4\}, \{(0, 2), (1, 2), (3, 2), (2, 4)\})$.

Step 5.

We obtain $X \setminus X^4 = \emptyset$; therefore, stop.

So, the optimal stationary strategies of the players are the following:

$$s^{1^*} : 0 \to \{1, 2\};\quad 1 \to \{2, 3\};\quad 2 \to \{4\};\quad 3 \to \{2\}.$$

$$s^{2^*} : 0 \to \{2, 4\};\quad 1 \to \{2, 4\};\quad 2 \to \{4\};\quad 3 \to \{2, 4\}.$$

The set of optimal stationary strategies in G generates the tree given in Fig. 3.20, i.e., the optimal strategies are as follows:

$$s^{1^*}(0), s^{2^*}(0) \text{ generate the transition } (0, 1).$$

$$s^{1^*}(1), s^{2^*}(1) \text{ generate the transition } (1, 2).$$

$$s^{1^*}(2), s^{2^*}(2) \text{ generate the transition } (2, 4).$$

$$s^{1^*}(3), s^{2^*}(3) \text{ generate the transition } (3, 4).$$

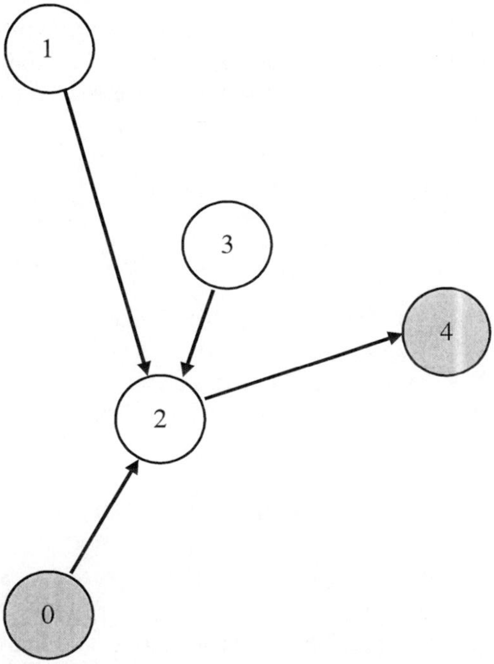

Fig. 3.20 Tree corresponding to optimal strategies of the problem

This tree gives the optimal stationary strategies of the players for an arbitrary starting position $x_0 = x \in X$. In this example, for $x_0 = 0$, we obtain a Stackelberg solution which is also a Nash equilibrium. If we fix $x_0 = 1$, this Stackelberg solution is not a Nash equilibrium.

3.18.4 An Algorithm for Hierarchical Control Problems

Based on the results from Sect. 3.18.3, we can propose an algorithm for solving the multi-objective hierarchical control problem from Sect. 3.15. First, we show that the hierarchical control problem from Sect. 3.15 can be reduced to a stationary hierarchical control problem on an auxiliary network $(\overline{G}, c^1, c^2, \ldots, c^m, y_0, y_T)$ for which Stackelberg stationary strategies should be found.

We construct the graph $\overline{G} = (Y, E)$ of the network as follows:

$$Y = Y^0 \cup Y^1 \cup Y^2 \cup \cdots \cup Y^{t_1} \cup Y^{t_1+1} \cup \cdots \cup Y^{t_2} \quad (t^k \cap Y^l = \emptyset,\ k \neq l),$$

where $Y^t = (X, t)$ corresponds to the set of states $x(t) \in X$ of system $\mathbb{L}$ at time moment t $(t = 0, 1, 2, \ldots, t_2)$:

$$E = E^0 \cup E^1 \cup E^2 \cup \cdots \cup E^{t_1} \cup E^{t_1+1} \cup \cdots \cup E^{t_2-1} \cup E^f,$$

where E^t, $t = 0, 1, 2, \ldots, t_2 - 1$ represents the set of directed edges in $\overline{G}$ that connects vertices from Y^t with vertices from Y^{t+1}.

We include an arbitrarily directed edge $((x,t), (y, t+1))$ in E^t, $t = 0, 1, 2, \ldots, t_2 - 1$ if in the control process at time moment t for a given state $x = x(t)$, there exist vectors of control parameters $u^1(t), u^2(t), \ldots, u^m(t)$ from corresponding feasible sets $U_t^1(x(t))$, $U_t^2(x(t))$, $\ldots$, $U_t^m(x(t))$ such that

$$y(t+1) = g_t(x(t), u^1(t), u^2(t), \ldots, u^m(t)).$$

In an analogous way, we define the set E^f. We include an arbitrary edge $((x,t), (x_f, t_2))$ in E^f, $t = t_1 - 1, t_1, t_1 + 1, \ldots, t_2 - 1$, if at time moments $t \in [t_1 - 1, t_2 - 1]$ for a state $x(t)$, there exist vectors of control parameters $u^1(t), u^2(t), \ldots, u^m(t)$ from corresponding feasible sets $U_t^1(x(t)), U_t^2(x(t)), \ldots, U_t^m(x(t))$ such that

$$x_f(t+1) = g_t(x(t), u^1(t), u^2(t), \ldots, u^m(t)).$$

Additionally, with each vertex (x,t), we associate a set of subsets $A^i(x,t) = \{X_i^j(x, t+1),\ j = 1, 2, \ldots, K_i(x)\}$, where an arbitrary set $X_i^j(x, t+1)$ represents the set of possible next states $x(t+1)$ if player i fixes a feasible vector of control parameters $u(t) \in U_t^i(x(t))$, i.e., $|A^i(x,t)| = |U_t(x(t))|$.

In $\overline{G}$, we define the cost functions $c^1, c^2, \ldots, c^m$ as follows:

With each edge $e_t = ((x,t), (y, t+1)) \in E^t$, we associate the costs

$$c^i_{e_t} = c^i_t(x(t), y(t+1)),\ i = 1, 2, \ldots, m,\ t = 0, 1, 2, \ldots, t_2 - 1$$

and with edge $e_t = ((x,t), (x_f, T_2)) \in E^f$, we associate the costs

$$c^i_{e_t} = c^i_t(x(t), x_f(t+1)),\ i = 1, 2, \ldots, m,\ t = t_1 - 1, t_1, t_1 + 1, \ldots, t_2 - 1.$$

After that, we use the algorithm from Sect. 3.18.3 and determine Stackelberg stationary strategies on $\overline{G}$ with a fixed starting state $y_0 = (x_0, t)$ and a final state $y_t = (x_f, t_2)$.

Taking into account that there exists a bijection between the set of Stackelberg stationary strategies of the players on $\overline{G}$ and the Stackelberg solution to the hierarchical control problem from Sect. 3.15, we can find a Stackelberg solution to the problem.

Reference

1. Alonso, P., Boratto, M., Peinado, J., Ibáñez, J., & Sastre, J. (2014). *On the evaluation of matrix polynomials using several GPGPUs*. Departament of Information Systems and Computation, Universitat Politècnica de València, Tech. Rep.
2. Alpern, S. (1991). Cycles in extensive form perfect information games. *Journal of Mathematical Analysis and Applications, 159*(1), 1–17.
3. Altman, E. (1999). *Constrained Markov decision processes: Stochastic modeling*. Routledge.
4. Aronson, J. E. (1989). A survey of dynamic network flows. *Annals of Operations Research, 20*(1), 1–66.
5. Ballard, G., Demmel, J., Holtz, O., Lipshitz, B., & Schwartz, O. (2012). Communication-optimal parallel algorithm for Strassen's matrix multiplication. In *Proceedings of the 24th ACM Symposium on Parallelism in Algorithms and Architectures* (pp. 193–204).
6. Bartlett, M. S. (1978). *An introduction to stochastic processes: With special reference to methods and applications*. CUP Archive.
7. Basar, T., & Olsder, G. J. (1999). Dynamic noncooperative game theory (2nd ed.). Classics in applied mathematics 23. Society for Industrial and Applied Mathematics.
8. Bauer, H. (1981). *Probability theory and elements of measure theory.* Academic Press.
9. Bellman, R. (1957). *Dynamic programming*. Princeton University Press.
10. Bellman, R. (1957). A Markovian decision process. *Journal of Mathematics and Mechanics, 6*, 679–684.
11. Bellman, R. (1959). Functional equations in the theory of dynamic programming—XI: Limit theorems. *Rendiconti del Circolo Matematico di Palermo, 8*(3), 343–345.
12. Bellman, R., & Kalaba, R. E. (1965). *Dynamic programming and modern control theory* (Vol. 81). Academic Press.
13. Bernstein, D. S. (2009). *Matrix mathematics: Theory, facts, and formulas* (2nd ed.). Princeton University Press.
14. Berman, A., & Plemmons, R. J. (1979). *Nonnegative matrices in the mathematical sciences*. Academic Press.
15. Bertsekas, D. P. (1987). *Dynamic programming: Deterministic and stochastic models*. Prentice Hall.
16. Bertsekas, D. P., & Shreve, S. E. (1978). *Stochastic optimal control: The discrete-time case*. Academic Press.
17. Blackwell, D. (1965). Discounted dynamic programming. *The Annals of Mathematical Statistics. 36*(1), 226–235.

D. Lozovanu, S. W. Pickl, *Markov Decision Processes and Stochastic Positional Games*, International Series in Operations Research & Management Science 349,
https://doi.org/10.1007/978-3-031-40180-0

18. Blackwell, D., & Ferguson, T. S. (1968). The big match. *The Annals of Mathematical Statistics, 39*(1), 159–163.
19. Boliac, R., Lozovanu, D., & Solomon, D. (2000). Optimal paths in network games with p players. *Discrete Applied Mathematics, 99*(3), 339–346.
20. Bollobás, B. (2001). *Random graphs* (2nd ed.). Cambridge Studies in Advanced Mathematics. Cambridge University Press.
21. Boltjanski, W. G. (1976). *Optimale Steuerung diskreter Systeme.* Akademische Verlagsgesellschaft Geest & Portig K. G.
22. Boros, E., & Gurvich, V. (2002). *On Nash-solvability in pure stationary strategies of finite games with perfect information which may have cycles.* DIMACS Technical Report, 18, 1–32.
23. Boyd, S., & Vandenberghe, L. (2004). *Convex optimization.* Cambridge University Press.
24. Brucher, P. (1972). Discrete parameter optimization problem and essential efficient points. *Operations Research, 16*(5), 189–197.
25. Burkard, R. E., Keiding, H., Pruzan, P. M., & Krarup, J. (1981). A relationship between optimality and efficiency in multicriteria 0–1 programming problems. *Computers & Operations Research, 8*(4), 241–247.
26. Burkard, R. E., Krarup, J., & Pruzan, P. M. (1982). Efficiency and optimality in minisum, minimax 0-1 programming problems. *Journal of the Operational Research Society, 33*(2), 137–151.
27. Butenko, S., Murphey, R., & Pardalos, P. M. (Eds.). (2003). *Cooperative control: Models, applications and algorithms.* Kluwer Academic Publishers.
28. Butkovic, P., & Cuninghame-Green, R. A. (1992). An $O(n^2)$ algorithm for the maximum cycle mean of an $n \times n$ bivalent matrix. *Discrete Applied Mathematics, 35*(2), 157–162.
29. Christofides, N. (1975). *Graph theory: An algorithmic approach.* Academic Press.
30. Chung, K. L. (1967). *Markov chains.* Springer.
31. Cinlar, E. (1975). *Introduction to stochastic processes.* Prentice Hall.
32. Condon, A. (1992). The complexity of stochastic games. *Information and Computation, 96*(2), 203–224.
33. Coppersmith, D., & Winograd, S. (1990). Matrix multiplication via arithmetic progressions. *Journal of Symbolic Computation, 9*(3), 251–280.
34. Cormen, T. H., Leiserson, C. E., Rivest, R. L., & Stein, C. (2009). *Introduction to algorithms.* The MIT Press.
35. D'Alberto, P., & Nicolau, A. (2009). Adaptive Winograd's matrix multiplications. *ACM Transactions on Mathematical Software (TOMS), 36*(1), 1–23.
36. Dasgupta, P., & Maskin, E. (1986). The existence of equilibrium in discontinuous economic games, part I (theory). *The Review of Economic Studies, 53*(1), 1–26.
37. Debreu, G. (1952). A social equilibrium existence theorem. *Proceedings of the National Academy of Sciences, 38*(10), 886–893.
38. Denardo, E. V. (1986). On linear programming in a Markov decision problem. *Management Science, 16,* 281–226.
39. Derman, C. (1970). *Finite state Markovian decision processes.* Academic Press.
40. Durinovic, S., Lee, H., Katehakis, M., & Filar, J. (1986). Multiobjective Markov decision process with average reward criterion. *Large Scale Systems: Theory and Applications, 10*(3), 215–226.
41. Dynkin, E. B., & Yushkevich, A. A. (1979). *Controlled Markov processes* (Vol. 235). Springer.
42. Ehrenfeucht, A., & Mycielski, J. (1979). Positional strategies for mean payoff games. *International Journal of Game Theory, 8*(2), 109–113.
43. Emelichev, V. A., & Perepelitsa, V. A. (1994). Complexity of discrete multicriteria problems. *Discrete Mathematics and Applications, 4*(2), 89–118.
44. Fabrikant, A., Papadimitriou, C., & Talwar, K. (2004). The complexity of pure Nash equilibria. In *Proceedings of the 36th annual ACM Symposium on Theory of Computing* (pp. 604–612). Chicago.

45. Fan, K. (1952). Fixed-point and minimax theorems in locally convex topological linear spaces. *Proceedings of the National Academy of Sciences, 38*(2), 121–126.
46. Fan, K. (1966). Applications of a theorem concerning sets with convex sections. *Mathematische Annalen, 163*(3), 189–203.
47. Federgruen, A., & Schweitzer, P. J. (1978). Discounted and undiscounted value-iteration in Markov decision problems: A survey. In: *Dynamic programming and its applications* (pp. 23–52). Academic Press.
48. Feller, W. (1957). *An Introduction to probability theory and its applications*. Wiley.
49. Filar, J. A. (1984). On stationary equilibria of a single-controller stochastic game. *Mathematical Programming, 30*(3), 313–325.
50. Filar, J. A. (1986). Quadratic programming and the single-controller stochastic game. *Journal of Mathematical Analysis and Applications, 113*(1), 136–147.
51. Filar, J. A., Kallenberg, L. C., & Lee, H. M. (1989). Variance-penalized Markov decision processes. *Mathematics of Operations Research, 14*(1), 147–161.
52. Filar, J. A., & Raghavan, T. E. S. (1984). A matrix game solution of the single-controller stochastic game. *Mathematics of Operations Research, 9*(3), 356–362.
53. Filar, J. A., Schultz, T. A., Thuijsman, F., & Vrieze, O. J. (1991). Nonlinear programming and stationary equilibria in stochastic games. *Mathematical Programming, 50*(1), 227–237.
54. Filar, J. A., & Vrieze, K. (1997). *Competitive Markov decision processes*. Springer.
55. Fink, A. M. (1964). Equilibrium in a stochastic *n*-person game. *Journal of Science of the Hiroshima University, Series A-I (Mathematics), 28*(1), 89–93.
56. Fleming, W. H., & Rishel, R. W. (1975). *Deterministic and stochastic optimal control*. Springer.
57. Flesch, J., Thuijsman, F., & Vrieze, K. (1997). Cyclic Markov equilibria in stochastic games. *International Journal of Game Theory, 26*(3), 303–314.
58. Fox, B. L., & Landi, D. M. (1968). Scientific applications: An algorithm for identifying the ergodic subchains and transient states of a stochastic matrix. *Communications of the ACM, 11*(9), 619–621.
59. Ford, L. R., Jr., & Fulkerson, D. R. (1958). Constructing maximal dynamic flows from static flows. *Operations Research, 6*(3), 419–433.
60. Garey, M. R., & Johnson, D. S. (1979). *Computers and intractability: A guide to the theory of NP-completeness*. W. H. Freeman.
61. Gheorghe, A. V. (1990). *Decision processes in dynamic probabilistic systems*. Kluwer Academic Publishers.
62. Ghosh, M. K. (1990). Markov decision processes with multiple costs. *Operations Research Letters, 9*(4), 257–260.
63. Gillette, D. (1957). Stochastic games with zero stop probabilities. *Contributions to the Theory of Games, 3*, 179–187.
64. Glicksberg, I. L. (1952). A further generalization of the Kakutani fixed point theorem, with application to Nash equilibrium points. *Proceedings of the American Mathematical Society, 3*(1), 170–174.
65. Gottlob, G., Greco, G., & Scarcello, F. (2005). Pure Nash equilibria: Hard and easy games. *Journal of Artificial Intelligence Research, 24*, 357–406.
66. Granas, A., & Dugundji, J. (2003). *Fixed point theory* (Vol. 14, pp. 15–16). Springer.
67. Gurvich, V. A., Karzanov, A. V., & Khachivan, L. G. (1988). Cyclic games and an algorithm to find minimax cycle means in directed graphs. *USSR Computational Mathematics and Mathematical Physics, 28*(5), 85–91.
68. Helmberg, G., Wagner, P., & Veltkamp, G. (1993). On Faddeev-Leverrier's method for the computation of the characteristic polynomial of a matrix and of eigenvectors. *Linear Algebra and Its Applications, 185*, 219–233.
69. Hordijk, A., & Kallenberg, L. C. M. (1979). Linear programming and Markov decision chains. *Management Science, 25*(4), 352–362.

70. Hordijk, A., & Kallenberg, L. C. M. (1980). On solving Markov decision problems by linear programming. In *Recent developments in Markov decision processes. International Conference on Markov Decision Processes*. Academic Press
71. Howard, R. A. (1960). *Dynamic programming and Markov processes*. Wiley.
72. Howard, R. A. (1963). Semi-Markovian decision processes. *Bulletin of the International Statistical Institute, 40*(2), 625–652.
73. Howard, R. A. (1972). *Dynamic probabilistic systems, Markov models*. Wiley.
74. Hu, Q., & Yue, W. (2007). *Markov decision processes with their applications* (Vol. 14). Springer.
75. Kakutani, S. (1941). A generalization of Brouwer's fixed point theorem. *Duke Mathematical Journal, 8*(3), 457–459.
76. Kallenberg, L. C. (1983). *Linear programming and finite Markovian control problems*. MC Tracts.
77. Kallenberg, L. C. (2011). *Markov decision processes*. Lecture Notes. University of Leiden, 2–5.
78. Karlin, S. (1969). *A first course in stochastic processes*. Academic Press.
79. Karp, R. M. (1978). A characterization of the minimum cycle mean in a digraph. *Discrete Mathematics, 23*(3), 309–311.
80. Karzanov, A. V., & Lebedev, V. N. (1993). Cyclical games with prohibitions. *Mathematical Programming, 60*(1), 277–293.
81. Keller-Gehrig, W. (1985). Fast algorithms for the characteristics polynomial. *Theoretical Computer Science, 36*, 309–317.
82. Kemeny, J. G., & Snell, J. L. (1960). *Finite Markov chains*. Van Nostrand, Princeton. Springer.
83. Kemeny, J. G., Snell, J. L., & Knapp, A. W. (1976). *Denumerable Markov chains*. Springer.
84. Khachiyan, L. G. (1980). Polynomial algorithms in linear programming. *USSR Computational Mathematics and Mathematical Physics, 20*(1), 53–72.
85. Khachiyan, L. G. (1982). On the exact solution of systems of linear inequalities and linear programming problems. *USSR Computational Mathematics and Mathematical Physics, 22*(4), 239–242.
86. Klinz, B., & Woeginger, G. J. (2004). Minimum-cost dynamic flows: The series-parallel case. *Networks: An International Journal, 43*(3), 153–162.
87. Kohlberg, E. (1974). Repeated games with absorbing states. *The Annals of Statistics, 2*, 724–738.
88. Krabs, W., & Pickl, S. (2003). Controllability of a time-discrete dynamical system with the aid of the solution of an approximation problem. *Control and Cybernetics, 32*(1), 57–74.
89. Krabs, W., & Pickl, S. (2003). *Analysis, controllability and optimization of time-discrete systems and dynamical games*. Springer.
90. Lamond, B. F., & Puterman, M. L. (1989). Generalized inverses in discrete time Markov decision processes. *Society for Industrial and Applied Mathematics. Journal on Matrix Analysis and Applications, 10*(1), 118–134.
91. Lawler, E. L. (1966). Optimal cycles in doubly weighted directed linear graphs. In P. Rosenstiehl (Ed.), *Theory of Graphs: International Symposium, Gordon and Breach*, New York, U.S.A., 1966 (pp. 209–213).
92. Lazari, A. (2010). Algorithms for determining the transient and differential matrices in finite Markov processes. *Buletinul Academiei de Ştiinţe a Republicii Moldova. Matematica, 2*(63), 84–99.
93. Lazari, A., & Lozovanu, D. (2020). New algorithms for finding the limiting and differential matrices in Markov chains. *Buletinul Academiei de Ştiinţe a Moldovei. Matematica, 92*(1), 75–88.
94. Le Gall, F. (2014). Powers of tensors and fast matrix multiplication. In: *Proceedings of the 39th International Symposium on Symbolic and Algebraic Computation* (pp. 296–303).
95. Liang, W., Baer, R., Saravanan, C., Shao, Y., Bell, A. T., & Head-Gordon, M. (2004). Fast methods for resumming matrix polynomials and Chebyshev matrix polynomials. *Journal of Computational Physics, 194*(2), 575–587.

96. Lozovanu, D. (1991). Algorithms to solve some classes of network minimax problems and their applications. *Cybernetics and Systems Analysis, 27*(1), 93–100.
97. Lozovanu, D. (1991). *Extremal-combinatorial problems and algorithms for its solving* (in Russian). Kishinev, Stiinta.
98. Lozovanu, D. (2004). Polynomial time algorithm for determining optimal strategies in cyclic games. In D. Bienstock, G. Nemhauser (Eds.), *Integer programming and combinatorial optimization. IPCO 2004*, New York. Lecture Notes in Computer Science (pp. 74–85). Springer.
99. Lozovanu, D. (2008). Multiobjective Control of time-discrete systems and dynamic games on networks. In A. Chinchuluun, P. M. Pardalos, A. Migdalas, & L. Pitsoulis (Eds.), *Pareto optimality, game theory and equilibria* (Vol 17, pp. 665–757).
100. Lozovanu, D. (2011). The game-theoretical approach to Markov decision problems and determining Nash equilibria for stochastic positional games. *International Journal of Mathematical Modelling and Numerical Optimisation, 2*(2), 162–174.
101. Lozovanu, D. (2018). Stationary Nash equilibria for average stochastic positional games. In *Frontiers of dynamic games: Game theory and management* (pp. 139–163). Springer.
102. Lozovanu, D. (2019). Pure and mixed stationary Nash equilibria for average stochastic positional games. In *Frontiers of dynamic games: Game theory and management* (pp. 131–155). Springer.
103. Lozovanu, D., & Fonoberova, M. (2006). Optimal dynamic multicommodity flows in networks. *Electronic Notes in Discrete Mathematics, 25*, 93–100.
104. Lozovanu, D., & Fonoberova, M. (2009). Optimal dynamic flows in networks and algorithms for finding them. In M. Dehmer & F. Emmert-Streb (Eds.), *Analysis of complex networks* (pp. 377–400). Wiley.
105. Lozovanu, D., & Lazari, A. (2010). An approach for determining the matrix of limiting state probabilities in discrete Markov processes. *Buletinul Academiei de Ştiinţe a Moldovei. Matematica, 62*(1), 77–91.
106. Lozovanu, D., & Petic, C. (1998). Algorithms for finding the minimum cycle mean in the weighted directed graph. *Computer Science Journal Moldova, 6*(1), 27–34.
107. Lozovanu, D., & Pickl, S. (2005). Nash equilibria for multiobjective control of time-discrete systems and polynomial-time algorithms for k-partite networks. *Central European Journal of Operations Research, 13*(2), 127–146.
108. Lozovanu, D., & Pickl, S. (2006). Nash equilibria conditions for cyclic games with p players. *Electronic Notes in Discrete Mathematics, 25*, 123–129.
109. Lozovanu, D., & Pickl, S. (2007). Algorithms and the calculation of Nash equilibria for multi-objective control of time-discrete systems and polynomial-time algorithms for dynamic c-games on networks. *European Journal of Operational Research, 181*(3), 1214–1232.
110. Lozovanu, D., & Pickl, S. (2007). Algorithms for solving multiobjective discrete control problems and dynamic c-games on networks. *Discrete Applied Mathematics, 155*(14), 1846–1857.
111. Lozovanu, D., & Pickl, S. (2007). Multiobjective hierarchical control of time-discrete systems and determining Stackelberg strategies. In *CTW-2007 Proceedings* (pp. 111–114).
112. Lozovanu, D., & Pickl, S. (2009). *Optimization and multiobjective control of time-discrete systems: Dynamic networks and multilayered structures*. Springer.
113. Lozovanu, D., & Pickl, S. (2009). Dynamic programming algorithms for solving stochastic discrete control problems. *Buletinul Academiei de Ştiinţe a Moldovei. Matematica, 60*(2), 73–90.
114. Lozovanu, D., & Pickl, S. (2009). Algorithmic solutions of discrete control problems on stochastic networks. In *CTW-2009 Proceedings* (pp. 221–224).
115. Lozovanu, D., & Pickl, S. (2009). Algorithms for solving discrete optimal control problems with infinite time horizon and determining minimal mean cost cycles in a directed graph as decision support tool. *Central European Journal of Operations Research, 17*(3), 255–264.
116. Lozovanu, D., & Pickl, S. (2009). Discrete control and algorithms for solving antagonistic dynamic games on networks. *Optimization, 58*(6), 665–683.

117. Lozovanu, D., & Pickl, S. (2009). An extension of a polynomial time algorithm for the calculation of the limit state matrix in a random graph. In D. M. Dubois (Ed.), *International journal of computing anticipatory systems* (pp. 92–97).
118. Lozovanu, D., & Pickl, S. (2010). Determining optimal stationary strategies for discounted stochastic optimal control problem on networks. In U. Faigle, R. Schrader, & D. Herrmann (Eds.), *CTW-2010 Proceedings* (pp. 115–118).
119. Lozovanu, D., & Pickl, S. (2010). Optimal stationary control of discrete processes and a polynomial time algorithm for stochastic control problem on networks. In *Proceedings of the International Conference on Computational Science, ICCS 2010, University of Amsterdam*. Procedia Computer Science (Vol. 1(1), pp. 1417–1426). Elsevier
120. Lozovanu, D., & Pickl, S. (2010). Algorithms for solving discrete optimal control problems with varying time of states' transactions of dynamical systems. *Dynamics of Continuous, Discrete and Impulsive Systems Series B: Applications and Algorithms, 17*(1), 101–111.
121. Lozovanu, D., & Pickl, S. (2011). Algorithms for determining the state-time probabilities and the limit matrix in Markov chains. *Buletinul Academiei de Ştiinţe a Moldovei. Matematica, 65*(1), 66–82.
122. Lozovanu, D., & Pickl, S. (2011). Discounted Markov decision processes and algorithms for solving stochastic control problem on networks. *CTW, 2011* (pp. 194–197).
123. Lozovanu, D., & Pickl, S. (2012). Determining the optimal strategies for antagonistic positional games in Markov decision processes. In *Operations Research Proceedings 2011: Selected Papers of the International Conference on Operations Research (OR 2011)*, Zurich, Switzerland (pp. 229–234). Springer.
124. Lozovanu, D., & Pickl, S. (2015). Determining the optimal strategies for discrete control problems on stochastic networks with discounted costs. *Discrete Applied Mathematics, 182*, 169–180.
125. Lozovanu, D., & Pickl, S. (2015). *Optimization of stochastic discrete systems and control on complex networks: Computational networks*. Springer.
126. Lozovanu, D., & Pickl, S. (2018). Nash equilibria in mixed stationary strategies for m-player mean payoff games on networks. *Contributions to Game Theory and Management, 11*, 103–112.
127. Lozovanu, D., & Pickl, S. (2019). Pure stationary Nash equilibria for discounted stochastic positional games. *Contributions to Game Theory and Management, 12*, 246–260.
128. Lozovanu, D., & Pickl, S. (2020). On the existence of stationary Nash equilibria in average stochastic games with finite state and action spaces. *Contributions to Game Theory and Management, 13*, 304–323.
129. Lozovanu, D., Pickl, S., & Kropat, E. (2011). Markov decision processes and determining Nash equilibria for stochastic positional games. *IFAC Proceedings Volumes, 44*(1), 13398–13403.
130. Lozovanu, D., Solomon, D., & Zelikovsky, A. (2005). Multiobjective games and determining Pareto-Nash equilibria. *Buletinul Academiei de Ştiinţe a Moldovei. Matematica, 49*(3), 115–122.
131. Lozovanu, D., & Stratila, D. (2001). The minimum-cost flow problem on dynamic networks and an algorithm for its solving. *Buletinul Academiei de Ştiinţe a Moldovei. Matematica, 37*(2), 38–56.
132. Lozovanu, D., & Stratila, D. (2003). Optimal flow in dynamic networks with nonlinear cost functions on edges. In *Analysis and Optimization of Differential Systems: IFIP TC7/WG7.2 International Working Conference on Analysis and Optimization of Differential Systems, 2002*, Constanta, Romania (pp. 247–258). Springer.
133. Mangasarian, O. L., & Stone, H. (1964). Two-person nonzero-sum games and quadratic programming. *Journal of Mathematical Analysis and Applications, 9*(3), 348–355.
134. Mertens, J. F., & Neyman, A. (1981). Stochastic games. *International Journal of Game Theory, 10*, 53–66.
135. Meyn, S. (2008). *Control techniques for complex networks*. Cambridge University Press.

136. Mine, H., & Osaki, S. (1968). Some remarks on a Markovian decision problem with an absorbing state. *Journal of Mathematical Analysis and Applications, 23*(2), 327–333.
137. Mine, H., & Osaki, S. (1970). *Markovian decision processes*. Elsevier.
138. Moulin, H. (1976). Prolongement des jeuxa deux joueurs de somme nulle. Une théorie abstraite des duels. *Mémoires de la Société Mathématique de France, 45*, 5–111.
139. Moulin, H. (1981). *Théorie des jeux pour l'économie et la politique*. Hermann.
140. Murphey, R., & Pardalos, P. M. (Eds.). (2002). *Cooperative control and optimization* (Vol. 66). Springer.
141. Nash, J. (1951). Non-cooperative games. *Annals of Mathematics, 54*(2), 286–295.
142. Neumann, J., & Morgenstern, O. (1953). *Theory of games and economic behavior*. Princeton University Press.
143. Neyman, A., & Sorin, S. (Eds.). (2003). *Stochastic games and applications* (Vol. 570). Springer.
144. Pareto, V. (1904). *Manuel d'Économie Politique*. V. Giard et E. Brière.
145. Pernet, C., & Storjohann, A. (2007). Faster algorithms for the characteristic polynomial. In *Proceedings of the 2007 International Symposium on Symbolic and Algebraic Computation* (pp. 307–314).
146. Pickl, S., & Lozovanu, D. (2009). Dynamic programming algorithms for solving stochastic discrete control problems. *Buletinul Academiei de Ştiinţe a Republicii Moldova. Matematica, 2*(60), 73–90.
147. Pickl, S., & Lozovanu, D. (2011). A linear programming approach for solving the discounted stochastic optimal control problem on certain networks. In *Proceedings of the Network and Electronic Commerce Research Conference NAEC 2011*, Riva del Garda, Italy.
148. Pickl, S., & Lozovanu, D. (2011). A dynamic programming approach for finite Markov processes and algorithms for the calculation of the limit matrix in Markov chains. *Optimization, 60*(10–11), 1339–1358.
149. Podinovskii, V. V., & Nogin, V. D. (1982). *Pareto-optimal'nye resheniya mnogokriterial'nykh zadach (in Russian) [Pareto-optimal solutions of multicriteria problems]*. Nauka.
150. Porteus, E. L. (1980). Overview of iterative methods for discounted finite Markov and semi-Markov decision chains. In *Recent developments in Markov decision processes* (pp. 1–20).
151. Puterman, M. L. (1990). *Markov decision processes. Handbooks in operations research and management science* (Vol. 2, pp. 331–434).
152. Puterman, M. L. (2014). *Markov decision processes: Discrete stochastic dynamic programming*. Wiley.
153. Raghavan, T. E. S. (2003). Finite-step algorithms for single-controller and perfect information stochastic games. In A. Neyman & S. Sorin (Eds.), *Stochastic games and applications*. NATO Science Series (Vol. 570, pp. 227–251). Springer.
154. Reny, P. J. (1999). On the existence of pure and mixed strategy Nash equilibria in discontinuous games. *Econometrica, 67*(5), 1029–1056.
155. Rogers, P. D. (1969). *Nonzero-sum stochastic games*. University of California.
156. Romanovski, I. V. (1967). Optimization of stationary control of a discrete deterministic process. *Cybernetics, 3*(2), 52–62.
157. Romanovski, I. V. (1973). On the solvability of Bellman's functional equation for a Markovian decision process. *Journal of Mathematical Analysis and Applications, 42*(2), 485–498.
158. Ross, S. M. (2014). *Introduction to probability models*. Academic Press.
159. Saad, Y. (2003). *Iterative methods for sparse linear systems*. Society for Industrial and Applied Mathematics.
160. Schultz, T. A. (1988). *Mathematical programming and stochastic games*. PhD Thesis, Johns Hopkins University.
161. Schweitzer, P. J. (1971). Iterative solution of the functional equations of undiscounted Markov renewal programming. *Journal of Mathematical Analysis and Applications, 34*(3), 495–501.
162. Schweitzer, P. J., & Federgruen, A. (1977). The asymptotic behavior of undiscounted value iteration in Markov decision problems. *Mathematics of Operations Research, 2*(4), 360–381.

163. Schweitzer, P. J., & Federgruen, A. (1978). Foolproof convergence in multichain policy iteration. *Journal of Mathematical Analysis and Applications, 64*(2), 360–368.
164. Schweitzer, P. J., & Federgruen, A. (1979). Geometric convergence of value-iteration in multichain Markov decision problems. *Advances in Applied Probability, 11*(1), 188–217.
165. Shapley, L. S. (1953). Stochastic games. *Proceedings of the National Academy of Sciences, 39*(10), 1095–1100.
166. Simon, L. K. (1987). Games with discontinuous payoffs. *The Review of Economic Studies, 54*(4), 569–597.
167. Sobel, M. J. (1971). Noncooperative stochastic games. *The Annals of Mathematical Statistics, 42*(6), 1930–1935.
168. Solan, E. (2009). Stochastic games. In R. Meyers (Ed.), *Encyclopedia of complexity and systems science*. Springer.
169. Solan, E., & Vieille, N. (2010). Computing uniformly optimal strategies in two-player stochastic games. *Economic Theory, 42*(1), 237–253.
170. Sorin, S. (1986). Asymptotic properties of a non-zero sum stochastic game. *International Journal of Game Theory, 15*, 101–107.
171. Stewart, W. J. (1995). *Introduction to the numerical solution of Markov chains*. Princeton University Press.
172. Stoer, J., & Bulirsch, R. (1993). *Introduction to numerical analysis*. Springer.
173. Strassen, V. (1969). Gaussian elimination is not optimal. *Numerische Mathematik, 13*(4), 354–356.
174. Takahashi, M. (1964). Equilibrium points of stochastic non-cooperative n-person games. *Journal of Science of the Hiroshima University, Series AI (Mathematics), 28*(1), 95–99.
175. Tijs, S. H., & Vrieze, O. J. (1986). On the existence of easy initial states for undiscounted stochastic games. *Mathematics of Operations Research, 11*(3), 506–513.
176. Thuijsman, F. (2003). The big match and the paris match. In A. Neyman & S. Sorin (Eds.), *Stochastic games and applications, NATO science series C, Mathematical and physical sciences* (Vol 570, pp. 195–204).
177. Thuijsman, F. (2003). Repeated games with absorbing states. In *Stochastic games and applications, NATO science series C, Mathematical and physical sciences* (Vol. 570, pp. 205–213).
178. Thuijsman, F., & Vrieze, O. J. (1987). The bad match; a total reward stochastic game. *Operations-Research-Spektrum, 9*(2), 93–99.
179. Thuijsman, F., & Vrieze, O. J. (1998). Total reward stochastic games and sensitive average reward strategies. *Journal of Optimization Theory and Applications, 98*, 175–196.
180. Thuijsman, F., & Raghavan, T. E. (1997). Perfect information stochastic games and related classes. *International Journal of Game Theory, 26*, 403–408.
181. Tsitsiklis, J. N. (2007). NP-hardness of checking the unichain condition in average cost MDPs. *Operations Research Letters, 35*(3), 319–323.
182. van der Wal, J. (1981). Stochastic dynamic programming: Successive approximations and nearly optimal strategies for Markov decision processes and Markov games. *Mathematical Center Tracts, 139*, Mathematisch Centrum.
183. Vieille, N. (2002). *Stochastic games: Recent results*. Handbook of game theory with economic applications (Vol. 3, pp. 1833–1850).
184. Vieille, N. (2003). *Two-player non-zero-sum games: A reduction*. Stochastic Games and Applications (pp. 281–292). Springer.
185. Vieille, N. (2000). Equilibrium in 2-person stochastic games I, II. *Israel Journal of Mathematics, 119*(1), 55–126.
186. Vöge, J., & Jurdziński, M. (2000). A discrete strategy improvement algorithm for solving parity games. In *Computer Aided Verification: Proceedings of the 12th International Conference, CAV 2000*, Chicago, IL, USA (pp. 202–215). Springer.
187. von Stackelberg, H. (1934). *Marktform und Gleichgewicht*. Springer.
188. Vrieze, O. J. (1981). Linear programming and undiscounted stochastic games in which one player controls transitions. *Operations-Research-Spektrum, 3*(1), 29–35.

189. Vrieze, O. J. (1987). Stochastic games with finite state and action spaces. *CWI Tracts, 33*, 1–221.
190. Vrieze, O. J., & Thuijsman, F. (1989). On equilibria in repeated games with absorbing states. *International Journal of Game Theory, 18*(3), 293–310.
191. Weber, G. W. (1999). Optimal control theory: On the global structure and connections with optimization. Part 1. *Journal of Computational Technologies, 4*(2), 3–26.
192. White, C. C., & Kim, K. W. (1980). Solution procedures for vector criterion Markov decision processes. *Large Scale Systems, 1*, 129–140.
193. White, C. C., & White, D. J. (1989). Markov decision processes. *European Journal of Operational Research, 39*(1), 1–16.
194. White, D. J. (1978). *Finite dynamic programming: An approach to finite Markov decision processes*. Wiley.
195. White, D. J. (1982). A multi-objective version of Bellman's inventory problem. *Journal of Mathematical Analysis and Applications, 87*(1), 219–227.
196. White, D. J. (1982). Multi-objective infinite-horizon discounted Markov decision processes. *Journal of Mathematical Analysis and Applications, 89*(2), 639–647.
197. White, D. J. (1985). Monotone value iteration for discounted finite Markov decision processes. *Journal of Mathematical Analysis and Applications, 109*(2), 311–324.
198. White, D. J. (1987). Decomposition in multi-item inventory control. *Journal of Optimization Theory and Applications, 54*, 383–401.
199. White, D. J. (1993). *Markov decision processes*. Wiley.
200. Williams, V. V. (2014). *Multiplying matrices in* $O(n^{2.373})$ *time* (pp. 1–73). Stanford University.
201. Yushkevich, A. A. (1982). On semi-Markov controlled models with an average reward criterion. *Theory of Probability & Its Applications, 26*(4), 796–803.
202. Zwick, U., & Paterson, M. (1996). The complexity of mean payoff games on graphs. *Theoretical Computer Science, 158*(1–2), 343–359.

Index

D. Lozovanu, S. W. Pickl, *Markov Decision Processes and Stochastic Positional Games*, International Series in Operations Research & Management Science 349,
https://doi.org/10.1007/978-3-031-40180-0